Conceptual foundations of quantum field theory

Quantum field theory is a powerful language for the description of the sub-atomic constituents of the physical world and the laws and principles that govern them. This book contains up-to-date in-depth analyses, by a group of eminent physicists and philosophers of science, of our present understanding of its conceptual foundations, of the reasons why this understanding has to be revised so that the theory can go further, and of possible directions in which revisions may be promising and productive.

These analyses will be of interest to students of physics who want to know about the foundational problems of their subject. The book will also be of interest to professional philosophers, historians and sociologists of science. It contains much material for metaphysical and methodological reflection on the subject; it will also repay historical and cultural analyses of the theories discussed in the book, and sociological analyses of the way in which various factors contribute to the way the foundations are revised. The authors also reflect on the tension between physicists on the one hand and philosophers and historians of science on the other, when revision of the foundations becomes an urgent item on the agenda.

This work will be of value to graduate students and research workers in theoretical physics, and in the history, philosophy and sociology of science, who have an interest in the conceptual foundations of quantum field theory.

T0171863

For Bob and Sam

Conceptual foundations of quantum field theory

EDITOR

TIAN YU CAO

Boston University

CAMBRIDGE
UNIVERSITY PRESS

PUBLISHED BY THE PRESS SYNDICATE OF THE UNIVERSITY OF CAMBRIDGE
The Pitt Building, Trumpington Street, Cambridge, United Kingdom

CAMBRIDGE UNIVERSITY PRESS
The Edinburgh Building, Cambridge CB2 2RU, UK
40 West 20th Street, New York NY 10011–4211, USA
477 Williamstown Road, Port Melbourne, VIC 3207, Australia
Ruiz de Alarcón 13, 28014 Madrid, Spain
Dock House, The Waterfront, Cape Town 8001, South Africa

http://www.cambridge.org

First published 1999
First paperback edition 2004

Typeset in 10/12½pt Times [wv]

A catalogue record for this book is available from the British Library

Library of Congress Cataloguing in Publication data

Conceptual foundations of quantum field theory / editor, Tian Yu Cao.
p.   cm.
Includes index.
ISBN 0 521 63152 1 hardback
1. Quantum field theory–Congresses.   I. Cao, Tian Yu, 1941–   .
QC174.45.A1C646   1999
530.14′3–dc21 98–24394 CIP

ISBN 0 521 63152 1 hardback
ISBN 0 521 60272 6 paperback

Contents

List of contributors   ix
Preface   xi
Photographs of the conference   xii

Introduction: Conceptual issues in quantum field theory   1
TIAN YU CAO

I Philosophers' interest in quantum field theory   **28**
1 Why are we philosophers interested in quantum field theory?   28
TIAN YU CAO

2 Quantum field theory and the philosopher   34
MICHAEL REDHEAD

II Three approaches to the foundations of quantum field theory   **41**
3 The usefulness of a general theory of quantized fields   41
A. S. WIGHTMAN

4 Effective field theory in condensed matter physics   47
R. SHANKAR

5 The triumph and limitations of quantum field theory   56
DAVID J. GROSS

6 Comments   68
SAM TREIMAN

Session discussions   70

III
7 Does quantum field theory need a foundation?   74
SHELDON LEE GLASHOW

IV Mathematics, statistics and quantum field theory   **89**
8 Renormalization group theory: its basis and formulation in statistical physics   89
MICHAEL E. FISHER

9 Where does quantum field theory fit into the big picture?   136
ARTHUR JAFFE

10 The unreasonable effectiveness of quantum field theory   148
ROMAN JACKIW

11 Comments: The quantum field theory of physics and of mathematics   161
HOWARD J. SCHNITZER

Session discussions   164

V **Quantum field theory and space-time**   **166**
Introduction   166
JOHN STACHEL

12 Quantum field theory and space-time – formalism and reality   176
BRYCE DEWITT

13 Quantum field theory of geometry   187
ABHAY ASHTEKAR and JERZY LEWANDOWSKI

14 'Localization' in quantum field theory: how much of QFT is
compatible with what we know about space-time?   207
CARLO ROVELLI

15 Comments   233
JOHN STACHEL

VI
16 What is quantum field theory, and what did we think it was?   241
STEVEN WEINBERG

17 Comments   252
LAURIE M. BROWN and FRITZ ROHRLICH

Session discussions   259

VII **Renormalization group**   **264**
18 What is fundamental physics? A renormalization group perspective   264
DAVID NELSON

19 Renormalization group: an interesting yet puzzling idea   268
TIAN YU CAO

VIII **Non-Abelian gauge theory**   **287**
20 Gauge fields, gravity and Bohm's theory   287
NICK HUGGETT and ROBERT WEINGARD

21 Is the Aharonov–Bohm effect local?   298
RICHARD HEALEY

Session discussions   310

IX **The ontology of particles or fields**   **314**
22 The ineliminable classical face of quantum field theory   314
PAUL TELLER

23 The logic of quanta   324
STEVEN FRENCH and DÉCIO KRAUSE

24 Do Feynman diagrams endorse a particle ontology? The roles of
   Feynman diagrams in *S*-matrix theory   343
   DAVID KAISER

25 On the ontology of QFT   357
   FRITZ ROHRLICH

X

26 Panel discussion   368
   STANLEY DESER (moderator), SIDNEY COLEMAN,
   SHELDON LEE GLASHOW, DAVID GROSS, STEVEN WEINBERG,
   ARTHUR WIGHTMAN

   Name index   387
   Subject index   391

Contributors

Abhay Ashtekar
Center for Gravitational Physics and Geometry, Department of Physics,
Penn State University, University Park, PA 16802, USA

Laurie M. Brown
Department of Physics and Astronomy, Northwestern University, Evenston,
IL 60208, USA

Tian Yu Cao
Department of Philosophy, Boston University, Boston, MA 02215, USA

Sidney Coleman
Department of Physics, Harvard University, Cambridge, MA 02138, USA

Stanley Deser
Department of Physics, Brandeis University, Waltham, MA 02254, USA

Bryce DeWitt
Department of Physics, University of Texas at Austin, Austin, TX 78712, USA

Michael E. Fisher
Institute for Physical Science and Technology, University of Maryland at College Park,
MD 20741-2431, USA

Steven French
Department of Philosophy, University of Leeds, Leeds, LS2 9JT, UK

Sheldon Lee Glashow
Department of Physics, Harvard University, Cambridge, MA 02138, USA; and
Department of Physics, Boston University, Boston, MA 02215, USA

David Gross
Department of Physics, Princeton University, Princeton, NJ 08544-0708, USA

Richard Healey
Department of Philosophy, University of Arizona, Tucson, AZ 85721, USA

Nick Huggett
Department of Philosophy, University of Illinois at Chicago, Chicago, IL 60607, USA

Roman Jackiw
Department of Physics, Massachusetts Institute of Technology, Cambridge,
MA 02139, USA

Arthur Jaffe
Departments of Physics and Mathematics, Harvard University, Cambridge, MA 02138, USA

David Kaiser
Department of History of Science, 235 Science Center, Harvard University, Cambridge, MA 02138, USA

Décio Krause
Department of Mathematics, Federal University of Parana, 81531-990, Curitiba, PR, Brazil

Jerzy Lewandowski
Institute of Theoretical Physics, University of Warsaw, Warsaw, Poland; and Max Planck Institut für Gravitationphysik, Schlaatzweg 1, 14473 Potsdam, Germany

David Nelson
Department of Physics, Harvard University, Cambridge, MA 02138, USA

Michael Redhead
Wolfson College, Cambridge University, Cambridge, CB3 9BB, UK

Fritz Rohrlich
Department of Physics, Syracuse University, Syracuse, NY 13210, USA

Carlo Rovelli
Department of Physics, University of Pittsburgh, Pittsburgh, PA 15260, USA

Howard Schnitzer
Department of Physics, Brandeis University, Waltham, MA 02254, USA

R. Shankar
Department of Physics, Yale University, PO Box 6666, New Haven, CT 06511-8167, USA

John Stachel
Department of Physics, Boston University, Boston, MA 02215, USA

Paul Teller
Department of Philosophy, University of California at David, CA 95616, USA

Sam Treiman
Department of Physics, Princeton University, Princeton, NJ 08544-0708, USA

Steven Weinberg
Department of Physics, University of Texas at Austin, Austin, TX 78712, USA

Robert Weingard
Department of Philosophy, Rutgers University, New Brunswick, NJ 08903, USA

Arthur S. Wightman
Departments of Mathematics and Physics, Princeton University, Princeton, NJ 08544-0708, USA

Preface

This volume is the result of a two-tier conference consisting of a two-day symposium followed by a one-day workshop, which was first conceived by a group of philosophers and historians of physics in the Greater Boston area, the core members of which were Babak Ashirafi of Massachusetts Institute of Technology, Ronald Anderson of Boston College, Tian Yu Cao of Boston University, David Kaiser of Harvard University and Silvan S. Schweber of Brandeis University, and then sponsored by the Center for Philosophy and History of Science, Boston University, and held at Boston University on March 1–3 1996, with financial support provided by the U.S. National Science Foundation and the Boston Philosophy of Science Association.

The intention was to offer an opportunity for a group of leading scholars to present their penetrating and in-depth analysis of various formulations and understandings of the foundations of quantum field theory, and to investigate philosophical and historical issues associated with these formulations, and also to provide a forum for the desirable, mutually beneficial but difficult exchange of views and ideas between physicists and mathematicians on the one side and philosophers and historians on the other. Although the experiment in dialogue was not completely successful, the publication of this volume will make the valuable contributions to this conference as well as interesting material about the tension between two groups of scholars accessible to a much wider audience for further theoretical, philosophical, historical, and sociological analysis.

During the long period of preparation for the conference, in addition to many planning meetings by our group, we also received advice and numerous suggestions from the prospective participants, and also from Professor Gerald Holton of Harvard University and Professor Robert S. Cohen of Boston University. We are grateful for their intellectual and spiritual support. Thanks also to Ms Corinne Yee and Ms Carolyn A. Fahlbeck, without whose effective handling of the complexities that constantly emerged in the process of meetings the conference would have been practically impossible.

Tian Yu Cao
Boston University

The 1996 Boston University Conference on the Foundations of Quantum Field
Theory

Stanley Deser, Sheldon Lee Glashow and David Gross

Silvan S. Schweber, David Gross, R. Shankar, Sam Treiman and Arthur
Wightman

Arthur Jaffe, Peter Galison, Roman Jackiw, Michael E. Fisher and Howard
Schnitzer

Sheldon Lee Glashow

Abhay Ashtekar, Carlo Rovelli, Bryce DeWitt and John Stachel

Steven Weinberg

Stanley Deser, David Gross, Sheldon Lee Glashow, Sidney Coleman, Steven
Weinberg and Arthur Wightman

Tian Yu Cao, Michael E. Fisher and David Nelson

Robert Weingard, Richard Healey, Ronald Anderson, Roman Jackiw and J. B. Kennedy

Steven French, Fritz Rohrlich, Paul Teller and David Kaiser

Laurie Brown and George Mackey

Jon Westling, Tian Yu Cao, Gerald Holton and Abner Shimony

R. Shankar, Francis Low and Sam Treiman

Laszlo Tisza, J. Scott Whitaker and Michael E. Fisher

Robert S. Cohen and Elena Mamchur

Introduction
Conceptual issues in quantum field theory

TIAN YU CAO

Quantum field theory (QFT) is a powerful language for describing the subatomic constituents of the physical world (quarks, leptons, gauge bosons, Higgs scalars, and so on) and the laws and principles that govern them. Not only has it provided a framework for understanding the hierarchical structure that can be built from these constituents, it has also profoundly influenced the development of contemporary cosmology and deeply penetrated into the current conception and imagination of the origin and evolution of the universe. For this reason, it has justifiably long been taken to be the foundation of fundamental physics: elementary particle physics and cosmology.

QFT reached its most spectacular success in the early 1970s in the standard model, which is able to describe the fundamental interactions of nature within a unified theoretical structure: the non-Abelian gauge theory. Ironically enough, however, this success brought about a long period of stagnation in its conceptual development: virtually nothing of physical importance occurred after the construction of the standard model except for its detailed experimental verification. This situation can be assessed in different ways, depending on what perspective we start with. For example, the stagnation could be taken to be an indication that the theoretical development of fundamental physics has come to an end. If we have already discovered all the fundamental laws, concepts and principles and integrated them into a coherent structure of the standard model, how can we find anything fundamentally new? Thus 'No new physics!' is the slogan of the apologist for the standard model.

Other physicists, such as string theorists, disagree. For them the success of the standard model, or even the adequacy of QFT as a framework for describing and understanding the nature and texture of the physical world, is far from perfect. First, there are too many empirical parameters that cannot be derived and understood from first principles. Second, the unification achieved is only partial: the electroweak theory and quantum chromodynamics (QCD) for the quark–gluon interactions are still separate pieces, let alone their unification with gravity. Moreover, the understanding and handling of physically meaningless infinities are far from satisfactory. But most importantly, there is no way to incorporate gravity into the framework of QFT. Thus the stagnation, from a string theorist's perspective, is only a temporary silence before the thunderstorm, i.e. another conceptual revolution, which, though it may not be related to the current pursuit of string theory, will radically revise many basic assumptions and concepts adopted by QFT, such as point excitations and the very basic ideas of space and time.

Still other physicists, mainly mathematical physicists pursuing the algebraic approach to QFT, feel that the stagnation reflects the profound crisis of traditional (Lagrangian and functional integral) approaches to QFT. According to the most critical among this category, the crisis has its roots in the inadequate pursuit of quantization and the unjustifiable neglect of the locality principle, a condensed form of the most important experience of 20th century physics. These led to the occurrence of infinities in the formalism, and to the superficial struggles for circumventing them, which have prevented us from gaining a deep understanding of the intrinsic meaning of local physics.[1]

Different assessments notwithstanding, the stagnation itself has provided physicists, mathematicians, and historians and philosophers of physics an opportunity to examine carefully and reflect thoroughly upon where QFT stands now and how it has evolved into the present situation. Conceptually, this kind of examination and reflection is indispensable for a proper understanding of the true meaning of QFT, and is also necessary for detecting and pursuing new directions in which theoretical physics may develop.

1 Reality, ontology and structural realism

In undertaking a historical and philosophical analysis of conceptual issues in QFT, different people naturally have different concerns. For mathematicians or mathematics-oriented physicists, the major concern is with the precise and rigorous proof of the existence and consistency of the symbolic system adopted by QFT. Thus plausible arguments adopted by practising physicists are far from satisfactory and convincing, and a large part of the conceptual development of QFT, seen from this perspective, seems to be restricted only to the so-called heuristic physics. For physicists who take QFT as part of empirical science, the physical interpretation of the adopted mathematical structure, and its power to provide a unified description of various observed or observable phenomena and make empirical predictions are far more important than the formal concern with existence and consistency, which they see as stifling their creative activities. For conceptual historians and in particular for philosophers, the major concern is with the presuppositions about the world adopted by QFT and the world picture suggested by its conceptual structure. They are also interested in understanding the successes and failures of QFT and the significant changes in QFT in terms of its basic entities and theoretical structure.

Different concerns lead to different opinions on various issues. But the basic dividing line that separates people is a philosophical one. If one takes an instrumentalist attitude towards scientific concepts and theories, then there is no room for any deep conceptual analysis. The only thing one can do is to compare concepts and theories with experience and to see how successful they are in making verifiable predictions. Once the link between theory and the world is cut and removed, the success or empirical adequacy becomes a primitive parameter and can have no explanation.

---

[1] See B. Schroer (1996): 'Motivations and Physical Aims of Algebraic QFT,' manuscript (July 1996).

Thus no further conceptual analysis of the theoretical structure can be done, or is even desirable.[2]

The realists, in contrast, make (and also try to justify) assumptions about the world in their conceptual structure. That is, they assume that some concepts are the representation of the physical reality, while acknowledging that others may merely be conventions. This entails a complexity in the theoretical structure, which is divided in broad terms into two parts: a representative part and a conventional part. The dividing line in most cases is vague, uncertain and shifting. And this requires a conceptual analysis to clarify the situation in a particular case at a particular moment of the conceptual evolution of a theory. For example, the answer given now to the question 'Are quarks real?' may be quite different from the answer given in 1964 when Gell-Mann first invented the concept of quarks. The situation with the Higgs particle is similar but with more uncertainty at this moment. Still more difficult is the question as to the reality of virtual particles: they can either be dismissed as an artifact of the perturbative expansion in terms of free particle states, or seen as entailed by the basic assumption of quantum fluctuations in QFT.

Another deep question involves the function played by conventions in a theory, and asks why it is possible for them to play such a function. Some realists assume that conventions are not merely conventions but encode within them some structural information about the entities under investigation. Conventions, such as particular ways of fixing the gauge, by definition can be replaced by other conventions. But the structural information encoded in them has to be retained, perhaps in a very complicated way. In short, the whole issue concerning the relationship between formalism and reality has its roots in the realistic attitude towards scientific theory and its ensuing division of a theory into representative and conventional parts.

Further complications in the conceptual analysis of the theoretical structure come from another assumption of the realists about the causal-hierarchical structure of the physical world. Some physicists assume that entities and phenomena of the world are causally connected and layered into quasi-autonomous domains. Usually a relatively upper level domain (or relatively restrictive phenomena) can be understood or derived, sometimes wholly, sometimes only partially, from a relatively deeper level (or relatively universal phenomena), which is assumed to be primitive in the derivation. While the relationship between domains at different levels, which are grasped by theoretical models at distinctive cognitive levels, is very complicated, involving both reducibility and emergence, the causal-hierarchical relationship within each domain in terms of the primitive and the derivative is universally assumed in scientific practice. Without this assumption, no theoretical discourse would be possible.

This assumed causal-hierarchical structure of the world is believed to be embodied in the hierarchy of conceptual structures of scientific theories. For the realists, a very

---

[2] On the 'divisive rhetoric' of this paragraph, Arthur Fine has made very interesting comments: 'I do not think that the issues you discuss about ontology would simply be dismissed by an instrumentalist, as you suggest. After all, as van Fraassen would say, the instrumentalist is committed to the QFT worldview even if she does not necessarily believe it. So, the instrumentalist too would like to know to what she is committed. I think that it is almost always a mistake when one says that the realist is interested in this or that feature of science but the instrumentalist not.' (Private exchange.) It would be very interesting to have a conceptual analysis of the ontological commitment and theoretical structure of a scientific theory, such as quantum field theory, from an instrumentalist's perspective. This would provide a testing ground for a proper judgment of the two competing philosophical positions: realism and instrumentalism.

important task in conceptual analysis, the so-called foundational analysis, is to analyze the logical relationship between the primitive concepts, which are assumed to be the representation of what are primitive in the domain, and the derivative ones in a conceptual structure. The latter have historically emerged from empirical investigations; and thus their logical relationship with the primitive ones is not explicitly clear.

Some physicists claim that they feel no need for foundations. This does not mean that their scientific reasoning is detached from the logical structure of concepts in their discipline, which represents the causal structure of the domain under investigation. What it means is only that their intuition at the heuristic level, which usually takes the accepted understanding of the foundations for granted, is enough for their daily researches. However, in a situation in which complicated conceptual problems cannot be understood properly at the heuristic level, or in a period of crisis when basic preconceptions have to be radically revised for the further development of the discipline, a clarification of the foundations is badly needed and cannot be avoided without hindering the discipline from going further. Nonetheless, the difficulties in grasping the questions in a mathematically precise way and in the conceptual analysis of the unfamiliar logical and ontological foundations would deter most practising physicists from doing so.

From a realist point of view, the clarification of what the basic ontology is in a given theory is a very important aspect of the foundational discussion. Here the basic ontology of a theory is taken to be the irreducible conceptual element in the logical construction of reality within the theory. In contrast to appearance or epiphenomena, and also opposed to mere heuristic and conventional devices, the basic ontology is concerned with real existence. That is, it is not only objective, but also autonomous in the sense that its existence is not dependent upon anything external, although its existence may be interconnected with other primary entities. As a representation of the deep reality, the basic ontology of a theory enjoys great explanatory power: all appearance or phenomena described by the theory can be derived from it as a result of its behavior.

It is obvious that any talk of a basic ontology involves a reductive connotation. Ontological and epistemological emergence notwithstanding, a reductive pursuit is always productive and fruitful in terms of simplification and unification in a scientific discourse, and thus is always highly desirable. Furthermore, the ontological commitment of a theory specifies the basic entities to be investigated, provides stimulation and guiding principles for further theoretical and experimental researches, and dictates the structure of the theory and its further development, which results in a series of theories, or a distinctive research program.

In order to clarify what the basic ontology of QFT is, we have to realize that any statement about an ontology refers to certain underlying particular entities and their intrinsic properties, mainly through reference to the structural characteristics of these entities. One can even argue that any ontological characterization of a system is always and exclusively structural in nature. That is, part of what an ontology is, is mainly specified or even constituted by the established structural properties and relations of the underlying entities. Moreover, this is the only part of the ontology that is accessible to scientific investigations through the causal chains that relate the structural assertions about the hypothetical entities to observable phenomena. The recognition that structural properties and relations are constitutive of an ontology is crucial in our understanding of the nature of space-time that, arguably, has underlain, or individuated, or at least indexed local fields, which in turn are arguably the basic ontology of the traditional QFT.

Various issues related to the place of space-time in the theoretical structure of QFT have been addressed by authors in this volume, and I will turn to some of them shortly. Here I only want to stress that in any discussion of the basic ontology of QFT, the distinctive theoretical context has to be clearly specified. For example, in various formulations of QFT that are based on the concept of space-time provided by the special theory of relativity (STR), we can find two categories of entities: local fields individuated by space-time points, and non-individual quanta[3] that are global in nature. If we take Kant's position that the individuation and identification of entities are necessary for our conception of the world as consisting of distinct entities,[4] then we may take local fields as the basic ontology and quanta as an appearance derived from the fields. As we shall discuss in the next section, there are good reasons to argue for a particle ontology instead. This position requires a rejection of the Kantian metaphysics, which, however, often cause some confusions that are not unrelated to the conflation of several theoretical contexts at different cognitive levels.

For example, it is legitimate to argue that a conceptual analysis of a complex concept (such as that of a local field) into various elements (such as its substantial stuff, the various properties attached to it, and individuating space-time points) does not generally ensure that these elements have their own existence. However, in the discussion of the ontology of STR-based QFT, this argument itself does not give sufficient grounds to reject the autonomous existence of space-time points, which are presupposed by STR as an irreducible structure underlying a field system. Some scholars argue, mainly by adopting Einstein's argument against the hole argument in the context of general theory of relativity (GTR),[5] that space-time points have no reality because they have to be individuated by the metric field that defines the spatio-temporal relations.[6] It is further argued that the classical metric field, as an agent for individuating space-time points that individuate local fields, itself is not a primitive entity, but only an appearance of some substratum that is quantal in nature: an argument for quantum gravity (QG). These arguments are interesting in their own right. They have improved our understanding of the nature of space-time, and pointed to a new direction for the further development of QFT, which requires a clarification of the inter-theoretical relationship of the concepts involved. But these arguments are irrelevant to the discussion of the ontology of STR-based QFT. In order to claim their relevance to the discussion about the ontology of QFT, we have to have mathematical formulations of GTR-based or QG-based QFT in the first place.

---

[3] For an analysis of why quanta can be regarded as physical entities without individuality, see Steven French and Décio Krause in this volume.

[4] This is compatible with the ontological commitment, in fact has provided the metaphysical foundation, of Cantor's set theory and all of classical mathematics.

[5] For a detailed historical investigation and insightful conceptual analysis about Einstein's hole argument against the idea of general covariance and his later argument against his own hole argument, see John Stachel (1989): 'Einstein's search for general covariance, 1912–1915,' in *Einstein and the History of General Relativity* (eds. D. Howard and J. Stachel, Birkhauser), 63–100.

[6] Although the lack of individuality of space-time points does entail their lack of reality, this entailment is not generally true because some non-individual entities, such as quanta, are also real.

This brings us to another issue of interest: the nature of mathematical concepts and formalisms. In our construction of a physical theory to approximate the structure of reality, we have to use mathematical concepts (such as the continuum, the vacuum, a local field, the bare charge, gauge symmetry, ghosts, etc.) as an idealization and exploit their logical ramifications to the ultimate conclusions, because this is the only window through which we can have access to reality. Then, in view of the shifting line separating representative and conventional parts of a theoretical structure, an important interpretive question arises: what is the criterion for the physical reality or partial reality of a mathematical concept?

Obviously only those mathematical concepts that are indispensable and independent of particular formalisms, or invariant under the transformations between equivalent formalisms, have a chance to claim to be part of reality, either as a basic ontology or as derivative structures. A case in point is the status of ghosts in non-Abelian gauge theories: they have appeared in some, but not in other, empirically equivalent formalisms, and thus cannot claim to be part of physical reality. It is debatable, however, whether gauge invariance should be taken as a criterion for reality, although it is generally accepted to be relevant to observability. A trivial counter-example is the accepted reality of the non-gauge-invariant charged fermion field (such as the electron or proton field). A justification for its reality, from a structural realist point of view, is that this non-gauge-invariant concept has become a universal carrier of structural relations in various experimental situations. The empirical and logical constraints are so tight that the acceptance of its existence becomes mandatory if we do not want to give up what we have achieved, since the scientific revolution in the 16th and 17th centuries, in scientific reasoning and hypothetico-deductive methodology in general. But its claim to reality can only be partial, extending only with respect to the structural information the concept carries with it so far.

2 Particles and fields

Historically, in the physical interpretation of STR-based QFT, particularly in the discussion of the field–particle relationship, the operator formalism of free fields and their Fock space structure has played a central role. The statement that particles are just the quanta of the fields makes sense only within this formalism, and many difficult interpretive questions hiding behind this statement become discernible when we take a closer look at the logical structure of the formalism. In order to address these questions, we have to pause and consider the core concept of the structure: quantization.

With regard to the idea of quantization, physicists as well as philosophers are divided into two camps. The first endorses the idea of active quantization, which presumes the necessity of finding a given classical reality, or a classical structure, or a classical theory, and then tries to establish a procedure to quantize it. The second camp, that of the quantum realist, rejects the two-step strategy and tries to formulate a quantum theory directly without quantizing a classical system.

Initially, the indispensability of classical structures for quantum theories was forcefully argued by Niels Bohr on three related grounds. First, the need for unambiguous communication requires classical concepts. Second, quantum theory makes sense only when the system it describes is measured by classical apparatus. Third, there is the correspondence principle: any quantum system, and thus quantum theory, must have its classical limit.

Bohr's correspondence principle provided the ultimate justification for the idea of active quantization. But it had foundered as early as 1928 when Jordan and Wigner tried to incorporate fermions into the operator formalism and to interpret fermions as quanta of a fermion field: there simply was no classical fermion field available to be quantized and the quantum fermion field they introduced had no classical limit. Thus, the measurement problem notwithstanding, the quantum realist feels confident that the microscopic world is quantum in nature, regardless of the existence of its classical limit, which may or may not exist at a certain level, although ultimately, at the macroscopic level, it should have classical manifestations.

The first attempt to formulate a relativistic quantum framework for the description of particles and free fields without recourse to the idea of active quantization was Wigner's representation theory,[7] which, however, can be shown to be compatible with and equivalent to canonical quantization. This, together with the fact that various formalisms of QFT are constructed upon classical space-time, or classical histories in the case of the path-integral formalism, has raised a question: 'What is the meaning of being quantal, in contrast with being classical, which may or may not have a classical counterpart?'

From its genesis (Planck, Einstein and Bohr), the term 'quantum' was related to the assignment and measurement of physical properties in the microscopic world and their discreteness. The condition for ensuring certain discreteness of the measurable properties was called by Bohr and Sommerfeld the quantization condition. In the context of matrix mechanics, the Bohr–Sommerfeld quantization condition was converted into the canonical commutation relation (CCR). But the function played by CCR remained just the same: to ensure the discreteness of physical properties in measurement.

Deeper implications of the concept of quantum were explored soon after the formulation of CCR, which set the limit for the simultaneous measurements of certain pairs of properties. The most profound consequence was of course Heisenberg's uncertainty relations. On the surface, the uncertainty relations address the epistemological question of measurement. However, for justifying or simply understanding experimentally well-established relations, an ontological presupposition is indispensable: we have to assume the existence of intrinsic (perhaps also primitive) fluctuations in the microscopic world, which are numerically controlled by the uncertainty relations. Fluctuations of what? Physical properties. What the uncertainty relations tell us is just this, no more. In particular, no question as to whose properties are fluctuating is addressed by the implications of the concept of quantum or uncertainty relations. It is a line of thought whose upshot is this: for a system to be quantal, the discreteness of its properties must be assured, the uncertainty relations must be obeyed, and the intrinsic fluctuations must be assumed. Does a quantum system require a classical structure to be its underpinning? Physically, it is not necessary: a system can be intrinsically quantum in nature without a classical substratum. But there is no incompatibility either; STR-based QFT is a case in point.

It should be clear now that the idea of quantum or quantization sheds no light at all on the ontological question as to whether the underlying substratum of QFT consists

[7] E. P. Wigner (1939): *Ann. Math.*, **40**: 149.

of particles or fields. In order to develop the idea of particles as field quanta from the concept of a quantum system, several steps have to be taken. First, a quantum field has to be assumed. It does not matter too much whether this quantum field is intrinsically quantal or is quantal only as a result of quantizing a classical field. The difference only refers to the difference in the existence of a classical limit, but has no impact on the nature of the quantum field whatsoever.

The exact meaning of a quantum field is difficult to specify at this stage. The most we can say is only that there are two constraints on the concept. First, devised as a substratum for generating substantial particles (meaning those that carry or are able to carry energy and momentum), the field itself cannot be merely a probability wave, similar to the Schrödinger wave function, but has to be substantial too. Second, as a quantum system, which by definition is discrete, it is different from a classical field, which is continuous in character.

However, as we shall see, this intrinsically discrete system is not equal to a collection of discrete particles. Thus no particle interpretation of a quantum field is guaranteed. Worse still, even the very existence of discrete particles is not entailed automatically by the concept of a quantum field. Thus a widespread misconception that we can get particles from a field through quantization should be dispelled. In order to extract particles from a quantum field, we have to take other steps. The most important among them is that we have to assume the excitability of the field as a substratum; the source of excitation can be the intrinsic fluctuations of the field, or external disturbances; and what is to be excited, just as what is to be fluctuating, is the property of the field, which characterizes the state of the field.

Before specifying the further conditions for extracting particles from a quantum field, let us look at the arguments against taking the field as the basic ontology of QFT. These arguments are interesting because otherwise it would be so natural to take the field ontology for granted, considering the fact that in this line of reasoning the field is taken to be the primary entity from which particles are extracted.

A strong case can be made that empirically only particles are observed, and fields, except for the classical fields, are not observable.[8] This suggests relegating the concept of the field to the status of a convention, a device for generating particles and mediating their interactions. The case becomes even stronger when physicists realize that from the viewpoint of particle interactions, as the theory of collision suggests, what matters is the asymptotic state of particles but not the interpolating field. Besides, as Borchers's theory shows, the S-matrix does not distinguish a particular interpolating field within an equivalent class. In the same spirit, the point-like local fields in the algebraic approach to QFT were only allowed to have a function similar to that of coordinates for the local algebra of observables.

However, in order to develop a particle interpretation of QFT, two steps have to be taken. First, conditions under which the notion of particle emerges from that of the field have to be specified. Second, it has to be shown that the physical content of QFT is in fact exhausted by the notion of particles. The first challenge is met by the construction of the Fock space representation, which permits a number operator and a unique vacuum state with an eigenvalue zero of the number operator. However, some inherent difficulties have already been built in. First, in order to define the number operator, a notion of energy gap between the vacuum and the first particle

---

[8] See remarks by Sam Treiman and Fritz Rohrlich in this volume.

state has to be presumed, which however is not applicable to massless particles. Second, in specifying the spectral condition, particularly in picking up a preferred vacuum state, the Poincaré group has played a key role, but this is only a symmetry group of a flat space-time. The recognition that space-time is in fact not flat has cast deep doubts upon the whole Fock space approach, although this would go beyond the context of STR-based QFT.

Thus, within and only within the Fock space representation, particles can be taken as associated with excited states of the field, or as the quanta of the fields, and the structure of Green's functions and the S-matrix can be analyzed in terms of particles. But the Fock space representation can only be defined for free fields. In the case of interacting fields, there are many unitarily inequivalent representations, among which only one equivalence class involves the Fock representation. The latter, however, can only be defined asymptotically, and thus is not generally definable.

Even within the Fock space representation of free fields, the notion of particles as quanta is difficult to define. In addition to the well-known lack-of-identity problem, which, as Steven French and Décio Krause suggest in this volume, requires a radically new formal framework for understanding the notion of quanta and its underlying metaphysics, there is the famous Unruh effect that has shaken the foundation of the very notion of particles: a uniformly accelerating observer in flat space-time feels himself to be immersed in a thermal bath of particles when the quantum field is in its vacuum state. This shows that how a particle detector responds in a given Fock space state depends both on the nature of the detector and on its state of motion. In general, the particle interpretation emerges only when the field is coupled to a simple system (a particle detector) and the effects of the interaction are investigated in flat space-time or in a curved but stationary space-time.[9]

The ontological primacy of particles is challenged by another consideration. In a system of relativistic particles in interaction, measurements of the position and momentum of individual particles can be made only in asymptotic regions. This suggests to some positivism-oriented scholars that only the S-matrix describing a physical process deals with reality, and that the notion of particles as part of the process emerges only in the asymptotic regions, and thus does not itself designate a basic ontology.

An advocate of the particle ontology may retort that this argument presupposes a direct linkage between reality and observation; but the linkage may be indirect and, in this case, the argument collapses. It can be further argued that in the path-integral formalism, which is empirically equivalent to the operator formalism, the notion of particles enters *ab initio*, thus enjoying an ontologically primary status. But even in this case, the physical content of QFT cannot be exhausted by the notion of particles. Here the key concept situated at the center of the ontological discussion of QFT is the concept of the vacuum.

If we take particles as the primary entities, then the vacuum can only mean a state of nothingness. Dirac's concept of the vacuum as filled with unobservable zero energy photons and negative energy electrons was dismissed as early as the 1930s as an unnecessary complication, and thus cannot claim to be physically real. On the other hand, if we take the fields as the primary entities, then the vacuum, as a

[9] See Robert M. Wald (1994): *Quantum Field Theory in Curved Spacetime and Black Hole Thermodynamics*, p. 3 (University of Chicago Press).

ground state of a quantum field, designates a state of a substantial substratum. As a state of a substantial substratum, the vacuum can be conceived, as Bryce DeWitt argues in this volume, as a cauldron of interacting virtual particles arising from fluctuations, and as carrying a blueprint for the whole field dynamics. Such a fluctuating vacuum was taken to be the physical basis for the calculations of renormalization effects in the 1930s and 1940s. More direct indication for the substantiality of the vacuum state was provided by the Casimir effect first observed in the late 1940s, which shows that the energy density of the vacuum is not zero, but negative.[10]

An advocate of the particle ontology may argue that the physical content of the vacuum conceived as the ground state of the field actually can only be captured in terms of particles, even if only in terms of virtual but not real particles. But this argument is inconsistent and cannot stand by itself. It is untenable not because of the unobservability and the lack of necessity (as artifacts of perturbation calculations) of virtual particles, but mainly because the very idea of fluctuations can only be applied to some state of some entities, or some substance, or some substratum, but not to the state of nothingness. The properties of something may be fluctuating, but 'nothing' cannot be fluctuating. The incompatibility between the concepts of fluctuations and the vacuum state within the framework of the particle ontology has removed the conceptual basis for the notion of virtual particles in that same framework; and the concept of a fluctuating vacuum, even though it still remains within the Fock space framework, goes in fact beyond the framework of a particle ontology.

Moreover, in the field ontology framework, the number of particles is only a parameter characterizing the state of a free field. Thus the appearance of virtual particles as a result of the fluctuating changes of the field state, and the physical effects (such as renormalization effects) they cause, are conceivable. Even the creation and annihilation of real particles are conceivable if enough changes can be brought in from external disturbances. But in the particle ontology framework, particles are themselves entities. Thus, considering the primacy and fundamentality of ontology, the fluctuations of the properties of particles, or any external disturbances, cannot be conceived as the cause for the creation of real or even virtual particles. On top of all these difficulties, there are two more persistent difficulties which have further weakened the particle's candidacy for the basic ontology of QFT. First, the particle as a global concept is not spatially localizable. And second, in the case of interactions, it is not well defined.

A consistent interpretation of QFT, suggested by the operator formalism, seems to be this. The basic ontology is the quantum field. The particles, or the quanta, as the manifestation of the excited states of the field, characterize the states of the field. They can be empirically investigated and registered, but they do not exhaust the physical content of the field. In a very deep sense, the concept of quanta, as objective but not primitive entities, as phenomenological indicators for the complicated structural features of the primitive field (or substratum) manifested in various situations, has well embodied the idea of structural realism.

---

[10] Although a different interpretation of the Casimir effect, not based on the notion of fluctuating vacuum and operator formalism in general, but based on, for example, Schwinger's source theory, is possible. See Svend E. Rugh, Henrik Zinkernagel and Tian Yu Cao: 'The Casimir Effect and the Interpretation of the Vacuum,' forthcoming in *Studies in History and Philosophy of Modern Physics*.

3 Locality and consistency

In 19th century physics, fields were conceived as the states of the ether that permeates or even coincides with the Newtonian absolute space, either in the form of the mechanical ether (Maxwell), or in the form of the demechanized ether (Lorentz). They were also conceived as the states of space itself, as the distribution of physical properties over various localities of space, an idea that was first speculated on by Faraday, and then picked up by Einstein in his early reflections on the foundations of physics. In either case, the ontological priority of space over fields was presumed.

The situation remains essentially the same even in quantum physics. In STR-based local quantum field theory, according to the operator field formalism, local quantum fields should be conceived as the local excitations of the vacuum. If we interpret the vacuum as space itself, we return to Faraday and take a substantivalist view of space. Even if we interpret the vacuum not as the void space itself, but as the ground state of a substratum, a substantial underlying field that is quantal in nature, the ontological priority of space-time (which replaces space in the context of STR) has not been challenged: the very idea of local excitations requires that the fields at least have to be indexed by space-time points, which presumes the prior (or at least inseparable) existence of space-time itself.

It is beyond dispute that the essential spirit of field theories is captured by the concept of locality, from which a large portion of their physical content can be derived. In the context of QFT, however, the concept of locality has two separate although not unrelated meanings: local fields and local interactions. Mathematically, local fields endowed with complicated physical properties are represented by spatially dimensionless and structureless point-like fields indexed by space-time points.[11] In terms of point-like fields, the causal structure of STR can be extended from the observables to the unobservable point-like fields and expressed by the vanishing of space-like commutators. The physical meaning of local interactions is the rejection of action at a distance. But in the theoretical system of point-like operator fields, the idea of local interactions requires that all primary quantities of interest should be expressed as the expectation values of operator products at the same space-time point. It is obvious that without the ontological assumption of the point-like fields, no concept of local interactions or local couplings can be properly defined.

The concept of local couplings is consequential and pivotal to the understanding of the structure of QFT. Since we have already assumed the existence of quantum fluctuations in the microscopic world, which is epistemically entailed but ontologically presumed and numerically controlled by the uncertainty relations, a local coupling means the coupling of infinitely fluctuating fields, which leads to ultraviolet divergences in QFT calculations. The occurrence of ultraviolet divergences was initially conceived to be an indication of the inconsistency of quantum electrodynamics (QED) in the 1930s and early 1940s, but has been properly recognized, first by Schwinger in 1948, as a sign of the limited validity domain of QED and of the existence of a new physics. In the same style of reasoning, the absence of ultraviolet divergences in quantum chromodynamics (QCD) suggested by the renormalization group calculations is taken to be an indication that QCD is a complete theory with

---

[11] The replacement of a point-like field by the smeared value of the field in a small neighborhood of the space-time point, as advocated by axiomatic field theorists, has changed nothing essential. In fact, it can only be taken as a mathematical manipulation on the basis of the concept of point-like fields.

unlimited validity domain, thus eliminating any possibility of a new physics.[12] Conceptually, however, it is not quite clear how to reconcile the renormalization group calculations with the concept of local couplings at short distances.[13] In addition, a new physics may still be required at least at large distances where the QCD coupling becomes very strong.

Physically, the ultraviolet infinities have their origin in the hidden assumption that the bare parameters are measured at zero distances or infinitely high energies. The recognition that physical parameters are only measurable at finite distances led to the idea of perturbative renormalization. That is, in the perturbative calculations, if we redefine the physical parameters as those we observe in laboratories by incorporating the ultraviolet infinities into non-observable bare parameters, and if we replace the bare parameters in the Lagrangian by formal power series in the renormalized parameters, then from the naive power counting point of view, the perturbative expansion of renormalizable theories in the renormalized parameters is finite to all orders without changing the formal structure of the Lagrangian. But this is impossible for nonrenormalizable theories. For this reason, the consistency of the QFT framework is claimed to be restored by renormalization, whereas nonrenormalizable theories had been rejected for a long time because of their inconsistency.

In low energy practical calculations, however, there is no big difference between renormalizable and nonrenormalizable theories. The difference is largely conceptual and theoretical: whereas renormalizable theories are insensitive to possible new physics at higher energies, that is not the case for nonrenormalizable theories. The difference should be taken seriously only when we take QFT as a fundamental framework and demand that all theories within this framework should be globally valid at all energy regimes. However, if we recognize that all quantum field theories are only an effective theory, valid only in a limited energy regime, then the difference, although it remains, seems almost irrelevant, and the very problem of ultraviolet infinities seems also to have evaporated. But this requires a change of perspective and a radical modification of the concept of fundamentality in physical theories (which will be discussed in later sections).

For some mathematics-oriented physicists, however, even the consistency of perturbatively renormalizable theories remains problematic: globally conceived, this consistency cannot be taken for granted, but rather has to be proven in a mathematically rigorous way. That is, it has to be shown that a system of nonlinear equations for interacting fields has exact solutions compatible with quantum mechanics. From this perspective, perturbatively renormalizable theories such as QED and QCD are judged to be inconsistent, and the constructive program is just a response to this consistency problem. The central theme of constructive QFT is the meaning of renormalization without perturbation theory, and the theory itself should be taken as a nonperturbative theory of renormalization for QFT models.

Crucial to the constructive program is the concept of phase cell localization, with which dominant degrees of freedom associated with a certain momentum range and simultaneously localized in a certain space-time volume (a phase cell) are

[12] A more careful analysis of this situation is suggested by Stephen Adler: 'I think that here you should distinguish between (a) divergences requiring renormalization, which are still present in QCD, and (b) presence or absence of an ultraviolet Landau ghost, which differentiates QED from QCD.' (Private correspondence.)
[13] See Cao's discussion in this volume.

identified and selected in functional integrals representing interactions. Since each phase cell has only a small number of variables, there is no ultraviolet problem within a given phase cell. Thus the global behavior of the interacting field system can be analyzed in terms of the dependence among different cells. In order to establish the effective decoupling among cells, a set of rules is designed for the inductive construction from one length scale to the next one. It consists essentially of integrating out those degrees of freedom localized in large momentum phase cells and leaving an effective interaction with resulting convergence factors associated with the remaining degrees of freedom.

With this method, in cases where the dimension of space-time is less than four, it can be shown that there is an exponential decay of the correlations among phase cells as a function of the scaled distance. This decay is indicative of a weaker sense of locality and establishes the effective decoupling of different cells. Thus a convergent expansion can be defined and the ultraviolet problem is solved. The case of four dimensions remains open, but Arthur Jaffe is optimistic and believes that technically a constructive solution to the ultraviolet problem of QCD and other Yang–Mills theories is within reach now.

Two features of the constructive theory are worth noting. First, with the help of phase cell localization, the global gauge fixing can be replaced by a more sensible local gauge fixing in each phase cell followed by patching local sections together. Second, the phase cell expansion bears some resemblance to Kadanoff's renormalization group analysis, but differs from the latter by being inductive rather than iterative.

4 Renormalization group

The concept of renormalization group, as the name suggests, has its origin in perturbative renormalization of QFT. In its later development, however, the richness of the concept has been explored in two different ways and has resulted in two different versions: the Gell-Mann–Low version mainly applied to QFT models and the Kadanoff–Wilson version mainly applied to models in condensed matter physics. In contemporary discussion, the deep connection and subtle difference between the two is a controversial subject.[14]

In the context of QFT, the recognition that bare parameters are defined at short distances and physical or renormalized parameters are defined and measurable at finite distances suggests a deep linkage of the behavior of fields across disparate (length or equivalently energy) scales. This linkage can be physically justified by the idea that the fields' behaviors at different scales are connected through their intrinsic quantum fluctuations owing to the coupling of the fluctuations at different scales. The difference and linkage of parameters at different scales can be summarized by the concept of scale dependence of parameters, the physical justification of which is usually provided by the screening or anti-screening effects of the fluctuating and polarized vacuum.

The idea of a continuous scale dependence of electric charge, or equivalently the idea that charge can be renormalized at various scales and that the variation of the renormalized charge is continuous and smooth, first suggested by Dirac in 1933

---

[14] See contributions to this volume by Fisher, Weinberg, Nelson and Cao. See also D. V. Shirkov (1993): 'Historical remarks on the renormalization group,' in *Renormalization: From Lorentz to Landau (and Beyond)* (ed. Laurie M. Brown, Springer), 167–186.

and then picked up by Dyson in his smoothed interaction representation in 1951, was fruitfully explored by Stueckelberg and Petermann in 1953 to suggest a group of infinitesimal operations associated with charge renormalizations at different scales. Independently, Gell-Mann and Low also exploited Dyson's renormalization transformations (on the basis of the idea of renormalizing charge at a sliding energy scale) and established functional equations for analyzing the variation of the physics with the change of scale at which we investigate the behavior of the system. These equations are particularly powerful in analyzing the short distance behavior of QED and, in particular, QCD interactions.

The power of the renormalization group equations thus established can be greatly enhanced if they are formulated, in accordance with the spirit of the algebraic approach to QFT (that physical reality of a theory can only be reconstructed from observables of the theory), not in terms of non-observable fields, but in terms of the local observables of the theory. The observable algebra can be extended to a field algebra, and each algebra has many representations. Thus by applying some selection criteria, such as a spectrum condition and the idea that elementary systems are local excitations of the vacuum, some physical states and symmetries (such as the states of quarks and gluons and the color symmetry) of the algebra of observables at arbitrarily small scales (the so-called ultra-particles and ultra-symmetries, which are different from the starting ones) can be uncovered as intrinsic features of the underlying theory. Although these ultra-particles may not be understood as Wigner particles with well-defined spin and mass and new superselection rules (liberated color), the physical reality of these structures, according to the criterion of the algebraic approach, is beyond doubt.[15]

In the context of condensed matter physics, the idea of renormalization group (RG) also describes the change of physics with scale. But what is meant by 'the change of physics with scale' in the Kadanoff–Wilson version of RG is much richer than the scale dependence of parameters in QFT. It mainly refers to the flow of parameters in the space of Hamiltonians with the reduction of energy scale (by integrating out unobservable short distance degrees of freedom). Here the units of interest and analysis are not parameters but Hamiltonians, and the nature of the space should be understood as a collection of all possible theories for describing certain physical problems of interest. Roughly speaking, there are two possibilities. In general, these Hamiltonians have different structures; but in some special cases, similar to the situation in QFT, the Hamiltonians connected by RG transformations may have the same parameter structure.

Crucial to the understanding of the nature of RG is the mechanism generating the flow, the RG transformations. In QFT, the RG transformations can be understood relatively easily or trivially in terms of the scale dependence of parameters. In condensed matter physics, it is much more complicated. In the original version of Kadanoff's self-similarity transformations of a system into itself (which are achieved by taking averages of spins in blocks followed by rescaling so that big blocks shrink

---

[15] See S. Doplicher and J. E. Roberts (1990): 'Why there is a field algebra with a compact gauge group describing the superselection structure in particle physics,' *Commun. Math. Phys.*, **131**: 51; R. Haag (1992): *Local Quantum Physics* (Springer); D. Buchholz (1994): 'On the manifestations of particles,' in *Mathematical Physics Towards the 21st Century* (eds. R. N. Sen and A. Gersten, Ben Gurion University Press); D. Buchholz and R. Verch (1995): 'Scaling algebra and renormalization group in algebraic quantum field theory,' preprint, DESY 95-004; D. Buchholz (1995): 'Quarks, gluons, color: facts of fiction?' hep-th/9511002.

down), the transformed or renormalized Hamiltonian has a form identical to the original one except for the renormalization of a single parameter. The flaw in Kadanoff's procedure is that there can be no justification for the idea of keeping only one effective renormalized coupling. In fact, as shown by Fisher in this volume, infinitely many new parameters will be generated by the transformations, which will inevitably carry the renormalized Hamiltonians out of the too small manifold of the original models.

In a more sophisticated version developed by Wilson, the flow is generated by systematically integrating out appropriate degrees of freedom, with the aim of reaching certain fixed points of the flow. However, in contrast to the Gell-Man–Low version in which the RG transformations have a conceptual basis in the scale dependence of parameters and can be carried out by differential operators (which are *a priori* deducible from the scale dependence of parameters), the Kadanoff–Wilson version has no standard recipe for designing effective RG transformations. Thus, as acknowledged by Fisher, the design of sensible and smooth RG transformations that will approach fixed points is an art more than a science, helped only by a general philosophy of attempting to eliminate those micro degrees of freedom of least direct importance to the macrophenomena while retaining those of most importance. If Fisher is right, then the RG in condensed matter physics can be regarded as merely a computational device, which does not represent the structural features of the physical reality, while the RG in QFT can be taken to be a representation of the scale dependence of parameters.

To be sure, the two versions are not incompatible. In fact it can be argued that each RG transformation in QFT, with which a renormalized coupling is defined, can be viewed as an implicit Kadanoff–Wilson operation of integrating out the degrees of freedom at energies higher than the energy scale at which the coupling is defined. But the difference between two versions runs quite deep and should not be ignored. The source of the difference lies in the definability of the concept of scale dependence of parameters. The concept is definable in QFT because there is only one unique and fixed theory with one or few parameters. This is not the case in condensed matter physics. Thus the concept is not definable there, or can only be understood as a shorthand for a simplified view of the RG flow. Another way of expressing the difference is to say that for QFT systems there is no characteristic scale because the intrinsic quantum fluctuations at all scales are coupled to each other and make equal contributions to physical processes. These result in renormalization effects on low energy physics due to high energy processes. In the systems described by condensed matter physics the fluctuations have a different origin: they are not necessarily coupled to each other, and the physics at a lower energy scale is essentially independent of the physics at higher energy scales.

If we look at the same situation from a different perspective, we may agree with Wilson and Fisher that the traditional view of scale dependence and renormalization, or even the Gell-Mann–Low version of RG, has made completely invisible the physics of many scales, which refers, not to the same Hamiltonian with changing parameters at different scales, but to different Hamiltonians at different scales.

The conceptual difference between the two versions also has its mathematical manifestations. Thus, in the Gell-Mann–Low version, the RG is really a group; that is, for each of its transformations there is a unique inverse operation. On the other hand, in the Kadanoff–Wilson version, the RG is not a true group but only a semi-group: its transformations normally have no unique inverse operation. The reason why generally no unique invertible transformation is possible in the Kadanoff—Wilson version is that a renormalized Hamiltonian generated by a RG transformation is only an approximation resulting from ignoring many details.

Nelson once argued that the RG in QFT also lacks a unique inverse in the sense that there are many conceivable high energy theories that would yield the same effective low energy physics. Thus there is no fundamental distinction between the two versions of RG in this regard.[16] This is true because when the same effective field theory is obtained through low energy approximation from different high energy theories, the forms of the high energy Hamiltonians must have changed by the approximations taken. But this shows only that the Kadanoff–Wilson version can also have applications to QFT systems,[17] not that it does not differ from the Gell-Mann–Low version in which the RG operates only within a fixed model. Essentially, the mathematical difference comes from the difference between the operation of an exact continuous group in one version and the operation of averaging out in the other.

In conclusion we may say that while the Gell-Mann–Low version of RG is exact and powerful but only narrowly defined, the Kadanoff–Wilson version is much richer and has a wider range of applications, although its operation is less well defined and relatively vague.

5 Effective field theories and beyond

RG as a defining principle for organizing physics, not graph by graph, but scale by scale, together with one of its profound implications (which itself is an essential feature of nature), namely the decoupling of low energy phenomena from high energy processes (which is characterized by mass scales associated with spontaneous symmetry breakings), has enabled us to understand not only the structure and dynamics of QFT models, but also the inter-theoretical relationship among these models and beyond. In particular, RG and the decoupling have helped us to understand why a description by renormalizable interactions seems selected by nature, as suggested by the success of the standard model. The explanation is clear and simple: at low energies, any general theory will appear as a renormalizable theory because all of its nonrenormalizable interactions will be suppressed by a power of the experimental energy divided by a fundamental energy scale (a heavy mass), as indicated by the decoupling theorem. Put another way, any renormalizable theory is only a low energy manifestation or an effective field theory (EFT) of a more general theory.

The above understanding of effective field theories, which first appeared in the mid-1970s, was soon developed into a new perspective, elevated to a new guiding principle, pursued as a new approach or a new research program, and advocated in recent years by Weinberg and others as a new language, a new framework, or even a new philosophy, for conceptualizing various issues in fundamental physics. Central to this new perspective is a conceptual reconciliation of presumed universality and accessible

[16] Private communication.
[17] An important application of this kind is Joseph Polchinski's 1984 paper 'Renormalization and effective Lagrangians,' *Nucl. Phys.*, **B231**: 269–295.

limited validity achievable by some physical mechanism for taking low energy approximations, such as the RG flow or decoupling.

This reconciliation makes the EFT approach superior to traditional ones both in model construction and in theory justification. In model construction, the EFT approach is more general and more flexible because it assumes no restrictions to simplicity and renormalizability and incorporates all possible terms consistent with the accepted principles and symmetries, including nonrenormalizable terms, into the most general possible theory. In theory justification, the EFT approach is more pertinent because only a limited consistency within its validity domain, rather than the unnecessary and irrelevant global consistency at short distances, is required. This stance is justified by the fact that each EFT has a well-defined validity domain delineated by a characteristic energy scale associated with a relevant spontaneous symmetry breaking, and thus any deliberation on its consistency beyond this domain becomes relatively irrelevant. From this perspective, the whole constructive program, the major concern of which is just the existence or nonperturbative renormalizability of QFT models or their global consistency, seems to be irrelevant to the acceptance of these models.

Weinberg asserts, in this volume and elsewhere, that S-matrix theory of hadron physics, which was popular in the 1960s, embodies the spirit of EFT philosophy, because it assumes all possible terms without any restrictions, and that QFT itself is only a practical and effective way of doing S-matrix theory. He even takes one further step and asserts that QFT is an inevitable result of assuming the general principles of quantum mechanics, relativity and cluster decomposition. Even if QFT is not the only possible result of these principles because of the possibility of string theory, at sufficiently low energy, he asserts, any relativistic quantum theory satisfying the cluster decomposition principle would look like a QFT. Weinberg's thesis has raised an interesting question concerning the place of general principles, as compared to concrete models, in the theoretical development of physics. While some physicists claim that implementing principles is more important in probing the secrets of nature because these principles are the condensed forms of all our past experience that should not be forgotten, others argue that the explorations of models are more fruitful because new features and new ideas about nature, such as gauge invariance or supersymmetry, would not automatically emerge from deliberations upon general principles.

Different opinions on the relative importance notwithstanding, a consensus among physicists is that the standard model can be taken as an effective theory of a deeper, if not final, theory. The reasons for taking the standard model as an effective theory are simple: its enormous empirical successes at the present energy scales and its consistency within its validity domain have justified it as a valid theory. But serious challenges have cried for a deeper and more general theory which would embrace all of its successes. Thus the standard model should be properly taken to be an effective theory of such a deeper theory, if we can find one.

Most serious among the challenges to the standard model is how to incorporate gravity. It has become clear since the 1970s that within the traditional QFT framework there is no way to have a renormalizable model for gravity. More importantly, any quantum theory of gravity has to be compatible with the classical theory of gravity, that is, with Einstein's general theory of relativity (GTR). But, as we shall see in the next section, taking GTR into consideration will remove most of the conceptual foundations of STR-based QFT on which the standard model is constructed.

Another serious challenge to the standard model is the lack of a rigorous proof of its consistency in its own validity domain, namely in the realm of the strong interactions. Here I am not referring to the short distance behavior, which ironically causes no trouble in QCD because of asymptotic freedom, but to the long distance behavior, that is, to the confinement of colored constituents. Many plausible arguments exist in terms of superconductivity models, condensation and duality, but rigorous proofs are still lacking. Here is perhaps the most pertinent area in which the constructive approach could contribute.

Still another challenge comes from some new developments in supersymmetric models. Central to the standard model is the idea of gauge invariance that dictates the dynamics of physical systems. Yet in some supersymmetric models, it can be shown that gauge symmetries are not fundamental, but rather artifacts that are stable only at long distances. Based on duality considerations, gauge particles, as 'magnetic' degrees of freedom, can be taken as composites or collective modes of the more elementary 'electric' degrees of freedom. Thus the underlying theory of 'electric' variables cannot have these 'magnetic' symmetries as its own symmetries.[18] Although these considerations have not rejected gauge symmetries as effective structures existing at low energies, with which the standard model is constructed, they do suggest the existence of a more fundamental theory from which gauge symmetries can be derived, and of which the standard model is only an effective theory.[19]

In addition to the above challenges, there are many other important questions to be addressed, such as the possibility of neutrino masses and oscillations, the existence of the Higgs particles and new lepton–quark families, the unification of the electroweak and strong interactions, the incorporation of the idea of supersymmetry, the mass hierarchy, the origin of families, the explanation of parameters, etc. Some of these questions, to use Glashow's terminology, are empirical, others are intrinsic, still others are emergent. But there are also some meta-questions that cannot be answered or even addressed within the context of the standard model.

As pointed out by Treiman in his comments, the standard model is flexible enough to adapt itself to many of these questions, and to many other possible big surprises as well. But this flexibility is not incompatible with the claim that the standard model is a mature model within the context of QFT, whose success has ruled out all other models except, perhaps, string theory (which, however, is claimed to have gone beyond the framework of QFT), because there are no new fundamental laws that can be discovered within the framework of QFT. Yet this maturity can be taken as an indication that the frontier of fundamental physics lies, not within the confines of QFT, but elsewhere, and the very existence of meta-questions seems to suggest that for the further development of fundamental physics we have to go beyond the confines of QFT.

6 Striving for new frameworks

In attempting to develop a consistent fundamental physical theory that goes beyond the standard model or the framework of QFT in general, physicists find that available theoretical resources for doing so are quite limited. Many important concepts and

[18] See N. Seiberg (1995): 'The power of duality: Exact results in 4-D SUSY field theory,' hep-th.9506077.
[19] Although, as noted by Stephen Adler (private correspondence), Seiberg's models (the more fundamental theories) 'all have large fundamental gauge symmetry groups.'

insights, such as those suggested by STR and quantum theory, have already been integrated into the old framework, except for those suggested by GTR, which is essentially a theory about the nature of space-time and gravity. In order to properly appreciate how radical it would be to assimilate the basic insights from GTR (as compared with those, such as supersymmetry, duality, Kaluza–Klein theory, or nonlocal excitations, that are not incompatible with STR's view that space-time has an independent existence) and to develop a GTR-based QFT, let us first look more closely at the place the concept of space-time occupies in the theoretical structure of STR-based QFT.

As we mentioned before, without considering GTR, space-time in field theories has at least played a role of indexing fields. This role seems to be seriously challenged by quantum theory: the limits to the measurability of fields set by uncertainty relations make it impossible to have a precise definition of a point-like space-time coincidence, which lies at the foundation of the usual concept of space-time points (or locality). This challenge, however, is partially met by axiomatic field theory, in which fields are defined not as entirely point-like local, but as operator-valued distributions with test functions having a finite compact support in space-time. Thus a weak definition of locality in terms of imprecise coincidences (coincidences not in terms of space-time points, but in terms of distributions over finite space-time regions) can be given and space-time can retain its foundational role of indexing fields. Secondary considerations of precision notwithstanding, the place of the concept of space-time in QFT remains unchanged: fields and their dynamic behavior are characterized by their locations and propagation in space-time, which as an absolute entity (a background manifold endowed with a fixed metric and causal structure) is independent of, and unaffected by, the existence of quantum fields. Essentially, this remains a Newtonian dualist view of physical reality: material content is distributed over an independently existing space-time.

A hidden assumption underlying the above conception is the substantivalist view of space-time, which however was challenged or even undermined by Einstein's interpretation of GTR. According to Einstein, space-time has no separate existence, but rather is a structural quality of the gravitational field.[20] As a derivative or phenomenological existence, space-time is nothing but spatio-temporal relations of physical entities, which are constituted by the gravitational field. Categorizing broadly, some philosophers call this interpretation of GTR a relationist view of space-time.

There is an old fashioned relationist view, according to which space-time is nothing but the totality of space-time relations that can only be defined in terms of (or constituted by) some coordinate system set up by material bodies.[21] Einstein's view shares some features of this view in the sense that it also rejects the independent existence of a background space-time, and defines the locality of dynamic entities only in terms of their relations with each other. But it has a new feature lacking in the old relationist view, namely that there is a special agent, the metric or gravitational field,[22] that is solely responsible for defining or constituting space-time relations. In

---

[20] See A. Einstein (1952): 'Relativity and the problem of space,' appendix 5 in the 15th edition of *Relativity: The Special and the General Theory* (Methuen, London, 1954).

[21] See Adolf Grünbaum (1977): 'Absolute and relational theories of space and space-time,' in *Foundations of Space-Time Theories* (eds. J. Earman, C. Glymore and J. Stachel, University of Minnesota Press), 303–373; and Paul Teller's contribution to this volume.

[22] There is a subtle difference between these two concepts; see John Stachel's clarification in this volume. But this difference is irrelevant to the discussion here.

contrast to the Newtonian space-time (which has an independent existence without physical interactions with other entities), and to the Leibnizian space-time relations (which are not entities at all), the gravitational field is a dynamic entity that interacts with all other physical entities. It is this universal coupling determining the relative motion of objects we want to use as rods and clocks that allows us to give the gravitational field a metrical interpretation, which in turn justifies its intrinsic role in constituting space-time relations. However, this constitutive role does not justify a widely circulated simple identification of the gravitational field with space-time, because this would be a categorical mistake, confusing the constituting agent with the space-time relations being constituted.

Taking Einstein's view of space-time seriously requires a radical revision of the foundations of QFT and a formulation of GTR-based QFT so that we can have a consistent fundamental theory. A simple replacement of the fixed and flat Minkowski space-time by a dynamic and curved, yet still classical, space-time as an indispensable structure for indexing quantum fields cannot claim to be a consistent fundamental theory. Even worse, it is doomed to fail because the very ideal of localization is not achievable in this context owing to the lack of a well-defined metric structure (which itself as a dynamic entity is in interaction with the quantum fields to be indexed). If we take the metric as a quantum entity, then the need for superposition (required by duality and the probability interpretation of the wave aspect of any quantum entity) makes it impossible to have any metric structure well defined for all quantum states. In both cases, the metric structure itself appears only as a special feature of certain states in the solutions of the theory, rather than a condition given for solving the theory. At most, then, the notion of localization in the context of GTR can only be defined, in a very weak sense, in terms of diffeomorphism invariance.

In such a diffeomorphism invariant QFT with no possibility of localizing fields in the traditional sense, most STR-based QFT axioms cannot even be formulated because of their dependence on the concept of space-time interval, which in turn depends on a well-defined metric structure. Consequentially, most key notions of QFT, such as the vacuum, particle states, Hamiltonian, time evolution and unitarity, simply cease to make sense. In a deep sense, a movement in this direction has indicated a crisis of the Newtonian conception of the structure of physical reality as space-time on which matter moves.

In a positive sense, a GTR-based QFT, in which space-time is eliminated as a primary entity and the focus is shifted to the gravitational fields that constitute space-time relations, suggests a quantum theory of gravitational fields. However, there are arguments for resisting quantizing gravity. For example, if we take the gravitational field not as a fundamental field, but only a description of gravitational effects, or an agent constituting the metrical properties of space-time, which are macroscopic in nature, then it would be meaningless to quantize gravity because then the concern would be microscopic in nature.

Philosophically, this is only a veiled version of the Kantian transcendental argument for the necessity of space and time: we need space-time to make sense of our experience (of local fields); a classical relationist view of space-time is tolerable because the macroscopic space-time relations, even though they are constituted by the gravitational fields, are still available for the construction of our experience of local fields; but the quantization of the gravitational fields will completely remove space-time and even space-time relations from fundamental discussions, making our experience or conception of local fields impossible, and this is unacceptable.

A similar argument prevails at a more fundamental level. According to the Copenhagen interpretation, the ontological primacy of a classical measuring system, the quantum nature of whose constituents is beyond doubt, is presumed so that quantum theory can make sense. Thus there is no reason to reject the ontological primacy of space-time or classical gravitational fields, which is transcendentally necessary for any sensible theory, and the quantization of gravity will not make the existing theory more fundamental (in the sense that all discussions are conducted at the same quantum level), if the Copenhagen interpretation is taken for granted, as is the case for most discussions of quantum gravity.

Thus in arguing for the necessity of quantizing gravity, we need reasons other than the metaphysical faith in a purely quantum theory, or the psychological desire to extend the successful quantum theories of fields to the case of gravity. This is particularly important because in fact there are no empirical data that require the analogical extension, or guide and constrain the construction of a quantum theory of gravity. All motivations in searching for a theory of quantum gravity, such as those for mathematical consistency and unification of fundamental interactions, are purely internal and speculative. Most convincing among them is that a consistent treatment of interactions between gravitational fields and other quantum fields (such as gravitational radiation) requires a quantum treatment of gravitational fields, whose classical limit will give rise to the metric structure for constructing the space-time relations. Otherwise, no formulations of GTR-based QFT can be taken to be a consistent fundamental theory.

Since at the classical level the gravitational fields are closely related to space-time relations, many physicists take quantum space-time or quantum geometry as a substitute for quantum gravity. There is even an attempt to apply the quantum principles directly to the very definition of space-time, and then to use spectra of space-time operators to capture the quantum (noncommutative) nature of the space-time in a diffeomorphism invariant way.[23] However, this diffeomorphism invariant theory is in harmony with GTR only in the sense of being a generally covariant theory. In its spirit, it is in direct contradiction with that of GTR, according to which space-time does not have an independent existence, and space-time relations are not the so-called nonsupervenient relations (which are neither determined by nor dependent on the monadic properties of the relata), but rather are constituted by the gravitational fields. Thus what should be quantized is not space-time itself, but the gravitational fields, and the 'quantum' nature of space-time should come as a result of quantizing the gravitational fields. But it is not at all clear that we can talk about quantum space-time or quantum geometry in a meaningful way if we remember that the meaning of quantum is exhausted by the discreteness in measurements that require the existence of 'classical' space-time as a precondition.

Of course, this consideration does not exclude the possibility of developing a new concept of quantization without any reference to space-time. Such a new concept is necessary for a quantum theory of gravity because here the existence of space-time is not presumed but only emerges in the classical limit of the theory. It is also possible, as Carlo Rovelli suggests in this volume, if the localization of gravitational fields is

---

[23] Alain Connes (1994): *Non-Commutative Geometry* (Academic Press).

defined only with respect to one another, whose physical states are labeled by graphs only without any reference to their space-time locations, that is, without assuming the existence of a background space-time or metric. Under these conditions, as the work by Ashtekar and his collaborators has shown, it is possible to develop a quantum theory in which dynamical variables constructed from the gravitational field have discrete spectra whereas their classical limits appear to be continuous observables, such as area and volume. Thus it is possible to obtain the metric structure, through coarse graining, from quantum operators (defined over a structureless manifold) and the related excitations in a diffeomorphism invariant theory without assuming a fixed background metric.

However, in Ashtekar's approach, a background differential manifold with a smooth topology is still assumed, and what is quantized are only the gravitational fields (or the connections) and other 'observables' constructed from them, but not the smooth manifold itself. And this has raised a serious question of profound metaphysical interest, concerning the ontological status of manifold. If a differential manifold structure is necessary even for theories like Ashtekar's in which no background space-time or metric structure is assumed, some philosophers would argue that then the manifold must be taken as the basic substance, as a kind of dematerialized ether, which is indispensable for the characterization of fields or for providing a substantial support to fields.[24]

Since a manifold without a metric is not space-time, and no dynamical variables are localized by a manifold because of diffeomorphic invariance, the manifold substantivalists, although they continue to adhere to the dualist conception of the physical world as divided into dynamical matter and a background stage, have given up much of the substance of space-time substantivalism. But even their last effort to keep a separate background stage is easy to defeat, as indicated by Stachel in this volume, if we pay attention to the actual process in which a manifold associated with a solution to the gravitational field equations is obtained: we first solve these equations locally, and then look for the maximal extension of this local patch to a global manifold that is compatible with certain selection criteria. Thus, the global structure of the manifold is not fixed, but depends on the dynamics of the theory. That is, it is not specifiable before the specification of a particular solution to the field equations.

In the absence of a background geometry, it would be impossible to introduce a Gaussian measure and the associated Fock space for defining free fields, particles and the vacuum state. Furthermore, in this case, the concepts of light-cone singularity (products of operators evaluated at the same point) and its regularization would also be meaningless because the ultimate justification of these concepts is the uncertainty relations, which would be unintelligible outside the context of measurements in space-time with a background geometry. But it is not easy, in using some concepts of QFT (such as symmetry and quantization) to construct a new theory without a background geometry, not to carry over some package of concepts from QFT. These concepts may have only implicit rather than explicit reference to the background geometry, and thus may not be explicitly provocative. But, still, they are unjustifiable in the new context because of their implicit dependence on the background geometry.

---

[24] For this kind of manifold substantivalism, see H. Field (1980): *Science without Numbers* (Princeton University Press), p. 35; and J. Earman (1989): *World Enough and Spacetime* (MIT Press), p. 155.

Thus, frequently we have encountered the adoption of illegitimate concepts, such as light-cone singularity or its equivalent and regularization, in the construction of a new fundamental theory of gravity. And this has revealed the inconsistency of the theoretical endeavor and puzzled readers who are seeking a consistent construction.

In a consistent formulation of GTR-based QFT, with its concomitant radical revision of the space-time underpinning, the foundations of STR-based QFT would be removed, and some fundamental concepts such as locality, causality, the vacuum and particles, would have to be defined in a radically different way. Thus the theory that would emerge might not be recognizable as a successor of the standard model, nor even as a new version of QFT.

In addition to the insights gained from GTR, there are other conceptual resources that can be employed to construct new frameworks that go beyond the standard model. Most important among them are the concepts of supersymmetry and strings.

It has become well known recently that in supersymmetric models, it can be shown that the idea of duality between electric and magnetic variables, which originated from the electromagnetic theory, can be extended to dualities between strong and weak couplings and between elementary and collective excitations. Thus there is no invariant distinction between the dual partners, and one can always be expressed in terms of the other. One insight suggested by duality is that perhaps some strong coupling phenomena, such as confinement, can have a weak coupling description. Another suggestion is that gauge fields (and thus gauge symmetries) are elementary only in the magnetic language, and may in fact be composites of the more elementary electric variables, which nevertheless are stable at long range. Thus they can only be taken to be long range artifacts.[25]

It is frequently argued that in order to accommodate newer symmetries, such as supersymmetry, a richer framework, namely string theory, is needed. Similar to our discussion above on the formulation of GTR-based QFT, in its conception and development string theory has received no evidential support or empirical constraints, but is only guided by the considerations of mathematical consistency and agreement with existing theories. In its first 'breakthrough' in 1984–85, the guiding idea was unification. Its second 'revolution', that of 1994–95, is said to be characterized by the convergence of such ideas as supersymmetry, Kaluza–Klein theory, minimum length, 'quantum' nature of space-time, nonlocality, duality, a finite theory without any ultraviolet divergences and a unique theory of no adjustable nondynamical parameters, features that were thought crucial for going beyond the standard model. But essentially these endeavors centered around string theory are still guided only by the consistency consideration.

The most radical new feature of string theory is that its basic degrees of freedom, strings, are nonlocal although they interact locally. However, the local interaction claim is somewhat weakened by the fact that there is no invariant way of specifying when and where the interaction takes place. The underlying conception of space-time advocated by string theorists is very interesting. Nobody disputes that the incorporation of the ideas of supersymmetry, Kaluza–Klein theory and quantization, although speculative, has radically enriched our imagination of space-time. Yet the conservative side of its conception is also striking: it takes the enlarged space-time coordinates themselves directly as physical degrees of freedom that have to be treated dynamically

---

[25] See note 18.

and quantum mechanically, thus providing the newest version of substantivalism. Different on the whole from what is envisioned by GTR-based QFT, string theory has not provided a consistent conception of space-time.

Furthermore, despite its radical claims, string theory remains a theoretical structure built on the foundations of QFT (such as an underlying fluctuating substratum, excitations, quantization, uncertainty relations, Fock space structure, symmetries, etc.) except for its basic quanta being extended entities rather than point-like fields, the possibility and necessity of which, however, is only claimed rather than being justified by any physical arguments.

Reservations regarding its claim of radicality notwithstanding, physically the mathematical manipulations of string theory are highly suggestive. First, it gives an existence proof that gravity can be treated quantum mechanically in a consistent way. Second, it has predicted many new physical degrees of freedom: strings, classical objects such as smooth solitons and singular black holes, and new types of topological defects such as D-branes. Third, it is able to count the number of states of certain black holes and thus gives a statistical mechanical interpretation to the Bekenstein–Hawking entropy. Fourth, it sheds much light on field theory duality and on the gauge-theoretical structure of the standard model. However, even at the speculative level, string theory continues to lack theoretical structures that are responsible for the structure of its substratum (vacuum) and for its dynamics.

7 Ontological synthesis, reductionism and the progress of physics

Nowadays many physicists take the standard model (or STR-based QFT in general) and GTR as effective field theories (EFTs) of a deeper theory, such as GTR-based QFT or string theory. The reason for such a conception of inter-theoretical relationships is easy to understand. Although QFT and GTR are empirically successful, there remain many interesting questions (such as the mass hierarchy and family structure of fundamental fermions, dynamics being dictated by some specific gauge groups, CP-violation and vanishing of the cosmological constant, etc., the so-called meta-questions), that are unanswerable and impossible to address within their frameworks. In addition, the established consistency and validity domains of both theories are limited by the incompatibility between STR-based QFT and GTR, both in terms of localizability of quantum states and in terms of the quantization of gravity. In order to have a theory with the widest possible validity domain, which is consistent with all physical principles established by GTR and QFT, and also driven by the desire to address meta-questions and to solve various puzzles related to them, physicists are searching for a deeper theory, which is in some sense an extension of both QFT and GTR so that in limiting cases it can be reduced to them, and thus is compatible with all empirical evidence supporting QFT and GTR. It is this reducibility that justifies calling QFT and GTR the EFTs of the deeper theory.

One discernible feature of any conceivable new deeper theory can be captured by the idea of an ontological synthesis of QFT and GTR. That is, it must take the quantal characteristic of the basic ontology from QFT (but discard its space-time underpinning as only a phenomenological description), and combine it with the basic ontological commitment of GTR that the universal space-time relations are ontologically constituted by the gravitational field (which, however, has to be quantized). The result of the synthesis is not just one new physical entity (the quantized gravitational field), but rather a new consistent theoretical structure with new conceptual foundations.

Looking at string theory from this perspective, we find that it is only a continuation and small extension of QFT. At the foundational level, string theory shares almost all the basic concepts of QFT, including (i) the general principles such as Lorentz invariance, causality, unitarity; (ii) the existence of an underlying fluctuating substratum, its space-time underpinning, its quantization, its ground state as the vacuum, and its real and virtual excitations that are constrained by the uncertainty relations; (iii) the role of gauge groups for fixing dynamics, and of other symmetry groups as constraints; and (iv) the couplings among various length scales, which are related to the idea of the renormalization group, and the hierarchy of couplings delineated by spontaneous symmetry breaking, which underlies the idea of effective theories; except for (v) the point-like excitations and the local interaction at a single space-time point. Here, although the radical difference between extended and point-like excitations have been stressed by the string theorists, we find no difference in the assumptions concerning what is to be excited and the mechanism responsible for the excitations (quantum fluctuations or external disturbances). Thus it is fair to say that string theory and QFT are two theories within the same conceptual framework, and the difference between them is not as radical as string theorists have usually asserted.

As we have shown above, the fact that both QFT and GTR (as particular languages for describing the physical world and as world views about how space-time and matter with its interactions are integrated into a conceptual whole) should be taken as EFTs of a deeper fundamental theory has raised an interesting question concerning the relationship between effective and fundamental theories.

If we take the idea of renormalization group seriously, that is, if we take physics as changing with scales, then there is no ground for saying that some theory is more fundamental than others if they are valid at different scales. In this case, they are equally fundamental, and each of them has its own validity domain and enjoys theoretical autonomy in the sense that it cannot be reduced to or derived from another theory. In particular, a theory valid at smaller length scales is not necessarily more fundamental than others because it may not be valid for phenomena described by other theories.

The above observation should not be taken as an argument against the idea of a fundamental theory. If a theory's validity domain covers the validity domains of several other theories, then this theory, with a consistent, well-defined substratum that serves as a foundation for a unified description of phenomena that were previously described separately, is certainly more fundamental than others. However, even in this case, a fundamental theory originally formulated at smaller scales (or lower levels) may not exhaust the empirical content of higher level phenomena because, although valid at higher levels, it may decouple from the complexities that emerge at higher levels.

This explains why the reductive pursuit of a fundamental theory, fruitful as it is in encouraging us to probe deeper levels of the world, which helps to give a partially unified description of phenomena under investigation, is unable to deal with the new complexities that emerge at higher levels. Furthermore, in the reconstruction of higher level complexity from lower level simplicity, more and more lower level complexities will be discovered and have to be appealed to for the success of the reconstruction. This development by itself is cognitively significant. But it also explains why ontological reductionism is doomed to fail. A telling case in point is the attempt at a reduction of hadrons to quarks and gluons. It involves the discovery of endless complications about the structure of the QCD vacuum and its fluctuations, the sought-for mass gap, renormalization group transformations, duality of strong and weak coupling phases, and so on. After the heroic efforts made by numerous physicists for more than three decades, we now find that no definite meaning can even be assigned to the most basic claim of ontological reductionism, that quarks and gluons are parts of hadrons.

If theory reduction is doomed to fail because of the doomed failure of ontological reduction, then there is simply no chance for the extreme version of reductionism, namely atomism, according to which the process of reduction will sooner or later come to an end, and all existing and logically possible theories will be reduced to a final theory, the ontological justification of which is that all physical entities and the laws and principles governing their behaviors should be reducible to the most elementary entities and the corresponding laws and principles governing their behaviors. In place of reductionism and atomism, as suggested by the ideas of renormalization group, decoupling and EFTs, is quasi-pluralism in theoretical ontology based on the idea of layered fundamentality. According to this position, causal connections between different layers exist, but each layer has its own basic entities and fundamental laws and principles that cannot be reduced to those at lower levels, and thus enjoys a quasi-autonomous status.

The implications of such an EFT perspective for our conceptions of how physics progresses is an interesting question. It may imply, as Nelson and Shankar contend in this volume, that progress is often made by moving away from the fundamental theory and looking for effective theories, and that there are good reasons to do so because decoupling makes the former irrelevant to the phenomena of interest. Furthermore, if physics, as Teller argues in this volume, can only give us a plethora of models, each one of which gives us a window on certain aspects, but not a whole panorama, of the way the world looks, and the progress is measured by answering questions that could not be answered before, then any new theory that can help us in this respect is acceptable even if it is inconsistent with other existing theories.

But such a phenomenological implication may not be the only one the EFT perspective can offer. In fact, this perspective is not incompatible with the view that physics progresses by eliminating inconsistencies and formulating more and more fundamental theories, each of which can be viewed as an EFT of the next fundamental theory. A more fundamental theory, such as GTR-based QFT and string theory, can be formulated by first detecting some vague logical connection among diverse physical ideas and various mathematical concepts by appealing to psychological associations, and then eliminating the marginal inconsistencies among them (such as that between the idea of gravity and of quantization) either by delineating the validity context of the concepts (such as quantization and space-time) or by widening perspectives (from STR-based QFT to GTR-based QFT), and finally formulating a consistent theory by ontological synthesis as discussed above. In this more fundamental theory with widened perspective, the previous meta-questions would turn into intrinsic and thus solvable questions, which is a tangible and thus a universally accepted indication for progress.

Glashow argues in this volume that with the discovery of the standard model the progress of fundamental physics has reached a mature stage: no meta-questions can be found in any other sciences but fundamental physics and Cosmology. Although many new surprises can be expected, answers to the remaining meta-questions can have no effect on our material (as opposed to moral, philosophical, etc.) lives and social well-being. As a reflection of this situation, he argues, most sciences are directed toward intrinsic and emergent questions, the solutions of which contribute to human welfare. By contrast, he further argues, the costly explorations of meta-questions in fundamental physics, although intellectually inspirational and culturally pleasurable, are practically irrelevant or even wasteful, because they absorb huge amounts of economic and human resources but add nothing to them.

Obviously, this view of fundamental physics has far-reaching implications for science policy. However, some physicists disagree and argue that fundamental physics is far from mature, and, as testimony to this assessment, note the intensive investigations in the direction of quantum gravity and string theory, which they claim have brought about rapid progress in understanding the so-called meta-questions. If this is the case, then surprises in fundamental understanding can be expected to occur, and the applications of fundamental discoveries to practical questions and their impact on human welfare cannot be precluded.[26] Clearly the implications of this view for science policy are quite different from those drawn from Glashow's view, and further discussions on this issue may be fruitful not only intellectually, but also practically and politically.

Acknowledgments

In preparing this introductory essay, which is intended only to provide a coherent context for properly appreciating the contributions to this volume, I have benefited from exchanges with Stephen Adler, James Cushing, Arthur Fine, Sheldon Lee Glashow, Joseph Polchinski, Silvan S. Schweber and Abner Shimony, and from reading manuscripts generously sent to me by Bert Schroer.

[26] Glashow insists that this kind of impact can be precluded, and is 'prepared to defend' his belief. (Private correspondence.)

Part One. Philosophers' interest in quantum field theory

1. Why are we philosophers interested in quantum field theory?

TIAN YU CAO

This two-tier conference signals a new phase of philosophers' interest in quantum field theory, which has been growing noticeably in the last few years. However, some prominent physicists have shown their deep suspicions against ignorant philosophers' intrusion into their profession, and have expressed their hostility quite openly. In the philosophy community, some prominent philosophers of physics have also expressed their suspicions against the rationale of moving away from the profound foundational problems raised by Einstein and Bohr, Bohm and Bell, such as those concerning the nature of space-time and measurement, possibility and implications of hidden variables and nonlocality, and stepping into the technical complexity of quantum field theory, which is only an application of quantum mechanics in general without intrinsically distinct philosophical questions to be explored. In order to dispel these suspicions, it is desirable to highlight certain aspects of quantum field theory which require philosophical reflections and deserve further investigations. This discussion intends to suggest that philosophers can learn many important lessons from quantum field theory, and may be of some help in clarifying its conceptual foundations. At this stage of crisis that quantum field theory is experiencing now, the clarification may contribute to the radical transformation of our basic concepts in theoretical physics, which is necessary for a happy resolution of the crisis and the emergence of a new promising fundamental physical theory.

Generally speaking, philosophers are interested in the metaphysical assumptions adopted by science and the world picture suggested by science. They are also interested in understanding the successes, failures and significant changes that have happened in science, in terms of its ontology and theoretical structure. Thus it is natural for philosophers to be interested in examining various scientific theories from these angles. But for historical and institutional reasons, many philosophers of physics have for many decades been preoccupied by questions raised by Newton and Leibniz, by Einstein and Bohr, and by Bohm and Bell, without properly appreciating novel features that have occurred in quantum field theory and statistical mechanics. They have ignored the fact that physics, both in terms of richness of its theoretical structure and in terms of the complicated world picture it suggests, has moved far away from Einstein and Bohr, and thus raised many new foundational issues to be clarified. It is true that the old profound puzzles of measurement and nonlocality continue to be a serious challenge to human intellect and deserve our reflections, just as the old puzzle of action-at-a-distance of Newton's gravity remained a serious challenge to the scholars musing during the period between Maxwell's theory of electromagnetic field and Einstein's general theory of relativity. But there is a possibility that serious reflections on the recent developments in contemporary physics may shed new light on the old puzzles, just as Einstein's theory of gravity did resolve the puzzle of Newton's action-at-a-distance. Even if it turns out not to be true for the 20th century puzzles, these recent developments in their own right remain a proper scientific context for our philosophical reflections upon science.

As a representation for describing subatomic entities and as a framework for the hierarchical structure of the microscopic world which can be built from these entities, quantum field theory embodies the reductionist view of science and is taken to be the foundation of fundamental physics: that is, the foundation of particle physics and cosmology. The reductionist pursuit reached its most spectacular success in the standard model of the 1970s, which, in addition to a unified representation of fundamental interactions, has also profoundly influenced the development of contemporary cosmology. Thus to a large extent, our present conception of nature, concerning the ultimate constituents of matter, the laws and principles that govern them, and the origin and evolution of the whole universe, is shaped by quantum field theory.

This explains why some philosophers take quantum field theory as the contemporary locus of metaphysical research[1] when they try to use current physical theory as a guide to detect and resolve metaphysical questions such as the following: 'Is the world ultimately continuous, or discrete, or dual in nature?' 'What is the meaning of the concepts of particle and field in the context of the microscopic world?' 'What is the nature of the vacuum and how should one conceive and understand the vacuum fluctuations in terms of substance that is supposed to obey the law of conservation?' 'What is the physical mechanism underlying cooperative correlations between the fluctuating local quantities at different length scales, which is assumed by quantum field theorists in their multiplicative renormalization procedure?' 'What justifies the idea of scale dependence of physical parameters that underlies the idea and technique of renormalization group in quantum field theory?' 'How does one reconcile global features, such as quanta, the vacuum and gauge conditions, with a local field theoretical framework, and what are their implications for our understanding of causal connections?' 'Is the world homogeneous or hierarchical in nature? If it is hierarchical, then what is the nature of the relationship between different levels in the hierarchy, reducible or not?' These and many other questions will be discussed in the following contributions to this volume.

Quantum field theory also provides rich material for epistemological discussion. Let me give just two examples. One concerns concepts, the other concerns theoretical structure. A persistent question in the philosophy of science is whether we should give theoretical concepts an instrumental interpretation or a realistic interpretation. The instrumentalist has difficulty in explaining the effectiveness, and the realist has difficulty with the elusiveness or underdetermination of the concepts in their relation to empirical data. Now in quantum field theory no one would deny that quantum particles and quantum fields are different from classical particles and classical fields. Thus the terms particle and field, with their inescapable classical connotation, in the context of quantum field theory, can only function as metaphors. But metaphors are useful only when they capture certain structural similarities with the

---

[1] See Howard Stein (1970): 'On the notion of field in Newton, Maxwell and beyond,' in *Historical and Philosophical Perspectives of Science* (ed. R. Stuewer, University of Minnesota Press), 264–287.

originals. That is, theoretical terms are useful only when they carry certain structural features of the objects they intend to describe. This attitude toward theoretical concepts seems to be accepted by almost all quantum field theorists. For example, the prominent physicist Rudolph Haag suggests that in the context of quantum field theory, the real function of the concept of particle is to indicate the charge structure, and the concept of field is to indicate the local nature of interactions.[2] This kind of understanding certainly supports a structural realistic interpretation of theoretical concepts, which was suggested by Henri Poincaré, Bertrand Russell, Rudolf Carnap and many other philosophers a long time ago, and has been adopted by some philosophers and conceptual historians of science, including myself, in arguing for the continuity and cumulativity of scientific development.[3]

The theoretical structure of the standard model is notoriously complicated, tightly constrained by various general requirements or principles, such as covariance, unitarity and renormalizability. One result of complying with these accepted principles is that, although a counter-example has been declared recently,[4] in most formulations of non-Abelian gauge theories we have to assume the existence of ghosts, even though we can manage to make them permanently unobservable. This has raised a serious interpretive question about the ontological status of the ghosts: should we take them only as a step in our mathematical manipulation, or give them more realistic interpretation, letting them enjoy a status similar to that of quarks, gluons and Higgs particles, if the latter are also permanently elusive to our detection? It has also raised a question about relative weights among the criteria for theory acceptance: should we take parsimony of ontology, or internal consistency and coherence, as the most important criterion for accepting a scientific theory?

Technically, the standard model is highly successful. Yet a proper understanding of its successes remains to be achieved. For example, the standard model would be impossible without the ideas of gauge invariance, spontaneous and anomalous symmetry breakings, asymptotic freedom and renormalizability. While our understanding of the first two ideas is relatively clear, no proper understanding of the other two ideas would be possible without a proper understanding of the idea of renormalization group. Now the question is what physical mechanism justifies the idea of a continuous scale dependence of physical parameters in the context of quantum field theory, without which the idea and technique of renormalization group adopted by quantum field theory would be devoid of their physical content, and thus be incomprehensible.

Besides, the success of the standard model has often been exaggerated. In fact, it has many conceptual difficulties. First, the description of low-energy pion–nucleon interactions by QCD, and the explanation of quark confinement in a four-dimensional space-time, have not been achieved (if they can be achieved at all) within the framework of the standard model. Second, the unification of the electroweak with the strong interactions has been attempted; these attempts, although suggestive, are still open to question. Most notably, the quantization of gravity and its unification with other interactions are generally believed to be unattainable goals within the framework of the standard model.

[2] Rudolph Haag (1992): *Local Quantum Physics* (Springer).
[3] See Tian Yu Cao (1997): *Conceptual Developments of 20th Century Field Theories* (Cambridge University Press).
[4] See Bryce DeWitt's contribution to this volume.

At a more fundamental level, traditionally the consistency of a quantum field theory is taken to be threatened by divergences and saved only by its renormalizability. But a rigorous proof of the renormalizability of a quantum field theory can only be obtained through the renormalization group approach. If a theory has a fixed point for the renormalization group transformations, then the theory is renormalizable, but not otherwise. Yet the proof of the existence of fixed points for the standard model in a four-dimensional space-time has not been achieved mathematically. Moreover, the idea of the renormalization group and the related ideas of decoupling and effective field theories have also suggested the legitimacy of nonrenormalizable interaction terms in a quantum field theory, and this has raised a serious question concerning the place of renormalizability in our understanding of the consistency of quantum field theory, and opened the door for alternative formulations.

Philosophically, the most interesting developments, dictated by the inner logic of the standard model, are a set of concepts, namely symmetry breakings, renormalization group and decoupling, which, ironically, are eroding the very foundation of the standard model: reductionism. A new world picture suggested by these advances is a hierarchy layered into quasi-autonomous domains, which are separated by mass scales associated with spontaneous symmetry breakings. Connections between layers exist and can be expressed by renormalization group equations. Yet the ontology and dynamics of each layer, according to the decoupling theorem, are quasi-stable, almost immune to whatever happens in other layers, and are describable by stable effective theories.

These developments in quantum field theory invite serious philosophical reflections so that their significance can be properly appreciated. Ontologically, the picture of a hierarchy of quasi-autonomous domains has endorsed the existence of objective emergent properties. This in turn has set an intrinsic limit to the reductionist methodology. Thus the development of quantum field theory as a global reductionist pursuit has reached its critical point at which its own conceptual foundation of reductionism has been undermined. Moreover, the historization of the laws of nature, suggested by the notion of spontaneous symmetry breakings and the related idea of cosmic phase transitions, seems to have undermined the idea of immutable fundamental laws of nature, another conceptual foundation of quantum field theory. Furthermore, the occurrence of a tower of quasi-stable effective theories has suggested that we need to revise the way we conceptualize the growth of scientific knowledge.

If we take these interpretations seriously, then some of the conceptual difficulties encountered by the standard model are unlikely to be normal puzzles which can be solved by the established methodology. What is required, it seems to some empiricists, is a drastic change of our conception of fundamental physics itself, a change from aiming at a fundamental theory (as the foundation of physics) to having effective theories valid at various accessible energy scales.

Many theorists have rejected this conception. For David Gross and Steven Weinberg, effective field theories are only the low energy approximations to a deeper theory, and can be obtained from it in a systematic way. However, an interesting point is worth noticing. Although both Gross and Weinberg believe that within the reductionist methodology, ways out of conceptual difficulties of the standard model can be found sooner or later, with the help of more sophisticated mathematics or novel physical ideas, both of them have lost their confidence in quantum field theory as the foundation of physics, and conceived a deeper theory or a final theory not as a field theory, but as a string theory or some other theory radically different from quantum field theory.[5] But string theory, or related speculations such as M-theory, can hardly be taken to be a physical theory at this stage; neither can it be taken as a mathematical theory because, as some mathematicians have argued, there are too many conjectures and too few rigorously proven results.

A question worth discussing is this: from a string theorist's point of view, what kind of defects in the foundations of quantum field theory have deprived it of its status as the foundation of physics? A related question is: should we take strings as a continuation of quantum field theory, or as a revolution, a radical breaking away from quantum field theory? For some more conservative mathematical physicists, these questions are non-existent because, they believe, by more and more mathematical elaborations, a consistent quantum field theory can be established and continue to serve as the foundation of physics. Thus at the present stage, there are essentially three approaches to the questions of the foundations of quantum field theory, of quantum field theory as the foundation of physics, and of the reasons why quantum field theory can no longer be taken as the foundation of physics: (i) an effective theory approach, (ii) a string theory approach, and (iii) a mathematical approach, the most active version of which now, as compared with the original axiomatic approach in the 1950s and 1960s and the constructive approach that was actively pursued from the late 1960s to the 1980s, is the algebraic approach initiated in the early 1960s. The assessment of these three approaches is closely related to the philosophical debate over reductionism versus emergentism.

The three approaches start from different basic presuppositions, give different over-all pictures of the progress of physics, and represent different visions about the central issues the present stage of fundamental physics is meant to address, namely, the possibility of a final unified and finite theory which would close the door for any new physics. Guided by different visions, physicists are pursuing different research strategies, adopting different intellectual arguments about fundamental physics, aside from the philosophical argument about reductionism and emergentism, as we heard last time at the workshop held on November 6, 1993, and sponsored by the American Academy of Arts and Sciences. A wonderful example of this sort is provided by Roman Jackiw's paper in this volume. While string theorists try to avoid the concept of a point-like local field and its consequence infinities, and to construct a finite theory, Jackiw finds infinities entailed by the concept of local fields useful for achieving the breakdown of symmetry, which, as he rightly indicates, is central to the present stage of fundamental physics. One of the duties of philosophers is to examine the credentials of basic assumptions adopted by each approach and the fertility of different visions.

In the debate between those who hold a vision of a final theory and those who hold a vision of many effective theories, many philosophers do not really believe in the possibility of a final theory. Nevertheless, some of them still join the physicists of the first category and support the illusory vision with the concept of regulative function.[6] These philosophers argue that the ideal of a final theory is heuristically and regulatively more productive because its esthetic value can ignite physicists

---

[5] See the contributions to this volume by David Gross and Steven Weinberg.
[6] See, for example, Michael Redhead's contribution to this volume.

with intellectual excitement, whereas pursuing effective theories, though more practical, is intellectually less exciting.

This argument is very popular among string theorists because it justifies their mental state. Yet it can be neutralized with two counter-arguments. First, the vision of effective field theories is not meant to take a purely phenomenological approach. Rather, it is fully compatible with a pluralist view of fundamentality: there can be many fundamental theories of physics, each of which is responsible for a certain level of complexity in the physical world; thus no one can claim to be more fundamental than the others. Yet within their own domain, theorists can still try to discover the ultimate and underlying order, pursue the esthetic value of unification, and enjoy all the intellectual excitement thereby, even though they realize that their unified fundamental theory is only of a limited validity. But this limited applicability is always the case and would not put a damper on their excitement. The reason for this claim is obvious: even the most dedicated final theorist would not imagine any application of his final theory to economics or poetry; and the intellectual excitement enjoyed by theorists working at various levels of complexities has also defied the exclusive privilege of the final theorists for intellectual excitement. Second, the vision of effective theories keeps our mind open to the future, while the vision of a final theory closes our eyes to any new physics, any future development, except for some not so exciting applications. Thus, even in terms of the heuristic or regulative role, the vision of a final theory does not enjoy superiority.

Finally, let me say a few words about this conference. At this stage, both physicists and philosophers feel that there is a tension between them: for physicists, philosophers are ignorant or even dangerous because they don't understand and tend to distort complicated concepts and theories in physics; for philosophers, physicists are naive about their own work, and rely mainly on handwaving arguments. While I agree with Michael Redhead who claims that philosophers can contribute to the clarification of conceptual foundations of quantum field theory, I also agree with Sidney Coleman who advises me that in order to discuss philosophical questions in physics, philosophers should learn physics first. This two-tier conference is an experiment, aiming to promote the desirable, mutually beneficial, yet difficult dialogue between philosophically interested physicists and scientifically informed philosophers. If both sides can throw away unnecessary closed-minded arrogance and hostility, then a productive dialogue may not be an illusion.

2. Quantum field theory and the philosopher

MICHAEL REDHEAD

The topic I want to address principally is what exactly philosophers can be expected to contribute to discussions of the foundations of QFT. This of course is part of the broader problem of the relation between science and philosophy generally. I will begin, then, with some general points before turning to more detailed issues with specific reference to QFT.

Philosophy is a second-order activity reflecting on the concepts, methods and fundamental presuppositions of other disciplines, art, politics, law, science and so on, but also, most importantly, examining reflexively its own arguments and procedures. To put it crudely other disciplines may refer to philosophy as a sort of 'court of appeal' to resolve foundational disputes (at a sufficient level of generality). But philosophy has to serve as its own court of appeal, to pull itself up by its own bootstraps so to speak.

To many, science is the paradigmatic example of objective, rational, empirically warranted knowledge. So it is to epistemology, the theory of knowledge, that we must turn to examine the credentials of the scientific enterprise. But the first thing to notice about philosophy as compared with science, and in particular physics, is that there is a striking lack of consensus about what knowledge is and how it can be achieved. There is an enormous gap between so-called analytic philosophy which broadly agrees that there is something special about science and scientific method, and tries to pin down exactly what it is, and the modern continental tradition which is deeply suspicious of science and its claims to truth and certainty, and generally espouses a cultural relativism that derives from Nietzsche and goes all the way to the postmodern extremes of Derrida and his ilk. There is also the social constructivism that sees scientific facts as mere 'social constructions' and is a direct offshoot of the continental relativism. But the analytic tradition itself divides sharply between instrumentalism and realism, the former seeing theories in physics as mathematical 'black boxes' linking empirical input with empirical output, while the latter seeks to go behind the purely observational data, and reveal something at least of the hidden 'theoretical' processes that lie behind and explain the observational regularities.

To my own taste the philosophical position which is, or should be, of most interest to the physicist is the realist one, although again one must be careful to distinguish different answers to the question, realist about what? Is it the entities, the abstract structural relations, the fundamental laws or what?

My own view is that the best candidate for what is 'true' about a physical theory is the abstract structural aspect. The main reason for this is that there is significantly greater continuity in structural aspects than there is in ontology, and one of the

34

principal arguments against realism is the abrupt about-turns in ontological commitment even in the development of classical physics. Of course the structural aspects also change, but there is a more obvious sense in which one structure grows out of another in a natural development or generalization, so that a cumulative model of scientific progress is much easier to maintain. Why do I reject instrumentalism? Simply because there is no clear way of demarcating the observational from the theoretical – all observations are theory-laden, as the slogan goes, so if we are going to be realists at all, we had better go directly to the theoretical level.

Now how does all this look from the QFT point of view? It must be recalled that QFT was developed, admittedly in rather a rudimentary form, very soon after the advent of quantum mechanics (QM) itself and as a result was infected with a jargon of observables and Copenhagenist complementarity-speak which showed the two major influences of positivism, the elimination of 'unobservable' metaphysics, and the dominance of Bohr's 'perspectival' views, on the interpretation of QM. That is not to say that one should conflate Bohrian philosophy with positivism, but at least they combine in prohibiting the philosopher-physicist from delving into a reality behind the phenomena, for example to explain the limitations set by the uncertainty principle or the weird combination of particle-like and wave-like properties subsumed under the phrase wave–particle duality.

Even in Dirac's first major paper on radiation theory published in 1927 (Dirac (1927)), he claimed that his method of second quantization resolved the wave–particle duality problem, roughly by showing how there were two equivalent routes to the quantized radiation field, one starting from a many-particle theory subject to the Bose–Einstein statistics, which when reformulated in an occupation-number (or Fock space) formalism was mathematically equivalent to another, treating the one-particle Schrödinger equation as a 'classical' field subjected to a quantization procedure adapted to a system with an infinite number of degrees of freedom, but with this 'second quantization' modelled exactly on the procedures employed in deriving the first-quantized Schrödinger equation from the Hamiltonian for a single classical particle.

When extended to massive fields and adapted to Fermion fields by substituting anticommutators for commutators in the quantization procedure, this led to the standard elementary text-book exposition of QFT as a theory of quantized excitations of the underlying field. Despite Dirac's claims, QFT by no means eliminated wave–particle duality or other manifestations of complementarity, since there were plenty of noncommuting observables around. Speaking very crudely, and referring for simplicity to neutral boson fields, there was a complementarity between the specification of a sharp *local* field amplitude (the field picture) and sharp *global* quantities representing the total momentum or energy in the field (the particle picture).

At this point we can bring in some of the insights of algebraic quantum field theory (AQFT) as it came to be developed in the 1960s. Attached to each bounded open set of space-time (we restrict ourselves to Minkowski space-time) there is an algebra of observables corresponding to quantities that are 'constructed' from the fields and can be measured by local operations confined to the relevant space-time region. Local propositions correspond to the projectors in these local algebras. These algebras and the associated representations of their symmetry properties are carried in a global Hilbert space, which, in the simplest free-field cases we shall be considering, is scaffolded by a unique translation-invariant global vacuum state and orthogonal one-particle, two-particle, etc., states of definite total energy and momentum. From this point of view the particle concept is a global one, not a local one. Crudely, to ask whether there is a particle *somewhere* in the field, means looking *everywhere*, and that is not a local thing to do! Now in nonrelativistic QFT, such as quantizing the Schrödinger equation, this can be nicely circumvented by defining local number operators, which specify the number of particles in a bounded region, but in the relativistic case, the global number operator cannot be broken down into a sum of commuting local operators in this way. So all we have is the global concept. In a truly local physics particles don't exist in the relativistic theories, except in an attenuated 'approximate' sense, which may be good enough for physicists but definitely not for philosophers trying to understand in absolute 'all or nothing' categories what QFT is about!

Of course the vacuum is a special sort of particle state, but the question 'Are we in the vacuum state?' cannot be given a local answer. Technically the projectors onto the vacuum state (or any of the many-particle states in the global Hilbert space) do not belong to any member of the net of local algebras. As a result all the local quantities exhibit the famous vacuum fluctuations. There is a nonvanishing probability in the vacuum state of answering 'yes' to any local question, such as 'Are there 100 coulombs of charge in this small volume element?'! Of course, the probabilities of such local events can be very, very small indeed, but it is a crucial consequence of the axioms of AQFT that they cannot vanish, and to the philosopher, unlike perhaps the physicist, there is all the difference in the world between something's being very small and its actually being zero!

Not only are there vacuum fluctuations, but even more remarkably, as we shall discuss in a moment, there are correlations even at space-like separations between the fluctuating local quantities. These correlations are of course what make possible the remarkable Reeh–Schlieder theorem (1961) in axiomatic QFT, expressing the cyclicity of the vacuum state with respect to *any* local algebra, for the global Hilbert space. These correlations are also of particular importance in attempting to adapt the Bell inequality arguments and the EPR argument to assessing the possibility of a thoroughgoing realist construal of QFF along the lines parallel to the very extensive discussion of these questions for ordinary QM in the philosophical literature. The extension of the Bell theorem to the case of vacuum fluctuations was carried out by Summers and Werner (1985) and by Landau (1987). They showed that maximal violation of the Bell inequality held in the QFT case, even in the vacuum state. This shows *prima facie* that a common cause explanation of the vacuum correlation is not possible. But the conclusion here is not of such decisive interest as in the ordinary QM case with correlated particles emitted from a source. The reason for this is that even if the Bell inequality had not been violated in the vacuum case, this would not have provided an acceptable common cause explanation, since the common cause at an earlier time would have to originate in the self-same vacuum state (because it is a time-translation invariant state), so exactly the same correlations are already present in the 'source' state to pursue the analogy with the ordinary QM case, and so the 'explanation' of the correlations has embarked on an infinite regress. In other words, whether or not the Bell inequality is violated, we have inexplicable correlations.

In order to discuss the violation of the Bell inequality it is necessary to consider correlations that are less than maximal, since for maximal correlations the inequality is *not* violated. On the other hand, if we want to formulate the EPR argument in the vacuum, it is essential to demonstrate *maximal* correlations at space-like separation.

It will be recalled that the EPR argument is designed to show that if we follow an antirealist option about possessed values of physical magnitudes in non-eigenstates of these magnitudes (and this of course is the orthodox view) then we can demonstrate an unacceptable action-at-a-distance which comprises creating elements of reality, in Einstein's phrase, by measurement procedures performed even at space-like separation with respect to the event consisting of the creation of the element of reality.

It is really quite surprising that it is only very recently that attempts have been made to formulate the EPR argument in the context of a fully relativistic AQFT. This required the demonstration of *maximal* correlations between local projectors in algebras attached to space-like separated regions of space-time. The necessary demonstration was given by myself in a paper entitled 'More Ado About Nothing', published in *Foundations of Physics* in 1995. What was actually shown was that for *any* projector in the one algebra there also existed a projector in the other algebra that was maximally correlated with it in the sense that the conditional probability of answering 'yes' to the proposition expressed by the second projector, given an answer 'yes' to the proposition expressed by the first projector, was itself unity. Notice carefully that maximal correlation in this sense does *not* mean a correlation *coefficient* attaining the value of one. We are maximizing the correlation coefficient subject to fixed and in general highly unequal values of the marginal probabilities (it is of course only when the marginals are equal that the maximal correlation coefficient can attain the value unity). Indeed Fredenhagen's theorem (1985) shows that the correlation coefficient falls off exponentially with the Lorentzian distance between the regions in question. But my result on maximal correlations in no way conflicts with Fredenhagen's theorem. The two theorems together just show that the *marginal probability* for the answer 'yes' to the projector whose existence is demonstrated to correlate maximally with a *given* projector in the space-like separated region, is what falls off exponentially with the Lorentz-distance between the regions.

Having demonstrated the maximal correlations, there is a deal more work to do in running a proper relativistic version of the EPR argument. This project has been tackled very recently by Ghirardi and Grassi (1994), who say in essence that the EPR claim to demonstrate the spooky creation of elements of reality at a distance cannot be made to go through in a situation where the lack of an invariant time order necessitates a counterfactual reformulation of the famous EPR sufficient condition for identifying elements of reality. The Ghirardi–Grassi argument has been criticized in its detail by myself and my student Patrick La Rivière (Redhead and La Rivière (1996)), basically on the grounds of employing an ambiguously stated locality principle and more importantly a tacit assumption of determinism. In fact by giving up the determinism assumption we are able to secure an argument for peaceful coexistence between QFT and special relativity, on the orthodox antirealist reading.

I have gone into this matter in some detail to point out the distinction between the interests of physicists and philosophers in discussing the foundation of QFT. To the physicists the locality issue is taken care of by the space-like commutation of the observables of AQFT, but this only has bearing on the transmission of *statistical* effects at a distance, essentially enabling the *no-signalling theorems* of QM to be translated into the QFT framework. But to the philosopher, issues of space-like

causation arise also at the case-by-case level as well as at the statistical level, and it is
here that the delicate reasoning of the EPR and Bell literature needs to be brought
into play. To the philosopher the putative tension between relativity and quantum
mechanics even in so-called relativistic quantum theories is an important concep-
tual-foundational issue that generally is sidestepped by the physicist with her more
operationally based approach to interpreting the physics. But arguably operationalism is not sidestepping the need for philosophical analysis, but is itself just bad
philosophy! So in these sorts of discussion I believe the philosopher can really
teach the physicist something about the interpretation and conceptual foundations
of QFT. QFT may be seen by many physicists just as an instrument for extraordinarily accurate numerical predictions, as witnessed by the g-2 calculations in QED, for
example, but this very success cries out for explanation with reference to validating
the structural framework of the theory at the very least, in broadly realist terms.
When I talked a moment ago of the possible advantages of interpreting QFT anti-
realistically, that was to argue against a classical-style realism of possessed values,
not against a broader realism of physical structure.

So far I have discussed QFT partly from an elementary point of view involving
essentially Fourier analysis of free fields, or interactions treated in the first nonvanishing order of perturbation, and partly from the axiomatic perspective. I now want to
turn for a moment to some of the more modern technical developments such as the
programme of effective field theories, which have been much promoted by Weinberg
and others during the past fifteen years or so. Full technical details are to be found in
the review by Cao and Schweber published in 1993. So I don't want to duplicate this
comprehensive discussion, but to offer some comments on their conclusions. However, for the sake of philosophers in the audience I will provide just the crudest of
sketches.

From its infancy QFT was plagued by the problem of infinite and hence literally
nonsensical predictions when attempts were made to calculate higher-order terms
in the perturbation expansion for interacting fields as a power series in the coupling
constant expressing the strength of the interaction. These difficulties were tackled in
the late 1940s by the development of renormalization theory by Bethe, Schwinger,
Tomonaga, Feynman and Dyson, to name just the principal figures in the story.
The idea was to absorb the infinities into the rescaling, or renormalization as it was
called, of a finite number of parameters, such as masses and coupling constants
(charges) entering into the theory. The philosopher Paul Teller has usefully distinguished three approaches to understanding what was going on (Teller (1989)). The
first was the cutoff or regularization approach in which short-distance modifications
were made to the theory in such a way as to produce finite results, but only as a
divergent function of the cutoff parameter. After renormalization this parameter
disappeared from the theory except in small correction terms which now vanished
as the cutoff limit was taken. So the prescription was to take the limit *after* the renormalization had been effected. The question, much discussed in the early days of this
approach, was whether the cutoff prescription could be interpreted physically, or was
just a mathematical artifice. With the development of the more comprehensive and
mathematically elegant concept of so-called dimensional regularization, this second
option of just a mathematical artifice came to be largely adopted. The second
approach, which Teller calls the real infinities approach, effected the infinite subtractions on the integrands of the divergent integrals, the residual integrands leading to

convergent results. The method achieved the same results as the cutoff approach with the limit taken after the renormalization of the finite integrals, but the methodology of infinite subtractions was even less transparent to most philosophers than in the case of the regularization approach.

To make physical sense of the whole renormalization procedure, it seemed clear that the high-frequency behaviour that led the divergent integrals must be modified by additional physical processes and that the point about a renormalizable theory was that its low-energy behaviour was relatively insensitive to the detail of how this modification was achieved, but not totally insensitive, because of the renormalization effects. The renormalization procedure produces a mask of ignorance, as Teller puts it (his third approach), hiding the details of the true and supposedly finite theory behind the bland screen of the renormalized masses and charges, to mix our metaphors a little. This idea of the mask of ignorance is taken a step further in the EFT programme. There are an endless series of quasi-autonomous domains set by different energy scales arranged in a hierarchical tower, each providing an effective theory within its own domain, but feeding renormalization effects across boundaries in the tower, from the higher to the lower levels. There is much more technical detail to fill out this picture, derived from analysis of the so-called renormalization group, which has played a major role in so much of the more recent development in QFT, but for our purposes we can see the EFT programme as a bottom-up approach, leading from low-energy to higher-energy regimes in a progressive exploration of the EFT tower. Of course, if there were a final finite theory of everything, a TOE, we could no doubt recover the hierarchy of effective field theories from the top down so to speak, and this would revert to something like the mask of ignorance approach referred to above. But, say Cao and Schweber, what if there is no TOE, but just the indefinitely extending tower; how should we assimilate this into our overall picture of the progress of physics?

First, let me make the point that the issue of choosing between a TOE and an indefinite prolongation of the EFT tower can never be settled observationally. In this respect the debate is entirely parallel to the case of atomism asserting in ancient Greek philosophy a terminus to the potentially infinite process of successive subdivision of matter. The Aristotelian continuum admits no such terminus and corresponds in this analogy to the EFT programme. Famously, atomism cannot be proved nor disproved. If atoms have not emerged at any stage of subdivision this does not prove that atoms would not emerge at a later stage, so atomism cannot be disproved by not finding the atoms, while equally if putative atoms have emerged the fact that they are the *ultimate* terminus of subdivision cannot be proved, since there is always a possibility that more extreme efforts to divide them will succeed! In this sense atomism is a purely metaphysical hypothesis, but that does not mean that it cannot provide plausible solutions to *philosophical* problems such as the possibility of change in the cosmos, and this after all was the role of the atomic hypothesis for the ancient atomists like Democritus and Leucippus. But, furthermore, atomism can serve a very important heuristic and regulative role in guiding the generation of testable physical theories, which conjoin the general untestable thesis, atoms exist, with the additional claim that protons, electrons or whatever are the ultimate atoms. Of course the history of physics is littered with putative atoms that turned out not to be atoms at all in the philosophical sense, the proton and its quark structure being a case in point.

The EFT programme is really a modern version of the unending sequence of Chinese boxes each contained within and supporting the structure of the previous member of this descending chain. But from the point of view of methodology of science a recurring theme has been the search for an *ultimate* underlying order characterized by simplicity and symmetry that lies behind and explains the confusing complexity of the phenomenal world. To subscribe to the new EFT programme is to give up on this endeavour and retreat to a position that is admittedly more cautious and pragmatic and closer to experimental practice, but is somehow less intellectually exciting. Perhaps one should not allow such considerations to enter into one's evaluation of a scientific programme, but to my own taste, the regulative ideal of an ultimate theory of everything remains a powerful aesthetic ingredient motivating the exercise of the greatest intellectual ingenuity in the pursuit of what may admittedly, in itself, be an illusory goal. But that after all is the proper function of regulative ideals.

References

Cao, T.Y. and Schweber, S.S. (1993): 'The conceptual foundations and the philosophical aspects of renormalization theory', *Synthese* **97**, 33–108.

Dirac, P.A.M. (1927): 'The quantum theory of the emission and absorption of radiation', *Proceedings of the Royal Society of London A* **114**, 243–65.

Fredenhagen, K. (1985): 'A remark on the cluster theorem', *Communications in Mathematical Physics* **97**, 461–3.

Ghirardi, G.C. and Grassi, R. (1994): 'Outcome prediction and property attribution: The EPR argument reconsidered', *Studies in History and Philosophy of Science* **24**, 397–423.

Landau, L.J. (1987): 'On the non-classical structure of the vacuum', *Physics Letters A* **123**, 115–18.

Redhead, M.L.G. (1995): 'More Ado About Nothing', *Foundations of Physics* **25**, 123–37.

Redhead, M.L.G. and La Rivière, P. (1996): 'The relativistic EPR argument', forthcoming in the *Festschrift for Abner Shimony* edited by R.S. Cohen.

Reeh, H. and Schlieder, S. (1961): 'Bemerkungen zur Unitäräquivalenz von Lorentzinvarianten Feldern', *Nuovo Cimento* **22**, 1051–68.

Summers, S.J. and Werner, R. (1985): 'The vacuum violates Bell's inequalities', *Physics Letters A* **110**, 257–9.

Teller, P. (1989): 'Infinite renormalization', *Philosophy of Science* **56**, 238–57.

Part Two. Three approaches to the foundations of quantum field theory

3. The usefulness of a general theory of quantized fields

A. S. WIGHTMAN

It is well known that people working at the frontiers of physics often don't know what they are talking about. This statement has at least two meanings. First, the experimental evidence available may be incomplete or ambiguous and confusing. Second, the theoretical concepts employed may not have received a precise formulation.

In this talk, I will give arguments for the position that it can be useful to refine a theoretical framework to the point at which it is a mathematically precise and consistent scheme, even if, as is sometimes the case, the resulting conceptual frame has physical deficiencies. I will offer several examples to illustrate the point. Of course, it is only occasionally that this sort of enterprise can help in the hunt for the Green Lion, the ultimate Lagrangian of the world.

1 The Stone–Von Neumann Theorems

In the time dependent Schrödinger equation

$$i\hbar \frac{\partial \Psi}{\partial t} = H\Psi \tag{1.1}$$

H is an operator acting on the wave function Ψ. Which Ψ are allowed? The usual elementary discussion of this question six decades ago was for H the Hamiltonian of an atom. It required that Ψ be twice differentiable with the lamentable result that the eigenfunctions of H might not be among the Ψ for which H was defined. For a time independent H the Theorems of Stone and Von Neumann clarified this paradox in a way that can be briefly summarized as follows.

What is needed physically is a continuous temporal development of the wave function that preserves probabilities:

$$\Psi(t) = V(t)\psi, \quad -\infty < t < \infty \tag{1.2}$$

$$V(t)V(t') = V(t + t') \tag{1.3}$$

where $V(t)$ is defined on every wave function in the Hilbert space of states and

$$V(t)V(t)^* = V(t)^* \cdot V(t) = 1 \tag{1.4}$$

The form of such a $V(t)$ was determined by Stone.

Theorem Let $V(t)$, $-\infty < t < \infty$ be a one-parameter family of everywhere defined operators on a Hilbert space, satisfying (1.3) and (1.4) there, and continuous in t.

41

Then

$$V(t) = \exp(iHt) \tag{1.5}$$

where H is a uniquely determined self-adjoint operator.

Now a self-adjoint operator has as part of its definition a uniquely determined linear set of wave functions that constitute its domain, so under the hypotheses of Stone's Theorem, the law of temporal evolution (1.2) uniquely determines the domain of definition of the Hamiltonians H.

The practical application of Stone's Theorem gives rise to a series of mathematical problems. For example, if the ψ is a function of the coordinates $\vec{x}_1, \vec{x}_2, \ldots, \vec{x}_N$ of N particles and H is the usual non-relativistic atomic Hamiltonian

$$H = -\sum_{i=1}^{N} \frac{\hbar^2}{2m_i} \Delta_{\vec{x}_i} + V(\vec{x}_i, \ldots, \vec{x}_N) \tag{1.6}$$

then, under suitable assumptions on V, H will be Hermitian when defined on the C^∞ functions of compact support, but it won't be self-adjoint. The question arises: can its domain be enlarged in such a way that the operator so defined is self-adjoint? If so, is that extension unique? Here Von Neumann's theory of the self-adjoint extensions of Hermitian operators is relevant; it classifies all possible self-adjoint extensions if such exist. Distinct extensions define different dynamical laws (1.2). There is a considerable lore on this subject. I refer to the book by M. Reed and B. Simon (1972–79) for details.

Now you may say this is all fancy mathematical talk of little physical interest. To which I would respond that in a well-ordered household somebody has to dispose of the garbage. There is a precise and elegant answer to the question: on what wave functions is the Schrödinger Hamiltonian defined?

2 Continuous unitary representation of the Poincaré group and the general theory of quantized fields

In ancient times, Wigner argued that any Poincaré invariant quantum theory has an associated continuous unitary representation of the (covering group of the) Poincaré group. (For simplicity, I will usually ignore the double-valued representations of the Poincaré group.) Further, he interpreted the infinitesimal space and time translations as operators of total momentum and of total energy, and therefore (later on) restricted his attention to representations in which the energy–momentum lies in or on the plus light-cone; that is the *spectral condition*. He classified the irreducible representations satisfying the spectral condition.

The representation of the Poincaré group in a theory that describes stable particles has to satisfy further conditions. For each kind of stable particle, there should be states of arbitrary numbers of particles transforming according to the appropriate tensor product representation of the Poincaré group. In particular, there should be a state with no particles, the vacuum state; it should be invariant. The representation of the Poincaré group for the whole theory must then be reducible and the multiplicities of the irreducible constituents must be such that the preceding conditions are satisfied. So much for the general treatment of relativistic quantum theory.

It is natural to approach the relativistic theory of quantized fields from a point of view that imposes as a requirement that the theory possess a physically acceptable representation of the Poincaré group compatible with the relativistic transformation properties of the fields. For example, a scalar field ϕ satisfies

$$V(a, \Lambda)\phi(x)V(a, \Lambda)^{-1} = \phi(\Lambda x + a) \tag{2.1}$$

where for a space-time translation, a, and a Lorentz transformation, Λ,

$$\{a, \Lambda\} \rightarrow V(a, \Lambda) \tag{2.2}$$

is the unitary representation of the Poincaré group. It is also reasonable to restrict attention to fields which are local in the sense that they commute at space-like separate points, i.e.

$$[\phi(x), \phi(y)] = 0 \quad \text{for} \quad (x - y) \cdot (x - y) < 0 \tag{2.3}$$

Finally it is useful to consider theories in which the fields are a complete description. That can be taken to mean that vectors of the form $P\Psi_0$ are dense in the Hilbert space of states, where P runs over all polynomials in the smeared fields

$$\int d^4x f_i(x)\phi_i(x) \tag{2.4}$$

for some test functions f_i and quantized fields ϕ_i. When this is true the vacuum state Ψ_0 is said to be *cyclic* for the fields on space-time. The assumed cyclicity of the vacuum has the consequence that one can compute the vacuum expectation values

$$(\Phi_0, \phi(x_1), \dots, \phi(x_n)\Phi_0) \tag{2.5}$$

They turn out to be boundary values of analytic functions in the variables

$$(x_2 - x_1), (x_3 - x_2), \dots, (x_n - x_{n-1}) \tag{2.6}$$

and this leads to one of the most characteristic features of relativistic quantum field theory, expressed in the following theorem.

Theorem (Reeh–Schlieder)
Under the above hypotheses, the vacuum is cyclic for polynomials in the smeared fields with test functions restricted to have their supports in an arbitrarily small fixed set of space-time.

It was one of the surprises of the 1950s that two of the most prized general results of Lagrangian quantum field theory (also distinguished by being well verified experimentally), the CPT Theorem and the Connection of Spin With Statistics, turned out to be theorems of the general theory of quantized fields with only the above-mentioned hypotheses. The proofs were easy consequences of a characterization of the theories in terms of vacuum expectation values.

Two further developments of this general theory must be mentioned: the Haag–Ruelle theory of collision states and Borchers's theory of equivalence classes of relatively local fields. The first constructs stationary in- and out-states for massive colliding particles by universal limiting processes, thereby showing that a uniquely determined scattering and particle production theory is contained in any of the quantum field theories that satisfy the above assumptions and have a mass gap.

The second asserts that two irreducible sets of relatively local fields in a Hilbert space that transform under the same representation of the Poincaré group necessarily yield the same S-matrix.

This completes my survey of the state of the general theory of quantized fields as of thirty-five years ago. More details can be found in the books by Jost (1965) and Streater and Wightman (1964).

I now turn to a parallel development which began at the same time but reached definite form in the 1960s and has since yielded striking and very deep results.

3 Localization in relativistic quantum field theory; local quantum theory

In non-relativistic second quantization there are field operators satisfying the commutation or anticommutation relations:

$$[\psi(\vec{x}, t), \psi(\vec{y}, t)]_\pm = 0 \tag{3.1}$$

$$[\psi(\vec{x}, t), \psi(\vec{y}, t)^*]_\pm = \delta(\vec{x} - \vec{y}) \tag{3.2}$$

The state vector

$$\psi(\vec{x}, t)^* \Phi_0 \tag{3.3}$$

where Φ_0 is the no-particle state, describes a state in which one particle is localized at $\vec{x}$ at time t. Furthermore

$$(\psi(\vec{x}, t)^* \Phi_0, \psi(\vec{y}, t)^* \Phi_0) = 0 \tag{3.4}$$

if

$$\vec{x} \neq \vec{y}$$

This gives meaning to the statement: if a particle is definitely here, it is definitely not there. On the other hand, the corresponding statement for a neutral scalar field, ϕ, in relativistic quantum field theory is false:

$$(\phi(\vec{x}, t)\Psi_0, \phi(\vec{y}, t)\Psi_0) \neq 0 \tag{3.5}$$

for $\vec{x} \neq \vec{y}$. However, if ϕ describes a theory in which there is a mass gap (no state except the vacuum with $\mathbf{p} \cdot \mathbf{p} < m^2$ for some $m > 0$), the left-hand side of (3.5) falls to zero at least as fast as $\exp(-m|(\vec{x} - \vec{y})|)$ as $|\vec{x} - \vec{y}| \to \infty$.

Thus in relativistic quantum field theory, a notion of localization that regards $\phi(\vec{x}, t)\Psi_0$ as a state localized at $\vec{x}$ at time t will have to be fuzzier than the corresponding notion in the non-relativistic theory.

In order to study this situation using the well-developed mathematical tools of the theory of operator algebras, Rudolph Haag and his coworkers Huzihiro Araki and Daniel Kastler considered the algebra $A(O)$ of operators generated by the bounded observables that can be measured in a space-time region O. The collection $A(O)$ where O runs over a suitable collection of space-time regions is called a *net of local algebras*.

The first indication that this would be a fruitful approach was that it gave a theory of super-selection rules. It is characteristic of a theory in which there are super-selection rules that the Hilbert space of states decomposes into a direct sum of subspaces with the property that the relative phases of different subspaces are unobservable. Haag and Kastler argued that the representations of the net of local algebras in the different subspaces of the direct sum are unitary inequivalent, so the representation theory of the net of local algebras determines whether there are super-selection rules. A technical question then arises. The net of local algebras may have many irreducible representations. How are the physically relevant ones to be determined? There are several interesting answers: Borchers's Selection Criterion, the Doplicher–Haag–Robertson Criterion, the Buchholz–Fredenhagen Criterion. To describe them and the theory to which they give rise would require me to recapitulate a sizable chunk of Rudolf Haag's book (1992). To entice you to read it, let me make a list of some of the questions to which you will find answers there:

How does the notion of particle emerge from the observables of the local net?
What is the statistics of those particles?
Can they have parastatistics?
Are there other kinds of statistics?
How do the answers to the preceding questions change when the dimension of space is 2 or 1?

I could continue but instead will offer a physics homework problem which I hope will convince you, if you do it, that local quantum theory is useful.

In 1932, in a well-known review article entitled 'Quantum theory of radiation' (see especially pp. 98–103), Enrico Fermi discussed questions of localization and causality in an elementary example. Given are two atoms A and B of the same species separated by a distance R. Initially A is excited and B is in its ground state. Fermi calculated the probability that at time t, B is excited and A is in its ground state. He showed that the probability is zero until a time R/c. Over the years, other authors have studied Fermi's example and have come to varied conclusions: Fermi's calculation is wrong but his conclusion is right; Fermi's calculation is wrong and his conclusion is wrong, etc. The latest contributions to the debate can be found in Hegerfeldt (1994) and Buchholz and Yngvason (1994). Who is right?

4 Quantum field theory in curved space-time

In the general theory of quantized fields defined on Minkowski space, the spectral condition plays an essential role. In quantum field theory on curved space-time, it is quite unclear how to formulate a substitute. There is certainly a large class of differential manifolds with a geometry that is locally Minkowski on which classical wave propagation takes place in a manner that is a physically natural generalization of that on Minkowski space (globally hyperbolic manifolds). The trouble is that generically such manifolds have no non-trivial isometries so that, for them, there is no analogue of the Poincaré group, and therefore no way of characterizing states satisfying the spectral condition in terms of their energy and momentum. In particular, there is no analogue of the vacuum state. Nevertheless, there is one feature of the Minkowski space theory that generalizes to globally hyperbolic smooth manifolds. That is the division of the problem into a part dealing with the construction of the field algebra on the manifold and a part in which physically acceptable states (positive linear forms on the algebra) are classified and constructed.

In the early 1980s John Dimock carried out the first step of constructing the algebra of fields for the special cases of a free scalar field, a free spinor field, and, later on, a free electromagnetic field. The second part of the problem, the determination of the physically acceptable states, is appreciably easier for free fields because the n-point expectation value in a state p can be expressed in terms of two-point expectation values. For example, the scalar field n-point is zero for n odd while for n even it is

$$p(\phi(x_1)\ldots\phi(x_n)) = [1\ldots n] \tag{4.1}$$

The right-hand side is the Hafnian defined recursively

$$[1\ldots n] = \sum_{j=2}^{n} [1j][\hat{1}2\ldots\hat{j}\ldots n] \tag{4.2}$$

where $\hat{l}$ means delete l, and

$$[jk] = p(\psi(x_j)\phi(x_k)) \tag{4.3}$$

is a two-point expectation value in the state p. Discussions of the admissible form of (4.3) go back more than a quarter of a century to the work of DeWitt and Brehme, who proposed the so-called Hadamard form.

In the last ten years there has been an intense effort to find alternative approaches and to relate them to the Hadamard form. In particular, Haag, Narnhofer, and Stein argued that the existence of a scaling limit permits one to construct a limiting theory in the tangent space of the manifold. There one can impose the usual spectral condition. Their approach has the virtue that it makes sense not only for free fields but also for interacting fields.

Quite a different approach has its roots in the work of Hörmander and Duistermaat on hyperbolic equations on manifolds. It uses a refined characterization of singularities of distributions, the theory of *wave front sets*, to track the singularities of the solutions of hyperbolic differential equations. In its later incarnation in the Ph.D. thesis of Radzikowski, the wave front set of the two-point function of a field was connected to the Hadamard form. Radzikowski's proposed WFSSC (wave front set spectral condition) was criticized by M. Köhler who suggested a refined form valid for a wider class of examples. The analysis was deepened and broadened by Brunetti, Fredenhagen and Köhler and the resulting spectral condition MSC (which stands for *microlocal spectral condition*) looks like a promising ingredient for a future general theory of quantized fields valid on both curved and flat spacetime. The papers referred to in this section can be tracked from the paper by Brunetti, Fredenhagen and Köhler (1995).

References

Detlev Buchholz and Jacob Yngvason, There are no causality problems for Fermi's two-atom system, *Physical Review Letters*, **73** (1994) 613–16.

R. Brunetti, K. Fredenhagen and M. Köhler, *The Microlocal Spectrum Condition and Wick Polynomials of Free Fields on Curved Spacetimes*, *DESY* (1995) 95–196.

E. Fermi, Quantum theory of radiation, *Reviews of Modern Physics*, **3** (1932) 87–132.

R. Haag, *Local Quantum Physics: Fields, Particles, Algebras*, Springer Verlag (1992).

Gerhard C. Hegerfeldt, Causality problems for Fermi's two-atom system, *Physical Review Letters*, **72** (1994) 596–9.

R. Jost, *The General Theory of Quantized Fields*, American Mathematical Society (1965).

M. Reed and B. Simon, *Methods of Modern Mathematical Physics*, Volume I Chapter VIII, Volume II Chapter X (1972–79).

R.F. Streater and A.S. Wightman, *PCT, Spin and Statistics and All That*, W.A. Benjamin (1964).

4. Effective field theory in condensed matter physics

R. SHANKAR

I am presumably here to give you my perspective on quantum field theory from the point of view of a condensed matter theorist. I must begin with a disclaimer, a warning that I may not be really representing anyone but myself, since I find myself today working in condensed matter after following a fairly tortuous route. I must begin with a few words on this, not only since it will allow you to decide who else, if any, I represent, but also because my past encounters with field theory will parallel that of many others from my generation.

I started life in electrical engineering, as Professor Schweber said in the introduction. As an electrical engineering student, the only field theorists I knew about were Bethe and Schwinger, who had done some basic work on wave guides. When I graduated, I switched to physics and was soon working with Geoff Chew at Berkeley on particle physics. In those days (sixties and early seventies), the community was split into two camps, the field theorists and the S-matricists, and Geoff was the high priest of the latter camp. The split arose because, unlike in QED, where everyone agreed that the electron and the proton go into the Lagrangian, get coupled to the photon and out comes the hydrogen atom as a composite object, with strong interactions the situation seemed more murky. Given any three particles, you could easily view any one as the bound state of the other two because of the large binding energies. The same particles also played the role of mediating the forces. So the problem was not just, 'What is the Lagrangian?', but 'Who goes into it and who comes out?' So Geoff and his camp decided to simply make no reference to Lagrangians and deal with just the S-matrix. The split was therefore not between field theory and effective field theory, but field theory and no field theory. Now Geoff was a very original thinker and deeply believed in his view. For example, he would refer to the pomeron, a pole in the complex angular momentum plane, as The Pomeron, in very concrete terms, but refer to the photon as the Photon Concept, even in a brightly lit room bursting with photons. I must emphasize that Geoff bore no animosity towards field theory or field theorists – he greatly admired Mandelstam who was a top-notch field theorist – but he had little tolerance towards anyone confusing S-matrix ideas with field theory ideas. As you will hear from Professor Kaiser, this was easy to do. So I had a choice: either struggle and learn field theory and run the risk of blurting out some four letter word like ϕ^4 in Geoff's presence or simply eliminate all risk by avoiding the subject altogether. Being a great believer in the principle of least action, I chose the latter route. Like Major Major's father in Catch 22, I woke up at the crack of noon and spent eight and even twelve hours a day not learning field theory and soon I had not learnt more field theory than anyone else in Geoff's group and was quickly moving to the top.

Soon it was time to graduate. By this time great changes had taken place. Clear evidence for quarks appeared at SLAC and with the discovery of asymptotic freedom, QCD emerged as a real possibility for strong interactions. Geoff then packed me off to Harvard, to the Society of Fellows. I am sure he knew I would be mingling with the field theorists, but it did not bother him, which is in accord with what I said earlier. I view his sending me off to Harvard as akin to Superman's father putting him in a capsule and firing him towards earth. However, unlike Superman who landed on a planet where he could beat the pants off everyone around him, I landed at Harvard, where everyone knew more field theory than I did. I remember one time when I was learning QED, the chap who came to fix the radiator also stopped to help me fix my Coulomb gauge. Anyway I had very good teachers, and many of my mentors from that time are at this conference. Pretty soon I went to the Coop, bought a rug for my office and was sweeping infinities under it – first a few logarithms and pretty soon, quadratic ones.

My stay was nearly over when one day Ed Witten said to me, 'I just learnt a new way to find exact S-matrices in two dimensions invented by Zamolodchikov and I want to extend the ideas to supersymmetric models. You are the S-matrix expert aren't you? Why don't we work together?' I was delighted. All my years of training in Berkeley gave me a tremendous advantage over Ed – for an entire week. Anyway, around that time I moved to Yale and learnt from Polyakov who was visiting the US that according to Zamolodchikov, these S-matrices were also the partition functions for Baxter-like models. I could not resist plunging into statistical mechanics and never looked back. Over the years I moved over to many-body physics, where I find myself now.

Despite all this moving, I never stopped admiring or using field theory and have a unique perspective that can only be attained by someone who has squandered his youth achieving mediocrity in an impressive number of areas. To me field theory does not necessarily mean a Lorentz invariant, renormalizable theory in the continuum, it means any theory that takes the form of a functional integral over an infinite number of degrees of freedom.

I will now proceed to give my views on effective field theories, by which one means a field theory which works only down to some length scale or up to some energy scale. Often the opponents of effective field theories or even its practitioners feel they are somehow compromising. I will now make my main point. *Even when one knows the theory at a microscopic level, there is often a good reason to deliberately move away to an effective theory.* Now everyone knows the microscopic Lagrangian in condensed matter: you just need electrons, protons and the Coulomb interaction. Everything we know follows from this. There is no point getting any more fundamental, it makes no difference that the photon is part of a bigger gauge group, or that the electron and its neutrino form a doublet which itself is part of a family, we have no family values. As long as you start with the right particle masses and charges, nothing else matters. On the other hand, progress is often made by moving away from the fundamental and looking for effective theories. I will illustrate my view with one example because I have been working on it for some years and because some particle physicists like Polchinski and Weinberg have been interested in some aspects of this problem. Towards the end I will mention a few more examples.

The technique I am going to use is the renormalization group (RG). You will hear about it in greater detail from Michael Fisher and others. I am going to give you a caricature. Imagine that you have some problem in the form of a partition function

$$Z(a, b) = \int dx \int dy \, e^{-a(x^2 + y^2)} e^{-b(x+y)^4} \tag{1}$$

where a, b are the parameters. Suppose now that you are just interested in x, say in its fluctuations. Then you have the option of integrating out y as follows:

$$z(a', b', \ldots) = \int dx \left[\int dy \, e^{-a(x^2 + y^2)} e^{-b(x+y)^4} \right]$$
$$\equiv \int dx \, e^{-a'x^2} e^{-b'x^4 - c'x^6 + \cdots} \tag{2}$$

where a', b', etc., define the parameters of the effective field theory for x. These parameters will reproduce exactly the same averages for x as the original ones. This evolution of parameters, with the elimination of uninteresting degrees of freedom, is what we mean these days by renormalization, and as such has nothing to do with infinities; you just saw it happen in a problem with just two variables.

Notice that to get the effective theory we need to do a non-Gaussian integral. This can only be done perturbatively. At the simplest 'tree level', we simply drop y and find $b' = b$. At higher orders, we bring down the nonquadratic exponential and integrate over y term by term and generate effective interactions for x. This procedure can be represented by Feynman graphs with the only difference that variables in the loop are limited to the ones being eliminated.

Why do we renormalize in this fashion? We do so because certain tendencies of x are not so apparent when y is around, but surface as we zero in on x. For example, a numerically small term in the action can grow in size as we eliminate the unwanted variables and guide us towards the correct low energy structure. This will now be demonstrated with an example.

Consider a system of nonrelativistic spinless fermions in two space dimensions. The one-particle Hamiltonian is

$$H = \frac{K^2}{2m} - \mu \tag{3}$$

where the chemical potential μ is introduced to make sure we have a finite density of particles in the ground state: all levels up to the Fermi surface, a circle defined by

$$K_F^2 / 2m = \mu \tag{4}$$

are now occupied and form a circle (see Figure 1).

Notice that this system has gapless excitations. You can take an electron just below the Fermi surface and move it just above, and this costs as little energy as you please. So one question one can ask is if this will be true when interactions are turned on.

We are going to answer this using the RG. Let us first learn how to apply the RG to noninteracting fermions. To understand the low energy physics, we take a band of width Λ on either side of the Fermi surface. This is the first great difference between this problem and the usual ones in relativistic field theory and statistical mechanics. Whereas in the latter examples low energy means small momentum, here it means small deviations from the Fermi surface. Whereas in these older problems we zero in on the origin in momentum space, here we zero in on a surface.

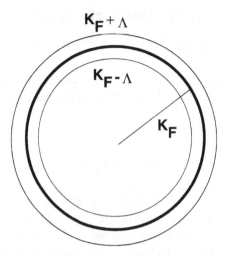

Fig. 1. The Fermi surface in two dimensions. The low energy modes lie within an annulus of thickness Λ on either side of it.

Let us cast the problem in the form of a Euclidean path integral:

$$Z = \int d\psi \, d\bar{\psi} \, e^{S_0} \tag{5}$$

where

$$S_0 = \int_0^{2\pi} d\theta \int_{-\infty}^{\infty} d\omega \int_{-\Lambda}^{\Lambda} dk \, \bar{\psi}(i\omega - vk)\psi \tag{6}$$

and v is the Fermi velocity, which gives the derivative of the energy with radial momentum on the Fermi surface and $k = K - K_F$.

Let us now perform mode elimination and reduce the cut-off by a factor s. Since this is a Gaussian integral, mode elimination just leads to a multiplicative constant we are not interested in. So the result is just the same action as above, but with $|k| \leq \Lambda$. Let us now perform the following additional transformations:

$$(\omega', k') = s(\omega, k)$$
$$(\psi', \bar{\psi}') = s^{-3/2}(\psi, \bar{\psi}) \tag{7}$$

When we do this, the action and the phase space all return to their old values. So what? Recall that our plan is to evaluate the role of quartic interactions in low energy physics as we do mode elimination. Now what really matters is not the absolute size of the quartic term, but its size relative to the quadratic term. By keeping fixed the size of the quadratic term and the phase space as we do mode elimination, this evaluation becomes very direct: if the quartic coupling grows it is relevant, if it decreases, it is irrelevant, and if it stays the same it is marginal.

Let us now turn on a generic four-Fermi interaction:

$$S_4 = \int \bar{\psi}(4)\bar{\psi}(3)\bar{\psi}(2)\bar{\psi}(1)u(4,3,2,1) \tag{8}$$

where $\int$ is a shorthand:

$$\int \equiv \prod_{i=1}^{3} \int d\theta_i \int_{-\Lambda}^{\Lambda} dk_i \int_{-\infty}^{\infty} d\omega_i \tag{9}$$

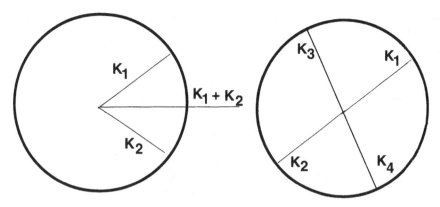

Fig. 2. Kinematical reasons why momenta are either conserved pairwise or restricted to the BCS channel.

At the tree level, we simply keep the modes within the new cut-off, rescale fields, frequencies and momenta, and read off the new coupling. We find

$$u'(k, \omega, \theta) = u(k/s, \omega/s, \theta) \qquad (10)$$

This is the evolution of the coupling function. To deal with coupling constants, we expand the function in a Taylor series (schematic)

$$u = u_0 + k u_1 + k^2 u_2 + \cdots \qquad (11)$$

where k stands for all the k's and ω's. An expansion of this kind is possible since couplings in the Lagrangian are nonsingular in a problem with short range interaction.

If we now make such an expansion and compare coefficients, we find that u_0 is marginal and the rest are irrelevant, as is any coupling of more than four fields. Now this is exactly what happens in ϕ_4^4. The difference here is that we still have dependence on the angles on the Fermi surface:

$$u_0 = u(\theta_1, \theta_2, \theta_3, \theta_4)$$

Therefore in this theory you are going to get coupling functions and not a few coupling constants.

Let us analyze this function. Momentum conservation should allow us to eliminate one angle. Actually it allows us to do more because of the fact that these momenta do not come from the entire plane, but a very thin annulus near K_F. Look at the left half of Figure 2. Assuming that the cut-off has been reduced to the thickness of the circle in the figure, it is clear that if two points 1 and 2 are chosen from it to represent the incoming lines in a four-point coupling, the outgoing ones are forced to be equal to them (not in their sum, but individually) up to a permutation, which is irrelevant for spinless fermions. Thus we have in the end just one function of the angular difference:

$$u(\theta_1, \theta_2, \theta_1, \theta_2) = F(\theta_1 - \theta_2) \equiv F(\theta) \qquad (12)$$

About forty years ago Landau came to the very same conclusion that a Fermi system at low energies would be described by one function defined on the Fermi surface. He did this without the benefit of the RG and for that reason, some of the leaps were hard to understand. Later, detailed diagrammatic calculations justified this picture. The RG provides yet another way to understand it. It also tells us other things, as we will now see.

The first thing is that the final angles are not slaved to the initial ones if the former are exactly opposite, as in the right half of Figure 2. In this case, the final ones can be anything, as long as they are opposite to each other. This leads to one more set of marginal couplings in the BCS channel,

$$u(\theta_1, -\theta_1, \theta_3, -\theta_3) = V(\theta_3 - \theta_1) \equiv V(\theta) \tag{13}$$

The next point is that since F and V are marginal at tree level, we have to go to one loop to see if they are still so. So we draw the usual diagrams shown in Figure 3. We eliminate an infinitesimal momentum slice of thickness $d\Lambda$ at $k = \pm\Lambda$, as shown by dark circles at the bottom of Figure 3.

These diagrams look like the ones in any quartic field theory, but each one behaves differently from the others and its traditional counterparts. Consider the first one (called ZS) for F. The external momenta have zero frequencies and lie on the Fermi surface since ω and k are irrelevant. The momentum transfer is exactly zero.

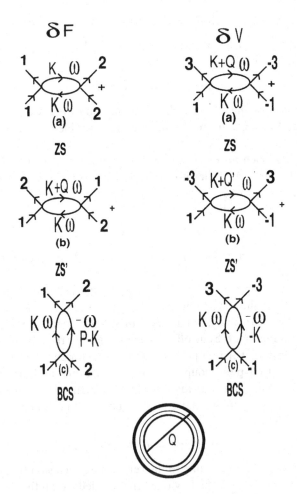

Fig. 3. One-loop diagrams for the flow of F and V.

So the integrand has the following schematic form:

$$\delta F \simeq \int d\theta \int dk \int d\omega \left(\frac{1}{(i\omega - \epsilon(K))} \frac{1}{(i\omega - \epsilon(K))} \right) \tag{14}$$

The loop momentum K lies in one of the two shells being eliminated. Since there is no momentum difference between the two propagators, they have the same energy $\epsilon(K)$ and the poles in ω lie in the same half-plane, and we get zero upon closing the contour in the other half-plane. In other words, this diagram can contribute if it is a particle–hole diagram, but given zero momentum transfer we cannot convert a hole at $-\Lambda$ to a particle at $+\Lambda$ or vice versa. In the ZS$'$ diagram, we have a large momentum transfer, called Q in the bottom of Figure 3. This is of order K_F and much bigger than the radial cut-off, a phenomenon unheard of in say ϕ^4 theory, where all momenta and transfers are of order Λ. This in turn means that the loop momentum is restricted not only in the radial direction to a sliver of width $d\Lambda$, but also in the angular direction, in order to be able to absorb this huge momentum and land up in the other shell being eliminated. So we have $du \simeq dt^2$, where $dt = d\Lambda/\Lambda$. The same goes for the BCS diagram. So F does not flow. Let us now turn to the renormalization of V. The first two diagrams are useless for the same reasons as before, but the last one, the BCS diagram, is special. Since the total incoming momentum is zero, the loop momenta are equal and opposite and no matter what direction K has, $-K$ is guaranteed to lie in the same shell. However, the loop frequencies are now equal and opposite so that the poles now lie in opposite directions. We now get a flow (dropping constants)

$$\frac{du(\theta_1 - \theta_3)}{dt} = -\int d\theta\, u(\theta_1 - \theta)u(\theta - \theta_3) \tag{15}$$

Here is an example of a flow equation for a coupling function. However, by expanding u in terms of angular momentum eigenfunctions we get an infinite number of flow equations

$$\frac{du_m}{dt} = -u_m^2 \tag{16}$$

one for each coefficient. What these equations tell us is that if the Cooper pair potential in angular momentum channel m is repulsive, it will get renormalized down to zero (a result derived many years ago by Anderson and Morel) while if it is attractive, it will run off, causing the BCS instability. This is the reason the V's are not a part of Landau theory which assumes we have no phase transitions.

Not only did Landau say we could describe Fermi liquids with an F, he also managed to compute the soft response functions of the system in terms of the F function even when it was large, say 10 in dimensionless units. The RG gives us a way to understand this. As we introduce a Λ and reduce it, a small parameter enters, namely Λ/K_F. This plays the role of $1/N$ in problems with N species of particles. Why do we have a large N here? Recall that the action of the noninteracting fermions had the form of an integral over one-dimensional fermions, one for each direction. How many fermions are there? The answer is not infinity since each direction has zero measure. It turns out that in any calculation, at any given Λ, the number of independent fermions is K_F/Λ. If you imagine dividing the annulus into patches of size Λ in the angular direction also, each patch carries an index and contributes to one species.

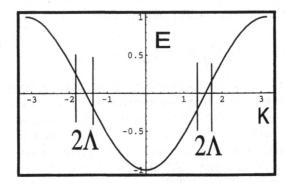

Fig. 4. The energy–momentum relation for free fermions on a linear lattice.

The same diagrams that dominate the usual $1/N$ expansions also dominate here and may be readily summed geometrically. Landau theory is just the $N = \infty$ limit. Thus the effective field theory has additional symmetries that allow for its solution.

A long paper of mine explains all this, as well as how it is to be generalized to anisotropic Fermi surfaces and Fermi surfaces with additional special features and consequently additional instabilities. The general scheme was proposed many years ago by Anderson in his book.

Polchinski independently analyzed the isotropic Fermi liquid (not in the same detail, since it was just a paradigm for him) and Weinberg independently derived the flow equations for the anisotropic superconductor.

I will now quickly discuss a few other examples of effective field theories in condensed matter. Consider spinless fermions on a one-dimensional lattice described by a Hamiltonian:

$$H = -\frac{1}{2}\sum_n \psi^\dagger(n+1)\psi(n) + H.C. \tag{17}$$

Upon Fourier transforming, we get the dispersion relation

$$E(K) = -\cos K \quad -\pi \leq K \leq \pi \tag{18}$$

depicted in Figure 4.

If we now linearize this near the Fermi points at $\pm K_F$ and introduce a k which measures $K - K_F$ (and not $|K| - K_F$) we find

$$H_{\text{eff}} = \int_{-\Lambda}^{\Lambda} dk[k\psi_R^\dagger(k)\psi_R(k) - k\psi_L^\dagger(k)\psi_L(k)]$$

which we recognize to be the beautiful one-dimensional Dirac Hamiltonian with $E = \pm k$. Thus from the original microscopic Hamiltonian Eqn. (17), which even its mother would not call beautiful, we have obtained a beautiful effective theory. We can add many interactions, and solve them by bosonization *à la* Luther and Peschel or Coleman.

Consider next the problem studied by Haldane, a one-dimensional spin antiferromagnetic Heisenberg chain with

$$H = \sum_n \vec{S}(n) \cdot \vec{S}(n+1)$$

This is already an effective interaction, which can be traced back to the Coulomb interaction and exchange forces. Haldane wanted to show that, contrary to naive expectations, this theory had two distinct limits as $S \to \infty$, depending on whether S was integer or half-integer. What Haldane did was to show that if we focus on the low energy modes of this model, we get the partition function for the nonlinear sigma model with a theta parameter equal to $2\pi S$:

$$Z = \int [\mathrm{d}n] \exp[-(\nabla n)^2 + 2\pi i S W]$$

where W is the winding number, which is necessarily integer. It follows that if S is an integer, the topological term makes no difference. On the other hand for half-integer values, it produces an alternating sign depending upon the instanton number.

I considered the effect of adding holes to this model. Now a hole is like a spinless fermion since you can only have zero or one hole per site, and the spinless fermion is equal to a Dirac field. So we get a sigma model coupled to a Dirac fermion. Just as the sigma model mimics QCD, this one mimics QCD coupled to massless quarks. In particular the massless fermions make the value of the topological coefficient irrelevant, which translates into the fact that these holes make all large S systems behave in one way independent of whether the spin is integer or half-integer.

Lastly, let us recall that Ising models near criticality can be represented by a Majorana field theory with action:

$$S = \int \bar{\psi}(i\partial + m)\psi \, \mathrm{d}^2 x \tag{19}$$

where m measures the deviation from criticality. If we now consider a random bond Ising model in which the bonds vary around the critical value, we get an Ising model with a position dependent mass term. To find the quenched averages in this model, one uses the replica trick and obtains the $N = 0$ component Gross–Neveu model, as shown by Dotsenko and Dotsenko. The amazing thing is that this model was originally invented for $N \to \infty$ and we see it being useful down to $N = 0$!

To conclude, very beautiful quantum field theories can arise in condensed matter physics as effective theories. In addition to their beauty, effective field theories are also very effective in answering certain questions that the more microscopic versions cannot.

References

References given below are not exhaustive and limited to those that were referred to in this talk. For the RG applied to fermions see R. Shankar, *Rev. Mod. Phys.*, **66**, 129, 1994. See also J. Polchinski, *Proceedings of the 1992 TASI Elementary Particle Physics*, editors J. Polchinski and J. Harvey, World Scientific, 1992, and S. Weinberg, *Nucl. Phys.*, **B413**, 567, 1994. For one-dimensional fermions see A. Luther and I. Peschel, *Phys. Rev.*, **B12**, 3908, 1975, S. Coleman, *Phys. Rev.*, **D11**, 2088, 1975. For the antiferromagnet see F.D.M. Haldane, *Phys. Lett.*, **93A**, 464, 1983, and for the doped case see R. Shankar, *Phys. Rev. Lett.*, **63**, 206, 1989. For the random bond Ising model see the introduction by V.S. Dotsenko and V.S. Dotsenko, *Adv. Phys.*, **32**, 129, 1983, and the review by B.N. Shalaev, *Phys. Reports*, **237**, 1994.

5. The triumph and limitations of quantum field theory

DAVID J. GROSS*

1 The triumph of quantum field theory

Although the title of this session is 'The foundations of quantum field theory', I shall talk, not of the foundations of quantum field theory (QFT), but of its triumphs and limitations. I am not sure it is necessary to formulate the foundations of QFT, or even to define precisely what it is. QFT is what quantum field theorists do. For a practising high energy physicist, nature is a surer guide as to what quantum field theory is as well to what might supersede it, than is the consistency of its axioms.

Quantum field theory is today at a pinnacle of success. It provides the framework for the standard model, a theory of all the observed forces of nature. This theory describes the forces of electromagnetism, the weak interaction responsible for radioactivity, and the strong nuclear force that governs the structure of nuclei, as consequences of local (gauge) symmetries. These forces act on the fundamental constituents of matter, which have been identified as pointlike quarks and leptons. The theory agrees astonishingly well with experiment to an accuracy of 10^{-6}–10^{-10} for electrodynamics, of 10^{-1}–10^{-4} for the weak interactions and of 1–10^{-2} for the strong interactions. It has been tested down to distances of 10^{-18} cm in some cases. We can see no reason why QFT should not be adequate down to distances of order the Planck length of 10^{-33} cm where gravity becomes important. If we note that classical electrodynamics and gravity are simply limiting cases of their quantum field theoretic generalizations, then quantum field theory works from the Planck length to the edge of the universe – over 60 orders of magnitude. No other theory has been so universally successful.

1.1 The problems of the past

It is hard for today's generation to remember the situation 25 years ago, when field theory had been abandoned by almost all particle physicists. The exhilaration following the development of quantum electrodynamics (QED) was short lived when the same methods were applied to the new field of mesons and nucleons. The proliferation of elementary particles and their strong couplings, as well as the misunderstanding and discomfort with renormalization theory, gave rise to despair with field theoretic models and to the conclusion that quantum field theory itself was at fault.

Renormalization was originally a response to the ultraviolet divergences that appeared in the calculations of radiative corrections in QED in a perturbative

* Supported in part by the National Science Foundation under grant PHY90-21984.

expansion in the fine structure constant. The basic observation was that if physical observables were expressed, not in terms of the bare parameters that entered the definition of the theory (like the bare mass of the electron) and refer to idealized measurements at infinitely small distances, but rather in terms of the physical parameters that are actually measurable at finite distances, then they would be finite, calculable functions of these. Feynman, Schwinger, Tomanaga and Dyson set forth the procedure for carrying out this renormalization to all orders in perturbation theory and proved that it yielded well-defined, finite results. Even though this program was very successful many physicists were uncomfortable with renormalization, feeling that it was merely a trick that swept the fundamental problem of ultraviolet divergences under the rug. Furthermore, there was great concern as to the consistency of quantum field theory at short distances. Most four-dimensional quantum field theories are not asymptotically free, so their short distance behavior is governed by strong coupling and thus not easily treatable. In the fifties it was suspected, especially by Landau and his school, that the nonperturbative ultraviolet behavior of QFT meant that these theories were inherently inconsistent, since they violated unitarity (which means that the total sum of probabilities for the outcomes of measurements of some physical processes was not unity). This is probably the case for most non-asymptotically free theories, which are most likely inconsistent as complete quantum field theories. The discovery of asymptotic freedom, however, has provided us with theories whose ultraviolet behavior is totally under control.

The disillusionment with QFT as a basis for the theory of elementary particles was also premature. What was missing was many ingredients, including the identification of the underlying gauge symmetry of the weak interactions, the concept of spontaneous symmetry breaking that could explain how this symmetry was hidden, the identification of the fundamental constituents of the nucleons as colored quarks, the discovery of asymptotic freedom which explained how the elementary colored constituents of hadrons could be seen at short distances yet evade detection through confinement, and the identification of the underlying gauge symmetry of the strong interactions. Once these were discovered, it was but a short step to the construction of the standard model, a gauge theory modeled on QED, which opened the door to the understanding of mesons and nucleons.

2 The lessons of quantum field theory

The development and successes of QFT have taught us much about nature and the language we should use to describe it. Some of the lessons we have learned may transcend QFT. Indeed they might point the way beyond QFT. The most important lessons, in my opinion, have to do with symmetry principles and with the renormalization group.

2.1 Symmetry

The most important lesson that we have learned in this century is that the secret of nature is symmetry. Starting with relativity, proceeding through the development of quantum mechanics and culminating in the standard model, symmetry principles have assumed a central position in the fundamental theories of nature. Local gauge symmetries provide the basis of the standard model and of Einstein's theory of gravitation.

Global symmetry principles express the invariance of physical laws under an actual transformation of the physical world. Local symmetry principles express the invariance of physical phenomena under a transformation of our description of them, yet local symmetry underlies dynamics. As Yang has stated: *Symmetry dictates interaction.* The first example of a gauge theory was general relativity where diffeomorphism invariance of space-time dictated the laws of gravity. In the standard model, non-Abelian gauge symmetry dictates the electroweak and strong forces. Today we believe that global symmetries are unnatural. They smell of action at a distance. We now suspect that all fundamental symmetries are local gauge symmetries. Global symmetries are either broken, or approximate, or they are the remnants of spontaneously broken local symmetries. Thus, Poincaré invariance can be regarded as the residual symmetry of general relativity in the Minkowski vacuum under changes of the space-time coordinates.

The story of symmetry does not end with gauge symmetries. In recent years we have discovered a new and extremely powerful new symmetry – supersymmetry – which might explain many mysteries of the standard model. We avidly await the experimental discovery of this symmetry. The search for even newer symmetries is at the heart of many current attempts to go beyond the standard model. String theory, for example, shows signs of containing totally new and mysterious symmetries with greater predictive power.

Another part of the lesson of symmetry is that, although the secret of nature is symmetry, much of the texture of the world is due to mechanisms of symmetry breaking. The spontaneous symmetry breaking of global and local gauge symmetries is a recurrent theme in modern theoretical physics. In quantum mechanical systems with a finite number of degrees of freedom global symmetries are realized in only one way. The laws of physics are invariant and the ground state of the theory is unique and symmetric. However, in systems with an infinite number of degrees of freedom a second realization of symmetry is possible, in which the ground state is asymmetric. This spontaneous symmetry breaking is responsible for magnetism, superconductivity, the structure of the unified electroweak theory and more. In such a situation the symmetry of nature's laws is hidden from us. Indeed, the search for new symmetries of nature is based on the possibility of finding mechanisms, such as spontaneous symmetry breaking or confinement, that hide the new symmetry.

There are two corollaries of the lesson of symmetry that are relevant to our understanding of QFT. First is the importance of special quantum field theories. A common strategy adopted years ago, say in constructive field theory, was to consider theories with only scalar fields. Their study, it was thought, would teach us the general principles of QFT and illuminate its foundations. This, to an extent, was achieved. But in the absence of vector or fermionic fields one cannot construct either gauge invariant or supersymmetric theories, with all of their special and rich phenomena. Today it might be equally foolhardy to ignore quantum gravity in the further development of QFT. Indeed the fact that QFT finds it so difficult to incorporate the dynamics of space-time suggests that we might search for more special theories.

Second, we have probably exhausted all possible symmetries of QFT. To find new ones we need a richer framework. Traditional quantum field theory is based on the principles of locality and causality, on the principles of quantum mechanics and on the principles of symmetry. It used to be thought that QFT, or even particular quantum field theories, were the unique way of realizing such principles. String theory provides us with an example of a theory that extends quantum field theory, yet embodies these same principles. It appears to contain new and strange symmetries that do not appear in QFT. If there are new organizing principles of nature, the framework of QFT may simply not be rich enough. We may need string theory, or even more radical theories, to deal with new symmetries, especially those of space-time.

2.2 The renormalization group

The second important lesson we have learned is the idea of the renormalization group and effective dynamics. The decoupling of physical phenomena at different scales of energy is an essential characteristic of nature. It is this feature of nature that makes it possible to understand a limited range of physical phenomena without having to understand everything at once. The renormalization group describes the change of our description of physics as we change the scale at which we probe nature. These methods are especially powerful in QFT which asserts control over physics at all scales.

Quantum field theories are most naturally formulated at short distances, where locality can be most easily imposed, in terms of some fundamental dynamical degrees of freedom (described by quantum fields). Measurement, however, always refers to physics at some finite distance. We can describe the low energy physics we are interested in by deriving an effective theory which involves only the low momentum modes of the theory. This procedure, that *of integrating out the high momentum modes of the quantum fields*, is the essence of the renormalization group, a transformation that describes the flow of couplings in the space of quantum field theories as we reduce the scale of energy.

The characteristic behavior of the solutions of the renormalization group equations is that they approach a finite dimensional sub-manifold in the infinite dimensional space of all theories. This defines an *effective* low energy theory, which is formulated in terms of a finite number of degrees of freedom and parameters and is largely independent of the high energy starting point. This effective low energy theory might be formulated in terms of totally different quantum fields, but it is equally *fundamental* to the original high energy formulation insofar as our only concern is low energy physics.

Thus, for example, QCD is the theory of quarks whose interactions are mediated by gluons. This is the appropriate description at energies of billions of electron volts. However, if we wish to describe the properties of ordinary nuclei at energies of millions of electron volts, we employ instead an *effective theory* of nucleons, composites of the quarks, whose interactions are mediated by other quark composites – mesons. Similarly, in order to discuss the properties of ordinary matter made of atoms at energies of a few electron volts we can treat the nuclei as pointlike particles, ignore their internal structure and take into account only the electromagnetic interactions of the charged nuclei and electrons.

The renormalization group influences the way we think about QFT itself. One implication is that there may be more than one, equally fundamental, formulation of a particular QFT, each appropriate for describing physics at a different scale of energy. Thus, the formulation of QCD as a theory of quarks and gluons is appropriate at high energies where, owing to asymptotic freedom, these degrees of freedom are weakly coupled. At low energies it is quite possible, although not yet realized in practice, that the theory is equivalent to a theory of strings, describing mesons as tubes of confined chromodynamic flux. Both formulations might be equivalent and complete, each appropriate to a different energy regime. Indeed, as this example suggests, a quantum field theory might be equivalent to a totally different kind of theory, such as a string theory.

The renormalization group has had a profound influence on how we think about renormalizability. Renormalizability was often regarded as a selection principle for QFT. Many quantum field theories, those whose couplings had dimensions of powers of an inverse mass (such as the Fermi theory of weak interactions), were not renormalizable. This meant that, once such interactions were introduced, it was necessary to specify an infinite number of additional interactions with an infinite number of free parameters in order to ensure the finiteness of physical observables. This seemed physically nonsensical, since such a theory has no predictive power and was taken to be the reason why theories of nature, such as QED, were described by renormalizable quantum field theories.

Our present view of things is quite different. The renormalization group philosophy can be applied to the standard model itself. Imagine that we have a unified theory whose characteristic energy scale, Λ, is very large or whose characteristic distance scale, $\hbar c/\Lambda$, is very small (say the Planck length of 10^{-33} cm). Assume further that just below this scale the theory can be expressed in terms of local field variables. As to what happens at the unification scale itself we assume nothing, except that just below this scale the theory can be described by a local quantum field theory. (String theory does provide us with an example of such a unified theory, which includes gravity and can be expressed by local field theory at distances much larger than the Planck length.) Even in the absence of knowledge regarding the unified theory, we can determine the most general quantum field theory. In absence of knowledge as to the principles of unification this theory has an infinite number of arbitrary parameters describing all possible fields and all possible interactions. We also assume that all the dimensionless couplings that characterize the theory at energy Λ are of order one (what else could they be?). Such a theory is useless to describe the physics at high energy, but at low energies, of order E, the effective dynamics, the effective Lagrangian that describes physics up to corrections of order E/Λ, will be parameterized by a finite number of couplings. The renormalization group describes how the various couplings run with energy. We start at Λ with whatever the final unified theory and then we can show that the low energy physics will be described by the most general renormalizable field theory consistent with the assumed couplings plus nonrenormalizable interactions that are suppressed by powers of the energy relative to the cutoff. If we demand further that the theory at the scale Λ contain the local gauge symmetry that we observe in nature, then the effective low energy theory will be described by the standard model up to terms that are negligible by inverse powers of the large scale compared to the energy that we observe. The extra interactions will give rise to weak effects, such as gravity or baryon decay. But these are very small and unobservable at low energy.

Nonrenormalizable theories were once rejected since, if they had couplings of order one at low energies, then their high energy behavior was uncontrollable unless one specified an infinite number of arbitrary parameters. This is now turned around. If all couplings are moderate at high energies, then nonrenormalizable interactions are unobservable at low energies. Furthermore, the standard model is the inevitable consequence of any unified theory, any form of the final theory, as long as it is local at the very high energy scale and contains the observed low energy symmetries. In some sense this is pleasing; we understand why the standard model emerges at low energy. But from the point of view of the unified theory that surely awaits us at very high energy it is disappointing, since our low energy theory tells us little about what the final theory can be. Indeed, the high energy theory need not be a QFT at all.

3 QCD as a perfect QFT

For those who ever felt uncomfortable with ultraviolet divergences, renormalization theory or the arbitrary parameters of quantum field theory, QCD offers the example of a perfect quantum field theory. By this I mean:

- This theory has no ultraviolet divergences at all. The local (*bare*) coupling vanishes, and the only infinities that appear are due to the fact that one sometimes expresses observables measured at finite distances in terms of those measured at infinitely small distances.
- The theory has no free, adjustable parameters (neglecting the irrelevant quark masses), and dimensional observables are calculable in terms of the dynamically produced mass scale of the theory $m = \Lambda \exp(-1/g_0^2)$, where g_0 is the *bare* coupling that characterizes the theory at high energies of order Λ.
- The theory shows no diseases when extrapolated to infinitely high energies. To the contrary, asymptotic freedom means that at high energies QCD becomes simple and perturbation theory is a better and better approximation.

Thus, QCD provides the first example of a complete theory with no adjustable parameters and with no indication within the theory of a distance scale at which it must break down.

There is a price to be paid for these wonderful features. The absence of adjustable parameters means that there are no small parameters in the theory. The generation of a dynamical mass scale means that perturbative methods cannot suffice for most questions. The flip side of asymptotic freedom is infrared slavery, so that the large distance properties of the theory, including the phenomenon of confinement, the dynamics of chiral symmetry breaking and the structure of hadrons, are issues of strong coupling.

What are the limitations of such a QFT? In traditional terms there are none. Yet, even if we knew not of the electroweak and gravitational interactions, we might suspect that the theory is incomplete. Not in the sense that it is inconsistent, but rather that there are questions that can be asked which it is powerless to answer; such as why is the gauge group $SU(3)$, or what dictates the dynamics of space-time?

4 The limitations of QFT

Quantum field theory is a mature subject; the frontier of fundamental physics lies elsewhere. Nonetheless there are many open problems in quantum field theory that should and will be addressed in the next decades.

First there are problems having to do with QCD, our most complete field theory. Much is understood, but much remains to be understood. These problems include the proof of the existence of QCD and of confinement; the development of analytic methods to control QCD in the infrared; and the development of numerical algorithms for Minkowski space and scattering amplitudes.

The second class of problems are more general than QCD, but would help in solving it as well. These include the development of large N methods; the formulation of a nonperturbative continuum regularization: the rigorous formulation of renormalization flow in the space of Hamiltonians; a first quantized path integral representation of gauge mesons and the graviton; the exploration of the phase structure of particular theories, particularly supersymmetric gauge theories; the complete classification and understanding of two-dimensional conformal field theories and integrable models; and the discovery and solution of special integrable quantum field theories.

You will have noticed that the big problems of high energy physics are not on the above list. These include: the unification of forces, the mass hierarchy problem (namely the question of why the scale of the electroweak symmetry breaking and the mass scale of the strong interactions are smaller than the Planck or unification scale by 14 to 18 orders of magnitude), the origin of lepton–quark families, the explanation of the parameters of the standard model, quantum gravity, the smallness or vanishing of the cosmological constant, the early history of the universe....

The reason I have not listed these is that I believe that their resolution does not originate in quantum field theory at all. To solve these we will have to go beyond quantum field theory to the next stage, for example to string theory. In this sense QFT has reached true maturity. Not only do we marvel at its success but we are aware of its boundaries. To truly understand a physical theory it is necessary to have the perspective of the next stage of physics that supersedes it. Thus we understand classical mechanics much better in the light of quantum mechanics, electrodynamics much better after QED. Perhaps the true understanding of QFT will only transpire after we find its successor.

The search for a replacement for QFT has been going on ever since its invention. Every conceptual and technical difficulty that was encountered was taken as evidence for a fundamental length at which QFT breaks down. With the success of QFT as embodied in the standard model the search for a fundamental length has been pushed down to the Planck length. There almost everyone believes that a new framework will be required, since many of the basic concepts of QFT are unclear once space-time fluctuates violently. The longstanding problem of quantizing gravity is probably impossible within the framework of quantum field theory. Einstein's theory of gravity appears to be an effective theory, whose dimensional coupling, Newton's constant G_N, arises from the scale of unification which might be close to the Planck mass M_P, i.e., $G_N \propto 1/M_P^2$. General relativity is then simply an incredibly weak force that survives at low energies and is only observable since it couples coherently, via long range forces, to mass, so that we can observe its effects on large objects. QFT has proved useless in incorporating quantum gravity into a consistent theory at the Planck scale. We need to go beyond QFT, to a theory of strings or to something else, to describe quantum gravity.

There are other indications of the limitations of QFT. The very success of QFT in providing us with an extremely successful theory of all the nongravitational forces of nature has made it clear that this framework cannot explain many of the features and parameters of the standard model which cry out for explanation. In the days before the standard model it was possible to believe that the requirement of renormalizability or symmetry would be sufficient to yield total predictive power (as in QCD). But today's understanding makes it clear that these principles are not sufficient.

Thus, the limitations of QFT are not those of consistency or incompleteness in its own terms, but rather of insufficiency and incompleteness in broader terms. If we restrict ourselves to effective field theories, or the use of QFT in dealing with non-relativistic many body systems, or fundamental theories of limited domain (such as QCD), then QFT is in fine shape. But if we are to come to grips with the quantization of the dynamics of space-time then QFT is, I believe, inadequate. I also believe that we will learn much about QFT itself from its successor. For example, there are certain features of special quantum field theories (such as the recently developed duality symmetries) whose deeper understanding might require the embedding of field theory within string theory.

5 Beyond QFT – string theory

We have one strong candidate for an extension of physics beyond QFT that does claim to be able to answer the questions that QFT cannot and more – string theory. String theory is a radically conservative extension of the principles of physics, in which one introduces fundamental degrees of freedom that are not pointlike, but rather have the structure of extended one-dimensional objects – strings – while leaving untouched (at least in the beginning) the other principles of causality, relativistic invariance and quantum mechanics. The structure of this theory, which appears to be unique and free of any nondynamical parameters, is quite remarkable. It yields a consistent theory of quantum gravity, at least in the perturbative, weak field domain, providing us with an existence proof that gravity and quantum mechanics are mutually consistent. In addition, it appears to possess all the ingredients that would be necessary to reproduce and explain the standard model. Most important, it is definitely an extension of the conceptual framework of physics beyond QFT.

There have been two major revolutions completed in this century: relativity (special and general) and quantum mechanics. These were associated with two of the three dimensional parameters of physics: $\hbar$, Planck's quantum of action, and c, the velocity of light. Both involved major conceptual changes in the framework of physics, but reduced to classical nonrelativistic physics when $\hbar$ or $1/c$ could be regarded as small. The last dimensional parameter we need in order to establish a set of funda-mental dimensional units of nature is Newton's gravitational constant, which sets the fundamental (Planck) scale of length or energy. Many of us believe that string theory is the revolution associated with this last of the dimensional parameters of nature. At large distances, compared with the string scale of approximately 10^{-33} cm, string theory goes over into field theory. At shorter distances it is bound to be very different, indeed it calls into question what we mean by distance or space-time itself.

The reason we are unable to construct predictive models based on string theory is our lack of understanding of the nonperturbative dynamics of string theory. Our present understanding of string theory is very primitive. It appears to be a totally con-sistent theory, which does away with pointlike structures and hints at a fundamental revision of the notions of space and time at short distances while at the same time reducing to field theory at large distances. It introduces a fundamental length in a way that had not been envisaged – not by, for example, discretizing space and time – but rather by replacing the fundamental pointlike constituents of matter with extended, nonlocal strings. The constituents are nonlocal but they interact locally; this is sufficient to preserve the usual consequences of locality – causality as expressed in the analyticity of scattering amplitudes.

To be more specific, string theory is constructed to date by the method of *first quantization*. Feynman's approach to QFT, wherein scattering amplitudes are constructed by summing over the trajectories of particles, with each history weighted by the exponential of (*i* times) the classical action given by the proper length of the path, is generalized to strings by replacing the length of the particle trajectory with the area swept out by the string as it moves in space-time. This yields a perturbative expansion of the amplitudes in powers of the string coupling, which is analogous to the Feynman diagram expansion of QFT. However, string theory exhibits profound differences from QFT. First, there is no longer any ambiguity, or freedom, in specifying the string interactions, since there is no longer an invariant way of specifying when and where the interaction took place. Consequently, the string coupling itself becomes a dynamical variable, whose value should ultimately be determined (in ways we do not yet understand). Furthermore, there are only a few, perhaps only one, consistent string theory. Finally, the issue of ultraviolet divergences is automatically solved, the smoothing out of world-lines to world-tubes renders the interactions extremely soft and ensures that string amplitudes are totally finite.

At low energies string theory goes over into field theory. That means that we can describe the scattering amplitudes by an effective field theory describing the light particles to any degree of approximation in powers of the momenta p/M_P. However, string theory is not just a complicated field theory. It exhibits features at short distances or high energies that are profoundly different from QFT; for example the Gaussian falloff of scattering amplitudes at large momenta. At the moment we are still groping towards an understanding of its properties for strong coupling and for short distances.

In our eventual understanding of string theory we might have to undergo a discontinuous conceptual change in the way we look at the world similar to that which occurred in the development of relativity and quantum mechanics. I think that we are in some sense in a situation analogous to where physics was in the beginning of the development of quantum mechanics, where one had a semiclassical approximation to quantum mechanics that was not yet part of a consistent, coherent framework. There was an enormous amount of confusion until quantum mechanics was finally discovered. What will this revolution lead to? Which of our concepts will have to be modified? There are many hints that our concepts of space-time, which are so fundamental to our understanding of nature, will have to be altered.

The first hint is based on a stringy analysis of the measurement of position, following Heisenberg's famous analysis in the quantum mechanics. Already in ordinary quantum mechanics space becomes somewhat fuzzy. The very act of measurement of the position of a particle can change its position. In order to perform a measurement of position x, with a small uncertainty of order Δx, we require probes of very high energy E. That is why we employ microscopes with high frequency (energy) rays or particle accelerators to explore short distances. The precise relation is that

$$\Delta x \approx \frac{\hbar c}{E},$$

where $\hbar$ is Planck's quantum of action and c is the velocity of light. In string theory, however, the probes themselves are not pointlike, but rather extended objects, and thus there is another limitation as to how precisely we can measure short distances. As energy is pumped into the string it expands and thus there is an additional uncertainty proportional to the energy. Altogether

$$\Delta x \approx \frac{\hbar c}{E} + \frac{GE}{c^5}.$$

Consequently it appears impossible to measure distances shorter than the Planck length.

The second hint is based on a symmetry of string theory known as duality. Imagine a string that lives in a world in which one of the spatial dimensions is a little circle of radius R. Such situations are common in string theory and indeed necessary if we are to accept the fact that the string theories are naturally formulated in nine spatial dimensions, so that if they are to look like the real world, six dimensions must be curled up, *compactified*, into a small space. Such perturbative solutions of realistic string theories have been found and are the basis for phenomenological string models. Returning to the simple example of a circle, duality states that the theory is identical in all of its physical properties to one that is compactified on a circle of radius $\bar{R} = L_P^2/R$, where L_P is the ubiquitous Planck length of 10^{-33} cm. Thus if we try to make the extent of one of the dimensions of space very small, by curling up one dimension into a circle of very small radius R, we would instead interpret this as a world in which the circle had a very large radius $\bar{R}$. The minimal size of the circle is of order L_P. This property is inherently stringy. It arises from the existence of stringy states that wind around the spatial circle and again suggests that spatial dimensions less than the Planck length have no meaning.

Another threat to our conventional view of space-time is the discovery that in string theory the very topology of space-time can continuously be altered. In perturbative string theory there are families of solutions labeled by various parameters. In some cases these solutions can be pictured as describing strings propagating on a certain curved spatial manifold. As one varies the parameters the shape and geometry of the background manifold vary. It turns out that by varying these parameters one can continuously deform the theory so that the essential geometry of the background manifold changes. Thus one can go smoothly from a string moving in one geometry to a string moving in another, although in between there is no simple space-time description. This phenomenon cannot be explained by ordinary quantum field theories.

Finally, during this past year, new developments have been made in the understanding of the structure of string theory. A remarkable set of conjectures have been formulated and tested that relate quite different string theories to each other (S, T, U dualities) for different values of their parameters and for different background space-time manifolds. Until now the methods we have employed to construct string theories have been quite conservative. To calculate string scattering amplitudes one used the method of 'first quantization', in which the amplitudes are constructed by summing over path histories of propagating strings, with each path weighted by the exponential of the classical action (the area of the world sheet swept out by the string as it moves in space-time). This approach, originally developed by Feynman for QED, is quite adequate for perturbative calculations. It was envisaged that to do better one would, as in QFT, develop a string field theory. However, these new developments suggest that in addition to stringlike objects, 'string theory' contains other extended objects of higher internal dimension, which cannot be treated by the same first quantized methods and for which this approach is inadequate. Even stranger, some of the duality symmetries of string theory connect theories whose couplings (g_1 and g_2) are inversely related, $g_1 = 1/g_2$. This is a generalization of electrodynamic duality, wherein the electric and magnetic fields and their charges (e and g, related by the Dirac quantization condition $eg = 2\pi\hbar$) are interchanged.

These developments hint that the ultimate formulation of string theory will be quite different than originally envisaged. It might be one in which strings do not play a fundamental role and it might be a theory that cannot be constructed as the quantization of a classical theory. Thus it appears that we are headed for a real theoretical crisis in the development of string theory. A welcome crisis, in my opinion, one that could force us to radically new ideas.

5.1 Lessons for QFT

What can we learn from string theory about quantum field theory? There are a few lessons that we can already extract, and I expect that many more will emerge in the future.

- First, this theory, used simply as an example of a unified theory at a very high energy scale, provides us with a vindication of the modern philosophy of the renormalization group and the effective Lagrangian that I discussed previously. Using string theory as the theory at the cutoff we can verify that at energies low compared with the cutoff (the Planck mass) all observables can be reproduced by an effective local quantum field theory and, most importantly, all dimensionless couplings in this effective, high energy, theory are of the same order of magnitude. Thus, we have an example of a theory which, as far as we can see, is consistent at arbitrarily high energies and reduces at low energy to quantum field theory. String theory could explain the emergence of quantum field theory in the low energy limit, much as quantum mechanics explains classical mechanics, whose equations can be understood as determining the saddlepoints of the quantum path integral in the limit of small $\hbar$.
- We also learn that the very same quantum field theories that play a special role in nature are those that emerge from the string. These include general relativity, non-Abelian gauge theories and (perhaps) supersymmetric quantum field theories. Thus, string theory could explain the distinguished role of these theories. From a practical point of view string theory can be used to motivate the construction of novel field theoretic models that include less familiar interactions as well as new kinds of particles, such as axions or dilatons, that are ubiquitous features of low energy string physics.
- Finally, it appears that some of the new and mysterious dualities of supersymmetric field theories have their natural explanation in the framework of string theory. To truly understand these features of QFT it may be necessary to consider a field theory as part of a string theory and use the latter to understand the former.

6 Conclusions

I believe that we are living in revolutionary times, where many of the basic principles of physics are being challenged by the need to go beyond QFT and in which many of our basic concepts will require fundamental revision. Will the exploration of the foundations or the philosophical meaning of QFT help us in these tasks? I admit that my prejudice is that the answer is no. The issues that face us now have little to do with those that were confronted in the struggle to make sense of QFT. Rather, it is the surprises that we unearth in the experimental exploration of nature as well as those that emerge in the theoretical exploration of our emerging theories that will force us to radical modifications of our basic preconceptions.

6. Comments

SAM TREIMAN

This commentator format is quite unfamiliar to me, and somewhat uncomfortable. I don't know whether I'm supposed to flatter the speakers, criticize them, grade them, or merely parrot them. Or should I go off on my own? This latter is tempting. If unleashed I could provide you with *true* answers to all the questions before us, philosophical, historical, physical, mathematical, physical-mathematical, and so on. But I'll bite my tongue and try to stick to the commentator role.

First, a few words on the topic of the opening session this afternoon – why are philosophers interested in quantum field theory? Quantum mechanics undoubtedly abounds in genuine, deep and still unresolved philosophical questions. These are usually posed, at least in popular accounts, in the context of finite systems of non-relativistic point particles. In preparation for the conference, I asked myself: does relativistic QFT introduce any really *distinctive* philosophical problems? There are several possibilities, which I can only express in low-brow form. For one thing, a field is really a collection of infinitely many degrees of freedom, one for each point in space. I can well suppose that infinity raises interesting questions for philosophers. Certainly for physicists, especially in the field theory context, it has been a great pre-occupation. As we'll hear later from Jackiw, infinity has its useful aspects. Next, for systems of relativistic particles in interaction, it is no longer so easily possible to speak of measurements of position and momentum of individual particles, except perhaps in asymptotic regions. An extreme version of this is that it is the S-matrix only that deals with reality. Then there's the question of virtual reality. Are real particles real when they are on the inside somewhere in a Feynman diagram? And what about quarks? There's the genuine physics question of whether quarks are truly confined, more precisely whether they don't exist in isolation. If so, what do the philosophers say? Do they – the quarks, that is – exist? Finally, I may remind you that the fields that enter into our theories are put there mainly for the purpose of making quanta – making particles, that is; and for intermediating particle interactions; or in some cases for acquiring nonzero vacuum expectation values to break symmetries; or to enter into local gauge transformations; and so on. But for the most part the field *observables* that we talk about aren't things that we really observe. The practical exceptions to this are of course the electromagnetic field and – still only in rather trivial ways – the gravitational field. For the rest, this persistent *non*-observation of field *observables* may provide some philosophical conundrums, though I must say that it doesn't bother me.

One seems to be driven to QFT by the general principles of quantum mechanics taken together with the requirement of inhomogeneous Lorentz invariance and cluster decomposition. However, in the nonrelativistic limit, for situations where particle number can be treated as conserved – and this covers vast territories of science – there is no need whatsoever for a field theoretic approach. The observables are all fully expressible in terms of the position, momentum and spin operators of the individual particles. Nevertheless, especially for systems involving a very large number of particles, it is greatly convenient to 'second quantize'; that is, to go over to a field theoretic description. Apart from its practical benefits within the many-body context, this has the virtue of fostering productive commerce between many-body physics and particle physics – with arrows moving in both directions. Some examples: the ideas of broken symmetry, critical phenomena, the renormalization group, effective field theory, and so on. We've heard something of this in the wonderful talk given by Professor Shankar, whose credo is: use any tools that work, provided they have a field theoretic flavor.

For the rest, let me focus on relativistic QFT. It will be clear that my two Princeton colleagues, Professors Wightman and Gross, have come at this topic today from quite different angles. Wightman's outlook was, in the main, generic. His talk was devoted, so to speak, to the theory of quantum field theories. There are many interesting findings and many open issues that are internal to that subject. But on a broad enough level, if you want to look for features common to *all* field theories, you of course cannot hope to extract many *phenomenological* predictions. But there *do* exist a few such predictions, as Wightman reminded us: the connection between spin and statistics and the CPT theorem. I would add the forward pi-nucleon dispersion relations, and various other bits and pieces of analyticity information. These things are all old hat by now; but they are nonetheless precious.

Gross comes at our subject from the other end. He's in pursuit of the *right* theory. As everyone knows, with the Standard Model we have in fact come a long way toward that right theory over the past several decades. Within the domain of currently accessible phenomenology, the Standard Model stands largely unchallenged; quantitatively, it is in fact spectacularly successful in areas where reliable calculations and reliable experiments happen to coincide. A definite model of course provides a broad target for specialists in falsification, and indeed there do exist some experimental hints of new physics that go beyond the model. But it must be kept in mind that the Standard Model is a somewhat flexible framework. It can be generalized to allow, say, for finite neutrino masses and oscillations, for additional Higgs particles, even for new families if we should need them, and so on. It can be extended in various ways to achieve unification between the strong and electroweak sectors; and most interestingly it can be enlarged to incorporate the ideas of supersymmetry. From Gross's Olympian perspective, the really essential features, the features that he sees as the ideals of a QFT, are *local* gauge symmetry, and symmetry breaking; with illumination provided by the renormalization group and the ideas of effective field theory. Global symmetry Gross dismisses as ugly, fortuitous at best. In a memorable phrase he describes global symmetry as *smelling of action at a distance*.

With all the success and potential of QFT, as embodied in the Standard Model and possible generalizations and extensions, you may ask: why does one even contemplate that there are limitations to QFT? My immediate response is, go ask Gross. But here is something of what he might tell you, what he did in fact tell us here. For one thing, the Standard Model, though there is no reason to think it inconsistent, makes choices that seem to be arbitrary; moreover, it is far from economical. Within QCD, for example, we don't know why the gauge group is SU(3) rather than some other symmetry group. There is the problem of the mass hierarchy, the large number of input parameters, etc. Gross is doubtful that these matters can be resolved tastefully within a QFT framework. For another and undoubtedly more decisive thing, there is the matter of quantum gravity. I don't know that it's a theorem, but there is certainly a widely held belief that quantum gravity cannot be successfully married to quantum field theory. But don't despair, Gross says: there is string theory. Whether or not that is the final answer, he says, it provides an existence proof that gravity *can* be made compatible with the principles of quantum mechanics.

Whatever successfully replaces QFT, if anything does, it seems pretty sure that it will effectively turn into a QFT somewhere below the Planck scale and well above the scale currently accessible to us. And although we won't directly reach the Planck scale very soon in our experiments, theoretical progress in that far-off domain may well illuminate the physics of our humdrum world. From Gross's perspective, the real theoretical frontier today lies in this post-QFT region. At any rate, that's where a lot of the action is now taking place in many quarters of the globe.

As for my own humble opinion, I find it very hard indeed to doubt that there must be striking new physics at the Planck scale. At the same time, I am perhaps a little more optimistic than Gross seems to be about the possibility of big surprises within QFT along the way to that remote territory.

Session discussions

DeWitt: If, as Dave suggests, we are in a period like that between the Bohr atom and Heisenberg and Schrödinger, why is it taking so long?

Gross: All historical analogies are crude. I think actually, as you know very well, since you've worked on quantum gravity for most of your life, that understanding the dynamics of space-time is a revolution which hasn't yet occurred. In my opinion, we're on the verge of this revolution but it might take 12 years or it might take 40 years. It is harder than the quantum mechanical revolution. One of the reasons that it's harder is that we are removed by 17, 18 orders of magnitude from the typical scale of gravity. We have no direct evidence for quantum gravity. We are enormously lucky that one of the forces from the Planck scale has survived 20 orders of magnitude – namely gravity. It is a force which is, in natural units of strength, 10^{-40}. Thus in an atom the gravitational energies are 10^{-40} of Coulomb energies. Yet gravity, luckily, is observable, since it couples to mass and we can put together big people and planets and scatter them together. So we're enormously lucky that we've found this remnant of an energy scale which is 20 orders of magnitude away. But we have no direct observation of that scale.

In some sense, the quantum mechanics revolution, as difficult and as wrenching as it was, might have been a lot easier than coming to grips with the dynamics of space-time. That's why it might take 120 years. I don't know. You can never predict intellectual developments, because they really are first order phase transitions. Also we might not be able to succeed. I think that string theory has only a 75% chance of being right.

It is very hard to extrapolate physics beyond phase transitions. But we can identify thresholds. And we've identified now the same Planckian scale from two different arguments. One argument follows from Newton's constant. That identifies a scale of 10^{19} Gev. The second argument, based on high precision experiments at low energies, predicts the same scale, roughly, within a factor of ten over 20 orders of magnitude. That's a remarkable coincidence. So particle physicists now have to confront the problems that you have been struggling with all your life. It might take an awful long time to solve these problems because they are very hard. The notions of space and time are quite fundamental to the way we've always thought about physics and having to change them is going to be very difficult. I think we're going to have to change them, and we're beginning to get hints of how.

Stachel: I don't think Dave's answer fully met the import of the question, as I understood it. The thing is we do have a clue. Even at the Newtonian level, gravitation is that clue. Properly understood, gravitation at the Newtonian level is already a gauge theory. And when you combine that gauge version of gravitation with special relativity, you get general relativity. Now then, you say, conventional methods of quantization won't quantize general relativity. Shouldn't one explore *nonconventional* methods of quantization before one gives up on general relativity and goes lusting after other gods, such as string theory?

Gross: Sam Treiman said in his talk that there is no theorem that you can't deal with quantum gravity using quantum field theory. You could imagine that you could control the theory for strong coupling and there are ultraviolet fixed points, asymptotic safety, as Steve Weinberg used to say. So you can't disprove the quantum field theory of gravity. That's not how science develops. Nobody ever proves that it is impossible to do something. The only thing that happens is people start doing other things because they're more successful. One of the reasons that many of us particle physicists, not just those doing string theory, believe that quantum field theory is insufficient, is because we've learned a lot about quantum field theory. We know that all the methods that we know how to use in quantum field theory don't work in quantum gravity. Even supersymmetry doesn't help. Furthermore, we have an example of something that does work: string theory.

Not only does it work, the thing that attracts string theorists who are interested in quantum gravity is that this is a theory that automatically contains gravity. Ed Witten likes to say that the one prediction of string theory is gravity. This isn't as silly a remark as it might sound. Imagine we happened to have been physicists living in some ocean, and that we were very small, and thus we couldn't feel the gravitational force. So we would never have been able to measure this force of 10^{-40} compared to the Coulomb interaction. We would never have invented it, right? But, if you had discovered string theory, it would have been an automatic consequence of string theory, in the same way, Ed likes to say, that antimatter is an automatic consequence of quantum field theory. One would have predicted it, and calculated Newton's constant.

So in that sense, we're not just happy to have an example where you can calculate gravitational radiative corrections and all the things we do in QED or QCD with a theory that contains gravity. But more than that; this is a theory in which gravity is essential, is an inherent part of the theory and is essentially unified with the other gauge interactions that play such a fundamental role in nature. That's why it's so attractive.

Deser: I happen to lean to the string theoretical side as underpinning quantum gravity. There's also another aspect, which is a supergravity side. The fact that space-time is no longer what it used to be, that's something you would not have discovered within normal gravity, and it may yet be the best of all possible worlds if there's a particular privileged supergravity, $d = 11$, which is closely related in fact to some of this vast duality of different pieces of string theory. The gravity people can say, 'See, we understand it at a local field theory level,' and the string theory people might say, 'No, this is just one corner of the big picture.' So I think that the debate has in any case gone beyond, what shall I say, classical quantum gravity and can never go back in that direction. What direction it will be may indeed take a few decades or centuries, or maybe a few months. But in any case I think that we have to understand the conceptual developments that come from string theory at the very least, given the compelling fact that there exists a closed string theory that has massless spin two. All of Einstein gravity is a consequence.

Fisher: There is a counter-argument to that, which was brought up by Ken Wilson, when he was strongly criticized for using *lattice* gauge theories. He said, 'Well, Poincaré invariance and similar symmetries of space-time that we observe could just be asymptotic properties valid on long length scales and time scales.' I would like to hear your comments on that. If one imagines one has any messy equation, let us say an algebraic equation, and if it happens that as you vary some parameters, two of the roots come together, then it is a very good bet, say \$99 to \$1, that they will meet together just like the two solutions of a quadratic equation, that is, in a completely symmetrical fashion. This will occur even if there is no symmetry at all in the original, underlying equation. So my question is: Is it not the case that the asymptotic fixed points of any large system are likely to show rather definite, well-defined symmetries? Given that the standard model happens to have a very large number of symmetries (or near symmetries) including some rather peculiar ones, what gives you the faith – maybe you did not actually profess it but you more or less implied it – that string theory is going to lead only to the standard model, rather than permit a large number of asymptotic 'solutions'? Perhaps we just happen to have fallen into the standard model in the universe in which we are sitting.

Gross: There are two views about symmetry, one of which Michael was expressing. I'll rephrase it as 'symmetry is not fundamental in the structure of physics'. It's just that as you let the renormalization group evolve, you get attracted to fixed points which are often more symmetric than where you started. This particular approach is taken to an extreme by Holga Nielson, who would like to start with nothing, no symmetry principles, not even any logical principles, and somehow get to a fixed point in which everything emerges. My view of this approach is, 'garbage in, beauty out'. I really don't like that point of view.

My point of view is exactly the opposite. 'Garbage around us, beauty up there.' That is the lesson that we've learned during the evolution of physics, by which I mean fundamental physics, the reductionist program. All our advances introduce bigger and bigger symmetries. I think it's beautiful and I like it, but most importantly, it seems to be a fact about the world, about nature. Therefore, we can use symmetry as one of our strategies for exploring nature. It's been a very useful strategy. Why is it so? That is a philosophical question, but an interesting question. I suppose you could ask equally, why is mathematics so effective in the description of physics? Mathematics is language, symmetries are relations, the world makes sense. It would be hard for me to imagine an ugly world of the type that then becomes simple at a fixed point.

Fisher: Well, all the asymptotic behavior and renormalization group fixed points that we look at in condensed matter theory seem to grow symmetries not necessarily reflecting those of the basic, underlying theory. In particular, I will show some experiments tomorrow, where, in fact, one knows for certain that the observed symmetry grows from a totally unsymmetric underlying physics. Although as a research strategy I think what you say about postulating symmetry is totally unarguable, one can remark, in opposition, that it is only the desperate man who seeks after symmetry! If we truly understand a theory, we should see symmetry coming out or, on the other hand, failing to appear. So I am certainly not criticizing you on strategy. But you recognize – you put it very nicely, and I was relieved to hear it – that the renormalization group principle works in a large space, there are many fixed points, and there are many model field theories. So I am still unclear as to the origin of your faith that string theory should give us the standard model rather than some other type of local universe.

Gross: I stated a theorem, but there was much hand-waving. To make the theory precise one would have to make many assumptions. One would have to assume certain symmetries and, of course, one would have to make all the usual assumptions that you can use the perturbative analysis of the renormalization group as a sufficient guide. Within those assumptions, one can derive the standard model from any unified theory. String theory is more than that. It's an example of a real theory, so you just calculate with it. Perturbatively, there are many perfectly consistent quantum mechanical solutions of string theory, with a well-defined semiclassical expansion, that actually look very much like the standard model. The only problem is that there are too many of them. There is enormous degeneracy of such solutions, and most don't break supersymmetry. So we don't know the dynamics that chooses the vacuum and breaks the symmetry. Maybe there isn't such a dynamics, but we hope there is.

Part Three

7. Does quantum field theory need a foundation?*

SHELDON LEE GLASHOW

Once upon a time, there was a controversy in particle physics. [Some physicists] searched for a self-consistent interpretation wherein all [particles] were equally elementary. Others ... insisted on the existence of a small number of fundamental constituents and a simple underlying force law ... Many recent experimental and theoretical developments seem to confirm the latter philosophy and lead toward a unique, unified, and remarkably simple and successful view of particle physics. [1]

1 Introduction

A consistent description of all observed phenomena of the microworld is at hand! The so-called standard model of elementary particle physics grew by fits and starts from an exceedingly complex interplay between experiment and theory. In recent decades, experimentalists, with immense fortitude and generous funding, have identified and studied what appear to be the basic building blocks of matter and the fundamental forces that govern their interactions. Meanwhile, theorists have laboriously created, refined and reformulated a mathematical framework – quantum field theory – in terms of which the standard model is expressed. Aside from the occasional appearance (and disappearance) of conflicting data,[1] the standard model has met every experimental test. And yet, too many deep questions remain unanswered for the standard model, and hence quantum field theory, to be the last word. Many theoretical physicists believe that an entirely new framework is needed – superstring theory or something like it – that will once again revise our understanding of space, time and matter.

In the course of these developments, the conceptual basis of the present theory has been obscured. Most physicists are too much concerned with the phenomena they explore or the theories they build to worry about the logical foundation or historical origin of their discipline. Most philosophers who explore the relationship between scientific knowledge and objective reality are ill-equipped to deal with the intricacies of quantum field theory. If we are to build a better theory, perhaps we should first know where we are and how we got here. The hope has been expressed that this conference will help enable physicists and philosophers to address the questions of what science is and how it has evolved.

---

* Research supported in part by NSF-PHYS-92-18167.
[1] The specialist will recall, among many examples, Rubbia's monojets, Simpson's 17 keV neutrino, the anomalon, the zeta particle, and today's anomalous values for R_b and R_c reported by the LEP collaborations at CERN.

74

Our focus is supposed to be on its foundations, but can QFT be disentangled from the rest of physics? QFT and its interpretation are intimately linked to its antecedents: classical mechanics, electromagnetism, quantum mechanics and special relativity; to its kissing cousins: statistical mechanics, general relativity and black-hole physics; and to its would-be successor: superstring theory. Indeed, the quantum workings of the microworld are recapitulated classically: as print-outs, photographs, and meter readings. Attempts to isolate QFT and discuss its foundations *per se* are likely to become paeans to its many triumphs, epitaphs for its many omissions, or mere metamathematical musings akin to the study of Newton's new math in his time or Einstein's in his.

Our host, Professor T.Y. Cao, studies the nature of scientific progress using the development of QFT as an illustrative example. Is each new layer of science placed firmly on the layer beneath? Or is science punctuated by revolutions by which the entire edifice must be built anew? With some qualifications, I believe that the former view prevails. Special relativity and quantum mechanics were certainly revolutionary, but it is wrong to say that they *overthrew* classical mechanics. Past successes are rarely abandoned: they are coopted. New disciplines define the bounds of validity of older ones. We now understand exactly in what sense classical mechanics is 'true' and why it will remain true for evermore. At the same time, new disciplines extend the arena of discourse to phenomena that could not be addressed before. In a similar way, superstring theory, if correct, will incorporate – but certainly not abolish! – quantum field theory and the standard model.

I shall approach the profound theme of this conference by discussing the sorts of questions about physics that may be addressed to a theory such as classical mechanics or quantum field theory. To do this, I first ask just what we may mean by physics:

- Many dictionaries ineptly describe physics as *the science of matter and energy and of the interactions between the two*. [2] The realization that matter and energy are one renders this definition senseless.
- Henry A. Rowland held that physics is *a science above all sciences, which deals with the foundation of the universe, with the constitution of matter of which everything in the universe is made, and with* [the space through] *which the various portions of the universe . . . affect each other*. [3] Mentioning the universe thrice, Rowland anticipated today's truly grand unification: the convergence of cosmology and particle physics, which must be regarded as the culmination of Newton's synthesis – the study of all features great and small.[2] And yet, what does it mean 'to deal with'?
- Philipp Frank claimed: *There are no boundaries between physics and philosophy, if one only formulates the task of physics in accordance with the doctrines of Ernst Mach using the words of Carnap: 'to order the perceptions systematically and from present perceptions to draw conclusions about perceptions to be expected'*. [4] We are reminded – as often we must – that physics has always been (and always must be) an experimental science.
- In another context, G.W. Hemming put the Carnap–Franck–Mach definition more succinctly: *The problem of billiards in all its generality is this: a ball impinges on a ball at rest or on a cushion – to find the subsequent path of each ball*. [5] The rest is commentary, albeit rather mathematical.

[2] I am indebted to Uri Sarid for this paraphrase of scripture.

A few personal remarks may clarify my standing to present a talk at this conference. Having learned a bit about QFT from Sam Schweber at Cornell, I applied for graduate study at Princeton, only to be rejected because my interviewers correctly discerned that I knew more of the language of physics than its substance. Instead, I attended Harvard where I had the good fortune to work with Julian Schwinger. His legendary mastery of mathematics never detracted him from his search for a simple and elegant description of the complexities of nature. Schwinger's can-do philosophy had much in common with that of our Harvard colleague, Percy Bridgman, who wrote:

Whatever may be one's opinion as to the simplicity of the laws of nature, there can be no question that the possessors of some such conviction have a real advantage in the race for physical discovery. Doubtless there are many simple connections still to be discovered, and he who has a strong conviction of the existence of these connections is much more likely to find them than he who is not at all sure that they are there. [6]

Schwinger insisted that the several features shared by weak interactions and electromagnetism indicated a common origin and the possible existence of a unified theory of both forces. His conclusion to a paper based on his Harvard lectures reads:

What has been presented here is an attempt to elaborate a complete dynamical theory of the elementary particles from a few general concepts. Such a connected series of speculations provides a convenient frame of reference in seeking a more coherent account of nature. [7]

It did! Schwinger's suggestions sent me on my way, although I did not then (and do not now) share his fluency in QFT, nor that of my colleagues in the audience. I participate reluctantly and with considerable trepidation because each of my three forays into formal QFT met with catastrophe.

- In 1959, I argued that softly broken gauge theories are renormalizable. [8] Abdus Salam and others immediately pointed out my egregious error. [9]
- In the 1960s, Sidney Coleman and I presented an analysis of the neutral kaon system that was either erroneous or had been done before (I forget which). T.D. Lee sought and obtained both a retraction and an apology.
- In 1970, John Iliopoulos and I discovered many remarkable cancellations among the expected divergences of a softly broken gauge theory. [10] Soon afterward, Tini Veltman told us that his student, Gerard 'tHooft, made our perfectly correct endeavor perfectly irrelevant.

Most of my successful research endeavors relate to the symmetries of a field-theoretic approach to nature, not to its dynamics. Along with my friend Gary Feinberg, I was more concerned with symmetry-related 'angular problems' than with the analytically more complex 'radial problems'.

QFT may be discussed at many levels: in terms of its relationship to observations, its mathematical consistency, or its emergence from earlier physical constructs. From the latter tack, we see QFT as the inevitable consequence of the demand for compatibility between quantum mechanics and special relativity. To oversimplify: Dirac's equation was relativistic, thereby implying the existence of antimatter and demanding a formalism which could describe the creation and annihilation of particles. Quantum electrodynamics emerged and became the paradigm for the description of all elementary particles. The dream was realized by today's standard model. (Although the origin of QFT lay in relativistic quantum mechanics, its non-relativistic version is now the instrument of choice in condensed-matter physics, fluid dynamics and elsewhere.) QFT is the well-entrenched language of particle physics, statistical mechanics, astrophysics and cosmology, encompassing and enabling their many successes. By many particle physicists, QFT is called upon without regard for its logical foundation, but as a useful tool with which to secure a predictive description of the building blocks of matter and the ways they interact – just as one drives without necessarily knowing how a car works.

A theory cannot become an established part of the scientific edifice until, first, its implications are shown to accord with experiment, and, second, its domain of applicability is established. Newtonian mechanics is absolutely true – within a well-defined envelope defined by c and $\hbar$. Similarly for classical electrodynamics and non-relativistic quantum mechanics. Like its predecessors, quantum field theory offers – and will always offer – a valid description of particle phenomena at energies lying within its own domain of applicability. This domain cannot extend all the way to the Planck scale, but its actual limits of applicability have not yet been probed. From this point of view, we are discussing the foundations of a theory that, whatever its successes, cannot be accepted as true.

In fact, QFT is just wrong! Quantum mechanics is all-encompassing. A correct theory must include quantum gravity and QFT is not up to the task. Furthermore, it is beset by divergences, both perturbative and non-perturbative. While not affecting the predictive power of QFT, divergences are disturbing. To such luminaries as Schwinger and Dirac, the appearance of divergences make QFT unacceptable as a final theory. These problems and others may be resolved if and when the standard model is shown to emerge from the superstring. One source of divergences – the discrete space-time points at which emissions and absorptions take place – is removed by the replacement of point-like particles by smooth and continuously evolving loops of string. Furthermore, superstring theory not only provides a quantum theory of gravity but it demands one. The only difficulty is difficulty itself: nobody has been able to wring from it any explicit and testable predictions, aside from the existence of gravity and the primacy of gauge theories. Furthermore, the superstring can never become a *Theory of Everything*: such a theory does not now exist, may never exist, and is probably not even a sensible concept.

2 Questions

Dirac's 'large number problem' appears in many guises. Why is there a spread of five orders of magnitude among the masses of fundamental fermions? Why are they so small compared with the Planck mass? Why is the cosmological constant tiny? Another problem with quantum field theory is that there are simply too many of these questions. Why are elementary particle forces described by gauge theories? Why is the gauge group what it is? There are many questions like this that simply cannot be answered within the context of the theory. I call questions of this kind *meta-questions*.

To clarify what I mean by a metaquestion, let me attempt to construct a four-fold classification of all of the questions that may be asked of nature in the context of a particular theoretical framework. By an *intrinsic question* is meant one that the theory can answer, whether in practice or in principle. The answers, when they can be found, must concur with experiment: otherwise the theory must be modified or rejected. An *emergent question* is one that expands the domain of applicability of a theory, thereby leading to essentially new phenomena. The distinction between intrinsic and emergent questions involves an element of surprise or novelty and may not always be clear.

A *metaquestion* is one that a theory should be able to answer but cannot, or one that cannot even be addressed. Metaquestions in physics are always unambiguously so, because our theories are explicit enough to say which questions they can or cannot answer. Finally, some questions are purely *empirical*. These lie beyond the framework of the theory and must be addressed directly by experiment or observation.[3] Some examples may clarify these notions:

- What are the constituents of nuclei?
  Empirical: This question was answered in 1932 with the discovery of the neutron. Today, we might regard the question as intrinsic to the standard model: nuclei with nucleon number A are the lightest states of $3A$ quarks.
- What if we consider an assemblage of $\sim 10^{57}$ nucleons?
  Emergent: We are led to the concept of a neutron star.
- Why is the neutron heavier than the proton?
  Intrinsic: Because the down quark is heavier than the up quark.
- Why is the down quark heavier than the up quark?
  Meta: God knows!
- What are cosmic rays?
  Empirical: Most primary cosmic particles were found to be energetic protons, along with some larger nuclei, photons and antiprotons. An ambitious experiment spearheaded by S.C.C. Ting – the Anti-Matter Spectrometer in Space – will search for a tiny admixture of antinuclei. A positive result would compel a major revision of our cosmology. [12]

Let's see if my classification is useful when applied to various ingredients of the standard model.

Classical mechanics: Intrinsic questions are involved in the prediction of eclipses, in space travel and in the solution to the celestial N-body problem. They abound in mechanics textbooks. Emergent questions led to the disciplines of acoustics, elasticity, fluid dynamics, and to both kinetic theory and its reconciliation with thermodynamics. (One of the most beautiful examples of an emergent question was posed centuries ago by Daniel Bernoulli: how would a gas behave if it were made up of tiny particles in motion making elastic collisions with the container?) Several old metasaws remain unanswered: What is mass? What is the origin of force? Why are gravitational and inertial masses equal? Today, these questions should be addressed to the successors of classical mechanics.

---

[3] John Horgan [11] proposes a class of what he finds to be unanswerable or *ironic* questions, such as, *Could quarks and electrons be composed of still smaller particles, ad infinitum?* As posed, this is an infinite series of questions. Without its coda, it is an important metaquestion. The issue is debated and an unambiguous answer is expected. *Just how inevitable was life's origin?* stands for a slew of wholly empirical questions now under active investigation: Was there once life on Mars? Are there other Earth-like planets in the Galaxy? Are intelligent beings from other solar systems trying to communicate with us? Nonetheless, I concede that 'Is there life on other galaxies?' is at present unanswerable, and that Horgan's metaphysical query, *What does quantum mechanics really mean?* is as ironic as any in literary criticism.

Classical E & M: 'Describe the radiation emitted by a given antenna' is intrinsic; 'How does an ionized gas behave?' is emergent and leads to plasma physics, or in its relativistic version, accelerator physics. Metaquestions about the electrodynamics of moving media led to special relativity, whence they became intrinsic: answerable and soon answered. Other metaquestions still vex us, such as: 'Is there an inner structure to the electron, and if so, what are the forces that hold it together?' and 'Does nature exhibit a magnetic analog to electric charge?'

Quantum mechanics: 'Find the binding energy of the helium atom' is intrinsic and (with considerable difficulty) answered. 'Find the complete spectrum of an iron atom' is intrinsic and answerable in principle but not in practice. 'How do high-T_c superconductors work?' or 'What is the mechanism of sonoluminescence?' are emergent questions. 'How and under what circumstances do nuclei undergo radioactive decay?' is a once-baffling metaquestion that became intrinsic and was answered by quantum field theory.

Quantum field theory: is where we are today. For decades, the unchallenged standard model of particle physics, based on QFT, has offered what appears to be a complete, consistent and correct description of all known particle phenomena. The standard model is a gauge theory founded on the Lie group $SU(3) \times SU(2) \times U(1)$. It involves three families of fundamental fermions and, in its simplest realization *à la* Weinberg and Salam, an as yet undetected Higgs boson. Here are a few once-metaquestions that are now intrinsic and answered: Why are there no strangeness-changing neutral currents? Why does the spectrum of a heavy quark bound to its antiquark resemble that of positronium? What is the origin of isotopic spin and unitary symmetry? Why do the lightest baryons form a representation of $SU(6)$? Why does deep-inelastic lepton scattering exhibit scaling behavior? The standard model grew like Topsy and it works! I will turn to its emergent and metaquestions after a brief digression.

Other sciences: What about the physics of condensed matter, plasmas, atoms, molecules, or light? Or, for that matter, chemistry, molecular biology, or geology? Many scientists argue vehemently that their disciplines are just as fundamental as particle physics, but there is a sense in which they are wrong. Most scientific disciplines are entirely free of metaquestions.[4] Any sensible question posed by the chemist or solid-state physicist is answerable in principle in terms of known and accepted microscopic and thermodynamic principles. None of their discoveries can change our view of the interactions and structure of atoms. There is no undiscovered vital principle underlying biology or even (in my view) consciousness. With no intent to disparage, I claim that the 'fundamental' laws (i.e., the language of discourse) governing most of the sciences have long ago been set in stone. Of course, this no more suggests an end to science than a knowledge of the rules of chess makes one a grand master.

3 Problems

All remaining metaquestions lurk at the extremes of the ladder of size: they are to be found exclusively in particle physics and cosmology. There is a flip side to this coin.

---

[4] Questions like 'How did lice evolve?', 'Can lice think?', and 'Why are there lice?' suggest that my classification should be used with care in the life sciences.

In olden days, answers to metaquestions led to extraordinary technological developments. This is no longer so. Most of modern science (and most of its support) is properly directed toward intrinsic and emergent questions that may, directly or indirectly, contribute to human welfare.

Metaquestions have become societally irrelevant. Their answers are unlikely to have any impact on the rest of science, or for that matter, on the workaday world. For example, particle physicists no longer study matter as found, but rather matter as made, and made at considerable cost. There will never be a practical device employing tau leptons or W bosons. Kaons have been seen for half a century, but have never been put to useful work. Although cosmologists should soon find out whether there will be a Big Chill or a Big Crunch in the distant future, the result will not change our way of life. The virtues of science at the meta-frontier are inspirational, intellectual and cultural, but they are anything but practical. Perhaps, as Michael V. Posner observed, [13] scientific research is not necessarily *an input into economic activity creating further prosperity,* but also can be *an output (like Scotch Whisky or visits to the grand opera) that is pleasurable to indulge in but which absorbs economic resources rather than adding to them.*

It is both remarkable and disconcerting that the standard model of particle physics has never been found to be in unambiguous conflict with experimental data. Carlo Rubbia, who once thought otherwise, endured five years as the director-general of a laboratory that generated (and generates) endless confirmations of the standard model. That is, many of the intrinsic questions of the standard model have explicit answers that conform to the phenomena. Other intrinsic questions (involving, say, the observed spectrum or decay patterns of hadrons) do not yet have explicit answers. Getting more and better answers to the intrinsic questions of the standard model is one of the challenges facing today's so-called phenomenological theorist – a qualification that must not be taken pejoratively.

The emergent questions associated with QFT range far and wide. QFT offers a language with which to discuss statistical mechanics, whose methods and techniques often complement those used in particle physics. QFT allows astrophysicists to describe the most violent events in the heavens, such as supernovae and active galactic nuclei. It is also the basis for neonatal cosmology, which offers the wondrous possibility of understanding the observed large-scale structure of the universe and its origin in the hot Big Bang. The study of black holes is an emergent (and immensely impractical) arena wherein both QFT and general relativity play crucial roles.

Although (or perhaps, because) the standard model works so well, today's particle physicists suffer a peculiar malaise. Imagine a television set with lots of knobs: for focus, brightness, tint, contrast, bass, treble, and so on. The show seems much the same whatever the adjustments, within a large range. The standard model is a lot like that. Who would care if the tau lepton mass were doubled or the Cabibbo angle halved? The standard model has about 19 knobs. They are not really adjustable: they have been adjusted at the factory. Why they have their values are 19 of the most baffling metaquestions associated with particle physics. With masses compared to the Planck mass, these are its dimensionless parameters:

- *Six* quark masses and *three* charged lepton masses,
- *Four* quark mixing parameters and *three* gauge coupling constants,
- *Two* Higgs parameters and *one* additional measure of CP violation.

Before asking why the muon is 200 times as heavy as the electron (or 20 powers of 10 less that the Planck mass), one might ask why there is a muon at all. Why is CP violation such a tiny effect? Why are the families of fundamental fermions so congruent, and why are there just three of them when one would seem to suffice? Why is the top quark so much heavier than other quarks? Why is the gauge group of the standard model what it is? Is there a Higgs boson? If so, is high-energy physics beyond 1 TeV bereft of essentially new phenomena? If not, what is the mechanism of spontaneous symmetry breaking? And of course, what are the limits of QFT?

With an unchallenged theory in hand and a host of puzzling metaquestions, what is today's particle theorist or high-energy experimenter to do? Some turn to intrinsic questions that are as yet unanswered. This endeavor is interesting in itself and may reveal flaws in the standard model. Perhaps the wild things lie in the heavy quark domain, where searches for CP violating effects will soon be carried out at laboratories throughout the world. Will the results simply confirm the standard model? In few other disciplines do their practitioners pray for a discrepancy between theory and experiment!

Another direction is the yet-unsuccessful search for a meaningful pattern among the parameters of the standard model. A third possibility is the appearance of surprises at LEP or the Tevatron Collider, i.e., data that are inexplicable by the standard model. Occasional unconvincing indications of anomalies lead to flurries of soon-to-be discarded theoretical papers involving all sorts of 'new physics'. An ambitious extension of the standard model – supersymmetry – makes just one quantitative and confirmed prediction, but it sufficed to make true believers of many of my colleagues.

Although some particle physicists anticipate a vast desert stretching from observed energies all the way to the Planck mass, I feel certain that the standard model is not quantum field theory's last stand. Neutrinos offer one reason for my optimism. In the minimal version of the standard model, they are regarded as massless particles. However, this treatment is both theoretically unsound and experimentally challenged. Neutrinos must acquire tiny masses, if not via a more elaborate gauge group (e.g., grand unified theory) then through the intervention of quantum gravity. And there are several experimental indications of neutrino oscillation, a phenomenon that demands neutrino mass.

Although it is straightforward to extend the standard model to accommodate neutrino masses, any such attempt vastly increases the number of arbitrary parameters of the theory: in place of 19, we end up with 31, 40 or more. No one can accept a theory with so many knobs! Furthermore, some of the neutrino-related parameters, while observable in principle and as fundamental as any others, may never be measured.[5] No way would nature leave us that way. A new, deluxe, possibly even supersymmetric version of the standard model will soon emerge as we patiently pave our portion of the long road toward the Planck mass, most of which we shall never travel save in our minds.

I close with a pastiche of unattributed snippets stolen from my fellow speakers. Quantum field theory, while it is unreasonably effective as a fantastic computational tool, has little to say about Fritz Rohrlich's cat. The standard model is one of the most successful theories ever created, but we are not approaching the end of physics. The

---

[5] This immensely sad observation was first made by J. Schechter and J.W.F. Valle [14].

fundamental forces may be unified, but then again they may not be. Although the ideal of an ultimate theory motivates the exercise of the greatest intellectual ingenuity in pursuit of what may be an illusory goal, in the end we shall not have arrived at reality. Nonetheless, the trip is wonderful and the landscape so breathtaking.

References

[1] A. De Rújula, H. Georgi & S.L. Glashow, *Physical Review* **D12** (1975) 147.

[2] *American Heritage Dictionary*, 2nd College Ed., Houghton-Mifflin (1985) Boston.

[3] Henry A. Rowland, *American Journal of Science* [4] **VIII** (1899) 401.

[4] Philipp Frank, *Between Physics and Philosophy*, Harvard University Press (1941) Cambridge, MA, p. 103.

[5] G.W. Hemming, QC, *Billiards Mathematically Treated*, Macmillan, 1899, London.

[6] P.W. Bridgman, *The Logic of Modern Physics*, Macmillan (1960) New York, p. 207.

[7] J. Schwinger, *Annals of Physics* **2** (1957) 407.

[8] S.L. Glashow, *Nuclear Physics* **10** (1959) 107.

[9] A. Salam, *Nuclear Physics* **18** (1960) 681; S. Kamefuchi, ibid., 691.

[10] S.L. Glashow & John Iliopoulos, *Physical Review* **D3** (1971) 1043.

[11] John Horgan, *The End of Science*, Addison Wesley (1996) Reading, MA, p.7.

[12] *An Antimatter Spectrometer in Space*: unpubl. Proposal by the Antimatter Study Group; see also: A. Cohen, A. De Rùjula & S.L. Glashow, *Astrophysical Journal* (in the press).

[13] Michael V. Posner, *Nature* **382** (1996) 123.

[14] J. Schechter & J.W.F. Valle, *Physical Review* **D23** (1981) 1666.

Discussions

Teller: I take great interest in seeing physicists disagreeing very heatedly among each other sometimes about what we're trying to accomplish.

Glashow: I think, phrased that way, there is a disagreement. There are those who sensed a final approach, so to speak. And there are others like Shankar with the view that Shankar expressed, although Shankar's view applied to condensed matter physics, that we should be satisfied with any small improvement of the structure we have. But traditionally most of us were trying to make the structure incrementally better, and pushing along in this way. We can no longer do that because, as I explained, we have this standard theory which admits no mucking around. It has its 19 parameters, it's otherwise more or less sacrosanct. You can't change it.

The age of model building is done, except if you want to go beyond this theory. Now the big leap is string theory, and they want to do the whole thing; that's very ambitious. But others would like to make smaller steps. And the smaller steps would be presumably, many people feel, strongly guided by experiment. That is to say the hope is that experiment will indicate some new phenomenon that does not agree with the theory as it is presently constituted, and then we just add bells and whistles insofar as it's possible. But as I said, it ain't easy. Any new architecture has, by its very nature, to be quite elaborate and quite enormous.

Low energy supersymmetry is one of the things that people talk about. It's a hell of a lot of new particles and new forces which may be just around the corner. And if they see some of these particles, you'll see hundreds, literally hundreds of people sprouting up, who will have claimed to predict exactly what was seen. In fact they've already sprouted up and claimed to have predicted various things that were seen and subsequently retracted. But you see you can't play this small modification game anymore. It's not the way it was when Sam and I were growing up and there were lots of little tricks you could do here and there that could make our knowledge better. They're not there any more, in terms of changing the theory. They're there in terms of being able to calculate things that were too hard to calculate yesterday. Some smart physicists will figure out how to do something slightly better, that happens. But we can't monkey around.

So it's either the big dream, the big dream for the ultimate theory, or hope to seek experimental conflicts and build new structures. But we are, everybody would agree that we have right now the standard theory, and most physicists feel that we are stuck with it for the time being. We're really at a plateau, and in a sense it really is a time for people like you, philosophers, to contemplate not where we're going, because we don't really know and you hear all kinds of strange views, but where we are. And maybe the time has come for you to tell us where we are. 'Cause it hasn't changed in the last 15 years, you can sit back and, you know, think about where we are.

Gross: In a relativistic theory with particle creation, it does not make sense to talk about localizing single particles. Quantum field theory is a local theory. This is sometimes formulated as saying that quantum field operators commute at space-like separation, so there can't be propagation of signals faster than the speed of light. The remark I made about string theory was that string theory introduced nonlocality into physics in a way which wasn't previously envisaged, by replacing particles by strings so that the fundamental dynamical objects are nonlocal. However, their interactions are local. Thus one can have nonlocal extended objects that communicate locally, and thereby giving rise to locality in the sense of causal behavior.

Tisza: Professor Glashow is reluctant to commit himself in matters of foundations. Yet my first impression was that he raised a number of cogent philosophical issues, an impression strengthened by the transcript of his talk. The Editor kindly agreed that I use the transcript to amplify my original comment to point out that Glashow's contribution does open a bridgehead toward the philosophical connection that this conference was intended to achieve.

The point of departure is the tension surrounding the standard model: none of its conclusions are contradicted by experiment, and having made important contributions to this model, Glashow notes this achievement with justified pride. However, he finds a great deal to criticize, mainly in what the model fails to tell us. Wishing to formulate questions to trigger improvements, he proceeds to establish a classification of the questions which can be posed in the context of a *single* theoretical structure. In a truly philosophical mindset he arrives at the interesting category of *meta-questions* which lie beyond the context of the theory. He lists many such questions of obvious interest, inquiring into what is behind the experimental facts. Unfortunately, we get no suggestion on how to answer them; such questions seem to generate only speculations.

I claim that this deadlock is avoidable and, at its best, mathematical physics does enable us to start from an available traditional knowledge and transcend it in terms of a thoroughly convincing argument. A striking example is Einstein's June 1905 paper on the *Electrodynamics of moving bodies*. Einstein pointed out a subtle conflict between the principles of the classical mechanics of point masses (CMP) on the one hand, and CED on the other. Instead of perpetuating this conflict as a paradox, he pointed a way to avoid it by suggesting a relativistic refinement for Newtonian absolute time. He demonstrated a metatheoretical analysis, somewhat, but not quite, in the sense of Glashow. The secret of turning metatheory into a practical enterprise appears to be the exploitation of a situation in which *two* theories grounded in distinct but overlapping experimental domains exhibit a marginal inconsistency. The elimination of the latter is the lever by which tradition is rejuvenated as obsolete concepts are exchanged for new and deeper ones.

Unfortunately, Einstein did not develop this pivotal idea into a deliberate strategy. Three months before the relativity paper, he had advanced his light quantum theory and set the stage for the wave–particle paradox, rather than abandoning the Newtonian point-particle concept in the quantum domain. It is an experimental fact that light has undulatory properties and that it manifests itself in discrete quanta of energy, momentum and angular momentum. The situation becomes paradoxical only if the quanta are assumed to be point-like in harmony with the Newtonian particle concept, as Einstein indeed assumed in his March paper and reiterated in subsequent reviews.

Einstein was not happy with this situation. In a revealing letter to Sommerfield (M. Eckert and W. Pricha, 1984, Eds., *Physikalische Blatter*, **40**, 29–40, see p. 33; reprinted in *Einstein Papers*, Vol. V, doc. 211), he considers the point representation of the light quantum only as an expediency which he was unable to avoid. The situation would have called for a *par excellence* metatheoretical analysis for which there was no precedent, however, and the removal of the point-particle paradox soon disappeared from the agenda.

What happened was that Niels Bohr and his circle succeeded in making an extraordinary fruitful use of this paradox by establishing QM, the deepest theory of mathematical physics. Yet QM has a questionable aspect and that is its interpretation, closely tied in with its foundation. There are important conceptual and mathematical differences between the original approaches of Heisenberg, Dirac, Schrödinger, and a measure of the success of the foundations is that the apparently disparate ingredients now coexist and are seen as indispensable in modern QM. This theory enables us to account for all naturally occurring material structures from protons, neutrons, electrons and photons. This is a highly cost-effective procedure even though the definition of these quasi-elementary ingredients is imperfect. The Heisenberg–Schrödinger controversy was never fully resolved, but was given legitimacy by making the 'wave–particle paradox' an acceptable ingredient of theory. This was a successful short range procedure, but now we are in the long range, and in order to facilitate the necessary widening of perspective, we might consider the 'wave–particle paradox' an acceptable *phenomenological* approximation in the context of the aforementioned structural problem. At present, however, we are interested in the internal excitations of particles, and their creation from the vacuum. This calls for a deepening of the theory by considering an 'intrinsic dynamics'. The first step toward this deepening is the recognition of the potential 'layering' of QM. I wish to point out one instance in which exactly this type of layering was expressed in the pioneering days. In August 1927 Pauli wrote a remarkable letter to Niels Bohr (Wolfgang Pauli, *Scientific Correspondence*, Eds. A. Hermann, K.V. Meyenne and V.F. Weisskopf, Springer Verlag,

1979, *Vol. 1: 1919–1929*, Letter [168]), with comments on the general state of QM. He suggests that one ought to distinguish two points of view, which are often not sufficiently kept apart. In the first and more phenomenological one, the nature of the particles is irrelevant, one may deal with electrons, or larger particles of matter, which may or may not carry a charge, and the electromagnetic field is ignored. The only thing that matters is the de Broglie wave associated with the particle. Pauli takes strong exception to people's expressing views, say, on a presumed organic relation between the de Broglie field and the electromagnetic field, without realizing that they have transcended the legitimate limits of the phenomenological theory. To address such problems one ought to assume the second point of view in which one deals with the charge and the spin of the elementary particles as a mark of an *intrinsic structure*. (My emphasis.) He concludes by noting that the means for such a deeper theory are still totally absent.

I may add that Pauli's letter is dated three months after the submission of his seminal paper on the electron spin, which he classified here as a phenomenological theory. Bohr answered Pauli's letter in a week, but responded only to other topics, without referring to the issue invoked.

I am not aware of anyone pursuing Pauli's program until I made a first step in this direction in L. Tisza, *Physical Review A* **40**, 6781 (1989). This was 60 years after Pauli's reflections, and I was able to involve the undulatory and spinning properties of photons and electrons to establish the beginning of the quantum formalism and interpret the state vector in terms of internal currents. I believe that the arguments of that paper are cogent, except for the last paragraph on the conjectured extension to further problems.

The foregoing scenario suggests a number of questions. (i) How did it happen that Einstein advanced substantially different methodologies within the same year? (ii) How did it happen that the less deep method became preponderant? And finally (iii) what is to be done to remove this subjectivity from the acquisition without losing what subjectivity has created?

Ad (i). Glashow made the key remark that 'all great physicists do very stupid things at some point in their life.' *Ad* (ii). The freedom to advance dumb ideas is useful, and maybe indispensable when experience confronts us with deeply puzzling situations. *Ad* (iii). We must make a sharp distinction between heuristic on the one hand, and consolidation on the other. In the former one must be bold enough to handle paradoxes which arise as new concepts are injected into a partially incorrect context. In the latter no paradox must be left unresolved as the old context is cleaned from overstatements.

When confronted with a 'dumb idea' of a great physicist, it behooves us to ask: what might have been his motivating misconception? As likely as not, the misconception entered the public domain to survive there long after the 'dumb idea' was forgotten. Einstein's misguided attempt to explain nuclear forces in terms of gravity, quoted by Glashow, is in line with his well-known preference for unity over diversity.

It is not far-fetched to surmise that Einstein inherited his unitary preference from Newton. The latter's insistence on considering sound propagation as a purely mechanical process must have prompted his fudging the formula of sound velocity, engagingly commented upon by Glashow. The correct theory involves the ratio of specific heats which in turn depends on the chemistry of the medium. Even the simple process of sound propagation calls for the joint application of mechanics and chemistry! How much more is this the case for the particulate constituents of matter which do account for both mechanics and chemistry.

It is a great achievement that entities with such dual properties have been experimentally discovered and mathematically described. What is missing is conceptual understanding, because the experimentally supported mechanical–chemical duality does not fit the expectation of Newtonian mechanical unification.

I suggest that we should reaffirm Newton's perennial insight that natural phenomena are susceptible to rigorous mathematical representation, but abandon his hubris that CMP is already the theory of everything (TOE). Instead, it should be considered as the point of departure of an evolutionary process. This program depends on a systematic development of the foundation of metatheory for the integration of theories of limited scope.

Once this key to the proper analysis is recognized, the program becomes readily manageable within a well-defined domain which is expanded stepwise from modest beginnings. The beginnings of a systematic metatheory is being published in a paper: 'The reasonable effectiveness of mathematics in the natural sciences.' It appears in a contribution to *Experimental Metaphysics – Quantum Mechanical Studies in Honor of Abner Shimony*, Kluwer, 1997.

Glashow: One of the reasons that Pauli might not have been especially proud of his introduction of spin matrices is that he, more than anyone else, knew that what he had done was not relativistic. And he was pleasantly surprised a year later when Dirac completed his job.

Tisza: The natural fusion of QM and SRT was a great discovery of Dirac. However, by doing this within the context of point particles, he aggravated Pauli's concern: in addition to spin and magnetic moment, the points carry now also lepton numbers. Besides, there is nothing wrong with the Pauli matrices, but their adaptation to special relativity hinges on the choice of the optimal spinor algebra used in my 1989 paper.

Fisher: I have a question for my field-theoretic colleagues, which again arises from the lines of demarcation between condensed matter and particle physics. In condensed matter physics, in one way or another, we do have action at a distance. Of course, we use (effective) field theories with gradient terms, usually just gradient-squared terms but more if we need them: however, we remain conscious that this is an approximation. But in quantum field theory we say that operators commute at spatially separated points and this represents locality; however, we would not get any further if we did not also have gradient terms of various sorts present. So it has always struck me that claiming that introducing a field gets rid of action at a distance is something of a swindle that physicists now accept. But to be honest to the philosophers, what we are really saying is that there *is* action at a distance, but it's a mighty small distance! We accept the Newtonian idealization of a derivative or a gradient.

That perspective has always seemed to me to offer an argument against point particles. Now I wonder, Shelly, if these issues resonate with you at all? One might say when we come to the theory of strings that we are looking at objects that have extension and are nonlocal. Then comes the point that David Gross made: when you actually look at how strings interact, it is somehow still local. Maybe that is a little less of a swindle. I don't know. Do you have any feelings on this?

Glashow: No, no, you're asking me questions on formal field theory, I explained my degree of incompetence.

Fisher: Maybe! But you have used it. And I am just asking you for your feelings about action at a distance, whether you really care about it?

Glashow: I thought it was settled: we have not had to talk about that any more for the last hundred years or so.

Fisher: Well, I am suggesting that the accepted resolution is really something of a swindle that most physicists have agreed to. But since philosophers are present, a little honesty might not be totally out of place! [laughter]

Gross: What actually do you mean by a swindle?

Fisher: I mean that there still seems to be a problem of how it is that infinitesimally nearby points interact, if one may say it that way. It depends again on one's view of the calculus. If one regards the continuum as God given (which historically it was not) one can still ask, when I put gradient-squared terms into a Hamiltonian or a Lagrangian, what am I doing? Certainly they are defined at points; but mathematically one cannot define the gradient of a point: one must have some sort of extension and take a limit. At the back of my own mind that has suggested some intrinsic sort of an approximation.

Gross: By the way, at what distance do you think we're making these swindles? You said you think it's a swindle. It's not a swindle at a distance of a centimeter, right?

Fisher: No, no, no! We know the calculus is extremely valuable, and therefore in neglecting to mention this issue to the philosophers, we are performing no swindle for any bridge or any airplane. But when you mention the tension in string theory and we recall historical pressures to introduce a fundamental length into fundamental physics, then I feel the question deserves more exposure in this forum.

Gross: I mean it is a totally experimental question at one level. We have tested locality to distances of order 10^{-18} cm. If you introduce a latticization of space, or some other of the old fashioned and rather boring and ugly ways of introducing cutoffs into the space-time structure, you'll run into conflicts with high precision experiments all over the place, unless you do it at distances at smaller than 10^{-16} to 10^{-18} cm. Beyond that you can do whatever you want. And string theory in some sense introduces a fundamental length, but in a much more beautiful way than just putting in some kind discrete space-time, and it doesn't violate Lorentz invariance.

Glashow: No, no. Maybe another way, there may be another way.

Gross: There might be lots of ways, but string theory is a very different way of doing that. But down to 10^{-18} cm it's no swindle, and it's tested, beyond that you can do whatever you want.

Fisher: When I suggested it was a swindle, I certainly did not imply disagreement with anything you say. In some sense our only job as physicists is to understand the accessible experiments. As you rightly stress, there is absolutely no evidence that we need to go beyond the continuity of space. But I am now observing that many of us are driven by various visions and puzzles. I take it from your comments that you have never been bothered about this issue. But if you actually look at what replaces action at a distance, it's the derivative! Okay? And if you ask where the derivative comes from, you have to talk about two separated points, and then you take a handy limit. It fits wonderfully with experiment, and so who am I to complain? I can win prizes on that basis! But, in my heart of hearts, I feel that if you were unhappy with action at a distance, you should still be a little unhappy.

Gross: Okay, it's actually quite interesting. Before this we had a disagreement on the origin of symmetry. You believed the symmetry is an accidental consequence of the flow of theories toward fixed points. Garbage in, beauty out. Whereas I believe that symmetry is fundamental. Now in the same spirit, I also have a prejudice about space-time, and it's exactly the opposite of yours. Coming from condensed matter physics or many body physics, it's probably more natural to believe in lattices and discreteness, and most people who have thought about fundamental lengths tend to think in that direction, that something is wrong with infinity or the continuum, and there really should be discrete points, and a derivative is just a difference. Actually I think you're giving up so much if you give up the continuum with all of its power and elegance that provides the framework for the symmetries that have played such an important role in physics. I would much rather see a more inventive solution to the problem, which by the way, I think string theory has done. It doesn't give up the kind of continuum used for calculus or anything like that. You lose none of the power that you get in continuum theory, yet at the same time, in an effective sense, it introduces a smallest possible length. It's a very novel way.

Fisher: I agree: I admire string theory and I am certainly not trying to criticize the approach of current fundamental theory. Rather, mainly for the benefit of the philosophers present, I am trying to bring out the degree to which matters that some physicists have certainly worried about in the past are still real worries. Personally, the issue has not worried me greatly. But anybody, like my colleague Ken Wilson and I myself, who would actually like to compute answers to some of the questions, or would like to solve QCD quantitatively, is forced to consider discrete lattices. Then we take pains to make the lattice spacings as small as we can; we are worried about T.D. Lee, because he introduces a random lattice but appears not to realize that randomness *per se* may be vitally important; and so on.

Thus I am entirely with you as a physicist but, in a philosophical context, would nevertheless say that the idea of action at a distance has been there for a long time and I would assert that physicists basically accept it. The best theories that we have finesse the issue in a very beautiful way. And we should not handicap ourselves by going to lattice theories or forcing a fundamental length. That would be ridiculous! The same remarks apply to symmetry: it is always an advantage if you find it or can introduce it.

But surely, from the philosopher's perspective, the issue is how much are these features intrinsic to what physicists are claiming, as against merely representing effective operational tools or guiding principles for getting sensible and successful theories. All professional physicists play the same game there!

Glashow: But you did suggest the possibility of abandoning the real numbers. You ask, 'Does nature know about the real numbers?' And I do recall some decades ago that David Finklestein made an attempt to create a quantum mechanics, or perhaps even a quantum field theory, which was purely algebraic. It was a fascinating attempt; of course, it didn't work.

Part Four. Mathematics, statistics and quantum field theory

8. Renormalization group theory: its basis and formulation in statistical physics

MICHAEL E. FISHER

The nature and origins of renormalization group ideas in statistical physics and condensed matter theory are recounted informally, emphasizing those features of prime importance in these areas of science in contradistinction to quantum field theory, in particular: critical exponents and scaling, relevance, irrelevance and marginality, universality, and Wilson's crucial concept of flows and fixed points in a large space of Hamiltonians.

Contents

Foreword
1 Introduction
2 Whence came renormalization group theory?
3 Where stands the renormalization group?
4 Exponents, anomalous dimensions, scale invariance and scale dependence
5 The challenges posed by critical phenomena
6 Exponent relations, scaling, and irrelevance
7 Relevance, crossover, and marginality
8 The task for renormalization group theory
9 Kadanoff's scaling picture
10 Wilson's quest
11 The construction of renormalization group transformations: the epsilon expansion
12 Flows, fixed points, universality and scaling
13 Conclusions
Acknowledgments
Selected bibliography
Appendix A Asymptotic behavior
Appendix B Unitarity of the renormalization group
Appendix C Nature of a semigroup

Foreword

It was a pleasure to participate in the Colloquium cosponsored by the Departments of Philosophy and of Physics at Boston University "On the Foundations of Quantum Field Theory." In the full title, this was preceded by the phrase: "A Historical Examination and Philosophical Reflections," which set the aims of the meeting. Naturally, the participants were mainly high-energy physicists, experts in field theories, and interested philosophers of science. I was called on to speak, essentially in a service role, because I had witnessed and had some hand in the development of renormalization group concepts and because I have played a role in applications where these ideas really mattered. It is hoped that this article, based on the talk I presented in Boston,[1] will prove a useful contribution to these Proceedings.

1 Introduction

It is held by some that the "Renormalization Group" – or, better, renormalization group*s* or, let us say, *Renormalization Group Theory* (or RGT) – is "one of the under-lying ideas in the theoretical structure of Quantum Field Theory." That belief sug-gests the potential value of a historical and conceptual account of RG theory and the ideas and sources from which it grew, as viewed from the perspective of statistical mechanics and condensed matter physics. Especially pertinent are the roots in the theory of critical phenomena.

The proposition just stated regarding the significance of RG theory for Quantum Field Theory (or QFT, for short) is certainly debatable, even though experts in QFT have certainly invoked RG ideas. Indeed, one may ask: How far is some concept only instrumental? How far is it crucial? It is surely true in physics that when we have ideas and pictures that are extremely useful, they acquire elements of reality in and of themselves. But, philosophically, it is instructive to look at the degree to which such objects are purely instrumental – merely useful tools – and the extent to which phy-sicists seriously suppose they embody an essence of reality. Certainly, many parts of physics are well established and long precede RG ideas. Among these is statistical mechanics itself, a theory *not* reduced and, in a deep sense, *not* directly reducible to lower, more fundamental levels without the introduction of specific, new postulates.

Furthermore, statistical mechanics has reached a stage where it is well posed math-ematically; many of the basic theorems (although by no means all) have been proved with full rigor. In that context, I believe it is possible to view the renormalization group as merely an instrument or a computational device. On the other hand, at one extreme, one might say: "Well, the partition function itself is really just a com-binatorial device." But most practitioners tend to think of it (and especially its logarithm, the free energy) as rather more basic!

Now my aim here is not to instruct those field theorists who understand these matters well.[2] Rather, I hope to convey to nonexperts and, in particular, to any with a philosophical interest, a little more about what Renormalization Group Theory is[3] – at least in the eyes of some of those who have earned a living by using it! One hopes such information may be useful to those who might want to discuss

---

[1] The present article is a slightly modified and extended version of the lecture: it was prepared for publication in *Reviews of Modern Physics* (**70**, 653–81, 1998) and is reproduced here with permission from the Amer-ican Physical Society. These Proceedings are listed in the Selected Bibliography below as Cao (1998).

[2] Such as D. Gross and R. Shankar (see Cao, 1998, and Shankar, 1994). Note also Bagnuls and Bervillier (1997).

[3] It is worthwhile to stress, at the outset, what a "renormalization group" is *not*! Although in many appli-cations the particular renormalization group employed may be invertible, and so constitute a continuous or discrete, group of transformations, it is, in general, only a *semigroup*. In other words a renormalization group is not necessarily invertible and, hence, cannot be 'run backwards' without ambiguity: in short it is *not* a "group". The misuse of mathematical terminology may be tolerated since these aspects play, at best, a small role in RG theory. The point will be returned to in Sections 8 and 11. See also Appendix C.

its implications and significance or assess how it fits into physics more broadly or into QFT in particular.

2 Whence came renormalization group theory?

This is a good question to start with: I will try to respond, sketching the foundations of RG theory in the *critical exponent relations* and crucial *scaling concepts*[4] of Leo P. Kadanoff, Benjamin Widom, and myself developed in 1963–66[5] – among, of course, other important workers, particularly Cyril Domb[6] and his group at King's College London, of which, originally, I was a member, George A. Baker, Jr., whose introduction of Padé approximant techniques proved so fruitful in gaining quantitative knowledge,[7] and Valeri L. Pokrovskii and A.Z. Patashinskii in the Soviet Union who were, perhaps, the first to bring field-theoretic perspectives to bear.[8] Especially, of course, I will say something of the genesis of the full RG concept – the systematic integrating out of appropriate degrees of freedom and the resulting RG flows – in the inspired work of Kenneth G. Wilson[9] as I saw it when he was a colleague of mine and Ben Widom's at Cornell University in 1965–72. And I must point also to the general, clarifying formulation of RG theory by Franz J. Wegner (1972a) when he was associated with Leo Kadanoff at Brown University: their focus on *relevant*, *irrelevant* and *marginal* 'operators' (or perturbations) has played a central role.[10]

But, if one takes a step back, two earlier, fundamental theoretical achievements must be recognized: the first is the work of Lev D. Landau, who, in reality, is the founder of systematic *effective field theories*, even though he might not have put it that way. It is Landau's *invention* – as it may, I feel, be fairly called – of the *order parameter* that is so important but often underappreciated.[11] To assert that there exists an order parameter in essence says: "I may not understand the microscopic phenomena at all" (as was the case, historically, for superfluid helium) "but I recognize that there is a microscopic level and I believe it should have certain general, overall properties especially as regards locality and symmetry: those then serve to govern the most characteristic behavior on scales greater than atomic." Landau and Ginzburg (a major collaborator and developer of the concept[12]) misjudged one or two of the important general properties, in particular the role of fluctuations and singularity; but that does not alter the deep significance of this way of looking at a complex, condensed matter system. Know the nature of the order parameter – suppose, for example, it is a

[4] Five influential reviews antedating renormalization-group concepts are Domb (1960), Fisher (1965, 1967b), Kadanoff *et al.* (1967) and Stanley (1971). Early reviews of renormalization group developments are provided by Wilson and Kogut (1974b) and Fisher (1974): see also Wilson (1983) and Fisher (1983). The first texts are Pfeuty and Toulouse (1975), Ma (1976), and Patashinskii and Pokrovskii (1979). The books by Baker (1990), Creswick *et al.* (1992), and Domb (1996) present retrospective views.

[5] See: Essam and Fisher (1963), Widom (1965a,b), Kadanoff (1966), and Fisher (1967a).

[6] Note Domb (1960), Domb and Hunter (1965), and the account in Domb (1996).

[7] See Baker (1961) and the overview in Baker (1990).

[8] The original paper is Patashinskii and Pokrovskii (1966); their text (1979), which includes a chapter on RG theory, appeared in Russian around 1975 but did not then discuss RG theory.

[9] Wilson (1971a,b), described within the QFT context in Wilson (1983).

[10] Note the reviews by Kadanoff (1976) and Wegner (1976).

[11] See Landau and Lifshitz (1958) especially Sec. 135.

[12] In particular for the theory of superconductivity: see V.L. Ginzburg and L.D. Landau, 1959, "On the theory of superconductivity," *Zh. Eksp. Teor. Fiz.* **20**, 1064; and, for a personal historical account, V.L. Ginzburg, 1997, "Superconductivity and superfluidity (what was done and what was not)," *Phys.-Uspekhi* **40**, 407–32.

complex number and like a wave function – then one knows much about the macroscopic nature of a physical system!

Significantly, in my view, Landau's introduction of the order parameter exposed a novel and unexpected *foliation* or level in our understanding of the physical world. Traditionally, one characterizes statistical mechanics as directly linking the *microscopic* world of nuclei and atoms (on length scales of 10^{-13} to 10^{-8} cm) to the *macroscopic* world of, say, millimeters to meters. But the order parameter, as a dynamic, fluctuating object, in many cases intervenes on an intermediate or *mesoscopic* level characterized by scales of tens or hundreds of angstroms up to microns (say, $10^{-6.5}$ to $10^{-3.5}$ cm). The advent of Wilson's concept of the renormalization group gave more precise meaning to the effective ("coarse-grained") Hamiltonians that stemmed from the work of Landau and Ginzburg. One now pictures the LGW – for Landau-Ginzburg-Wilson – Hamiltonians as true but significantly renormalized Hamiltonians in which finer microscopic degrees of freedom have been integrated-out. (See below for more concrete and explicit expressions.) Frequently, indeed, in modern condensed matter theory one *starts* from this intermediate level with a physically appropriate LGW Hamiltonian *in place* of a true (or, at least, more faithful or realistic) microscopic Hamiltonian; and *then* one brings statistical mechanics to bear in order to understand the macroscopic level. The derivation and validity of the many types of initial, LGW Hamiltonians may then be the object of separate studies to relate them to the atomic level.[13]

Landau's concept of the order parameter, indeed, brought light, clarity, and form to the general theory of phase transitions, leading eventually, to the characterization of multicritical points and the understanding of many characteristic features of ordered states.[14] But in 1944 a bombshell struck! Lars Onsager, by a mathematical *tour de force*, deeply admired by Landau himself,[15] computed exactly the partition function and thermodynamic properties of the simplest model of a ferromagnet or a fluid.[16] This model, the *Ising model*, exhibited a sharp critical point: but the explicit properties, in particular the nature of the critical singularities, disagreed profoundly – as I will explain below – with essentially all the detailed predictions of the Landau theory (and of all foregoing, more specific theories). From this challenge, and from experimental evidence pointing in the same direction,[17] grew the ideas of *universal* but nontrivial *critical exponents*,[18] special *relations* between different *exponents*,[19] and then, *scaling descriptions* of the region of a critical point.[20] These insights served as stimulus and inspiration to Kenneth Wilson in his pursuit of an understanding of quantum field theories.[21] Indeed, once one understood the close mathematical analogy between doing statistical mechanics with effective Hamiltonians and doing

[13] These issues have been discussed further by the author in "Condensed matter physics: Does quantum mechanics matter?" in *Niels Boher: Physics and the World*, edited by H. Feshbach, T. Matsui and A. Oleson, 1988 (Harwood Academic, Chur) pp. 177–83.

[14] See Landau and Lifshitz (1958).

[15] As I know by independent personal communications from Valeri Pokrovskii and from Isaak M. Khalatnikov.

[16] Onsager (1944), Kaufman and Onsager (1949), Onsager (1949).

[17] See, e.g., Fisher (1965), Stanley (1971).

[18] Domb (1960, 1996) was the principal pioneer in the identification and estimation of critical exponents: see also the preface to Domb (1996) by the present author.

[19] Advanced particularly in Essam and Fisher (1963).

[20] Widom (1965a,b), Domb and Hunter (1965), Kadanoff (1966), and Patashinskii and Pokrovskii (1966).

[21] Wilson (1971a,b; 1983).

quantum field theory (especially with the aid of Feynman's path integral) the connections seemed almost obvious. Needless to say, however, the realization of the analogy did not come overnight: in fact, Wilson himself was, in my estimation, the individual who first understood clearly the analogies at the deepest levels. And they are being exploited, to mutual benefit to this day.

In 1971, then, Ken Wilson, having struggled with the problem for four or five years,[22] was able to cast his renormalization group ideas into a conceptually effective framework – effective in the sense that one could do certain calculations with it.[23] And Franz Wegner, very soon afterwards,[24] further clarified the foundations and exposed their depth and breadth. An early paper by Kadanoff and Wegner (1971) showing when and how universality could *fail* was particularly significant in demonstrating the richness of Wilson's conception.

So our understanding of "anomalous," i.e., nonLandau-type but, in reality, standard critical behavior was greatly enhanced. And let me stress that my *personal aim* as a theorist is to *gain* insight and understanding. What that may truly mean is, probably, a matter for deep philosophical review: after all, what constitutes an *explanation*? But, on the other hand, if you work as a theoretical physicist in the United States, and wish to publish in the *Physical Review*, you had better *calculate* something concrete and interesting with your new theory pretty soon! For *that* purpose, the *epsilon expansion*, which used as a small, perturbation parameter the deviation of the spatial dimensionality, d, from four dimensions, namely, $\epsilon = 4 - d$, provided a powerful and timely tool.[25] It had the added advantage, if one wanted to move ahead, that the method looked something like a cookbook – so that "any fool" could do or check the calculations, whether they really understood, at a deeper level, what they were doing or not! But in practice that also has a real benefit in that a lot of calculations do get done, and some of them turn up new and interesting things or answer old or new questions in instructive ways. A few calculations reveal apparent paradoxes and problems which serve to teach one and advance understanding since, as Arthur Wightman has observed, one asks: "Maybe we should go back and think more carefully about what we are actually doing in implementing the theoretical ideas?" So that, in outline, is what I want to convey in more detail, in this exposition.

3 Where stands the renormalization group?

Beyond sketching the origins, it is the breadth and generality of RG theory that I wish to stress. Let me, indeed, say immediately that the full RG theory should no more be regarded as based on QFT perturbative expansions – despite that common claim – than can the magnificent structure of Gibbsian statistical mechanics be viewed as founded upon ideal classical gases, Boltzmannian kinetic theory, and the virial and cluster expansions for dilute fluids! True, this last route was still frequently retravelled in text books more than 50 years after Gibbs' major works were published; but it deeply misrepresents the power and range of statistical mechanics.

---

[22] See below and the account in Wilson (1983).
[23] As we will explain: see Wilson (1971a,b).
[24] Wegner (1972a,b).
[25] Wilson and Fisher (1972).

The parallel mischaracterizations of RG theory may be found, for example, in the much cited book by Daniel Amit (1978), or in Chapter 5 of the later text on *Statistical Field Theory* by Itzykson and Drouffe (1989), or, more recently, in the lecture notes entitled *Renormalization Group* by Benfatto and Gallavotti (1995), "dedicated to scholars wishing to reflect on some details of the foundations of the modern renormalization group approach." There we read that the authors aim to expose how the RG looks to them as physicists, namely: "this means the achievement of a coherent perturbation theory based on second order (or lowest-order) calculations." One cannot accept that! It is analogous to asking "What does statistical mechanics convey to a physicist?" and replying: "It means that one can compute the second-virial coefficient to correct the ideal gas laws!" Of course, historically, that is not a totally irrelevant remark; but it is extremely misleading and, in effect, insults one of America's greatest theoretical physicists, Josiah Willard Gibbs.

To continue to use Benfatto and Gallavotti as strawmen, we find in their preface that the reader is presumed to have "some familiarity with classical quantum field theory." That surely, gives one the impression that, somehow, QFT is necessary for RG theory. Well, it is totally unnecessary![26] And, in particular, by implication the suggestion overlooks entirely the so-called "real space RG" techniques,[27] the significant Monte Carlo RG calculations,[28] the use of *functional* RG methods,[29] etc. On the other hand, if one wants to do certain types of calculation, then familiarity with quantum field theory and Feynman diagrams can be very useful. But there is *no necessity*, even though many books that claim to tell one about renormalization group theory give that impression.

I do not want to be unfair to Giovanni Gallavotti, on whose lectures the published notes are based: his book is insightful, stimulating and, accepting his perspective as a mathematical physicist[30] keenly interested in field theory, it is authoritative. Furthermore, it forthrightly acknowledges the breadth of the RG approach citing as examples of problems implicitly or explicitly treated by RG theory:[31]

(i)   The KAM (Kolmogorov-Arnold-Moser) theory of Hamiltonian stability
(ii)   The constructive theory of Euclidean fields
(iii) Universality theory of the critical point in statistical mechanics
(iv)   Onset of chaotic motions in dynamical systems (which includes Feigenbaum's period-doubling cascades)
(v)   The convergence of Fourier series on a circle
(vi)   The theory of the Fermi surface in Fermi liquids (as described in the lecture by Shankar)

To this list one might well add:

(vii) The theory of polymers in solutions and in melts
(viii) Derivation of the Navier-Stokes equations for hydrodynamics

[26] See, e.g., Fisher (1974, 1983), Creswick, Farach and Poole (1992), and Domb (1996).
[27] See the reviews in Niemeijer and van Leeuwen (1976), Burkhardt and van Leeuwen (1982).
[28] Pioneered by Ma (1976) and reviewed in Burkhardt and van Leeuwen (1982). For a large scale calculation, see: Pawley, Swendsen, Wallace, and Wilson (1984).
[29] For a striking application see: D.S. Fisher and Huse (1985).
[30] The uninitiated should note that for a decade or two the term "mathematical physicist" has meant a theorist who provides *rigorous proofs* of his main results. For an account of the use of the renormalization group *in* rigorous work in mathematical physics, see Gawędski (1986).
[31] Benfatto and Gallavotti (1995), Chap. 1.

(ix) The fluctuations of membranes and interfaces

(x) The existence and properties of 'critical phases' (such as superfluid and liquid-crystal films)

(xi) Phenomena in random systems, fluid percolation, electron localization, etc.

(xii) The Kondo problem for magnetic impurities in nonmagnetic metals.

This last problem, incidentally, was widely advertised as a significant, major issue in solid state physics. However, when Wilson solved it by a highly innovative, *numerical* RG technique[32] he was given surprisingly little credit by that community. It is worth noting Wilson's own assessment of his achievement: "This is the most exciting aspect of the renormalization group, the part of the theory that makes it possible to solve problems which are unreachable by Feynman diagrams. The Kondo problem has been solved by a nondiagrammatic computer method."

Earlier in this same passage, written in 1975, Wilson roughly but very usefully divides RG theory into four parts: (a) the formal theory of fixed points and linear and nonlinear behavior near fixed points where he especially cites Wegner (1972a, 1976), as did I, above; (b) the diagrammatic (or field-theoretic) formulation of the RG for critical phenomena[33] where the ϵ expansion[34] and its many variants[35] plays a central role; (c) QFT methods, including the 1970–71 Callan-Symanzik equations[36] and the original, 1954 Gell-Mann-Low RG theory – restricted to systems with only a single, marginal variable[37] – from which Wilson drew some of his inspiration and which he took to name the whole approach.[38] Wilson characterizes these methods as efficient calculationally – which is certainly the case – but applying only to Feynman diagram expansions and says: "They completely hide the physics of many scales." Indeed, from the perspective of condensed matter physics, as I will try to explain below, the chief drawback of the sophisticated field-theoretic techniques is that they are safely applicable only when the basic physics is already well understood. By contrast, the general formulation (a), and Wilson's approach (b), provide insight and understanding into quite fresh problems.

Finally, Wilson highlights (d) "the construction of nondiagrammatic RG transformations, which are then solved numerically." This includes the real space, Monte Carlo, and functional RG approaches cited above and, of course, Wilson's own brilliant application to the Kondo problem (1975).

---

[32] Wilson (1975); for the following quotation see page 776, column 1.

[33] Wilson (1972), Brézin, Wallace and Wilson (1972), Wilson and Kogut (1974), Brézin, Le Guillou and Zinn-Justin (1976).

[34] Wilson and Fisher (1972), Fisher and Pfeuty (1972).

[35] Especial mention should be made of $1/n$ expansions, where n is the number of components of the vector order parameter (Abe, 1972, 1973; Fisher, Ma, and Nickel, 1972; Suzuki, 1972; and see Fisher, 1974, and Ma, 1976a) and of coupling-constant expansions in fixed dimension: see Parisi (1973, 1974); Baker, Nickel, Green, and Meiron (1976); Le Guillou and Zinn-Justin (1977); Baker, Nickel, and Meiron (1978): For other problems, dimensionality expansions have been made by writing $d = 8 - \epsilon$, $6 - \epsilon$, $4 + \frac{1}{2}m - \epsilon$ $(m = 1, 2, \cdots)$, $3 - \epsilon$, $2 + \epsilon$, and $1 + \epsilon$.

[36] The Callan–Symanzik equations are described, e.g., in Amit (1978) and Itzykson and Drouffe (1989). The coupling-constant expansions in fixed dimension (Parisi, 1973, 1974; Baker *et al.*, 1976) typically use these equations as a starting point and are usually presented purely formally in contrast to the full Wilson approach (b).

[37] See Wilson (1975), page 796, column 1. The concept of a "marginal" variable is explained briefly below: see also Wegner (1972a, 1976), Fisher (1974, 1983), and Kadanoff (1976).

[38] See Wilson (1975, 1983).

4 Exponents, anomalous dimensions, scale invariance and scale dependence

If one is to pick out a single feature that epitomizes the power and successes of RG theory, one can but endorse Gallavotti and Benfatto when they say "it has to be stressed that the *possibility of nonclassical critical indices* (i.e., *of nonzero anomaly* η) *is probably the most important achievement of the renormalization group.*"[39] For nonexperts it seems worthwhile to spend a little time here explaining the meaning of this remark in more detail and commenting on a few of the specialist terms that have already arisen in this account.

To that end, consider a locally defined microscopic variable which I will denote $\psi(\mathbf{r})$. In a ferromagnet this might well be the local magnetization, $\vec{M}(\mathbf{r})$, or spin vector, $S(\mathbf{r})$, at point $\mathbf{r}$ in ordinary d-dimensional (Euclidean) space; in a fluid it might be the deviation $\delta\rho(\mathbf{r})$, of the fluctuating density at $\mathbf{r}$ from the mean density. In QFT the local variables $\psi(\mathbf{r})$ are the basic *quantum fields* which are 'operator valued.' For a magnetic system in which quantum mechanics was important, $\vec{M}(\mathbf{r})$ and $\vec{S}(\mathbf{r})$ would, likewise, be operators. However, the distinction is of relatively minor importance so that we may, for ease, suppose $\psi(\mathbf{r})$ is a simple classical variable. It will be most interesting when ψ is closely related to the order parameter for the phase transition and critical behavior of concern.

By means of a scattering experiment (using light, x rays, neutrons, electrons, etc.) one can often observe the corresponding *pair correlation function* (or basic 'two-point function')

$$G(\mathbf{r}) = \langle \psi(\mathbf{0})\psi(\mathbf{r}) \rangle, \qquad (1)$$

where the angular brackets $\langle \cdot \rangle$ denote a statistical average over the thermal fluctuations that characterize all equilibrium systems at nonzero temperature. (Also understood, when $\psi(\mathbf{r})$ is an operator, are the corresponding quantum-mechanical expectation values.)

Physically, $G(\mathbf{r})$ is important since it provides a direct measure of the influence of the leading microscopic fluctuations at the origin $\mathbf{0}$ on the behavior at a point distance $r = |\mathbf{r}|$ away. But, almost by definition, in the vicinity of an appropriate critical point – for example the Curie point of a ferromagnet when $\psi \equiv \vec{M}$ or the gas–liquid critical point when $\psi = \delta\rho$ – a strong "ordering" influence or correlation spreads out over, essentially, macroscopic distances. As a consequence, precisely *at* criticality one rather generally finds a *power-law decay*, namely,

$$G_c(\mathbf{r}) \approx D/r^{d-2+\eta} \quad \text{as} \quad r \to \infty, \qquad (2)$$

which is characterized by the *critical exponent* (or critical *index*) $d - 2 + \eta$.

Now all the theories one first encounters – the so-called 'classical' or Landau-Ginzburg or van der Waals theories, etc.[40] – predict, quite unequivocally, that η *vanishes*. In QFT this corresponds to the behavior of a free massless particle. Mathematically, the reason underlying this prediction is that the basic functions entering the theory have (or are assumed to have) a smooth, *analytic*, *nonsingular* character so that, following Newton, they may be freely differentiated and thereby expanded

---

[39] See Benfatto and Gallavotti (1995) page 64.

[40] Note that 'classical' here, and in the quote from Benfatto and Gallavotti above means 'in the sense of the ancient authors'; in particular, it is *not* used in contradistinction to 'quantal' or to allude in any way to quantum mechanics (which has essentially no relevance for critical points at nonzero temperature: see the author's article cited in Footnote 13).

in Taylor series with positive integral powers[41] even at the critical point. In QFT the classical exponent value $d - 2$ (implying $\eta = 0$) can often be determined by naive dimensional analysis or 'power counting': then $d - 2$ is said to represent the 'canonical dimension' while η, if nonvanishing, represents the 'dimensional anomaly.' Physically, the prediction $\eta = 0$ typically results from a neglect of fluctuations or, more precisely as Wilson emphasized, from the assumption that only fluctuations on much smaller scales can play a significant role: in such circumstances the fluctuations can be safely incorporated into *effective* (or *renormalized*) parameters (masses, coupling constants, etc.) with no change in the basic character of the theory.

But a power-law dependence on distance implies a *lack* of a definite length scale and, hence, a *scale invariance*. To illustrate this, let us rescale distances by a factor b so that

$$\mathbf{r} \Rightarrow \mathbf{r}' = b\mathbf{r}, \tag{3}$$

and, at the same time, rescale the order parameter ψ by some "covariant" factor b^ω where ω will be a critical exponent characterizing ψ. Then we have

$$G_c(\mathbf{r}) = \langle \psi(\mathbf{0})\psi(\mathbf{r})\rangle_c \Rightarrow$$

$$G_c'(b\mathbf{r}) = b^{2\omega}\langle \psi(\mathbf{0})\psi(b\mathbf{r})\rangle_c \approx b^{2\omega}D/b^{d-2+\eta}r^{d-2+\eta}. \tag{4}$$

Now, observe that if one has $\omega = \frac{1}{2}(d - 2 + \eta)$, the factors of b drop out and the form in Eq. (2) is recaptured. In other words $G_c(\mathbf{r})$ is *scale-invariant* (or covariant): its variation reveals no characteristic lengths, large, small, or intermediate!

Since power laws imply scale invariance and the *absence* of well separated scales, the classical theories should be suspect at (and near) criticality! Indeed, one finds that the "anomaly" η does *not* normally vanish (at least for dimensions d less than 4, which is the only concern in a condensed matter laboratory!). In particular, from the work of Kaufman and Onsager (1949) one can show analytically that $\eta = \frac{1}{4}$ for the $d = 2$ Ising model.[42] Consequently, the analyticity and Taylor expansions presupposed in the classical theories are *not* valid.[43] Therein lies the challenge to theory! Indeed, it proved hard even to envisage the nature of a theory that would lead to $\eta \neq 0$. The power of the renormalization group is that it provides a conceptual and, in many cases, a computational framework within which anomalous values for η (and for other exponents like ω and its analogs for all local quantities such as the energy density) arise naturally.

In applications to condensed matter physics, it is clear that the power law in Eq. (2) can hold only for distances relatively large compared to atomic lengths or lattice spacings which we will denote a. In this sense the scale invariance of correlation functions is only *asymptotic* – hence the symbol $\approx$, for "asymptotically equals,"[44] and the proviso $r \to \infty$ in Eq. (2). A more detailed description would account for the effects of nonvanishing a, at least in leading order. By contrast, in QFT the microscopic distance a represents an "ultraviolet" cutoff which, since it is in general unknown,

---

[41] The relevant expansion variable in scattering experiments is the square of the scattering wave vector, $\mathbf{k}$, which is proportional to $\lambda^{-1}\sin\frac{1}{2}\theta$ where θ is the scattering angle and λ the wavelength of the radiation used. In the description of near-critical thermodynamics, Landau theory assumes (and mean-field theories lead to) Taylor expansions in powers of $T - T_c$ and $\Psi = \langle \psi(\mathbf{r})\rangle$, the equilibrium value of the order parameter.

[42] Fisher (1959); see also Fisher (1965, Sec. 29; 1967b, Sec. 6.2), Fisher and Burford (1967).

[43] Precisely the same problem undermines applications of catastrophe theory to critical phenomena; the assumed expressions in powers of $(T - T_c)$ and $\Psi = \langle \psi \rangle$ are simply not valid.

[44] See Appendix A for a discussion of appropriate conventions for the symbols $\simeq$, $\approx$, and $\sim$.

one normally wishes to remove from the theory. If this removal is not done with surgical care – which is what the renormalization program in QFT is all about – the theory remains plagued with infinite divergencies arising when $a \to 0$, i.e., when the "cutoff is removed." But in statistical physics one always anticipates a short-distance cutoff that sets certain physical parameters such as the value of T_c; infinite terms *per se* do not arise and certainly do *not* drive the theory as in QFT.

In current descriptions of QFT the concept of the *scale dependence of parameters* is often used with the physical picture that the typical properties of a system measured at particular length (and/or time) scales change, more-or-less slowly, as the scale of observation changes. From my perspective this phraseology often represents merely a shorthand for a somewhat simplified view of RG *flows* (as discussed generally below) in which only one variable or a single trajectory is followed,[45] basically because one is interested only in one, unique theory – the real world of particle physics. In certain condensed matter problems something analogous may suffice or serve in a first attack; but in general a more complex view is imperative.

One may, however, provide a more concrete illustration of scale dependence by referring again to the power law Eq. (2). If the exponent η vanishes, or equivalently, if ψ has its canonical dimension, so that $\omega = \omega_{can} = \frac{1}{2}(d-2)$, one may regard the amplitude D as a fixed, measurable parameter which will typically embody some real physical significance. Suppose, however, η does *not* vanish but is nonetheless relatively *small*: indeed, for many $(d = 3)$-dimensional systems, one has $\eta \simeq 0.035$.[46] Then we can introduce a "renormalized" or "scale-dependent" parameter

$$\tilde{D}(R) \approx D/R^\eta \quad \text{as} \quad R \to \infty, \tag{5}$$

and rewrite the original result simply as

$$G_c(r) = \tilde{D}(r)/r^{d-2}. \tag{6}$$

Since η is small we see that $\tilde{D}(R)$ varies slowly with the scale R on which it is measured. In many cases in QFT the dimensions of the field ψ (*alias* the order parameter) are subject only to marginal perturbations (see below) which translate into a $\log R$ dependence of the renormalized parameter $\tilde{D}(R)$; the variation with scale is then still weaker than when $\eta \neq 0$.

5 The challenges posed by critical phenomena

It is good to remember, especially when discussing theory and philosophy, that physics is an experimental science! Accordingly, I will review briefly a few experimental findings[47] that serve to focus attention on the principal theoretical challenges faced by, and rather fully met by, RG theory.

In 1869 Andrews reported to the Royal Society his observations of carbon dioxide sealed in a (strong!) glass tube at a mean overall density, ρ, close to $0.5 \, \text{g cm}^{-3}$. At room temperatures the fluid breaks into two phases: a liquid of density $\rho_{liq}(T)$ that coexists with a lighter vapor or gas phase of density $\rho_{gas}(T)$ from which it is separated

---

[45] See below and, e.g., Wilson and Kogut (1974), Bagnuls and Bervillier (1997).

[46] See, e.g., Fisher and Burford (1967), Fisher (1983), Baker (1990), and Domb (1996).

[47] Ideally, I should show here plots of impressive experimental data and, in particular, dramatic color pictures of carbon dioxide passing through its critical point. [See Stanley (1971) for black and white photographs.] It is not, however, feasible to reproduce such figures here; instead the presentation focuses on the conclusions as embodied in the observed power laws, etc.

by a visible meniscus or interface; but when the temperature, T, is raised and reaches a sharp critical temperature, $T_c \simeq 31.04\,°C$, the liquid and gaseous phases become identical, assuming a common density $\rho_{liq} = \rho_{gas} = \rho_c$ while the meniscus disappears in a "mist" of "critical opalescence." For all T above T_c there is a complete "continuity of state," i.e., no distinction whatsoever remains between liquid and gas (and there is no meniscus). A plot of $\rho_{liq}(T)$ and $\rho_{gas}(T)$ – as illustrated somewhat schematically in Fig. 1(d) – represents the so-called gas–liquid *coexistence curve*: the two halves,

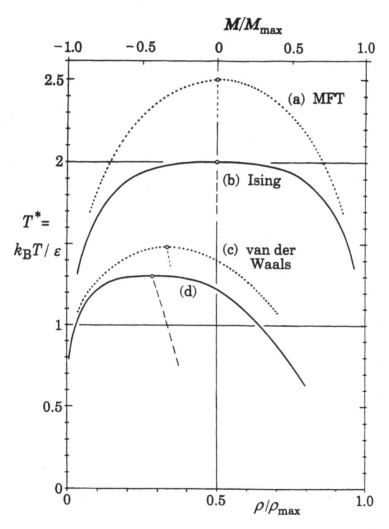

Fig. 1. Temperature variation of gas–liquid coexistence curves (temperature, T, versus density, ρ) and corresponding spontaneous magnetization plots (magnetization, M, versus T). The solid curves, (b) and (d), represent (semi-quantitatively) observation and modern theory, while the dotted curves (a) and (c) illustrate the corresponding 'classical' predictions (mean-field theory and van der Waals approximation). These latter plots are parabolic through the critical points (small open circles) instead of obeying a power law with the universal exponent $\beta \simeq 0.325$: see Eqs. (9) and (11). The energy scale ε, and the maximal density and magnetization, ρ_{max} and M_{max}, are nonuniversal parameters particular to each physical system; they vary widely in magnitude.

$\rho_{\text{liq}} > \rho_c$ and $\rho_{\text{gas}} < \rho_c$, meet smoothly at the *critical point* (T_c, ρ_c) – shown as a small circle in Fig. 1: the dashed line below T_c represents the *diameter* defined by $\bar{\rho}(T) = \frac{1}{2}[\rho_{\text{liq}}(T) + \rho_{\text{gas}}(T)]$.

The same phenomena occur in all elemental and simple molecular fluids and in fluid mixtures. The values of T_c, however, vary widely: e.g., for helium-four one finds 5.20 K while for mercury $T_c \simeq 1764$ K. The same is true for the critical densities and concentrations: these are thus "nonuniversal parameters" directly reflecting the atomic and molecular properties, i.e., the physics on the scale of the cutoff a. Hence, in Fig. 1, ρ_{max} (which may be taken as the density of the corresponding crystal at low T) is of order $1/a^3$, while the scale of $k_B T_c$ is set by the basic microscopic potential energy of attraction denoted ε. While of considerable chemical, physical, and engineering interest, such parameters will be of marginal concern to us here. The point, rather, is that the *shapes* of the coexistence curves, $\rho_{\text{liq}}(T)$ and $\rho_{\text{gas}}(T)$ versus T, become asymptotically *universal* in character as the critical point is approached.

To be more explicit, note first an issue of symmetry. In QFT, symmetries of many sorts play an important role: they may (or must) be built into the theory but can be "broken" in the physically realized vacuum state(s) of the quantum field. In the physics of fluids the opposite situation pertains. There is no real physical symmetry between coexisting liquid and gas: they are just different states, one a relatively dense collection of atoms or molecules, the other a relatively dilute collection – see Fig. 1(d). However, if one compares the two sides of the coexistence curve, gas and liquid, by forming the ratio

$$R(T) = [\rho_c - \rho_{\text{gas}}(T)]/[\rho_{\text{liq}}(T) - \rho_c], \tag{7}$$

one discovers an extraordinarily precise *asymptotic* symmetry. Explicitly, when T approaches T_c from below or, introducing a convenient notation,

$$t \equiv (T - T_c)/T_c \to 0-, \tag{8}$$

one finds $R(T) \to 1$. This simply means that the physical fluid builds for itself an exact mirror symmetry in density (and other properties) *as* the critical point is *approached*. And this is a universal feature for all fluids near criticality. (This symmetry is reflected in Fig. 1(d) by the high, although *not* absolutely perfect, degree of asymptotic *linearity* of the coexistence-curve diameter, $\bar{\rho}(T)$ – the dashed line described above.)

More striking than the (asymptotic) symmetry of the coexistence curve is the universality of its shape close to T_c, visible in Fig. 1(d) as a flattening of the graph relative to the parabolic shape of the corresponding classical prediction – see plot (c) in Fig. 1, which is derived from the famous van der Waals equation of state. Rather generally one can describe the shape of a fluid coexistence curve in the critical region via the power law

$$\Delta\rho \equiv \frac{1}{2}[\rho_{\text{liq}}(T) - \rho_{\text{gas}}(T)] \approx B|t|^\beta \quad \text{as} \quad t \to 0-, \tag{9}$$

where B is a *non*universal amplitude while the critical exponent β takes the *universal* value

$$\beta \simeq 0.325 \tag{10}$$

(in which the last figure is uncertain). To stress the point: β is a nontrivial number, not known exactly, but it is the *same* for all fluid critical points! This contrasts starkly with the classical prediction $\beta = \frac{1}{2}$ [corresponding to a parabola: see Fig. 1(c)]. The value in Eq. (10) applies to $(d = 3)$-dimensional systems. Classical theories make the same predictions for all d. On the other hand, for $d = 2$, Onsager's work (1949) on the square-lattice Ising model leads to $\beta = \frac{1}{8}$. This value has since been confirmed experimentally by Kim and Chan (1984) for a "two-dimensional fluid" of methane (CH_4) adsorbed on the flat, hexagonal-lattice surface of graphite crystals.

Not only does the value in Eq. (10) for β describe many types of fluid system, it also applies to anisotropic *magnetic materials*, in particular to those of Ising-type with one "easy axis." For that case, in vanishing magnetic fields, H, below the Curie or critical temperature, T_c, a ferromagnet exhibits a spontaneous magnetization and one has $M = \pm M_0(T)$. The sign, $+$ or $-$, depends on whether one lets H approach zero from positive or negative values. Since, in equilibrium, there is a full, natural physical symmetry under $H \Rightarrow -H$ and $M \Rightarrow -M$ (in contrast to fluid systems) one clearly has $M_c = 0$: likewise, the asymptotic symmetry corresponding to Eq. (7) is, in this case, exact for all T (see Fig. 1, plots (a) and (b)). Thus, as is evident in Fig. 1, the *global shape* of a spontaneous magnetization curves does not closely resemble a normal fluid coexistence curve. Nevertheless, in the asymptotic law

$$M_0(T) \approx B|t|^\beta \quad \text{as} \quad t \to 0-, \tag{11}$$

the exponent value in Eq. (10) still applies for $d = 3$: see Fig. 1(b); the corresponding classical "mean-field theory" in plot (a) again predicts $\beta = \frac{1}{2}$. For $d = 2$ the value $\beta = \frac{1}{8}$ is once more valid!

And, beyond fluids and anisotropic ferromagnets, many other systems belong – more correctly their critical behavior belongs – to the "Ising universality class." Included are other magnetic materials (antiferromagnets and ferrimagnets), binary metallic alloys (exhibiting order–disorder transitions), certain types of ferroelectrics, and so on.

For each of these systems there is an appropriate order parameter and, via Eq. (2), one can then define (and usually measure) the correlation decay exponent η which is likewise universal. Indeed, essentially any measurable property of a physical system displays a universal critical singularity. Of particular importance is the exponent $\alpha \simeq 0.11$ (Ising, $d = 3$) which describes the divergence to infinity of the specific heat via

$$C(T) \approx A^\pm / |t|^\alpha \quad \text{as} \quad t \to 0\pm \tag{12}$$

(at constant volume for fluids or in zero field, $H = 0$, for ferromagnets, etc.). The amplitudes A^+ and A^- are again *non*universal; but their *dimensionless ratio*, A^+/A^-, is *universal*, taking a value close to 0.52. When $d = 2$, as Onsager (1944) found, $A^+/A^- = 1$ and $|t|^{-\alpha}$ is replaced by $\log |t|$. But classical theory merely predicts a jump in specific heat, $\Delta C = C_c^- - C_c^+ > 0$, for all d!

Two other central quantities are a divergent isothermal compressibility $\chi(T)$ (for a fluid) or isothermal susceptibility, $\chi(T) \propto (\partial M/\partial H)_T$ (for a ferromagnet) and, for all systems, a *divergent correlation length*, $\xi(T)$, which measures the growth of the 'range of influence' or of correlation observed, say, via the decay of the correlation function $G(R; T)$ – see Eq. (1) above – to its long-distance limit. For these functions we write

$$\chi(T) \approx C^\pm / |t|^\gamma \quad \text{and} \quad \xi(T) \approx \xi_0^\pm / |t|^\nu, \tag{13}$$

as $t \to 0\pm$, and find, for $d = 3$ Ising-type systems,

$$\gamma \simeq 1.24 \quad \text{and} \quad \nu \simeq 0.63 \tag{14}$$

(while $\gamma = 1\frac{3}{4}$ and $\nu = 1$ for $d = 2$).

As hinted, there are other universality classes known theoretically although relatively few are found experimentally.[48] Indeed, one of the early successes of RG theory was delineating and sharpening our grasp of the various important universality classes. To a significant degree one found that only the vectorial or tensorial character of the relevant order parameter (e.g., scalar, complex number *alias* two-component vector, three-component vector, etc.) plays a role in determining the universality class. But the whys and the wherefores of this self-same issue represent, as does the universality itself, a prime challenge to any theory of critical phenomena.

6 Exponent relations, scaling and irrelevance

By 1960–62 the existence of universal critical exponents disagreeing sharply with classical predictions may be regarded as well established theoretically and experimentally.[49] The next theoretical step was the discovery of *exponent relations*, that is, simple algebraic equations satisfied by the various exponents *independently* of the universality class. Among the first of these were[50]

$$\gamma = (2 - \eta)\nu \quad \text{and} \quad \alpha + 2\beta + \gamma = 2. \tag{15}$$

[48] See, e.g., the survey in Fisher (1974b) and Aharony (1976).

[49] This retrospective statement may, perhaps, warrant further comment. First, the terms "universal" and "universality class" came into common usage only after 1974 when (see below) the concept of various types of renormalization-group fixed point had been well recognized (see Fisher, 1974b). Kadanoff (1976) deserves credit not only for introducing and popularizing the terms but especially for emphasizing, refining, and extending the concepts. On the other hand, Domb's (1960) review made clear that all (short-range) Ising models should have the same critical exponents irrespective of lattice structure but depending strongly on dimensionality. The excluded-volume problem for polymers was known to have closely related but *distinct* critical exponents from the Ising model, depending similarly *on* dimensionality but *not* lattice structure (Fisher and Sykes, 1959). And, as regards the Heisenberg model – which possesses what we would now say is an ($n = 3$)-component vector or O(3) order parameter – there were strong hints that the exponents were again different (Rushbrooke and Wood, 1958; Domb and Sykes, 1962).

On the experimental front matters might, possibly be viewed as less clear-cut: indeed, for ferromagnets, nonclassical exponents were unambiguously revealed only in 1964 by Kouvel and Fisher. However, a striking experiment by Heller and Benedek (1962) had already shown that the order parameter of the *anti*-ferromagnet MnF_2, namely, the *sublattice magnetization* $M_0^\dagger(T)$, vanishes as $|t|^\beta$ with $\beta \simeq 0.33_5$. Furthermore, for fluids, the work of the Dutch school under Michels and the famous analysis of coexistence curves by Guggenheim (1949) allowed little doubt – see Rowlinson (1959), Chap. 3, especially pp. 91–95 – that all reasonably simple atomic and molecular fluids displayed the same but *non*classical critical exponents with $\beta \simeq \frac{1}{3}$: And, also well before 1960, Widom and Rice (1955) had analyzed the *critical isotherms* of a number of simple fluids and concluded that the corresponding critical exponent δ (see, e.g., Fisher, 1967b) took a value around 4.2 in place of the van der Waals value $\delta = 3$. In addition, evidence was in hand showing that the *consolute point* in binary fluid mixtures was similar (see Rowlinson, 1959, pp. 165–6).

[50] See Fisher (1959; 1962; 1964, see Eq. (5.7); 1967b) for the first relation here; the second relation was advanced in Essam and Fisher (1963) where the now generally accepted notation for the thermodynamic critical exponents was also introduced. See, in addition, Fisher (1967a) based on a lecture given in March 1965. Actually the initial proposal was written as $\alpha' + 2\beta + \gamma' = 2$, where the primes denote exponents defined *below* T_c. This distinction, although historically important, is rarely made nowadays since, in general, scaling (see below) implies the $T \gtrless T_c$ equalities $\alpha' = \alpha$, $\gamma' = \gamma$, $\nu' = \nu$, etc. [also mentioned in Essam and Fisher (1967a)]. Moved by the suggested thermodynamic exponent equality, Rushbrooke (1963) quickly showed that for magnetic systems (with $H \Rightarrow -H$ symmetry) the positivity of specific heats implied by the Second Law of Thermodynamics could be used to prove rigorously the *inequality* $\alpha' + 2\beta + \gamma' \geq 2$. His proof was soon extended to fluid systems (Fisher, 1964), see Eq. (2.20). Corresponding to the first equality in Eq. (15), the inequality $\gamma \leq (2 - \eta)\nu$ was proven rigorously in Fisher (1969). Other valuable exponent inequalities encompassing "scaling laws" for the exponents as the limiting case of equality were proved by Griffiths (1965, 1972) for thermodynamic exponents and Buckingham and Gunton (1969) for correlation exponents.

As the reader may check from the values quoted above, these relations hold *exactly* for the $d = 2$ Ising models and are valid when $d = 3$ to within the experimental accuracy or the numerical precision (of the theoretical estimates[51]). They are even obeyed exactly by the classical exponent values (which, today, we understand[52] as valid for $d > 4$).

The first relation in Eq. (15) pertains just to the basic correlation function $G(\mathbf{r}; T) = \langle \psi(0)\psi(\mathbf{r}) \rangle$ as defined previously. It follows from the assumption,[53] supported in turn by an examination of the structure of Onsager's matrix solution to the Ising model,[54] *that in the critical region all lengths* (much larger than the lattice spacing a) *scale like the correlation length* $\xi(T)$ – introduced in Eq. (13). Formally one expresses this principle by writing, for $t \to 0$ and $r \to \infty$,

$$G(T; \mathbf{r}) \approx \frac{D}{r^{d-2+\eta}} \mathcal{G}\left(\frac{r}{\xi(T)}\right), \tag{16}$$

where, for consistency with (2), the *scaling function*, $\mathcal{G}(x)$, satisfies the normalization condition $\mathcal{G}(0) = 1$. Integrating $\mathbf{r}$ over all space yields the compressibility/susceptibility $\chi(T)$ and, thence, the relation $\gamma = (2 - \eta)\nu$. This *scaling law* highlights the importance of the correlation length ξ in the critical region, a feature later stressed and developed further, especially by Widom (1965), Kadanoff (1966, 1976), and Wilson (1983).[55] It is worth remarking that in QFT the inverse correlation length, ξ^{-1}, is basically equivalent to the *renormalized mass* of the field ψ: *masslessness* then equates with *criticality* since $\xi^{-1} \to 0$.

The next theoretical question was: "How can one construct an *equation of state* for a system which has nonclassical critical exponents?" The "equation of state" – for concreteness let us say, for a ferromagnet – is an equation relating the magnetization, M, the temperature, T, the magnetic field, H, and perhaps some further variable, say P, such as, for example, the overall pressure or, more interestingly, the strength of the direct electromagnetic, dipole–dipole couplings. More generally, one wants to know the free energy $F(T, H, P)$ from which all the thermodynamic properties follow[56] – or, better still, the full correlation function $G(\mathbf{r}; T, H, P)$ (where previously we had supposed $H = 0$ and $P = P_0$, fixed) since this gives more insight into the "structure" of the system.

The equation of state is crucial knowledge for any applications but, at first sight, the question appears merely of somewhat technical interest. Classical theory provides a simple answer – basically just a power series expansion in $(T - T_c)$, $(M - M_c)$, and $(P - P_c)$, etc.; but that always *enforces* classical exponent values! It transpires, therefore, that the mathematical issues are much more delicate. For convenience, let us focus on the *singular part* of the free-energy density, namely,[57]

$$f_s(t, h, g) \equiv -\Delta F(T, H, P)/V k_B T, \tag{17}$$

---

[51] See, e.g., Fisher (1967b), Baker (1990), Domb (1996).

[52] See Wilson and Fisher (1972), Wilson and Kogut (1974), Fisher (1974, 1983).

[53] See Fisher (1959, 1962).

[54] Onsager (1944), Kaufman and Onsager (1949).

[55] See also Wilson and Kogut (1974).

[56] Thus, for example, the equation of state is given by $M = -(\partial F/\partial H)_{T,P}$; the specific heat is $C = -T(\partial^2 F/\partial T^2)_{H=0,P}$.

[57] The "singular part," ΔF in Eq. (17), is found by subtracting from F analytic terms: $F_0(T, H, P) = F_c + F_1(T - T_c) + F_2 H + \cdots$. In Eq. (17) the volume V of the physical system is shown but a conceptually crucial theoretical issue, namely the *taking* of the *thermodynamic limit*, $V \to \infty$, has, for simplicity, been ignored. In Eq. (18), μ_B denotes the Bohr magneton, so that h is dimensionless.

as a function of the physically appropriate reduced variables

$$t = (T - T_c)/T_c, \qquad h = \mu_B H / k_B T, \qquad g = P / k_B T. \qquad (18)$$

Now, not only must $f(t, h, g)$ reproduce all the correct critical singularities when $t \to 0$ (for $h = 0$, etc.), it must *also* be *free* of singularities, i.e. "analytic," *away* from the critical point (and the phase boundary $h = 0$ below T_c).

The solution to this problem came most directly via Widom's (1965b) *homogeneity* or, as more customarily now called, *scaling hypothesis* which *embodies* a minimal number of the critical exponents. This may be written

$$f_s(t, h, g) \approx |t|^{2-\alpha} \mathcal{F}\left(\frac{h}{|t|^{\Delta}}, \frac{g}{|t|^{\phi}}\right), \qquad (19)$$

where α is the specific heat component introduced in Eq. (12) while the new exponent, Δ, which determines *how h scales with t*, is given by

$$\Delta = \beta + \gamma. \qquad (20)$$

Widom observed, incidentally, that the classical theories themselves obey scaling: one then has $\alpha = 0$, $\Delta = 1\frac{1}{2}$, $\phi = -\frac{1}{2}$.

The second new exponent, ϕ, did *not* appear in the original critical-point scaling formulations;[58] neither did the argument $z = g/|t|^{\phi}$ appear in the *scaling function* $\mathcal{F}(y, z)$. It is really only with the appreciation of RG theory that we know that such a dependence should in general be present and, indeed, that a full spectrum $\{\phi_j\}$ of such higher-order exponents with $\phi \equiv \phi_1 > \phi_2 > \phi_3 > \cdots$ must normally appear![59]

But how could such a spectrum of exponents be overlooked? The answer – essentially as supplied by the general RG analysis[60] – is that g and all the higher-order "coupling constants," say g_j, are *irrelevant* if their associated exponents ϕ_j are *negative*. To see this, suppose, as will typically be the case, that $\phi \equiv \phi_1 = -\theta$ is negative (so $\theta > 0$). Then, on approach to the critical point we see that

$$z = g/|t|^{\phi} = g|t|^{\theta} \to 0. \qquad (21)$$

Consequently, $\mathcal{F}(y, z)$ in Eq. (19) can be replaced simply by $\mathcal{F}(y, 0)$ which is a function of just a *single variable*. Furthermore, asymptotically when $T \to T_c$ we get the *same* function *whatever* the actual value of g – clearly[61] this is an example of *universality*.

Indeed, within RG theory this is the general mechanism of universality: in a very large (generally infinitely large) space of Hamiltonians, parametrized by t, h, and all the g_j, there is a controlling critical point (later seen to be a *fixed point*) about which each variable enters with a characteristic exponent. All systems with Hamiltonians differing only through the values of the g_j (within suitable bounds) will exhibit

---

[58] Widom (1965), Domb and Hunter (1965), Kadanoff (1966) Patashinskii and Pokrovskii (1966); and see Fisher (1967b) and Stanley (1971).

[59] See Wilson (1971a) and, for a very general exposition of scaling theory, Fisher (1974a).

[60] Wegner (1972, 1976), Fisher (1974a), Kadanoff (1976).

[61] Again we slide over a physically important detail, namely, that T_c, for example, will usually be a function of any irrelevant parameter such as g. This comes about because, in a full scaling formulation, the variables t, h, and g appearing in Eq. (19) must be replaced by *nonlinear scaling fields* $\tilde{t}(t, h, g)$, $\tilde{h}(t, h, g)$ and $\tilde{g}(t, h, g)$ which are smooth functions of t, h, and g (Wegner, 1972, 1976; Fisher, 1983). By the same token it is usually advantageous to introduce a prefactor A_0 in Eq. (19) and "metrical factors" E_j in the arguments $y \equiv z_0$ and z_j (see, e.g., Fisher, 1983).

the *same critical behavior* determined by the same free-energy scaling function $\mathcal{F}(y)$, where now we drop the irrelevant argument(s). Different universality classes will be associated with different controlling critical points in the space of Hamiltonians with, once one recognizes the concept of RG *flows*, different "domains of attraction" under the flow. All these issues will be reviewed in greater detail below.

In reality, the expectation of a general form of scaling[62] is frequently the most important consequence of RG theory for the practising experimentalist or theorist. Accordingly, it is worth saying more about the meaning and implications of Eq. (19). First, (i) it very generally *implies* the thermodynamic exponent relation Eq. (15) connecting α, β and γ; and (ii) since all leading exponents are determined entirely by the two exponents α and $\Delta(=\beta+\gamma)$, it predicts similar exponent relations for any other exponents one might define – such as δ specified *on* the critical isotherm[63] by $H \sim M^{\delta}$. Beyond that, (iii) if one fixes P (or g) and similar parameters and observes the free energy or, in practice, the equation of state, the data one collects amount to describing a function, say $M(T, H)$, of *two variables*. Typically this would be displayed as *sets of isotherms*: i.e., many plots of M vs. H at various closely spaced, fixed values of T near T_c. But according to the scaling law Eq. (19) if one plots the *scaled variables* $f_s/|t|^{2-\alpha}$ or $M/|t|^{\beta}$ against the scaled field $h/|t|^{\Delta}$, for appropriately chosen exponents and critical temperature T_c, one should find that all these data "collapse" (in Stanley's (1971) picturesque terminology) onto a single curve, which then just represents the scaling function $x = \mathcal{F}(y)$ itself!

Indeed, this dramatic collapse is precisely found in fitting experimental data. Furthermore, the same "collapse" occurs for different systems since the scaling function $\mathcal{F}(y)$ itself *also* proves to be *universal* (when properly normalized), as first stressed by Kadanoff (1976). A particularly striking example of such data collapse, yielding the same scaling function for a range of irrelevant parameter values, may be found in the recent work by Koch *et al.* (1989).[64] They studied a quite different physical problem, namely, the proposed "vortex-glass" transition in the high-T_c superconductor YBCO. There the voltage drop, E, across the specimen, measured over 4 or 5 decades, plays the role of M; the current density J, measured over a similar range, stands in for h, while the external magnetic field, H, acting on the sample, provides the irrelevant parameter P. The scaling function was finally determined over 10 decades in value and argument and seen to be universal!

7 Relevance, crossover, and marginality

As mentioned, the scaling behavior of the free energy, the equation of state, the correlation functions, and so on, always holds only in some *asymptotic sense* in condensed matter physics (and, indeed, in most applications of scaling). Typically, scaling becomes valid when $t \sim (T - T_c)$ becomes small, when the field H is small, and when the microscopic cutoff a is much smaller than the distances of interest. But one often needs to know: "How small is small enough?" Or, put in another

---

[62] Allowing for irrelevant variables, nonlinear scaling fields, and universality, as indicated in Eq. (19) and the previous footnote.

[63] See also Footnote 49 above.

[64] The scaling function, as plotted in this reference, strikes the uninitiated as two distinct functions, one for $T \geq T_c$, another for $T \leq T_c$. However, this is due just to the presentation adopted: scaling functions like $\mathcal{F}(y)$ in Eq. (19) are typically single functions *analytic through* $T = T_c$ for $y < \infty$ (i.e., $h \neq 0$) and can be re-plotted in a way that exhibits that feature naturally and explicitly.

language, "What is the nature of the leading corrections to the dominant power laws?" The "extended scaling" illustrated by the presence of the second argument $z = g/|t|^\phi$ in Eq. (19) provides an answer via Eq. (21) – an answer that, phenomeno-logically, can be regarded as independent of RG theory *per se*[65] but, in historical fact, essentially grew from insights gained via RG theory.[66]

Specifically, if the physical parameter $P \propto g$ is irrelevant then, by definition, $\phi = -\theta$ is negative and, as discussed, $z = g|t|^\theta$ becomes small when $|t| \to 0$. Then one can, fairly generally, hope to expand the scaling function $\mathcal{F}(y, z)$ in powers of z. From this one learns, for example, that the power law Eq. (11) for the spontaneous magnetization of a ferromagnet should, when t is no longer very small, be modified to read

$$M_0(T) = B|t|^\beta (1 + b_\theta |t|^\theta + b_1 t + \cdots), \tag{22}$$

where $b_\theta (\propto g)$ and b_1 are nonuniversal.[67] The exponent θ is often called the "*correction-to-scaling*" exponent – of course, it is universal.[68] It is significant because when θ is smaller than unity and b_θ is of order unity, the presence of such a singular correction hampers the reliable estimation of the primary exponent, here β, from experimental or numerical data.

Suppose, on the other hand, that ϕ is *positive* in the basic scaling law Eq. (19). Then when $t \to 0$ the scaled variable $z = g/|t|^\phi$ grows larger and larger. Consequently the behavior of $\mathcal{F}(y, z)$ for z small or vanishing becomes of less and less interest. Clearly, the previous discussion of asymptotic scaling fails! When that happens one says that the physical variable P represents a *relevant perturbation* of the original critical behavior.[69] Two possibilities then arise. *Either* the critical point may be *destroyed* altogether. This is, in fact, the effect of the magnetic field, which must itself be regarded as a relevant perturbation since $\phi_0 \equiv \Delta = \beta + \gamma > 0$. *Alternatively*, when z grows, the true, asymptotic critical behavior may *crossover*[70] to a new, quite *distinct* universality class with different exponents and a new asymptotic scaling function, say, $\mathcal{F}_\infty(y)$.[71]

The crossover scenario is, in fact, realized when the physical system is a ferromag-net with microscopic spin variables, say $\vec{S}(\mathbf{r})$, coupled by *short-range* "exchange" interactions while P measures the strength of the additional, *long-range* magnetic dipole–dipole coupling mediated by the induced electromagnetic fields.[72] Interested theorists had *felt* intuitively that the long-range character of the dipole–dipole coup-ling *should* matter, i.e., P should be *relevant*. But theoretically there seemed no feasible way of addressing the problem and, on the other hand, the experimentally observed critical exponents (for an important class of magnetic materials) seemed quite independent of the dipole–dipole coupling P.

[65] See Fisher (1974a).

[66] Wegner (1972) and Fisher (1974).

[67] See Wegner (1972, 1976) and Fisher (1974, 1983).

[68] For $d = 3$ Ising-type systems one finds $\theta \simeq 0.54$: see Chen *et al.* (1982), Zinn and Fisher (1996).

[69] Wegner (1972, 1976), Kadanoff (1976): see also Fisher (1983).

[70] See the extensive discussion of crossover in Fisher (1974b) and Aharony (1976).

[71] Formally, one might write $\mathcal{F}_\infty(y) = \mathcal{F}(y, z \to z_\infty)$ where z_∞ is a critical value which could be ∞; but a more subtle relationship is generally required since the exponent α in the prefactor in Eq. (19) changes.

[72] A "short-range" interaction potential, say $J(\mathbf{r})$, is usually supposed to decay with distance as $\exp(-r/R_0)$ where R_0 is some microscopic range, but certainly must decay *faster* than $1/r^{d+2}$; the dipole–dipole potential, however, decays more slowly, as $1/r^d$, and has a crucially important angular dependence as well.

The advent of RG theory changed that. First, it established a general framework within which the relevance or irrelevance of some particular perturbation P_j could be judged – essentially by the positive or negative sign of the associated exponent ϕ_j, with especially interesting *nonscaling* and *nonuniversal* behavior likely in the *marginal* case $\phi_j = 0$.[73] Second, for many cases where the $P_j = 0$ problem was well understood, RG theory showed how the *crossover exponent* ϕ could be determined exactly or perturbatively. Third, the ϵ expansion allowed calculation of ϕ *and* of the new critical behavior to which the crossover occurred.[74] The dipole–dipole problem for ferromagnets was settled via this last route: the dipole perturbation is *always* relevant; *however*, the new, dipolar critical exponents for typical ferromagnets like iron, nickel and gadolinium are numerically so close in value to the corresponding short-range exponents[75] that they are almost indistinguishable by experiment (or simulation)!

On the other hand, in the special example of *anisotropic*, easy-axis or Ising-type ferromagnets in $d = 3$ dimensions the dipolar couplings behave as *marginal* variables at the controlling, *dipolar* critical point.[76] This leads to the prediction of *logarithmic* modifications of the classical critical power laws (by factors diverging as $\log |T - T_c|$ to various powers). The predicted logarithmic behavior has, in fact, been verified experimentally by Ahlers *et al.* (1975). In other cases, especially for $d = 2$, marginal variables lead to continuously variable exponents such as $\alpha(g)$, and to quite different thermal variation, like $\exp(\tilde{A}/|t|^{\tilde{\nu}})$; such results have been checked both in exactly solved statistical mechanical models and in physical systems such as superfluid helium films.[77]

I have entered into these relatively detailed and technical considerations – which a less devoted reader need only peruse – in order to convey something of the flavor of how the renormalization group is used in statistical physics and to bring out those features for which it is so valued; because of the multifaceted character of condensed matter physics these are rather different and more diverse than those aspects of RG theory of significance for QFT.

8 The task for renormalization group theory

Let us, at this point, recapitulate briefly by highlighting, from the viewpoint of statistical physics, what it is one would wish RG theory to accomplish. First and foremost, (i) it should explain the ubiquity of power laws at and near critical points: see Eqs. (2), (9), (11)–(13). I sometimes like to compare this issue with the challenge to atomic physics of explaining the ubiquity of sharp spectral lines. Quantum mechanics responds, crudely speaking, by saying: "Well, (a) there is some wave – or a *wave function* ψ – needed to describe electrons in atoms, and (b) to fit a wave into a confined space the wave length must be quantized: hence (c) only certain definite energy levels are allowed and, thence, (d) there are sharp, spectral transitions between them!"

[73] See the striking analysis of Kadanoff and Wegner (1971).

[74] Fisher and Pfeuty (1972), Wegner (1972b).

[75] Fisher and Aharony (1973).

[76] Aharony (1973, 1976).

[77] See Kadanoff and Wegner (1971) and, for a review of the extensive later developments – including the Kosterlitz-Thouless theory of two-dimensional superfluidity and the Halperin-Nelson-Kosterlitz-Thouless-Young theory of two-dimensional melting – see Nelson (1983).

Of course, that is far from being the whole story in quantum mechanics; but I believe it captures an important essence. Neither is the first RG response the whole story: but, *to anticipate*, in Wilson's conception RG theory crudely says: "Well, (a) there is a *flow* in some *space*, $\mathbb{H}$, *of Hamiltonians* (or "coupling constants"); (b) the critical point of a system is associated with a *fixed point* (or stationary point) of that flow; (c) the flow operator – technically the *RG transformation*,[78] $\mathbb{R}$ – can be *linearized* about that fixed point; and (d) typically, such a linear operator (as in quantum mechanics) has a spectrum of discrete but nontrivial eigenvalues, say λ_k; then (e) each (asymptotically independent) exponential term in the flow varies as $e^{\lambda_k \ell}$, where ℓ is the *flow* (or renormalization) *parameter* and corresponds to a physical power law, say $|t|^{\phi_k}$, with critical exponent ϕ_k proportional to the eigenvalue λ_k." How one may find suitable transformations $\mathbb{R}$, and why the flows matter, are the subjects for the following chapters of our story.

Just as quantum mechanics does much more than explain sharp spectral lines, so RG theory should also explain, at least in principle, (ii) the values of the leading thermodynamic and correlation exponents, α, β, γ, δ, ν, η, and ω (to cite those we have already mentioned above) and (iii) clarify why and how the classical values are in error, including the existence of borderline dimensionalities, like $d_\times = 4$, above which classical theories become valid. Beyond the leading exponents, one wants (iv) the correction-to-scaling exponent θ (and, ideally, the higher-order correction exponents) and, especially, (v) one needs a method to compute crossover exponents, ϕ, to check for the relevance or irrelevance of a multitude of possible perturbations. Two central issues, of course, are (vi) the understanding of *universality* with nontrivial exponents and (vii) a derivation of *scaling*: see (16) and (19).

And, more subtly, one wants (viii) to understand the *breakdown* of universality and scaling in certain circumstances – one might recall continuous spectra in quantum mechanics – and (ix) to handle effectively logarithmic and more exotic dependences on temperature, etc.

An important further requirement as regards condensed matter physics is that RG theory should be firmly related to the science of statistical mechanics as perfected by Gibbs. Certainly, there is no need, and should be no desire, to replace standard statistical mechanics as a basis for describing equilibrium phenomena in pure, homogeneous systems.[79] Accordingly, it is appropriate to summarize briefly the demands of statistical mechanics in a way suitable for describing the formulation of RG transformations.

We may start by supposing that one has a set of microscopic, fluctuating, mechanical variables: in QFT these would be the various quantum fields, $\psi(\mathbf{r})$, defined – one supposes – at all points in a Euclidean (or Minkowski) space. In statistical physics we

---

[78] As explained in more detail in Sections 11 and 12 below, a specific renormalization transformation, say $\mathbb{R}_b$, acts on some 'initial' Hamiltonian $\mathcal{H}^{(0)}$ in the space $\mathbb{H}$ to transform it into a new Hamiltonian, $\mathcal{H}^{(1)}$. Under repeated operation of $\mathbb{R}_b$ the initial Hamiltonian "flows" into a sequence $\mathcal{H}^{(\ell)}$ ($\ell = 1, 2, \cdots$) corresponding to the iterated RG transformation $\mathbb{R}_b \cdots \mathbb{R}_b$ (ℓ times) which, in turn, specifies a new transformation $\mathbb{R}_{b^\ell}$. These "products" of repeated RG operations serve to define a *semigroup* of transformations that, in general, does *not* actually give rise to a group: see Footnote 3 above, the discussion below in Section 11 associated with Eq. (35) and Appendix C.

[79] One may, however, raise legitimate concerns about the adequacy of customary statistical mechanics when it comes to the analysis of random or impure systems – or in applications to systems far from equilibrium or in metastable or steady states – e.g., in fluid turbulence, in sandpiles and earthquakes, etc. And the use of RG ideas in chaotic mechanics and various other topics listed above in Section 3, clearly does *not* require a statistical mechanical basis.

will, rather, suppose that in a physical system of volume V there are N discrete "degrees of freedom." For classical fluid systems one would normally use the coordinates $\mathbf{r}_1, \mathbf{r}_2, \cdots, \mathbf{r}_N$ of the constituent particles. However, it is simpler mathematically – and the analogies with QFT are closer – if we consider here a set of "*spins*" $s_{\mathbf{x}}$ (which could be vectors, tensors, operators, etc.) associated with discrete lattice sites located at uniformly spaced points $\mathbf{x}$. If, as before, the lattice spacing is a, one can take $V = Na^d$ and the density of degrees of freedom in d spatial dimensions is $N/V = a^{-d}$.

In terms of the basic variables $s_{\mathbf{x}}$, one can form various "local operators" (or "physical densities" or "observables") like the local magnetization and energy densities

$$M_{\mathbf{x}} = \mu_{\mathrm{B}} s_{\mathbf{x}}, \qquad \mathcal{E}_{\mathbf{x}} = -\tfrac{1}{2} J \sum_{\delta} s_{\mathbf{x}} s_{\mathbf{x}+\delta, \cdots} \tag{23}$$

(where μ_{B} and J are fixed coefficients while δ runs over the nearest-neighbor lattice vectors). A physical system of interest is then specified by its Hamiltonian $\mathcal{H}[\{s_{\mathbf{x}}\}]$ – or energy function, as in mechanics – which is usually just a spatially uniform sum of local operators. The crucial function is the *reduced Hamiltonian*

$$\overline{\mathcal{H}}[s; t, h, \cdots, h_j, \cdots] = -\mathcal{H}[\{s_{\mathbf{x}}\}; \cdots, h_j, \cdots]/k_{\mathrm{B}} T, \tag{24}$$

where s denotes the set of all the microscopic spins $s_{\mathbf{x}}$ while $t, h, \cdots, h_j, \cdots$ are various "*thermodynamic fields*" (in QFT – the coupling constants): see Eq. (18). We may suppose that one or more of the thermodynamic fields, in particular the temperature, can be controlled directly by the experimenter; but others may be "given" since they will, for example, embody details of the physical system that are "fixed by nature."

Normally in condensed matter physics one thus focuses on some specific form of $\overline{\mathcal{H}}$ with at most two or three variable parameters – the Ising model is one such particularly simple form with just two variables, t, the reduced temperature, and h, the reduced field. An important feature of Wilson's approach, however, is to regard any such "physical Hamiltonian" as merely specifying a subspace (spanned, say, by "coordinates" t and h) in a very large space of possible (reduced) Hamiltonians, $\mathbb{H}$: see the schematic illustration in Fig. 2. This change in perspective proves crucial to the proper formulation of a renormalization group: in principle, it enters also in QFT although in practice, it is usually given little attention.

Granted a microscopic Hamiltonian, statistical mechanics promises to tell one the thermodynamic properties of the corresponding macroscopic system! First one must compute the partition function

$$Z_N[\overline{\mathcal{H}}] = \mathrm{Tr}_N^s \{ e^{\overline{\mathcal{H}}[s]} \}, \tag{25}$$

where the *trace operation*, $\mathrm{Tr}_N^s \{ \cdot \}$, denotes a summation or integration[80] over all the possible values of all the N spin variables $s_{\mathbf{x}}$ in the system of volume V. The *Boltzmann factor*, $\exp(\overline{\mathcal{H}}[s])$, measures, of course, the probability of observing the microstate specified by the set of values $\{s_{\mathbf{x}}\}$ in an equilibrium ensemble at temperature T. Then the thermodynamics follow from the total free-energy density, which is

---

[80] Here, for simplicity, we suppose the $s_{\mathbf{x}}$ are classical, commuting variables. If they are operator-valued then, in the standard way, the trace must be defined as a sum or integral over diagonal matrix elements computed with a complete basis set of N-variable states.

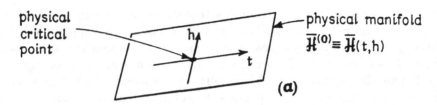

Fig. 2. Schematic illustration of the space of Hamiltonians, $\mathbb{H}$, having, in general, infinitely many dimensions (or coordinate axes). A particular physical system or model representing, say, the ferromagnet, iron, is specified by its reduced Hamiltonian $\overline{\mathcal{H}}(t, h)$, with $t = (T - T_c)/T_c$ and $h = \mu_B H/k_B T$ defined for that system: but in $\mathbb{H}$ this Hamiltonian specifies only a submanifold – the physical manifold, labelled (a), that is parametrized by the 'local coordinates' t and h. Other submanifolds, (b), (c), ..., located elsewhere in $\mathbb{H}$, depict the physical manifolds for Hamiltonians corresponding to other particular physical systems, say, the ferromagnets nickel and gadolinium, etc.

given by[81]

$$f[\overline{\mathcal{H}}] \equiv f(t, h, \cdots, h_j, \cdots) = \lim_{N,V \to \infty} V^{-1} \log Z_N[\overline{\mathcal{H}}]; \qquad (26)$$

this includes the singular part $f_s[\overline{\mathcal{H}}]$ near a critical point of interest: see Eq. (17). Correlation functions are defined similarly in standard manner.

To the degree that one can actually perform the trace operation in Eq. (25) for a particular model system and take the "thermodynamic limit" in Eq. (26) one will obtain the precise critical exponents, scaling functions, and so on. This was Onsager's (1944) route in solving the $d = 2$, spin-$\frac{1}{2}$ Ising models in zero magnetic field. At first sight one then has no need of RG theory. That surmise, however, turns out to be far from the truth. The issue is "simply" one of understanding! (Should one ever achieve truly high precision in simulating critical systems on a computer – a prospect which still seems some decades away – the same problem would remain.) In short, while one knows for sure that $\alpha = 0$ (log), $\beta = \frac{1}{8}$, $\gamma = 1\frac{3}{4}$, $\nu = 1$, $\eta = \frac{1}{4}$, $\cdots$ for the planar Ising models one does not know *why* the exponents have these values or *why* they satisfy the exponent relations Eqs. (15) or why the scaling law Eq. (16) is obeyed. Indeed, the seemingly inevitable mathematical complexities of solving even

---

[81] In Eq. (26) we have explicitly indicated the thermodynamic limit in which N and V become infinite maintaining the ratio $V/N = a^d$ fixed: in QFT this corresponds to an infinite system with an ultraviolet lattice cutoff.

such physically oversimplified models exactly[82] serve to conceal almost all traces of general, underlying mechanisms and principles that might "explain" the results. Thus it comes to pass that even a rather crude and approximate solution of a two-dimensional Ising model by a real space RG method can be truly instructive.[83]

9 Kadanoff's scaling picture

The year from late-1965 through 1966 saw the clear formulation of scaling for the thermodynamic properties in the critical region and the fuller appreciation of scaling for the correlation functions.[84] I have highlighted Widom's (1965) approach since it was the most direct and phenomenological – a bold, new thermodynamic hypothesis was advanced by generalizing a particular feature of the classical theories. But Domb and Hunter (1965) reached essentially the same conclusion for the thermodynamics based on analytic and series-expansion considerations, as did Patashinskii and Pokrovskii (1966) using a more microscopic formulation that brought out the relations to the full set of correlation functions (of all orders).[85]

Kadanoff (1966), however, derived scaling by introducing a completely new concept, namely, the *mapping* of a critical or near-critical system onto itself by a reduction in the effective number of degrees of freedom.[86] This paper attracted much favorable notice since, beyond obtaining all the scaling properties, it seemed to lay out a direct route to the actual *calculation* of critical properties. On closer examination, however, the implied program seemed – as I will explain briefly – to run rapidly into insuperable difficulties and interest faded. In retrospect, however, Kadanoff's scaling picture embodied important features eventually seen to be basic to Wilson's conception of the full renormalization group. Accordingly, it is appropriate to present a sketch of Kadanoff's seminal ideas.

For simplicity, consider with Kadanoff (1966) a lattice of spacing a (and dimensionality $d > 1$) with $S = \frac{1}{2}$ Ising spins $s_\mathbf{x}$ which, by definition, take only the values $+1$ or

---

[82] See the monograph by Rodney Baxter (1982).

[83] See Niemeijer and van Leeuwen (1976), Burkhardt and van Leeuwen (1982), and Wilson (1975, 1983) for discussion of real-space RG methods.

[84] Although one may recall, in this respect, earlier work (Fisher, 1959, 1962, 1964) restricted (in the application to ferromagnets) to zero magnetic field.

[85] It was later seen (Kiang and Stauffer, 1970; Fisher, 1971, Sec. 4.4) that thermodynamic scaling with general exponents (but particular forms of scaling function) was embodied in the "droplet model" partition function advanced by Essam and Fisher (1963) from which the exponent relations $\alpha' + 2\beta + \gamma' = 2$, etc., were originally derived. (See Eq. (15), Footnote 49, and Fisher, 1967b, Sec. 9.1; 1971, Sec. 4.)

[86] Novelty is always relative! From a historical perspective one should recall a suggestive contribution by M.J. Buckingham, presented in April 1965, in which he proposed a division of a lattice system into cells of geometrically increasing size, $L_n = b^n L_0$, with controlled intercell couplings. This led him to propose "the *existence* of an asymptotic 'lattice problem' such that the description of the nth order in terms of the $(n-1)$th is the same as that of the $(n+1)$th in terms of the nth." This is practically a description of "scaling" or "self similarity" as we recognize it today. Unfortunately, however, Buckingham failed to draw any significant, correct conclusions from his conception and his paper seemed to have little influence despite its presentation at the notable international conference on *Phenomena in the Neighborhood of Critical Points* organized by M.S. Green (with G.B. Benedek, E.W. Montroll, C.J. Pings, and the author) and held at the National Bureau of Standards, then in Washington, D.C. The Proceedings, complete with discussion remarks, were published, in December 1966, under the editorship of Green and J.V. Sengers (1966). Nearly all the presentations addressed the rapidly accumulating experimental evidence, but many well known theorists from a range of disciplines attended including P.W. Anderson, P. Debye, C. de Dominicis, C. Domb, S.F. Edwards, P.C. Hohenberg, K. Kawasaki, J.S. Langer, E. Lieb, W. Marshall, P.C. Martin, T. Matsubara, E.W. Montroll, O.K. Rice, J.S. Rowlinson, G.S. Rushbrooke, L. Tisza, G.E. Uhlenbeck, and C.N. Yang; but B. Widom, L.P. Kadanoff, and K.G. Wilson are *not* listed among the participants.

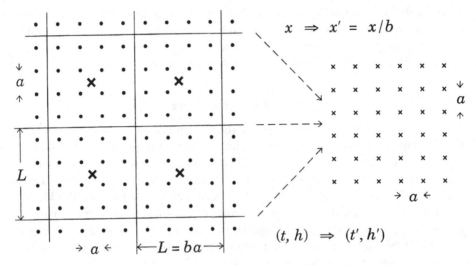

Fig. 3. A lattice of spacing a of Ising spins $s_x = \pm 1$ (in $d = 2$ dimensions) marked by solid dots, divided up into Kadanoff blocks or cells of dimensions $(L = ba) \times (L = ba)$ each containing a block spin $s'_{x'} = \pm 1$, indicated by a cross. After a rescaling, $x \Rightarrow x' = x/b$, the lattice of block spins appears identical with the original lattice. However, one supposes that the temperature t, and magnetic field h, of the original lattice can be renormalized to yield appropriate values, t' and h', for the rescaled, block-spin lattice: see text. In this illustration the spatial rescaling factor is $b = 4$.

−1: see Fig. 3. Spins on nearest-neighbor sites are coupled by an energy parameter or coupling constant, $J > 0$, which favors parallel alignment [see, e.g., Eq. (23) above]. Thus at low temperatures the majority of the spins point "up" ($s_x = +1$) or, alternatively, "down" ($s_x = -1$); in other words, there will be a spontaneous magnetization, $M_0(T)$, which decreases when T rises until it vanishes at the critical temperature $T_c > 0$: recall (11).

Now divide the lattice up into (disjoint) blocks, of dimensions $L \times L \times \cdots \times L$ with $L = ba$ so that each block contains b^d spins: see Fig. 3. Then associate with each block, say $\mathcal{B}_{x'}$ centered at point x', a new, effective *block spin*, $s'_{x'}$. If, finally, we *rescale* all spatial coordinates according to

$$x \Rightarrow x' = x/b, \tag{27}$$

the new lattice of block spins $s'_{x'}$ looks just like the original lattice of spins s_x. Note, in particular, the density of degrees of freedom is unchanged: see Fig. 3.

But if this appearance is to be more than superficial one must be able to relate the new or "renormalized" coupling J' between the block spins to the original coupling J, or, equivalently, the renormalized temperature deviation t' to the original value t. Likewise one must relate the new, renormalized magnetic field h' to the original field h.

To this end, Kadanoff supposes that b is large but less than the ratio, ξ/a, of the *correlation length*, $\xi(t, h)$, to the lattice spacing a; since ξ diverges at criticality – see Eq. (13) – this allows, asymptotically, for b to be chosen *arbitrarily*. Then Kadanoff notes that the total coupling of the magnetic field h to a block of b^d spins is equivalent to a coupling to the average spin

$$\bar{s}_{\mathbf{x}'} \equiv b^{-d} \sum_{\mathbf{x} \in \mathcal{B}_{\mathbf{x}'}} s_{\mathbf{x}} \cong \zeta(b) s_{\mathbf{x}'}', \tag{28}$$

where the sum runs over all the sites $\mathbf{x}$ in the block $\mathcal{B}_{\mathbf{x}'}$, while the "asymptotic equivalence" to the new, Ising block spin $s_{\mathbf{x}'}'$ is, Kadanoff proposes, determined by some "spin rescaling or renormalization factor" $\zeta(b)$. Introducing a similar thermal renormalization factor, $\vartheta(b)$, leads to the *recursion relations*

$$t' \approx \vartheta(b)t \quad \text{and} \quad h' \approx \zeta(b)h. \tag{29}$$

Correspondingly, the basic correlation function – compare with Eqs. (1), (4), and (16) – should renormalize as

$$G(\mathbf{x}; t, h) \equiv \langle s_0 s_{\mathbf{x}} \rangle \approx \zeta^2(b) G(\mathbf{x}'; t', h'). \tag{30}$$

In summary, under a spatial scale transformation and the integration out of all but a fraction b^{-d} of the original spins, the system asymptotically *maps back into itself* although at a renormalized temperature and field! However, the map is *complete* in the sense that *all* the statistical properties should be related by similarity.

But how should one choose – or, better, determine – the renormalization factors ζ and ϑ? Let us consider the basic relation Eq. (30) *at criticality*, so that $t = h = 0$ and, by Eq. (29), $t' = h' = 0$. Then, if we accept the observation/expectation Eq. (2) of a power law decay, i.e., $G_c(\mathbf{x}) \sim 1/|\mathbf{x}|^{d-2+\eta}$, one soon finds that $\zeta(b)$ must be just a power of b. It is natural, following Kadanoff (1966), then to propose the forms

$$\zeta(b) = b^{-\omega} \quad \text{and} \quad \vartheta(b) = b^{\lambda}, \tag{31}$$

where the two exponents ω and λ characterize the critical point under study while b is an essentially unrestricted *scaling parameter*.

By capitalizing on the freedom to choose b as $t, h \to 0$, or, more-or-less equivalently, by *iterating* the recursion relations Eqs. (29) and (30), one can, with some further work, show that all the previous scaling laws hold, specifically Eqs. (15), (16), and (19) although with $g \equiv 0$. Of course, all the exponents are now determined by ω and λ: for example, one finds $\nu = 1/\lambda$ and $\beta = \omega\nu$. Beyond that, the analysis leads to new exponent relations, namely, the so-called *hyperscaling laws*[87] which explicitly involve the spatial dimensionality: most notable is[88]

$$d\nu = 2 - \alpha. \tag{32}$$

But then Kadanoff's scaling picture is greatly strengthened by the fact that this relation holds *exactly* for the $d = 2$ Ising model! The same is true for all other exactly soluble models when $d < 4$.[89]

Historically, the careful numerical studies of the $d = 3$ Ising models by series expansions[90] for many years suggested a small but significant deviation from Eq. (32) as allowed by pure scaling phenomenology.[91] But, in recent years, the

[87] See Fisher (1974a) where the special character of the hyperscaling relations is stressed.
[88] See Kadanoff (1966), Widom (1965a) and Stell (1965, unpublished, quoted in Fisher, 1969; and 1968).
[89] See, e.g., Fisher (1983) and, for the details of the exactly solved models, Baxter (1982).
[90] For accounts of series expansion techniques and their important role see: Domb (1960, 1996), Baker (1961, 1990), Essam and Fisher (1963), Fisher (1965, 1967b), and Stanley (1971).
[91] As expounded systematically in Fisher (1974a) with hindsight enlightened by RG theory.

accumulating weight of evidence critically reviewed has convinced even the most cautious skeptics.[92]

Nevertheless, all is not roses! Unlike the previous exponent relations (all being independent of d) hyperscaling fails for the classical theories unless $d = 4$. And since one knows (rigorously for certain models) that the classical exponent values are valid for $d > 4$, it follows that hyperscaling cannot be generally valid. Thus something is certainly missing from Kadanoff's picture. Now, thanks to RG insights, we know that the breakdown of hyperscaling is to be understood via the second argument in the "fuller" scaling form Eq. (19): when d exceeds the appropriate borderline dimension, $d_\times$, a "dangerous irrelevant variable" appears and must be allowed for.[93] In essence one finds that the scaling function limit $\mathcal{F}(y, z \to 0)$, previously accepted without question, is no longer well defined but, rather, diverges as a power of z: asymptotic scaling survives but $d^* \equiv (2 - \alpha)/\nu$ sticks at the value 4 for $d > d_\times = 4$.

However, the issue of hyperscaling was *not* the main road block to the analytic development of Kadanoff's picture. The principal difficulties arose in explaining the *power-law* nature of the rescaling factors in Eqs. (29)–(31) and, in particular, in justifying the idea of a *single* effective, renormalized coupling J' between adjacent block spins, say $s'_{x'}$ and $s'_{x'+\delta'}$. Thus the interface between two adjacent $L \times L \times L$ blocks (taking $d = 3$ as an example) separates two block faces each containing b^2, strongly interacting, original lattice spins s_x. Well below T_c all these spins are frozen, "up" or "down," and a single effective coupling could well suffice; but at and above T_c these spins must fluctuate on many scales and a single effective-spin coupling seems inadequate to represent the inherent complexities.[94]

One may note, also that Kadanoff's picture, like the scaling hypothesis itself, provides no real hints as to the origins of universality: the rescaling exponents ω and λ in Eq. (31) might well change from one system to another. Wilson's (1971a) conception of the renormalization group both answered the problem of the "lost microscopic details" of the original spin lattice and provided a natural explanation of universality.

10 Wilson's quest

Now because this account has a historical perspective, and since I was Ken Wilson's colleague at Cornell for some 20 years, I will say something about how his search for a deeper understanding of quantum field theory led him to formulate renormalization group theory as we know it today. The first remark to make is that Ken Wilson is a markedly independent and original thinker and a rather private and reserved person. Secondly, in his 1975 article, in *Reviews of Modern Physics*, from which I have already quoted, Ken Wilson gave his considered overview of RG theory which, in my judgement, still stands well today. In 1982 he received the Nobel Prize and in his Nobel lecture, published in 1983, he devotes a section to "Some History Prior to 1971" in which he recounts his personal scientific odyssey.

---

[92] See Fisher and Chen (1985) and Baker and Kawashima (1995, 1996).

[93] See Fisher in (Gunton and Green, 1974, p. 66) where a "dangerous irrelevant variable" is characterized as a "hidden relevant variable"; and (Fisher, 1983, App. D).

[94] In hindsight, we know this difficulty is profound: in general, it is *impossible* to find an adequate single coupling. However, for certain special models it does prove possible and Kadanoff's picture goes through: see Nelson and Fisher (1975) and Fisher (1983). Further, in defense of Kadanoff, the condition $b \ll \xi/a$ was supposed to "freeze" the original spins in each block sufficiently well to justify their replacement by a simple block spin.

He explains that as a student at Caltech in 1956–60, he failed to avoid "the default for the most promising graduate students [which] was to enter elementary-particle theory." There he learned of the 1954 paper by Gell-Mann and Low "which was the principal inspiration for [his] own work prior to Kadanoff's (1966) formulation of the scaling hypothesis." By 1963 Ken Wilson had resolved to pursue quantum field theories as applied to the strong interactions. Prior to summer 1966 he heard Ben Widom present his scaling equation of state in a seminar at Cornell "but was puzzled by the absence of any theoretical basis for the form Widom wrote down." Later, in summer 1966, on studying Onsager's solution of the Ising model in the reformulation of Lieb, Schultz and Mattis,[95] Wilson became aware of analogies with field theory and realized the applicability of his own earlier RG ideas (developed for a truncated version of fixed-source meson theory[96]) to critical phenomena. This gave him a scaling picture but he discovered that he "had been scooped by Leo Kadanoff." Thereafter Ken Wilson amalgamated his thinking about field theories on a lattice and critical phenomena learning, in particular, about Euclidean QFT[97] and its close relation to the transfer matrix method in statistical mechanics – the basis of Onsager's (1944) solution.

That same summer of 1966 I joined Bed Widom at Cornell and we jointly ran an open and rather wide-ranging seminar loosely centered on statistical mechanics. Needless to say, the understanding of critical phenomena and of the then new scaling theories was a topic of much interest. Ken Wilson frequently attended and, perhaps partially through that route, soon learned a lot about critical phenomena. He was, in particular, interested in the series expansion and extrapolation methods for estimating critical temperatures, exponents, amplitudes, etc., for lattice models that had been pioneered by Cyril Domb and the King's College, London group.[98] This approach is, incidentally, still one of the most reliable and precise routes available for estimating critical parameters. At that time I, myself, was completing a paper on work with a London University student, Robert J. Burford, using high-temperature series expansions to study in detail the correlation functions and scattering behavior of the two- and three-dimensional Ising models.[99] Our theoretical analysis had already brought out some of the analogies with field theory revealed by the transfer matrix approach. Ken himself undertook large-scale series expansion calculations in order to learn and understand the techniques. Indeed, relying on the powerful computer programs Ken Wilson developed and kindly made available to us, another one of my students, Howard B. Tarko, extended the series analysis of the Ising correlations functions to temperatures below T_c and to all values of the magnetic field.[100] Our results have lasted rather well and many of them are only recently being revised and improved.[101]

[95] See Schultz *et al.* (1964).

[96] See Wilson (1983).

[97] As stressed by Symanzik (1966) the Euclidean formulation of quantum field theory makes more transparent the connections to statistical mechanics. Note, however, that in his 1966 article Symanzik did not delineate the special connections to critical phenomena *per se* that were gaining increasingly wide recognition; see, e.g., Patashinskii and Pokrovskii (1966), Fisher (1969, Sec. 12) and the remarks below concerning Fisher and Burford (1967).

[98] See the reviews Domb (1960), Fisher (1965, 1967b), Stanley (1971).

[99] Fisher and Burford (1967).

[100] Tarko and Fisher (1975).

[101] See Zinn and Fisher (1996), Zinn, Lai, and Fisher (1996), and references therein.

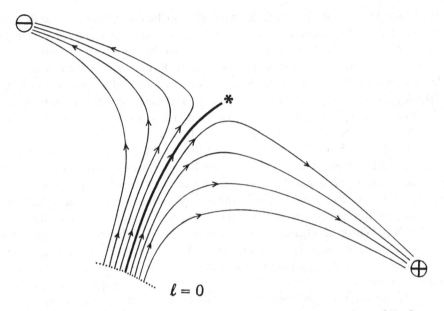

$\ell = 0$

Fig. 4. A 'vision' of flows in some large space inspired by a seminar of K. G. Wilson in the period 1967–70. The idea conveyed is that initially close, smoothly connected points at the start of the flow – the locus $\ell = 0$ – can eventually separate and run to far distant regions representing very different 'final' physical states: the essence of a phase transition. In modern terms the flow is in the space $\mathbb{H}$ of Hamiltonians; the intersection of the separatrix, shown bolder, with the initial locus ($\ell = 0$) represents the physical critical point; $*$ denotes the controlling fixed point, while $\oplus$ and $\ominus$ represent asymptotic high-T, disordered states and low-T, ordered states, respectively.

Typically, then, Ken Wilson's approach was always "hands on" and his great expertise with computers was ever at hand to check his ideas and focus his thinking.[102] From time to time Ken would intimate to Ben Widom or myself that he might be ready to tell us where his thinking about the central problem of explaining scaling had got to. Of course, we were eager to hear him speak at our seminar although his talks were frequently hard to grasp. From one of his earlier talks and the discussion afterwards, however, I carried away a powerful and vivid picture of *flows* – flows in a large space. And the point was that at the initiation of the flow, when the "time" or "flow parameter" ℓ, was small, two nearby points would travel close together; see Fig. 4. But as the flow developed a point could be reached – a bifurcation point (and hence, as one later realized, a stationary or fixed point of the flow) – beyond which the two originally close points could separate and, as ℓ increased, diverge to vastly differ-ent destinations: see Fig. 4. At the time, I vaguely understood this as indicative of how a sharp, nonanalytic phase transition could grow from smooth analytic initial data.[103]

But it was a long time before I understood the nature of the space – the space $\mathbb{H}$ of Hamiltonians – and the *mechanism* generating the flow, that is, a renormalization group transformation. Nowadays, when one looks at Fig. 4, one sees the locus of initial points, $\ell = 0$, as identifying the manifold corresponding to the original or

[102] See his remarks in Wilson (1983) on page 591, column 1.
[103] See the (later) introductory remarks in Wilson (1971a) related to Fig. 1 there.

"bare" Hamiltonian (see Fig. 2) while the trajectory leading to the bifurcation point represents a locus of critical points. The two distinct destinations for $\ell \to \infty$ then typically correspond to a high-temperature, fully disordered system and to a low-temperature fully ordered system: see Fig. 4.

In 1969 word reached Cornell that two Italian theorists, C. Di Castro and G. Jona-Lasinio, were claiming[104] that the "multiplicative renormalization group," as expounded in the field-theory text by Bogoliubov and Shirkov (1959), could provide "a microscopic foundation" for the scaling laws (which, by then, were well established phenomenologically). The formalism and content of the field-theoretic renormalization group was totally unfamiliar to most critical-phenomena theorists: but the prospect of a microscopic derivation was clearly exciting! However, the articles[105] proved hard to interpret as regards concrete progress and results. Nevertheless, the impression is sometimes conveyed that Wilson's final breakthrough was somehow anticipated by Di Castro and Jona-Lasinio.[106]

Such an impression would, I believe, be quite misleading. Indeed, Di Castro was invited to visit Cornell where he presented his ideas in a seminar that was listened to attentively. Again I have a vivid memory: walking to lunch at the Statler Inn after the seminar I checked my own impressions with Ken Wilson by asking, "Well, did he really say anything new?" (By "new" I meant some fresh insight or technique that carried the field forward.) The conclusion of our conversation was "No". The point was simply that none of the problems then outstanding – see the "tasks" outlined above (in Section 8) – had been solved or come under effective attack. In fairness, I must point out that the retrospective review by Di Castro and Jona-Lasinio themselves (1976) is reasonably well balanced. One accepted a scaling hypothesis and injected that as an ansatz into a general formalism; then certain insights and interesting features emerged; but, in reality, only scaling theory had been performed; and, in the end, as Di Castro and Jona-Lasinio say, "Still one did not see how to perform explicit calculations." Incidentally, it is also interesting to note Wilson's sharp criticism[107] of the account presented by Bogoliubov and Shirkov (1959) of the original RG ideas of Stueckelberg and Petermann (who, in 1953, coined the phrase "groupes de normalization") and of Gell-Mann and Low (1954).

One more personal anecdote may be permissible here. In August 1973 I was invited to present a tutorial seminar on renormalization group theory while visiting the Aspen Center for Physics. Ken Wilson's thesis advisor, Murray Gell-Mann, was in the audience. In the discussion period after the seminar Gell-Mann expressed his appreciation for the theoretical structure created by his famous student that I had set out in its generality, and he asked, "But tell me, what has all that got to do with the work Francis Low and I did so many years ago?"[108] In response, I explained the connecting thread and the far-reaching intellectual inspiration: certainly there is a thread but – to echo my previous comments – I believe that its length is comparable to that reaching from Maxwell, Boltzmann, and ideal gases to Gibbs' general conception of ensembles, partition functions, and their manifold inter-relations.

[104] The first published article was Di Castro and Jona-Lasinio (1969).

[105] See the later review by Di Castro and Jona-Lasinio (1976) for references to their writings in the period 1969–72 prior to Wilson's 1971 papers and the ϵ-expansion in 1972.

[106] See, for example, Benfatto and Gallavotti (1995) on page 96 in *A Brief Historical Note*, which is claimed to represent only the authors' personal "cultural evolution through the subject."

[107] See, especially, Wilson (1975) on page 796, column 1, and Footnote 10 in Wilson (1971a).

[108] That is, in Gell-Mann and Low (1954).

11 The construction of renormalization group transformations: the epsilon expansion

In telling my story I have purposefully incorporated a large dose of hindsight by emphasizing the importance of viewing a particular physical system – or its reduced Hamiltonian, $\overline{\mathcal{H}}(t, h, \cdots)$: see Eq. (24) – as specifying only a relatively small manifold in a large space, $\mathbb{H}$, of possible Hamiltonians. But why is that more than a mere formality? One learns the answer as soon as, following Wilson (1975, 1983), one attempts to implement Kadanoff's scaling description in some concrete, computational way. In Kadanoff's picture (in common with the Gell-Mann–Low, Callan–Symanzik, and General QFT viewpoints) one *assumes* that after a "rescaling" or "renormalization" the new, renormalized Hamiltonian (or, in QFT, the Lagrangian) has the *identical form* except for the renormalization of a single parameter (or coupling constant) or – as in Kadanoff's picture – of at most a small *fixed* number, like the temperature t and field h. That assumption is the dangerous and, unless one is especially lucky,[109] the *generally false* step! Wilson (1975, p. 592) has described his "liberation" from this straitjacket and how the freedom gained opened the door to the systematic design of RG transformations.

To explain, we may state matters as follows: Gibbs' prescription for calculating the partition function – see Eq. (25) – tells us to sum (or to integrate) over the allowed values of *all* the N spin variables $s_{\mathbf{x}}$. But this is very difficult! Let us, instead, adopt a strategy of "divide and conquer", by separating the set $\{s_{\mathbf{x}}\}$ of N spins into two groups: first, $\{s_{\mathbf{x}}^{<}\}$, consisting of $N' = N/b^d$ spins which we will leave as untouched fluctuating variables; and, second, $\{s_{\mathbf{x}}^{>}\}$ consisting of the remaining $N - N'$ spin variables over which we will integrate (or sum) so that they drop out of the problem. If we draw inspiration from Kadanoff's (or Buckingham's[110]) block picture we might reasonably choose to integrate over all but one central spin in each block of b^d spins. This process, which Kadanoff has dubbed "decimation" (after the Roman military practice), preserves translational invariance and clearly represents a concrete form of "coarse graining" (which, in earlier days, was typically cited as a way to derive, "in principle", mesoscopic or Landau-Ginzburg descriptions).

Now, after taking our partial trace we must be left with some new, *effective Hamiltonian*, say, $\overline{\mathcal{H}}_{\mathrm{eff}}[s^{<}]$, involving only the preserved, unintegrated spins. On reflection one realizes that, in order to be faithful to the original physics, such an effective Hamiltonian must be defined via its Boltzmann factor: recalling our brief outline of statistical mechanics, that leads directly to the explicit formula

$$e^{\overline{\mathcal{H}}_{\mathrm{eff}}[s^{<}]} = \mathrm{Tr}_{N-N'}^{s^{>}}\{e^{\overline{\mathcal{H}}[s^{<} \cup s^{>}]}\}, \qquad (33)$$

where the "union" $s^{<} \cup s^{>}$, simply stands for the full set of original spins $s \equiv \{s_{\mathbf{x}}\}$. By a spatial rescaling, as in Eq. (27), and a relabelling, namely, $s_{\mathbf{x}}^{<} \Rightarrow s_{\mathbf{x}'}'$, we obtain the "renormalized Hamiltonian", $\overline{\mathcal{H}}'[s'] \equiv \overline{\mathcal{H}}_{\mathrm{eff}}[s^{<}]$. Formally, then, we have succeeded in defining an *explicit renormalization transformation*. We will write

$$\overline{\mathcal{H}}'[s'] = \mathbb{R}_b\{\overline{\mathcal{H}}[s]\}, \qquad (34)$$

where we have elected to keep track of the spatial rescaling factor, b, as a subscript on the RG operator $\mathbb{R}$.

---

[109] See Footnote 94 above and Nelson and Fisher (1975) and Fisher (1983).
[110] Recall Footnote 86 above.

Note that if we complete the Gibbsian prescription by taking the trace over the renormalized spins we simply get back to the desired partition function, $Z_N[\mathcal{H}]$.[111] Thus nothing has been lost: the renormalized Hamiltonian retains all the thermo-dynamic information. On the other hand, experience suggests that, rather than try to compute Z_N directly from $\overline{\mathcal{H}}'$, it will prove more fruitful to *iterate* the transformation so obtaining a sequence, $\overline{\mathcal{H}}^{(\ell)}$, of renormalized Hamiltonians, namely,

$$\overline{\mathcal{H}}^{(\ell)} = \mathbb{R}_b[\overline{\mathcal{H}}^{(\ell-1)}] = \mathbb{R}_{b^\ell}[\overline{\mathcal{H}}], \qquad (35)$$

with $\overline{\mathcal{H}}^{(0)} \equiv \overline{\mathcal{H}}$, $\overline{\mathcal{H}}^{(1)} = \overline{\mathcal{H}}'$. It is these iterations that give rise to the *semigroup* character of the RG transformation.[112]

But now comes the crux: thanks to the rescaling and relabeling, the microscopic variables $\{s'_{\mathbf{x}'}\}$ are, indeed, completely equivalent to the original spins $\{s_{\mathbf{x}}\}$. However, when one proceeds to determine the nature of $\overline{\mathcal{H}}_{\text{eff}}$, and thence of $\overline{\mathcal{H}}'$, by using the formula (33), one soon discovers that one *cannot* expect the original form of $\overline{\mathcal{H}}$ to be reproduced in $\overline{\mathcal{H}}_{\text{eff}}$. Consider, for concreteness, an initial Hamiltonian, $\overline{\mathcal{H}}$, that describes Ising spins ($s_{\mathbf{x}} = \pm 1$) on a square lattice in zero magnetic field with just nearest-neighbor interactions of coupling strength $K_1 = J_1/k_{\text{B}}T$: in the most conservative Kadanoff picture there must be *some* definite recursion relation for the renormalized coupling, say, $K'_1 = \mathcal{T}_1(K_1)$, embodied in a definite function $\mathcal{T}(\cdot)$. But, in fact, one finds that $\overline{\mathcal{H}}_{\text{eff}}$ must actually contain *further* nonvanishing spin couplings, K_2, between second-neighbor spins, K_3, between third-neighbors, and so on up to *indefinitely* high orders. Worse still, four-spin coupling terms like $K_\square s_{\mathbf{x}_1} s_{\mathbf{x}_2} s_{\mathbf{x}_3} s_{\mathbf{x}_4}$ appear in $\overline{\mathcal{H}}_{\text{eff}}$, again for *all* possible arrangements of the four spins! And also six-spin couplings, eight-spin couplings, $\cdots$. Indeed, for any given set Q of $2m$ Ising spins on the lattice (and its translational equivalents), a nonvanishing coupling constant, K'_Q, is generated and appears in $\overline{\mathcal{H}}'$!

The only saving grace is that further iteration of the decimation transformation Eq. (33) cannot (in zero field) lead to anything worse! In other words the space $\mathbb{H}_{\text{Is}}$ of Ising spin Hamiltonians in zero field may be specified by the infinite set $\{K_Q\}$, of all possible spin couplings, and is *closed* under the decimation transformation Eq. (33). Formally, one can thus describe $\mathbb{R}_b$ by the full set of *recursion relations*

$$K'_P = \mathcal{T}_P(\{K_Q\}) \quad (\text{all } P). \qquad (36)$$

Clearly, this answers our previous questions as to what becomes of the complicated across-the-faces-of-the-block interactions in the original Kadanoff picture: they actually carry the renormalized Hamiltonian *out* of the (too small) manifold of near-est-neighbor Ising models and introduce (infinitely many) further couplings. The resulting situation is portrayed schematically in Fig. 5: the renormalized manifold for $\overline{\mathcal{H}}'(t', h')$ generally has no overlap with the original manifold. Further iterations, and *continuous* [see Eq. (40) below] as against discrete RG transformations, are suggested by the flow lines or "trajectories" also shown in Fig. 5. We will return to some of the details of these below.

In practice, the naive decimation transformation specified by Eq. (33) generally fails as a foundation for useful calculations.[113] Indeed, the design of effective RG

---

[111] The formal derivation, for those interested, is set out in Appendix B.
[112] This point is enlarged upon in Appendix C.
[113] See Kadanoff and Niemeijer in Gunton and Green (1974), Niemeijer and van Leeuwen (1976), Fisher (1983).

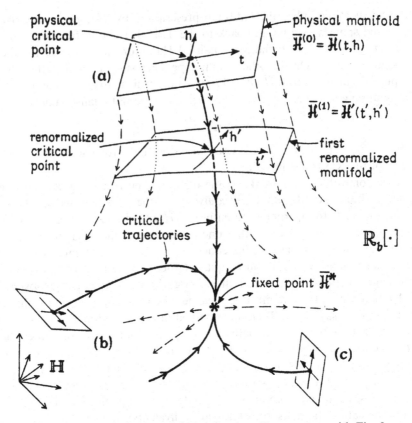

Fig. 5. A depiction of the space of Hamiltonians $\mathbb{H}$ – compare with Fig. 2 – showing initial or physical manifolds [labelled (a), (b), ..., as in Fig. 2] and the flows induced by repeated application of a discrete RG transformation $\mathbb{R}_b$ with a spatial rescaling factor b (or induced by a corresponding continuous or differential RG). Critical trajectories are shown bold: they all terminate, in the region of $\mathbb{H}$ shown here, at a fixed point $\overline{\mathcal{H}}^*$. The full space contains, in general, other nontrivial, critical fixed points, describing multicritical points and distinct critical-point universality classes; in addition, trivial fixed points, including high-temperature 'sinks' with no outflowing or relevant trajectories, typically appear. *Lines of fixed points* and other more complex structures may arise and, indeed, play a crucial role in certain problems. [After Fisher (1983).]

transformations turns out to be an art more than a science: there is no standard recipe! Nevertheless, there are guidelines: the general philosophy enunciated by Wilson and expounded well, for example, in a recent lecture by Shankar treating fermionic systems,[114] is to attempt to *eliminate* first those microscopic variables or degrees of freedom of "least direct importance" to the macroscopic phenomenon under study, while *retaining* those of most importance. In the case of ferromagnetic or gas–liquid critical points, the phenomena of most significance take place on long length scales – the correlation length, ξ, diverges; the critical point correlations, $G_c(\mathbf{r})$, decay slowly at long distances; long-range order sets in below T_c.

Thus in his first, breakthrough articles in 1971, Wilson used an ingenious "phase-space cell" decomposition for continuously variable scalar spins (as against ± 1 Ising

---

[114] See R. Shankar in Cao (1998) and Shankar (1994).

spins) to treat a lattice Landau–Ginzburg model with a general, single-spin or 'on-site' potential $V(s_{\mathbf{x}})$ acting on each spin, $-\infty < s_{\mathbf{x}} < \infty$. Blocks of cells of the smallest spatial extent were averaged over to obtain a single, renormalized cell of twice the linear size (so that $b = 2$). By making sufficiently many simplifying approximations Wilson obtained an explicit *nonlinear, integral recursion relation* that transformed the ℓ-times renormalized potential, $V^{(\ell)}(\cdot)$, into $V^{(\ell+1)}(\cdot)$. This recursion relation could be handled by computer and led to a *specific numerical estimate* for the exponent ν for $d = 3$ dimensions that was *quite different* from the classical value $\frac{1}{2}$ (and from the results of any previously soluble models like the spherical model[115]). On seeing that result, I knew that a major barrier to progress had been overcome!

I returned from a year's sabbatical leave at Stanford University in the summer of 1971, by which time Ken Wilson's two basic papers were in print. Shortly afterwards, in September, again while walking to lunch as I recall, Ken Wilson discussed his latest results from the nonlinear recursion relation with me. Analytical expressions could be obtained by expanding $V^{(\ell)}(s)$ in a power series:

$$V^{(\ell)}(s) = r_\ell s^2 + u_\ell s^4 + v_\ell s^6 + \cdots. \tag{37}$$

If truncated at quadratic order one had a soluble model – the Gaussian model (or free-field theory) – and the recursion relation certainly worked *exactly* for that! But to have a nontrivial model, one had to start not only with r_0 (as, essentially, the temperature variable) but, as a minimum, one also had to include $u_0 > 0$: the model then corresponded to the well-known $\lambda\varphi^4$ field theory. Although one might, thus, initially set $v_0 = w_0 = \cdots = 0$, all these higher-order terms were immediately generated under renormalization; furthermore, there was no reason for u_0 to be small and, for this reason and others, the standard field-theoretic perturbation theories were ineffective.

Now, I had had a long-standing interest in the effects of the spatial dimensionality d on singular behavior in various contexts:[116] so that issue was raised for Ken's recursion relation. Indeed, d appeared simply as an explicit parameter. It then became clear that $d = 4$ was a special case in which the leading-order corrections to the Gaussian model vanished. Furthermore, above $d = 4$ dimensions classical behavior reappeared in a natural way (since the parameters $u_0, v_0, \cdots$ all then became irrelevant). These facts fitted in nicely with the known special role of $d = 4$ in other situations.[117]

For $d = 3$, however, one seemed to need the infinite set of coefficients in Eq. (37) which all coupled together under renormalization. But I suggested that maybe one could treat the dimensional deviation, $\epsilon = 4 - d$, as a small, *nonintegral* parameter in analyzing the recursion relations for $d < 4$. Ken soon showed this was effective! Furthermore, the recursion relations proved to be *exact* to leading order in ϵ (so that if one replaced $b = 2$ by a general value, the expected universal results were, indeed, independent of b). A paper, entitled by Ken, "Critical exponents in 3.99 dimensions" was shortly written, submitted, and published:[118] it contained the first general formula for a nonclassical exponent, namely, $\gamma = 1 + \frac{1}{6}\epsilon + \mathrm{O}(\epsilon^2)$.

---

[115] For accounts of the critical behavior of the spherical model, see Fisher (1966a), where long-range forces were also considered, and, e.g., Stanley (1971), Baxter (1982), and Fisher (1983).
[116] Fisher and Gaunt (1964), Fisher (1966a,b; 1967c; 1972).
[117] See references in the previous footnote and Larkin and Khmel'nitskii (1969), especially Appendix 2.
[118] Wilson and Fisher (1972). The first draft was written by Ken Wilson who graciously listed the authors in alphabetical order.

It transpired, however, that the perturbation parameter ϵ provided more – namely, a systematic way of ordering the infinite set of discrete recursion relations not only for the expansion coefficients of $V^{(\ell)}(s)$ but also for further terms in the appropriate full space $\mathbb{H}$, involving spatial gradients or, equivalently but more usefully, the momenta or wave vectors $\mathbf{q}_i$ labelling the spin variables $\hat{s}_{\mathbf{q}}$, now re-expressed in Fourier space. With that facility in hand, the previous approximations entailed in the phase-space cell analysis could be dispensed with. Wilson then saw that he could precisely implement his *momentum-shell renormalization group*[119] – subsequently one of the most-exploited tools in critical phenomena studies!

In essence this transformation is like decimation[120] except that the division of the variables in Eq. (33) is made in momentum space: for ferromagnetic or gas–liquid type critical points the set $\{\hat{s}_{\mathbf{q}}^<\}$ contains those 'long-wavelength' or 'low-momentum' variables satisfying $|\mathbf{q}| \leq q_\Lambda/b$, where $q_\Lambda = \pi/a$ is the (ultraviolet) momentum cutoff implied by the lattice structure. Conversely, the 'short-wavelength', 'high-momentum' spin components $\{\hat{s}_{\mathbf{q}}^>\}$ having wave vectors lying in the momentum-space *shell*, $q_\Lambda/b < |\mathbf{q}| \leq q_\Lambda$, are integrated out. The spatial rescaling now takes the form

$$\mathbf{q} \Rightarrow \mathbf{q}' = b\mathbf{q}, \tag{38}$$

as follows from Eq. (27); but in analogy to $\zeta(b)$ in Eq. (28), a *nontrivial spin rescaling factor* ("multiplicative-wave function renormalization" in QFT) is introduced via

$$\hat{s}_{\mathbf{q}} \Rightarrow \hat{s}'_{\mathbf{q}'} = \hat{s}_{\mathbf{q}}/\hat{c}[b, \overline{\mathcal{H}}]. \tag{39}$$

The crucially important rescaling factor $\hat{c}$ takes the form $b^{d-\omega}$ and must be *tuned* in the critical region of interest [which leads to $\omega = \frac{1}{2}(d - 2 + \eta)$: compare with Eq. (4)]. It is also worth mentioning that by letting $b \to 1+$, one can derive a *differential* or continuous flow RG and rewrite the recursion relation Eq. (34) as[121]

$$\frac{\mathrm{d}}{\mathrm{d}\ell}\overline{\mathcal{H}} = \mathbb{B}[\overline{\mathcal{H}}]. \tag{40}$$

Such continuous flows are illustrated in Figs. 4 and 5. (If it happens that $\overline{\mathcal{H}}$ can be represented, in general only approximately, by a single coupling constant, say, g, then $\mathbb{B}$ reduces to the so-called beta-function $\beta(g)$ of QFT.)

For deriving ϵ expansions on the basis of the momentum shell RG, Feynman-graph perturbative techniques as developed for QFT prove very effective.[122] They enter basically because one can take $u_0 = O(\epsilon)$ small and they play a role both in efficiently

---

[119] See Wilson and Fisher (1972) Eq. (18) and the related text.

[120] A considerably more general form of RG transformation can be written as

$$\exp(\overline{\mathcal{H}}'[s']) = \mathrm{Tr}_N^s\{\mathcal{R}_{N',N}(s'; s)\exp(\overline{\mathcal{H}}[s])\},$$

where the trace is taken over the full set of original spins s. The $N' = N/b^d$ renormalized spins $\{s'\}$ are introduced via the RG kernel $\mathcal{R}_{N',N}(s'; s)$ which incorporates spatial and spin rescalings, etc., and which should satisfy a trace condition to ensure the partition-function-preserving property (see Footnote 111) which leads to the crucial free-energy flow equation: see Eq. (43) below. The decimation transformation, the momentum-shell RG, and other transformations can be written in this form.

[121] See Wilson (1971a) and Footnote 112 above: in this form the RG semigroup can typically be extended to an Abelian group (MacLane and Birkhoff, 1967); but as already stressed, this fact plays a negligible role.

[122] See Wilson (1972), Brézin, Wallace, and Wilson (1972), Wilson and Kogut (1974), the reviews Brézin, Le Guillou, and Zinn-Justin (1976), and Wallace (1976), and the texts Amit (1978) and Itzykson and Drouffe (1989).

organizing the calculation and in performing the essential integrals (particularly for systems with simple propagators and vertices).[123] Capitalizing on his field-theoretic expertise, Wilson obtained, in only a few weeks after submitting the first article, *exact expansions* for the exponents ν, γ, and ϕ to order ϵ^2 (and, by scaling, for all other exponents).[124] Furthermore, the anomalous dimension – defined in Eq. (2) at the beginning of our story – was calculated exactly to order ϵ^3: I cannot resist displaying this striking result, namely,

$$\eta = \frac{(n+2)}{2(n+8)^2}\epsilon^2 + \frac{(n+2)}{2(n+8)^2}\left[\frac{6(3n+14)}{(n+8)^2} - \frac{1}{4}\right]\epsilon^3 + \mathrm{O}(\epsilon^4), \tag{41}$$

where the symmetry parameter n denotes the number of components of the microscopic spin vectors, $\vec{s}_\mathbf{x} \equiv (s_\mathbf{x}^\mu)_{\mu=1,\cdots,n}$, so that one has just $n = 1$ for Ising spins.[125] Over the years these expansions have been extended to order ϵ^5 (and ϵ^6 for η)[126] and many further related expansions have been developed.[127]

12 Flows, fixed points, universality and scaling

To complete my story – and to fill in a few logical gaps over which we have jumped – I should explain how Wilson's construction of RG transformations in the space $\mathbb{H}$ enables RG theory to accomplish the "tasks" set out above in Section 8. As illustrated in Fig. 5, the recursive application of an RG transformation $\mathbb{R}_b$ induces a *flow* in the space of Hamiltonians, $\mathbb{H}$. Then one observes that "sensible," "reasonable," or, better, "well-designed" RG transformations are *smooth*, so that points in the original physical manifold, $\overline{\mathcal{H}}^{(0)} = \overline{\mathcal{H}}(t,h)$, that are close, say in temperature, remain so in $\overline{\mathcal{H}}^{(1)} \equiv \overline{\mathcal{H}}'$, i.e., under renormalization, and likewise as the flow parameter ℓ increases, in $\overline{\mathcal{H}}^{(\ell)}$. Notice, incidentally, that since the spatial scale renormalizes via $\mathbf{x} \Rightarrow \mathbf{x}' = b^\ell \mathbf{x}$ one may regard

$$\ell = \log_b(|\mathbf{x}'|/|\mathbf{x}|) \tag{42}$$

as measuring, logarithmically, the scale on which the system is being described – recall the physical *scale-dependence of parameters* discussed in Section 4; but note that, in general, the *form* of the Hamiltonian is also changing as the "scale" is changed or ℓ increases. Thus a partially renormalized Hamiltonian can be expected to take on a more-or-less generic, mesoscopic form: hence it represents an appropriate candidate to give meaning to a Landau-Ginzburg or, now, LGW effective Hamiltonian: recall the discussion of Landau's work in Section 2.

Thanks to the smoothness of the RG transformation, if one knows the free energy $f_\ell \equiv f[\overline{\mathcal{H}}^{(\ell)}]$ at the ℓ-th stage of renormalization, then one knows the *original* free

---

[123] Nevertheless, many more complex situations arise in condensed matter physics for which the formal application of graphical techniques without an adequate understanding of the appropriate RG structure can lead one seriously astray.

[124] See Wilson (1972) which was received on 1 December 1971 while Wilson and Fisher (1972) carries a receipt date of 11 October 1971.

[125] See, e.g., Fisher (1967b, 1974b, 1983), Kadanoff *et al.* (1967), Stanley (1971), Aharony (1976), Patashinskii and Pokrovskii (1979).

[126] See Gorishny, Larin, and Tkachov (1984) but note that the O(ϵ^5) polynomials in n are found accurately but some coefficients are known only within uncertainties.

[127] Recall Footnote 35.

energy $f[\overline{\mathcal{H}}]$ and its critical behavior: explicitly one has[128]

$$f(t, h, \cdots) \equiv f[\overline{\mathcal{H}}] = b^{-d\ell} f[\overline{\mathcal{H}}^{(\ell)}] \equiv b^{-d\ell} f_\ell(t^{(\ell)}, h^{(\ell)}, \cdots). \tag{43}$$

Furthermore, the smoothness implies that all the universal critical properties are preserved under renormalization. Similarly one finds[129] that the critical point of $\overline{\mathcal{H}}^{(0)} \equiv \overline{\mathcal{H}}$ maps onto that of $\overline{\mathcal{H}}^{(1)} \equiv \overline{\mathcal{H}}'$, and so on, as illustrated by the bold flow lines in Fig. 5. Thus it is instructive to follow the *critical trajectories* in $\mathbb{H}$, i.e., those RG flow lines that emanate from a physical critical point. In principle, the topology of these trajectories could be enormously complicated and even chaotic: in practice, however, for a well-designed or "apt" RG transformation, one most frequently finds that the critical flows terminate – or, more accurately, come to an asymptotic halt – at a *fixed point* $\overline{\mathcal{H}}^*$, of the RG: see Fig. 5. Such a fixed point is defined, via Eqs. (34) or (40), simply by

$$\mathbb{R}_b[\overline{\mathcal{H}}^*] = \overline{\mathcal{H}}^* \quad \text{or} \quad \mathbb{B}[\overline{\mathcal{H}}^*] = 0. \tag{44}$$

One then searches for fixed-point solutions: the role of the fixed-point equation is, indeed, roughly similar to that of Schrödinger's equation $\mathcal{H}\Psi = E\Psi$, for stationary states Ψ_k of energy E_k in quantum mechanics.

Why are the fixed points so important? Some, in fact, are not, being merely trivial, corresponding to *no interactions* or to *all spins frozen*, etc. But the *non*trivial fixed points represent critical states; furthermore, the nature of their criticality, and of the free energy in their neighborhood, must, as explained, be *identical* to that of all those distinct Hamiltonians whose critical trajectories converge to the same fixed point! In other words, a particular fixed point *defines a universality class* of critical behavior which "governs," or "attracts," all those systems whose critical points eventually map onto it: see Fig. 5.

Here, then, we at last have the natural explanation of *universality*: systems of quite different physical character may, nevertheless, belong to the domain of attraction of the *same* fixed point $\overline{\mathcal{H}}^*$ in $\mathbb{H}$. The distinct sets of inflowing trajectories reflect their varying physical content of associated irrelevant variables and the corresponding nonuniversal rates of approach to the asymptotic power laws dictated by $\overline{\mathcal{H}}^*$: see Eq. (22).

From each critical fixed point, there flow at least two "unstable" or outgoing trajectories. These correspond to one or more *relevant* variables, specifically, for the case illustrated in Fig. 5, to the temperature or thermal field, t, and the magnetic or ordering field, h. See also Fig. 4. If there are further relevant trajectories then, as discussed in Section 7, one can expect *crossover* to different critical behavior. In the space $\mathbb{H}$, such trajectories will then typically lead to distinct fixed points describing (in general) completely new universality classes.[130]

But what about *power laws* and *scaling*? The answer to this question was already sketched in Section 8; but we will recapitulate here, giving a few more technical

[128] Recall the partition-function preserving property set out in Footnote 111 above which actually implies the basic relation Eq. (43).

[129] See Wilson (1971a), Wilson and Kogut (1974), and Fisher (1983).

[130] A skeptical reader may ask: "But what if no fixed points are found?" This can well mean, as it has frequently meant in the past, simply that the chosen RG transformation was poorly designed or "not apt." On the other hand, a fixed point represents only the simplest kind of asymptotic flow behavior: other types of asymptotic flow may well be identified and translated into physical terms. Indeed, near certain types of trivial fixed point, such procedures, long ago indicated by Wilson (1971a, Wilson and Kogut, 1974), *must* be implemented: see, e.g., D.S. Fisher and Huse (1985).

details. However, trusting readers or those uninterested in the analysis are urged to *skip to the next section*!

That said, one must start by noting that the smoothness of a well-designed RG transformation means that it can always be expanded locally – to at least some degree – in a Taylor series.[131] It is worth stressing that it is this very property that fails for free energies in a critical region: to regain this ability, the large space of Hamiltonians is crucial. Near a fixed point satisfying Eq. (43) we can, therefore, rather generally expect to be able to *linearize* by writing

$$\mathbb{R}_b[\overline{\mathcal{H}}^* + g\mathcal{Q}] = \overline{\mathcal{H}}^* + g\mathbb{L}_b\mathcal{Q} + \mathrm{o}(g) \tag{45}$$

as $g \to 0$, or in differential form,

$$\frac{\mathrm{d}}{\mathrm{d}\ell}(\overline{\mathcal{H}}^* + g\mathcal{Q}) = g\mathbb{B}_1\mathcal{Q} + \mathrm{o}(g). \tag{46}$$

Now $\mathbb{L}_b$ and $\mathbb{B}_1$ are *linear operators* (albeit acting in a large space $\mathbb{H}$). As such we can seek eigenvalues and corresponding "eigenoperators", say $\mathcal{Q}_k$ (which will be "partial Hamiltonians"). Thus, in parallel to quantum mechanics, we may write

$$\mathbb{L}_b\mathcal{Q}_k = \Lambda_k(b)\mathcal{Q}_k \quad \text{or} \quad \mathbb{B}_1\mathcal{Q}_k = \lambda_k\mathcal{Q}_k, \tag{47}$$

where, in fact (by the semigroup property), the eigenvalues must be related by $\Lambda_k(b) = b^{\lambda_k}$. As in any such linear problem, knowing the spectrum of eigenvalues and eigenoperators or, at least, its dominant parts, tells one much of what one needs to know. Reasonably, the $\mathcal{Q}_k$ should form a basis for a general expansion

$$\overline{\mathcal{H}} \cong \overline{\mathcal{H}}^* + \sum_{k \geq 1} g_k \mathcal{Q}_k. \tag{48}$$

Physically, the expansion coefficient $g_k(\equiv g_k^{(0)})$ then represents the thermodynamic field[132] conjugate to the "critical operator" $\mathcal{Q}_k$ which, in turn, will often be close to some combination of *local* operators. Indeed, in a characteristic critical-point problem one finds two *relevant operators*, say $\mathcal{Q}_1$ and $\mathcal{Q}_2$ with $\lambda_1, \lambda_2 > 0$. Invariably, one of these operators can, say by its symmetry, be identified with the local energy density, $\mathcal{Q}_1 \cong \mathcal{E}$, so that $g_1 \cong t$ is the thermal field; the second then characterizes the order parameter, $\mathcal{Q}_2 \cong \Psi$ with field $g_2 \cong h$. Under renormalization each g_k varies simply as $g_k^{(\ell)} \approx b^{\lambda_k \ell} g_k^{(0)}$.

Finally,[133] one examines the flow equation (43) for the free energy. The essential point is that the degree of renormalization, b^ℓ, can be *chosen* as large as one wishes. When $t \to 0$, i.e., in the critical region which it is our aim to understand, a good choice proves to be $b^\ell = 1/|t|^{1/\lambda_1}$, which clearly diverges to ∞. One then finds that Eq. (43) leads to the *basic scaling relation* Eq. (19) which we will rewrite here in greater generality as

$$f_s(t, h, \cdots, g_j, \cdots) \approx |t|^{2-\alpha} \mathcal{F}\left(\frac{h}{|t|^\Delta}, \cdots, \frac{g_j}{|t|^{\phi_j}}, \cdots\right). \tag{49}$$

This is the essential result: recall, for example, that it leads to the "collapse" of equation-of-state data as described in Section 6.

---

[131] See Wilson (1971a), Wilson and Kogut (1974), Fisher (1974b), Wegner (1972, 1976), Kadanoff (1976).
[132] Reduced, as always, by the factor $1/k_\mathrm{B}T$: see, e.g., Eq. (18).
[133] See references in Footnote 131.

126   *Michael E. Fisher*

Now, however, the critical exponents can be expressed directly in terms of the RG eigenexponents λ_k (for the fixed point in question). Specifically one finds

$$2 - \alpha = \frac{d}{\lambda_1}, \qquad \Delta = \frac{\lambda_2}{\lambda_1}, \qquad \phi_j = \frac{\lambda_j}{\lambda_1}, \quad \text{and} \quad \nu = \frac{1}{\lambda_1}. \tag{50}$$

Then, as already explained in Sections 6 and 7, the sign of a given ϕ_j and, hence, of the corresponding λ_j determines the *relevance* (for $\lambda_j > 0$), *marginality* (for $\lambda_j = 0$), or *irrelevance* (for $\lambda_j < 0$) of the corresponding critical operator $\mathcal{Q}_j$ (or "perturbation") and of its conjugate field g_j: this field might, but for most values of j will *not*, be under direct experimental control. As explained previously, all exponent relations (15), (20), etc., follow from scaling, while the first and last of the equations (50) yield the *hyperscaling relation* Eq. (32).

When there are no marginal variables and the least negative ϕ_j is larger than unity in magnitude, a simple scaling description will usually work well and the Kadanoff picture almost applies. When there are *no* relevant variables and only one or a few *marginal variables*, field-theoretic perturbative techniques of the Gell-Mann-Low (1954), Callan-Symanzik[134] or so-called "parquet diagram" varieties[135] may well suffice (assuming the dominating fixed point is sufficiently simple to be well understood). There may then be little incentive for specifically invoking general RG theory. This seems, more or less, to be the current situation in QFT and it applies also in certain condensed matter problems.[136]

13 Conclusions

My tale is now told: following Wilson's 1971 papers and the introduction of the ϵ-expansion in 1972 the significance of the renormalization group approach in statistical mechanics was soon widely recognized[137] and exploited by many authors interested in critical and multicritical phenomena and in other problems in the broad area of condensed matter physics, physical chemistry, and beyond. Some of these successes have already been mentioned in order to emphasize, in particular, those features of the

---

[134] See Wilson (1975), Brézin *et al.* (1976), Amit (1978), Itzykson and Drouffe (1989).

[135] Larkin and Khmel'nitskii (1969).

[136] See, e.g., the case of dipolar Ising-type ferromagnets in $d = 3$ dimensions investigated experimentally by Ahlers, Kornblit, and Guggenheim (1975) following theoretical work by Larkin and Khmel'nitskii (1969) and Aharony (see 1976, Sec. 4E).

[137] Footnote 86 drew attention to the first international conference on critical phenomena organized by Melville S. Green and held in Washington in April 1965. Eight years later, in late May 1973, Mel Green, with an organizing committee of J.D. Gunton, L.P. Kadanoff, K. Kawasaki, K.G. Wilson, and the author, mounted another conference to review the progress in theory in the previous decade. The meeting was held in a Temple University Conference Center in rural Pennsylvania. The proceedings (Gunton and Green, 1974) entitled *Renormalization Group in Critical Phenomena and Quantum Field Theory*, are now mainly of historical interest. The discussions were recorded in full but most papers only in abstract or outline form. Whereas in the 1965 conference the overwhelming number of talks concerned experiments, now only J.M.H. (Anneke) Levelt Sengers and Guenter Ahlers spoke to review experimental findings in the light of theory. Theoretical talks were presented, in order, by P.C. Martin, Wilson, Fisher, Kadanoff, B.I. Halperin, E. Abrahams, Niemeijer (with van Leeuwen), Wegner, Green, Suzuki, Fisher and Wegner (again), E.K. Riedel, D.J. Bergman (with Y. Imry and D. Amit), M. Wortis, Symanzik, Di Castro, Wilson (again), G. Mack, G. Dell-Antonio, J. Zinn-Justin, G. Parisi, E. Brézin, P.C. Hohenberg (with Halperin and S.-K. Ma) and A. Aharony. Sadly, there were no participants from the Soviet Union but others included R. Abe, G.A. Baker, Jr., T. Burkhardt, R.B. Griffiths, T. Lubensky, D.R. Nelson, E. Siggia, H.E. Stanley, D. Stauffer, M.J. Stephen, B. Widom and A. Zee. As the lists of names and participants illustrates, many active young theorists had been attracted to the area, had made significant contributions, and were to make more in subsequent years.

full RG theory that are of general significance in the wide range of problems lying beyond the confines of quantum field theory and fundamental high-energy physics. But to review those developments would go beyond the mandate of this Colloquium.[138]

A further issue is the relevance of renormalization group concepts to quantum field theory. I have addressed that only in various peripheral ways. Insofar as I am by no means an expert in quantum field theory, that is not inappropriate; but perhaps one may step back a moment and look at QFT from the general philosophical perspective of understanding complex, interacting systems. Then, I would claim, statistical mechanics is a central science of great intellectual significance – as just one reminder, the concepts of "spin-glasses" and the theoretical and computational methods developed to analyze them (such as "simulated annealing") have proved of interest in physiology for the study of neuronal networks and in operations research for solving hard combinatorial problems. In that view, even if one focuses only on the physical sciences, the land of statistical physics is broad, with many dales, hills, valleys and peaks to explore that are of relevance to the real world and to our ways of thinking about it. Within that land there is an island, surrounded by water: I will not say "by a moat" since, these days, more and broader bridges happily span the waters and communicate with the mainland! That island is devoted to what was "particle physics" and is now "high-energy physics" or, more generally, to the deepest lying and, in that sense, the "most fundamental" aspects of physics. The reigning theory on the island is quantum field theory – the magnificent set of ideas and techniques that inspired the symposium[139] that led to this Colloquium. Those laboring on the island have built most impressive skyscrapers reaching to the heavens!

Nevertheless, from the global viewpoint of statistical physics – where many degrees of freedom, the ever-present fluctuations, and the diverse spatial and temporal scales pose the central problems – quantum field theory may be regarded as describing a rather special set of statistical mechanical models. As regards applications they have been largely restricted to $d = 4$ spatial dimensions [more physically, of course, to $(3 + 1)$ dimensions] although in the last decade *string theory* has dramatically changed that! The practitioners of QFT insist on the pre-eminence of some pretty special symmetry groups, the Poincaré group, $SU(3)$, and so on, which are not all so "natural" at first sight – even though the role of gauge theories as a unifying theme in modeling nature has been particularly impressive. But, of course, we know these special features of QFT are not matters of choice – rather, they are forced on us by our explorations of Nature itself. Indeed, as far as we know at present, there is only one high-energy physics; whereas, by contrast, the ingenuity of chemists, of materials scientists, and of Life itself, offers a much broader, multifaceted and varied panorama of systems to explore both conceptually and in the laboratory.

From this global standpoint, renormalization group theory represents a theoretical tool of depth and power. It first flowered luxuriantly in condensed matter physics, especially in the study of critical phenomena. But it is ubiquitous because of its

---

[138] Some reviews already mentioned that illustrate applications are Fisher (1974b), Wilson (1975), Wallace (1976), Aharony (1976), Patashinskii and Pokrovskii (1979), Nelson (1983), and Creswick *et al.* (1992). Beyond these, attention should be drawn to the notable article by Hohenberg and Halperin (1977) that reviews dynamic critical phenomena, and to many articles on further topics in the Domb and Lebowitz series *Phase Transitions and Critical Phenomena*, Vols. **7** and beyond (Academic Press Inc., London, 1983–).

[139] See Cao (1998).

potential for linking physical behavior across disparate scales; its ideas and techniques play a vital role in those cases where the fluctuations on many different physical scales truly interact. But it provides a valuable perspective – through concepts such as "relevance," "marginality" and "irrelevance" – even when scales are well separated! One can reasonably debate how vital renormalization group concepts are for quantum field theory itself. Certain aspects of the full theory do seem important because Nature teaches us, and particle physicists have learned, that quantum field theory is, indeed, one of those theories in which the different scales are connected together in nontrivial ways via the intrinsic quantum-mechanical fluctuations. However, in current quantum field theory, only certain facets of renormalization group theory play a pivotal role.[140] High-energy physics did not have to be the way it is! But, even if it were quite different, we would still need renormalization group theory in its fullest generality in condensed matter physics and, one suspects, in further scientific endeavors in the future.

Acknowledgments

I thank Professor Alfred I. Tauber, Director, and, especially, Professor Tian Yu Cao of the Center for Philosophy and History of Science at Boston University for their roles in arranging and organizing the symposium on the foundations of quantum field theory and for inviting me to participate. Daniel M. Zuckerman kindly read and commented on the draft manuscript (based on my lecture at the symposium) and assisted with checking and completing the Bibliography; Caricia Claremont patiently typed and retyped the manuscript. Comments from Stephen G. Brush, N.D. Mermin, R. Shankar, David J. Wallace, B. Widom, and Kenneth G. Wilson have been appreciated. Much of the scientific work by the author and his associates described or cited in this review has been supported by the National Science Foundation: for that support,* now entering its 32nd year, I am truly grateful.

Selected bibliography

The reader is cautioned that this article is not intended as a systematic review of renormalization group theory and its origins. Likewise, this bibliography makes no claims of completeness; however, it includes those contributions of most significance in the personal view of the author. The reviews of critical phenomena and RG theory cited in footnote 4 above contain many additional references. Further review articles appear in the series *Phase Transitions and Critical Phenomena*, edited by C. Domb and M.S. Green (later replaced by J.L. Lebowitz) and published by Academic Press Inc., London (1972–): some are listed below. Introductory accounts in an informal lecture style are presented in Fisher (1965, 1983).

Abe, R., 1972, "Expansion of a critical exponent in inverse powers of spin dimensionality," *Prog. Theoret. Phys.* **48**, 1414–15.

---

[140] It is interesting to look back and read in Gunton and Green (1973) pp. 157–160, Wilson's thoughts in May 1973 regarding the "Field theoretic implications of the renormalization group" at a point just before the ideas of *asymptotic freedom* became clarified for non-Abelian gauge theory by Gross and Wilczek (1973) and Politzer (1973).

* Most recently under grant CHE 96-14495.

Abe, R., 1973, "Expansion of a critical exponent in inverse powers of spin dimensionality," *Prog. Theoret. Phys.* **49**, 113–28.

Aharony, A., 1973, "Critical Behavior of Magnets with Dipolar Interactions. V. Uniaxial Magnets in *d* Dimensions," *Phys. Rev.* B **8**, 3363–70; 1974, erratum ibid. **9**, 3946.

Aharony, A., 1976, "Dependence of universal critical behavior on symmetry and range of interaction," in *Phase Transitions and Critical Phenomena*, edited by C. Domb and M.S. Green (Academic Press Inc., London), Vol. **6**, pp. 357–424.

Ahlers, G., A. Kornblit and H.J. Guggenheim, 1975, "Logarithmic corrections to the Landau specific heat near the Curie temperature of the dipolar Ising ferromagnet LiTbF$_4$," *Phys. Rev. Lett.* **34**, 1227–30.

Amit, D.J., 1978, *Field Theory, the Renormalization Group and Critical Phenomena* (McGraw-Hill Inc., London); see also 1993, the expanded Revised 2nd Edition (World Scientific, Singapore).

Bagnuls, C. and C. Bervillier, 1997, "Field-theoretic techniques in the study of critical phenomena," *J. Phys. Studies* **1**, 366–82.

Baker, G.A., Jr., 1961, "Application of the Padé approximant method to the investigation of some magnetic properties of the Ising model," *Phys. Rev.* **124**, 768–74.

Baker, G.A., Jr., 1990, *Quantitative Theory of Critical Phenomena* (Academic, San Diego).

Baker, G.A., Jr. and N. Kawashima, 1995, "Renormalized coupling constant for the three-dimensional Ising model," *Phys. Rev. Lett.* **75**, 994–7.

Baker, G.A., Jr. and N. Kawashima, 1996, "Reply to Comment by J.-K. Kim," *Phys. Rev. Lett.* **76**, 2403.

Baker, G.A., Jr., B.G. Nickel, M.S. Green, and D.I. Meiron, 1976, "Ising model critical indices in three dimensions from the Callan–Symanzik equation," *Phys. Rev. Lett.* **36**, 1351–4.

Baker, G.A., Jr., B.G. Nickel, and D.I. Meiron, 1978, "Critical indices from perturbation analysis of the Callan–Symanzik equation," *Phys. Rev.* B **17**, 1365–74.

Baxter, R.J., 1982, *Exactly Solved Models in Statistical Mechanics* (Academic Press Inc., London).

Benfatto, G. and G. Gallavotti, 1995, *Renormalization Group*, Physics Notes, edited by P.W. Anderson, A.S. Wightman, and S.B. Treiman (Princeton University, Princeton, NJ).

Bogoliubov, N.N. and D.V. Shirkov, 1959, *Introduction to the Theory of Quantized Fields* (Interscience, New York), Chap. VIII.

Brézin, E., J.C. Le Guillou, and J. Zinn-Justin, 1976, "Field theoretical approach to critical phenomena," in *Phase Transitions and Critical Phenomena*, edited by C. Domb and M.S. Green (Academic Press Inc., London), Vol. **6**, pp. 125–247.

Brézin, E., D.J. Wallace, and K.G. Wilson, 1972, "Feynman-graph expansion for the equation of state near the critical point," *Phys. Rev.* B **7**, 232–9.

Buckingham, M.J. and J.D. Gunton, 1969, "Correlations at the Critical Point of the Ising Model," *Phys. Rev.* **178**, 848–53.

Burkhardt, T.W. and J.M.J. van Leeuwen, 1982, Eds., *Real Space Renormalization* (Springer Verlag, Berlin).

Cao, T.Y., 1998, Ed., *The Conceptual Foundations of Quantum Field Theory* (Cambridge University Press, Cambridge).

Chen, J.-H., M.E. Fisher, and B.G. Nickel, 1982, "Unbiased estimation of corrections to scaling by partial differential approximants," *Phys. Rev. Lett.* **48**, 630–34.

Creswick, R.J., H.A. Farach, and C.P. Poole, Jr., 1992, *Introduction to Renormalization Group Methods in Physics* (John Wiley & Sons, Inc., New York).

De Bruijn, N.G., 1958, *Asymptotic Methods in Analysis* (North Holland, Amsterdam).

Di Castro, C. and G. Jona-Lasinio, 1969, "On the microscopic foundation of scaling laws," *Phys. Lett.* **29A**, 322–3.

Di Castro, C. and G. Jona-Lasinio, 1976, "The renormalization group approach to critical phenomena," in *Phase Transitions and Critical Phenomena*, edited by C. Domb and M.S. Green (Academic Press Inc., London), Vol. **6**, pp. 507–58.

Domb, C., 1960, "On the theory of cooperative phenomena in crystals," *Adv. Phys. (Phil. Mag. Suppl.)* **9**, 149–361.

130   *Michael E. Fisher*

Domb, C., 1996, *The Critical Point: A historical introduction to the modern theory of critical phenomena* (Taylor and Francis, London).

Domb, C. and D.L. Hunter, 1965, "On the critical behavior of ferromagnets," *Proc. Phys. Soc.* (London) **86**, 1147–51.

Domb, C. and M.F. Sykes, 1962, "Effect of change of spin on the critical properties of the Ising and Heisenberg models," *Phys. Rev.* **128**, 168–73.

Essam, J.W. and M.E. Fisher, 1963, "Padé approximant studies of the lattice gas and Ising ferromagnet below the critical point," *J. Chem. Phys.* **38**, 802–12.

Fisher, D.S. and D.A. Huse, 1985, "Wetting transitions: a functional renormalization group approach," *Phys. Rev.* B **32**, 247–56.

Fisher, M.E., 1959, "The susceptibility of the plane Ising model," *Physica* **25**, 521–4.

Fisher, M.E., 1962, "On the theory of critical point density fluctuations," *Physica* **28**, 172–80.

Fisher, M.E., 1964, "Correlation functions and the critical region of simple fluids," *J. Math. Phys.* **5**, 944–62.

Fisher, M.E., 1965, "The nature of critical points," in *Lectures in Theoretical Physics*, Vol. VIIC (University of Colorado Press, Boulder) pp. 1–159.

Fisher, M.E., 1966a, "Notes, definitions and formulas for critical point singularities," in M.S. Green and J.V. Sengers, 1966, Eds., cited below.

Fisher, M.E., 1966b, "Shape of a self-avoiding walk or polymer chain," *J. Chem. Phys.* **44**, 616–22.

Fisher, M.E., 1967a, "The theory of condensation and the critical point," *Physics* **3**, 255–83.

Fisher, M.E., 1967b, "The theory of equilibrium critical phenomena," *Repts. Prog. Phys.* **30**, 615–731.

Fisher, M.E., 1967c, "Critical temperatures of anisotropic Ising lattices. II. General upper bounds," *Phys. Rev.* **162**, 480–85.

Fisher, M.E., 1969, "Phase Transitions and Critical Phenomena," in *Contemporary Physics*, Vol. 1, *Proc. International Symp., Trieste, 7–28 June 1968* (International Atomic Energy Agency, Vienna) 19–46.

Fisher, M.E., 1971, "The theory of critical point singularities," in *Critical Phenomena*, *Proc. 1970 Enrico Fermi Internat. Sch. Phys., Course No. 51, Varenna, Italy*, edited by M.S. Green (Academic Press Inc., New York) pp. 1–99.

Fisher, M.E., 1972, "Phase transitions, symmetry and dimensionality," *Essays in Physics*, Vol. **4** (Academic Press Inc., London) pp. 43–89.

Fisher, M.E., 1974a, "General scaling theory for critical points," *Proc. Nobel Symp. XXIV, Aspenäsgården, Lerum, Sweden, June 1973*, in *Collective Properties of Physical Systems*, edited by B. Lundqvist and S. Lundqvist (Academic Press Inc., New York), pp. 16–37.

Fisher, M.E., 1974b, "The renormalization group in the theory of critical phenomena," *Rev. Mod. Phys.* **46**, 597–616.

Fisher, M.E., 1983, "Scaling, universality and renormalization group theory," in *Lecture Notes in Physics*, Vol. 186, *Critical Phenomena*, edited by F.J.W. Hahne (Springer, Berlin), pp. 1–139.

Fisher, M.E. and A. Aharony, 1973, "Dipolar interactions at ferromagnetic critical points," *Phys. Rev. Lett.* **30**, 559–62.

Fisher, M.E. and R.J. Burford, 1967, "Theory of critical point scattering and correlations. I. The Ising model," *Phys. Rev.* **156**, 583–622.

Fisher, M.E. and J.-H. Chen, 1985, "The validity of hyperscaling in three dimensions for scalar spin systems," *J. Physique* (Paris) **46**, 1645–54.

Fisher, M.E. and D.S. Gaunt, 1964, "Ising model and self-avoiding walks on hypercubical lattices and 'high density' expansions," *Phys. Rev.* **133**, A224–39.

Fisher, M.E., S.-K. Ma, and B.G. Nickel, 1972, "Critical exponents for long-range interactions," *Phys. Rev. Lett.* **29**, 917–20.

Fisher, M.E. and P. Pfeuty, 1972, "Critical behavior of the anisotropic *n*-vector model," *Phys. Rev.* B **6**, 1889–91.

Fisher, M.E. and M.F. Sykes, 1959, "Excluded volume problem and the Ising model of ferromagnetism," *Phys. Rev.* **114**, 45–58.

Gawędski, K., 1986, "Rigorous renormalization group at work," *Physica* **140A**, 78–84.

Gell-Mann, M. and F.E. Low, 1954, "Quantum electrodynamics at small distances," *Phys. Rev.* **95**, 1300–12.

Gorishny, S.G., S.A. Larin, and F.V. Tkachov, 1984, "ϵ-Expansion for critical exponents: The $O(\epsilon^5)$ approximation," *Phys. Lett.* **101A**, 120–23.

Green, M.S. and J.V. Sengers, Eds., 1966, *Critical Phenomena: Proceedings of a Conference held in Washington, D.C. April 1965*, N.B.S. Misc. Publ. **273** (U.S. Govt. Printing Off., Washington, 1 December 1966).

Griffiths, R.B. 1965, "Thermodynamic inequality near the critical point for ferromagnets and fluids," *Phys. Rev. Lett.* **14**, 623–4.

Griffiths, R.B., 1972, "Rigorous results and theorems," in *Phase Transitions and Critical Phenomena*, edited by C. Domb and M.S. Green (Academic Press Inc., London) Vol. **1**, pp. 7–109.

Gross, D. and F. Wilczek, 1973, "Ultraviolet behavior of non-Abelian gauge theories," *Phys. Rev. Lett.* **30**, 1343–6.

Guggenheim, E.A., 1949, *Thermodynamics* (North Holland, Amsterdam), Fig. 4.9.

Gunton, J.D. and M.S. Green, Eds., 1974, *Renormalization Group in Critical Phenomena and Quantum Field Theory: Proc. Conference*, held at Chestnut Hill, Pennsylvania, 29–31 May 1973 (Temple University, Philadelphia).

Heller, P. and G.B. Benedek, 1962, "Nuclear magnetic resonance in MnF_2 near the critical point," *Phys. Rev. Lett.* **8**, 428–32.

Hille, E., 1948, *Functional Analysis and Semi-Groups* (American Mathematical Society, New York).

Hohenberg, P.C. and B.I. Halperin, 1977, "Theory of dynamic critical phenomena," *Rev. Mod. Phys.* **49**, 435–79.

Itzykson, D. and J.-M. Drouffe, 1989, *Statistical Field Theory* (Cambridge University Press, Cambridge)

Jeffreys, H. and B.S. Jeffreys, 1956, *Methods of Mathematical Physics*, 3rd Edn. (Cambridge University Press, London).

Kadanoff, L.P., 1966, "Scaling laws for Ising models near T_c," *Physics* **2**, 263–72.

Kadanoff, L.P., 1976, "Scaling, universality and operator algebras," in *Phase Transitions and Critical Phenomena*, edited by C. Domb and M.S. Green (Academic Press Inc.), Vol. **5a**, pp. 1–34.

Kadanoff, L.P. *et al.*, 1967, "Static phenomena near critical points: Theory and experiment," *Rev. Mod. Phys.* **39**, 395–431.

Kadanoff, L.P. and F.J. Wegner, 1971, "Some critical properties of the eight-vertex model," *Phys. Rev.* B **4**, 3989–93.

Kaufman, B. and L. Onsager, 1949, "Crystal statistics. II. Short-range order in a binary lattice," *Phys. Rev.* **76**, 1244–52.

Kiang, C.S. and D. Stauffer, 1970, "Application of Fisher's droplet model for the liquid–gas transition near T_c," *Z. Phys.* **235**, 130–39.

Kim, H.K. and M.H.W. Chan, 1984, "Experimental determination of a two-dimensional liquid–vapor critical-point exponent," *Phys. Rev. Lett.* **53**, 170–73.

Koch, R.H., V. Foglietti, W.J. Gallagher, G. Koren, A. Gupta and M.P.A. Fisher, 1989, "Experimental evidence for vortex-glass superconductivity in Y-Ba-Cu-O," *Phys. Rev. Lett.* **63**, 1511–14.

Kouvel, J.S. and M.E. Fisher, 1964, "Detailed magnetic behavior of nickel near its Curie point," *Phys. Rev.* **136**, A1626–32.

Landau, L.D. and E.M. Lifshitz, 1958, *Statistical Physics*, Vol. 5 of *Course of Theoretical Physics* (Pergamon Press, London) Chap. XIV.

Larkin, A.I. and D.E. Khmel'nitskii, 1969, "Phase transition in uniaxial ferroelectrics," *Zh. Eksp. Teor. Fiz.* **56**, 2087–98: *Soviet Physics – JETP* **29**, 1123–8.

Le Guillou, J.C. and J. Zinn-Justin, 1977, "Critical exponents for the *n*-vector model in three dimensions from field theory," *Phys. Rev. Lett.* **39**, 95–8.

Ma, S.-K., 1976a, *Modern Theory of Critical Phenomena* (W.A. Benjamin, Inc., Reading, Mass.)

Ma, S.-K., 1976b, "Renormalization group by Monte Carlo methods," *Phys. Rev. Lett.* **37**, 461–74.

MacLane, S. and G. Birkhoff, 1967, *Algebra* (The Macmillan Co., New York) Chap. II, Sec. 9, Chap. III, Secs. 1, 2.

Nelson, D.R., 1983, "Defect-mediated phase transitions," in *Phase Transitions and Critical Phenomena*, edited by C. Domb and M.S. Green (Academic Press Inc., London), Vol. 7, pp. 1–99.

Nelson, D.R. and M.E. Fisher, 1975, "Soluble renormalization groups and scaling fields for low-dimensional spin systems," *Ann. Phys. (N.Y.)* **91**, 226–74.

Niemeijer, Th. and J.M.J. van Leeuwen, 1976, "Renormalization theory for Ising-like systems," in *Phase Transitions and Critical Phenomena*, edited by C. Domb and M.S. Green, Vol. **6** (Academic Press Inc., London) pp. 425–505.

Nickel, B.G., 1978, "Evaluation of simple Feynman graphs," *J. Math. Phys.* **19**, 542–8.

Onsager, L., 1944, "Crystal statistics. I. A two-dimensional model with an order–disorder transition," *Phys. Rev.* **62**, 117–49.

Onsager, L., 1949, Discussion remark following a paper by G.S. Rushbrooke at the International Conference on Statistical Mechanics in Florence, *Nuovo Cim. Suppl.* (9) **6**, 261.

Parisi, G., 1973, "Perturbation expansion of the Callan–Symanzik equation," in Gunton and Green (1974), cited above.

Parisi, G., 1974, "Large-momentum behaviour and analyticity in the coupling constant," *Nuovo Cim.* A **21**, 179–86.

Patashinskii, A.Z. and V.L. Pokrovskii, 1966, "Behavior of ordering systems near the transition point," *Zh. Eksper. Teor. Fiz.* **50**, 439–47: *Soviet Phys. – JETP* **23**, 292–7.

Patashinskii, A.Z. and V.L. Pokrovskii, 1979, *Fluctuation Theory of Phase Transitions* (Pergamon Press, Oxford).

Pawley, G.S., R.H. Swendsen, D.J. Wallace, and K.G. Wilson, 1984, "Monte Carlo renormalization group calculations of critical behavior in the simple-cubic Ising model," *Phys. Rev.* B **29**, 4030–40.

Pfeuty, P. and G. Toulouse, 1975, *Introduction au Group des Renormalisation et à ses Applications* (Univ. de Grenoble, Grenoble); translation, *Introduction to the Renormalization Group and to Critical Phenomena* (John Wiley & Sons Inc., London, 1977).

Politzer, H.D., 1973, "Reliable perturbative results for strong interactions?" *Phys. Rev. Lett.* **30**, 1346–9.

Riesz, F. and B. Sz.-Nagy, 1955, *Functional Analysis* (F. Ungar Publ. Co., New York) Chap. X, Secs. 141, 142.

Rowlinson, J.S., 1959, *Liquids and Liquid Mixtures* (Butterworths Scientific Publications, London).

Rushbrooke, G.S., 1963, "On the thermodynamics of the critical region for the Ising problem," *J. Chem. Phys.* **39**, 842–3.

Rushbrooke, G.S. and P.J. Wood, 1958, "On the Curie points and high-temperature susceptibilities of Heisenberg model ferromagnetics," *Molec. Phys.* **1**, 257–83.

Schultz, T.D., D.C. Mattis and E.H. Lieb, 1964, "Two-dimensional Ising model as a soluble problem of many fermions," *Rev. Mod. Phys.* **36**, 856–71.

Shankar, R., 1994, "Renormalization-group approach to interacting fermions," *Rev. Mod. Phys.* **66**, 129–92.

Stanley, H.E., 1971, *Introduction to Phase Transitions and Critical Phenomena* (Oxford University Press, New York).

Stell, G., 1968, "Extension of the Ornstein–Zernike theory of the critical region," *Phys. Rev. Lett.* **20**, 533–6.

Suzuki, M., 1972, "Critical exponents and scaling relations for the classical vector model with long-range interactions," *Phys. Lett.* **42A**, 5–6.

Symanzik, K., 1966, "Euclidean quantum field theory. I. Equations for a scalar model," *J. Math. Phys.* **7**, 510–25.

Tarko, H.B. and M.E. Fisher, 1975, "Theory of critical point scattering and correlations. III. The Ising model below T_c and in a field," *Phys. Rev.* B **11**, 1217–53.

Wallace, D.J., 1976, "The ϵ-expansion for exponents and the equation of state in isotropic systems," in *Phase Transitions and Critical Phenomena*, edited by C. Domb and M.S. Green (Academic Press Inc., London), Vol. **6**, pp. 293–356.

Wegner, F.J., 1972a, "Corrections to scaling laws," *Phys. Rev.* B **5**, 4529–36.

Wegner, F.J., 1972b, "Critical exponents in isotropic spin systems," *Phys. Rev.* B **6**, 1891–3.

Wegner, F.J., 1976, "The critical state, general aspects," in *Phase Transitions and Critical Phenomena*, edited by C. Domb and M.S. Green (Academic Press Inc., London), Vol. **6**, pp. 7–124.

Widom, B., 1965a, "Surface tension and molecular correlations near the critical point," *J. Chem. Phys.* **43**, 3892–7.

Widom, B., 1965b, "Equation of state in the neighborhood of the critical point," *J. Chem. Phys.* **43**, 3898–905.

Widom, B. and O.K. Rice, 1955, "Critical isotherm and the equation of state of liquid–vapor systems," *J. Chem. Phys.* **23**, 1250–55.

Wilson, K.G., 1971a, "Renormalization group and critical phenomena. I. Renormalization group and Kadanoff scaling picture," *Phys. Rev.* B **4**, 3174–83.

Wilson, K.G., 1971b, "Renormalization group and critical phenomena. II. Phase-space cell analysis of critical behavior," *Phys. Rev.* B **4**, 3184–205.

Wilson, K.G., 1972, "Feynman-graph expansion for critical exponents," *Phys. Rev. Lett.* **28**, 548–51.

Wilson, K.G., 1975, "The renormalization group: critical phenomena and the Kondo problem," *Rev. Mod. Phys.* **47**, 773–840.

Wilson, K.G., 1983, "The renormalization group and critical phenomena," (1982 Nobel Prize Lecture) *Rev. Mod. Phys.* **55**, 583–600.

Wilson, K.G. and M.E. Fisher, 1972, "Critical exponents in 3.99 dimensions," *Phys. Rev. Lett.* **28**, 240–43.

Wilson, K.G. and J. Kogut, 1974, "The renormalization group and the ϵ expansion," *Phys. Repts.* **12**, 75–200.

Zinn, S.-Y. and M.E. Fisher, 1996, "Universal surface-tension and critical-isotherm amplitude ratios in three dimensions," *Physica* A **226**, 168–80.

Zinn, S.-Y., S.-N. Lai, and M.E. Fisher, 1996, "Renormalized coupling constants and related amplitude ratios for Ising systems," *Phys. Rev.* E **54**, 1176–82.

Appendix A. Asymptotic behavior

In specifying critical behavior (and asymptotic variation more generally) a little more precision than normally used is really called for. Following well-established custom, I use $\simeq$ for "approximately equals" in a rough and ready sense, as in $\pi^2 \simeq 10$. But to express "$f(x)$ varies like x^λ when x is small and positive," i.e., just to specify a critical exponent, I write:

$$f(x) \sim x^\lambda \quad (x \to 0+). \tag{A1}$$

Then the precise implication is

$$\lim_{x \to 0+} [\ln |f(x)| / \ln x] = \lambda; \tag{A2}$$

see Fisher (1967b, Sec. 1.4; 1983, Sec. 2.4). To give more information, specifically a *critical amplitude* like D in Eq. (2), I define $\approx$ as "asymptotically equals" so that

$$f(x) \approx g(x) \quad \text{as} \quad x \to 0+ \tag{A3}$$

implies

$$\lim_{x \to 0+} f(x)/g(x) = 1. \tag{A4}$$

Thus, for example, one has $(1 - \cos x) \approx \frac{1}{2}x^2 \sim x^2$ when $x \to 0$. See Fisher (1967b, Secs. 6.2, 6.3 and 7.2) but note that in Eqs. (6.2.6)–(6.3.5) the symbol $\simeq$ should read $\approx$; note also De Bruijn's (1958) discussion of $\approx$ in his book *Asymptotic Methods in Analysis*. The thoughtless "strong recommendation" (of the American Institute of Physics and American Physical Society) to use $\approx$ as "approximately equals" is to be, and has been strongly decried![141] It may also be remarked that few physicists, indeed, use $\sim$ in the precise mathematical sense originally introduced by Poincaré in his pioneering analysis of asymptotic series: see, e.g., Jeffreys and Jeffreys (1956) Secs. 17·02, 23·082, and 23·083. De Bruijn and the Jeffreys also, of course, define the $O(\cdot)$ symbol which is frequently misused in the physics literature: thus $f = O(g)$ $(x \to 0)$, should mean $|f| < c|g|$ for some constant c and $|x|$ small enough, so that, e.g., $(1 - \cos x) = O(x)$ is correct even though less informative than $(1 - \cos x) = O(x^2)$.

Appendix B. Unitarity of the renormalization transformation

We start with the definition Eq. (33) and recall Eq. (25) to obtain

$$Z_{N'}[\overline{\mathcal{H}}'] \equiv \mathrm{Tr}_{N'}^{s'}\{e^{\overline{\mathcal{H}}[s']}\} = \mathrm{Tr}_{N'}^{s^<}\{e^{\overline{\mathcal{H}}_{\mathrm{eff}}[s^<]}\} = \mathrm{Tr}_{N'}^{s^<}\mathrm{Tr}_{N-N'}^{s^>}\{e^{\overline{\mathcal{H}}[s^< \cup s^>]}\}$$
$$= \mathrm{Tr}_N^s\{e^{\overline{\mathcal{H}}[s]}\} = Z_N[\overline{\mathcal{H}}],$$

from which the free energy $f[\overline{\mathcal{H}}]$ follows via Eq. (26).

Appendix C. Nature of a semigroup

We see from the discussion in the text and from the relation (35), in particular, that successive decimations with scaling factors b_1 and b_2 yield the quite general relation

$$\mathbb{R}_{b_2}\mathbb{R}_{b_1} = \mathbb{R}_{b_2 b_1},$$

which essentially defines a unitary *semigroup* of transformations. See Footnotes 3 and 78 above, and the formal algebraic definition in MacLane and Birkhoff (1967): a unitary semigroup (or 'monoid') is a set M of elements, $u, v, w, x, \cdots$, with a binary operation, $xy = w \in M$, which is associative, so $v(wx) = (vw)x$, and has a unit u, obeying $ux = xu = x$ (for all $x \in M$) – in RG theory, the unit transformation corresponds simply to $b = 1$. Hille (1948) and Riesz and Sz-Nagy (1955) describe semigroups within a continuum, functional analysis context and discuss the existence of an infinitesimal generator when the flow parameter ℓ is defined for continuous values $\ell \geq 0$: see Eq. (40) and Wilson's (1971a) introductory discussion.

Discussions

Jackiw: Neither a question nor an objection, just an addition for those who might be interested in the development of Ken Wilson's ideas. One should take into account that he began as a particle physicist and he was interested in the renormalization group in his thesis. His attempts, always as a particle physicist, were to use the renormalization group to solve some kind of particle physics/field theoretic

[141] More recently an internal AIP/APS memorandum (dated June 1994) states: "*Approximate equality.* $\simeq$ is preferred." Furthermore, this well advised change of policy may soon see the official light of day in newly revised guidelines to be published. (Private communication, March 1998.)

problem. He was forever interested in Gell-Mann-Low's paper. He was also interested in Bogoliubov's presentation of the renormalization group. Something that I don't understand is that in that period, most particle physicists, including the creators of the renormalization group, were indifferent to attempts at using it, especially those described in Bogoliubov's book. However, Wilson kept thinking that he could get something out of it. I don't think he actually succeeded for particle physics, and then he went off into condensed matter physics.

Fisher: I believe that what you say is reasonably close to the truth. However, Wilson certainly did get something out of his path of research! One can reread his Nobel Prize lecture [*Rev. Mod. Phys.* **55**, 583 (1983)]: granting the fact that everybody distorts even their own intellectual history (as the historians of science often demonstrate) one sees that your description of his *quest* is absolutely true. But it is worth emphasizing that Ken Wilson always had in mind the concrete calculational aspects of dealing with field theories. If, in reality, there was a cutoff, so be it! By that line of thought he started chopping up the continuum of degrees of freedom into discrete sets, then trying to remove those on the shortest scales, then rescaling, and so on. Those basic ideas are really very general.

Jackiw: But when I say he got something out of it, or he didn't get anything out of it, I'm thinking of particle physics. I don't think there's actually a particle physics experiment that is explained by his renormalization group ideas. He thought for example that deep-inelastic scaling would show power laws, and it didn't. And that's more or less when he stopped using the renormalization group in particle physics.

Fisher: I agree with that summary. But I feel it is a little too neat and tidy as a final verdict even for particle physics. Certainly, as I stressed, one always starts with certain prejudices and the expectation of power laws was rather natural given then understood facts about critical phenomena. But one of the nice things in condensed matter physics is the range of examples it provides. Let us consider two-dimensional critical systems: there are well-known cases where the asymptotic behavior is *not* described by simple power laws; but, nevertheless, Ken Wilson's ideas work. One can cite Kosterlitz-Thouless systems and the Kondo problems.

Within particle physics, the phenomenon of asymptotic freedom surely also fits the broad renormalization group approach? So I would say Ken Wilson was a little overly pessimistic. But he is a very focused person, and, as you say, power laws *per se* failed in particle physics and he moved on. In some real sense I do not think he should have felt pressed to; but that was his choice.

[Added note: In further response to Professor Jackiw's points, the reader is referred back to Footnote 107 (and associated text) as regards Bogoliubov's presentation of RG theory, and to Footnotes 137 and 139 and the text of the concluding section in connection with the relevance of RG theory to particle physics.]

9. Where does quantum field theory fit into the big picture?

ARTHUR JAFFE

Looking back on foundations

For centuries we have said that mathematics is the language of physics. In 1960 Eugene Wigner wrote the famous article expounding the 'unreasonable effectiveness of mathematics in the natural sciences' [Wig]. He concludes that lecture by stating, '... the miracle of the appropriateness of the language of mathematics for the formulation of the laws of physics is a wonderful gift which we neither understand nor deserve. We should be grateful for it and hope that it will remain valid in the future and that it will extend ... to wide branches of learning.'

Two basic questions have driven mathematical physics. First, what mathematical framework (basically what equations) describes nature? Once we decide on the equations, one can ask the second question: what are the mathematical properties of their solutions? We would like to know both their qualitative and quantitative properties. While it appears reasonable to answer these questions in order, they may be intimately related.

In that vein, let us inquire about the foundations of quantum electrodynamics. In this area of physics we encounter the most accurate quantitative and qualitative predictions about nature known to man. The Maxwell–Dirac equations are well accepted in their individual domains. But combined as a nonlinear system of equations, they lead to agreement between perturbation theoretic rules on the one hand – and to experiment on the other – that startles the imagination. In many parts of physics, accuracy to one part per thousand is standard; in a few areas of physics, measurement to a part per million remains a possibility; but in quantum electrodynamics one finds the possibility of many more orders of magnitude. Physicists have rules to predict and they can measure the anomalous magnetic moment of the electron to eleven decimal places: $(g - 2)/2 = 0.001159652200(40)$.

Yet in spite of these great successes, we do not know if the equations of relativistic quantum electrodynamics make mathematical sense. In fact the situation poses a dilemma. As a consequence of renormalization group theory, most physicists today believe that the equations of quantum electrodynamics in their simple form are inconsistent; in other words, we believe that the equations of electrodynamics have no solution at all! Stated differently, a majority of physicists have the opinion that if one can find a mathematically self-consistent theory of quantum field theory in four space-time dimensions, then it must encompass not only electrodynamics, but also other interactions and other forces, such as non-Abelian gauge theory, or QCD – or else the laws of physics must be posed in terms of a different framework from field theory, such as string theory. Some physicists maintain that only a 'string' theory will give a self-consistent description of nature, with quantum field theory emerging as a limiting case. Clearly this is appealing, but as yet we do not know whether this provides the answer.

In fact, only relatively recently has any progress at all been made on the much more modest question of whether any quantum field theory makes mathematical sense. Some years ago, one did not believe that the simplest examples of nonlinear wave equations were compatible with these two basic highlights of mathematical physics: relativity and quantum theory. A closely related question is whether special relativity is compatible with the Heisenberg commutation relations. But here concrete progress has been made. The question of alternative frameworks is clearly more ambitious. Some physicists believe that only one set of equations (the correct set of equations that describes nature) is mathematically self-consistent. I take a more conservative view. I believe that nature is more flexible, and that hard work will reveal the possibility of several mathematically consistent theories and associated equations, each of which has application in different domains of quantum physics.

Mathematical physics

From the time of Newton to the advent of quantum field theory, almost every well-known contributor to physics also had an impact on mathematics. In fact, borders between these subjects grew up after the development of quantum theory, and theoretical physics moved away from the frontiers of mathematics. Basically problems in quantum physics, especially field theory, seemed too difficult to handle by known mathematical tools. This became evident in the 1940s with the recognition that local singularities were a basic barrier to understanding the equations of physics as differential equations in an ordinary sense. The singularities of the nonlinear equations require 'renormalization'. This heroic route to redefining the constants in the equations arose as a set of rules necessary to understand perturbation theory. There was no real expectation at the time that these rules could be put within a mathematical framework encompassing a complete and self-contained theory of quantum fields – an exact theory, not defined solely in terms of the first few terms of a series expansion in a charge parameter.

In this atmosphere, physics and mathematics gradually drifted apart. Persons recognized in both areas, like Wigner, were the exception not the rule. Mathematical physics, which at one time was synonymous with all of theoretical physics, has since that time become a name attached to a specialized subfield. From the side of physics there was a general antipathy to mathematics and hence to foundations.

To a philosopher, this situation might appear strange; but Freeman Dyson aptly described it in his 1972 lecture, entitled 'Missed Opportunities'. Dyson began, 'I happen to be a physicist who started life as a mathematician. As a working physicist, I am acutely aware of the fact that the marriage between mathematics and physics, which was so enormously fruitful in past centuries, has recently ended in divorce.... I shall examine in detail some examples of missed opportunities, occasions on which mathematicians and physicists lost chances of making discoveries by neglecting to talk to each other' [D]. Dyson went on to describe how he had himself 'missed the opportunity of discovering a deeper connection between modular forms and Lie algebras, just because the number theorist Dyson and the physicist Dyson were not speaking to each other.'

Actually there were already signs about 1955 or 1960, some years earlier than Dyson's lecture, that this divorce between mathematics and physics was headed toward reconciliation. People in Princeton, Zürich, Paris, Göttingen, Rome and some other places were crossing the boundaries. It was during this period that Arthur Wightman and others began the school of axiomatic quantum field theory which has had so much influence.

But in addition to the traditional picture of mathematics as part of physics, the new line of communication (including the incident described by Dyson) included a large component of work which arose from ideas originating in physics that led to new mathematics. This too was an ancient story, as physics has been one of the richest sources for mathematical inspiration. But what was not clear at that time in the 1950s – or in the 1960s and 1970s – was the extent to which this intercourse would dominate the mainstream of mathematics. This has in fact turned out to be the case in the 1980s and 1990s. This approach has achieved sufficient acceptance that we can conclude not only that mathematics is the language of physics, but that in this area, physics has become the language of mathematics. Wigner would certainly be delighted, were he alive today, to learn that just 35 years after his famous lecture we are confronted with attempting to explain what we might call **Wigner's converse**: we marvel today at the *'unreasonable effectiveness of theoretical physics in mathematics.'*

At our meeting there seems to be great trust in the future. I share David Gross's optimism that we are on the road to finding the next level of Olympian beauty. I would add that physics really should not lose sight of the goal to formulate fundamental laws as mathematical statements – in other words, to put the laws into a framework of logic. As we understand physics today, this has not been achieved. However, one amazing aspect of today's research is that practically every branch of abstract mathematics appears relevant to some physicists – especially those who study string theory or related areas.

Constructive quantum field theory

If one believes in quantum fields, then one needs to ask whether the nonlinear equations of quantum field theory have a mathematical basis. For example, does a favorite wave equation, such as

$$\Box\varphi + \frac{\partial V(\varphi)}{\partial\varphi} = 0, \tag{1}$$

have a solution compatible with the Heisenberg condition

$$\left[\varphi(x,t), \frac{\partial\varphi(y,t)}{\partial t}\right] = iZ\delta(x-y), \tag{2}$$

where Z is an appropriate constant.[1] We are using units in which the velocity of light c and Planck's constant $\hbar$ both equal 1. Here

$$\Box = \frac{\partial^2}{\partial t^2} - \nabla^2,$$

---

[1] This is the natural generalization to a field $\varphi(x,t)$ depending continuously on the spatial variable x, of the standard equal-time Heisenberg relations for coordinates $q_k(t)$ indexed by k. In non-relativistic quantum theory, the coordinates q_k and conjugate momenta p_k satisfy $[q_k(t), p_l(t)] = i\delta_{kl}$. We choose the constant Z above to give an appropriate normalization to the scattering matrix determined by φ.

and $V = V(\varphi)$ is a potential function which defines the nonlinearity; we often choose V to be a polynomial.

We also require that φ be compatible with ordinary quantum theory. What this means is that the dependence $\varphi = \varphi(x, t)$ on time is given by an energy operator H for which

$$\varphi(x, t) = e^{itH} \varphi(x, 0)\, e^{-itH}. \tag{3}$$

If $\partial V / \partial \varphi = m^2 \varphi$, then the equation is linear. In that case, solutions describing particles of a mass m are well known, and these fields are also covariant under special relativity. These solutions describe particles moving freely, without interaction, following trajectories with a fixed momentum. Such free field solutions were written down by Wentzel, Dirac, and others in the early days of quantum field theory. On the other hand, if V is anything but a quadratic polynomial, then the ancient problem remains whether there are any fields that satisfy the nonlinear equations, are covariant, and satisfy the Heisenberg relations.

If we choose for V a cubic polynomial, then we are in trouble. In fact such a potential leads to an energy H which may be dominated by the nonlinearity, and which either for large positive φ or large negative φ becomes arbitrarily negative. This gives rise to unphysical states with arbitrarily negative energy densities focused in regions where the potential function itself becomes very negative. In fact, without using the positivity of the energy, and in particular of the potential function V, one can not rule out singularities of the solutions. Even in the classical (unquantized) wave equation this is a disaster and produces singularities in the solution, as was shown in 1957 by J. Keller [K], when the energy is unbounded from below.

Thus the simplest physical example of a nonlinear wave equation is the quartic wave equation for a particle of unit mass,

$$\Box \varphi + \varphi + \lambda \varphi^3 = 0, \tag{4}$$

where $\lambda \geq 0$ is a coupling constant (parameter). Efforts to solve such nonlinear problems went on over many years, and became known in the 1960s as 'constructive quantum field theory' (CQFT). From the point of view of physics, the difficulties arose because the Heisenberg condition (2) is singular. For that reason, the nonlinear term in the equation can only be defined after a 'renormalization' of the function of φ. Likewise, the nonlinear terms in the energy expression appear to be singular, and the renormalization appears to jeopardize the positivity of the energy.

On top of all these difficulties, scaling arguments make it clear that the larger the dimension of space-time, the more singular the nonlinear terms in equation (4) become. Thus the most regular field theories from this point of view appear to arise from interactions that are polynomials of low degree, and in the minimal space-time dimension – namely space-time dimension 2.

Even in two-dimensional space-time, there are certain renormalization divergences in the equations, though in that case physicists believe they understand them completely using the procedure called 'normal ordering' or 'Wick ordering'. On the other hand, this understanding only extends to a treatment of the equations in a perturbative sense, that is, the study of the proposed solution to the equation as a power series expansion in λ. It is well known that such series do not converge [J2], so such a study only makes sense if one knows by other means that a solution exists.

In space-time dimension 3, the problems become more serious. While the perturbation theory again makes sense for this equation (taken order by order), the equation requires both an infinite renormalization of the mass and of the ground state energy.[2] The three-dimensional equation (1) still lies in the so-called *super-renormalizable* category; it does not require renormalization of the coupling constant λ.

The first results for non-explicit examples of a nonlinear, relativistic quantum field were established in space-time dimension 2 [GJ1]. They include solutions to equations of the form (4). Relativistic covariance was demonstrated shortly afterwards [CJ], by constructing a local generator of Lorentz transformations and demonstrating compatibility between this generator and the equations of motion. We soon obtained corresponding global results [GJ3–4, GJS1–2]. Among other things, these yielded an example of the Wightman axioms [GJS1]. In addition, qualitative properties of the examples became within reach. For example, we could show that the particles described by the field had a definite, isolated mass $m = 1 + O(\lambda^2)$ [GJS2].

Some careful thinkers wonder whether the solutions constructed with mathematical precision by workers in this field are really the same examples considered from a heuristic point of view in mainstream physics text books. The straightforward answer yes follows from showing that these solutions have scattering matrix elements $S_{f,i}$ which are exactly the ones computed by the perturbation expansion based on Feynman diagrams and the text-book methods. In fact, for the equation under question, this has been established by Osterwalder and Séneór [OS], who have shown that the perturbation expansion of $S_{f,i}$ as a function of λ is *asymptotic*. This means that for any N, as $\lambda \to 0$,

$$S_{f,i} = \sum_{n=0}^{N} \lambda^n S_{f,i}(n) - O(\lambda^{N+1}). \tag{5}$$

Here $S_{f,i}(n)$ is the perturbation coefficient given by Feynman diagrams, as explained in the standard texts. On the basis of the constructive field theory methods, Dimock, Eckmann, Spencer, and Zirilli have understood low-energy scattering and bound states as well; see [GJ] for references.

Three dimensions and other improvements

Once it was understood that the simplest theories existed, it became clear that in fact they gave complete lower-dimensional examples of all the various axiom systems: Haag–Kastler axioms for algebraic quantum field theory, Wightman axioms for the vacuum expectation values and their general properties, Haag–Ruelle axioms for the theory of single particle states and for the resulting scattering matrix, and Osterwalder–Schrader axioms for relating Euclidean and Minkowski versions of the theories.

---

[2] This three-dimensional case also requires a change of the local unitary equivalence class of representations of the Heisenberg relations. In fact, the representation for the $d = 3$ problem is λ-dependent. We will not underscore this detail, but we do point out that in the two-dimensional problem one has a different result. The representation of the Heisenberg relations for the solution to the nonlinear wave equation, restricted to any bounded, space-time region, *is* locally unitarily equivalent to the representation for the free fields [GJ3]. Technically the local equivalence of representations is known as the 'locally Fock' property. This deep result cannot be extended to establish global unitary equivalence; in fact this negative global result was known for some time by the name 'Haag's theorem'. On the other hand, local unitary equivalence is closely tied to the lack of *wave-function renormalization* for the states in the Hilbert space of the field φ. The absence or presence of wave-function renormalization is the qualitative feature that distinguishes from each other the examples which are, or which are not, locally Fock.

In addition, it became clear that the methods also work in other examples: the first of these were boson–fermion equations in two dimensions, then for the equation (1) in space-time dimension 3, later for Abelian Higgs models coupling scalar and gauge theories in two and three dimensions, and finally for preliminary study of Yang–Mills theory in three and four space-time dimensions.

Once all these models came into being, a whole new realm of questions came within reach. These questions relate to the qualitative properties of the models: for example, do phase transitions occur (because of vacuum degeneracy for certain values of the coupling constants), do bound states occur, are the forces attractive or repulsive? And other questions also came into play: do scaling limits exist, do soliton states influence the structure of the mass spectrum, is there an effective maximum coupling, is there broken symmetry? Then there were more complicated structural questions, such as: what is the nature of the n-particle analysis of scattering amplitudes? Can one gain information about phase transitions? Are symmetries broken? Can one bound critical exponents for the mass, the magnetization, or the susceptibility? And other questions arose about supersymmetry: do examples of field theories with supersymmetry have solutions? Do some of these examples have equations without renormalization, justifying the old-fashioned way to look at field equations? Does supersymmetry get broken in any of these examples? And so forth. Other questions about the detailed nature of the low-energy behavior of the equations were also asked, and some of them were answered. Many, but not all, of these questions (and others) can be related to the wealth of concepts proposed without complete justification in the physics literature. See [GJ, JL1, JL2] for references to results of such investigations.

Phase cell localization

The method we used to tackle the analytic described above goes under the name 'phase cell localization and cluster expansions'. The fundamental question common to all such examples is to what extent different regions of space-time localization are independent. Since the answer to this question also depends on a localization in Fourier modes, we call this an *analysis of phase cell localization*. If different phase cells were exactly independent, then the expectations would factor into products of contributions from each cell. The result would be that correlations would vanish except at coinciding points. This is not the case, but at least in examples for which the smallest mass is strictly positive, we expect that correlations decay exponentially. Analogously, we expect that correlations between different phase cells also decay. In the lower-dimensional models, we can analyze this in configuration space alone, while in cases for which singularities of the Green's functions become more singular on the diagonal, true localization in phase cells becomes necessary.

The natural setting for phase cell localization is the functional integral representation of expectations. We use a representation of the heat kernel as a Wiener integral. In this framework, operators of quantum (field) theory become represented as classical integrals (in the case of bosons) or integrals over a Grassmann algebra (in the case of fermions). In order to implement the proof of decoupling, we devised a method based on an iterative construction: integrate out those degrees of freedom localized in particular large momentum phase cells, leaving an effective interaction with resulting convergence factors associated with the remaining degrees of freedom [GJ3–5].

This procedure can be organized inductively, and it defines an expansion. Thereby we define a set of rules to expand a given term into a sum of terms, with each step, or each collection of steps, giving rise to a small factor. These factors result in the convergence of the expansion. Certain steps in the phase cell expansion involve resummation of terms, so the expansion itself is recursively defined.

The basic ideas involved in our phase cell expansion overlap in some ways the ideas involved in the renormalization group analysis developed by Kadanoff and Wilson. On the one hand, the phase cell expansion has the flexibility of being *inductive*; in contrast, the renormalization group methods have the simplicity of being *iterative*, and of giving a simple differential equation to summarize the renormalization group flow. Both methods were developed about the same time. Except by some experts in constructive quantum field theory, not much attention was focused on the phase cell localization method; even within this community knowledge of the method is limited. The reason is that the phase cell localization method cannot be applied blindly as a 'black box': rather, the use of this method requires familiarity with its every detail.

In spite of its complexity, the results obtained by this method include many of the first constructive results in two and three dimensions, as well as certain results on the long distance behavior in lattice models. In the recent past, the methods used to attack the constructive field theory and lattice statistical mechanics problems have been a combination of phase cell localization method with the renormalization group method. Combining the best features of both methods allows the greatest flexibility.

One of the first uses of phase cell localization methods in constructive quantum field theory to demonstrate localization arose in the paper [GJ3] to analyze the locally Fock property as a consequence of a finite vacuum energy shift per unit volume. Another early result was the paper [GJ4] to demonstrate that the local perturbation of the Hamiltonian by the field in a fixed region leads to a bounded perturbation of the energy. For example, let $E(f)$ denote the ground state energy of the Hamiltonian $H(f) = H + \varphi(f)$. Here H is chosen to have ground state energy equal to 0, and $\varphi(f) = \int \varphi(x)f(x)\,dx$. We show that

$$|E(f)| \leq \int |f(x)|\,dx.$$

Similarly, if f_1 and f_2 vanish outside sets separated by distance d, then

$$|E(f_1 + f_2) - E(f_1) - E(f_2)| \leq o(1),$$

where o(1) decays sufficiently rapidly as $d \to \infty$.

This problem provided an elementary example on which to develop the method of phase cell localization (in space). It actually became the testing ground for the development of the proof that the Hamiltonian for a three-dimensional, φ^4 quantum field is bounded from below [GJ5]. This in turn was the first step to establishing the construction of a fully-fledged nonlinear example of quantum field theory in three-dimensional space-time [FO].

In order to analyze the phase cell localization for the φ^4 field theory, we used phase cell localization in small space-time cubes Δ. We choose cubes that have sufficiently small volume $|\Delta|$ that the most divergent renormalization (the vacuum energy shift) is finite. In this problem, this requires a volume $|\Delta| \leq \kappa^{-1}$, where κ denotes the momentum scale. Phase cell localization is possible in a cube Δ whose length ℓ satisfies $\ell \times \kappa \gg 1$, so these two conditions are compatible. In this framework, the definition of the average (low momentum) field in Δ is

$$\bar{\varphi} = \frac{1}{|\Delta|} \int_\Delta \varphi(x) \, dx,$$

and the fluctuation (high momentum) field in Δ is

$$\delta\varphi(x) = \varphi(x) - \bar{\varphi}.$$

The correlations of the high momentum parts of φ decay with the separation d,

$$|\langle f(\delta\varphi(x))g(\delta\varphi(y))\rangle - \langle f(\delta\varphi(x))\rangle\langle g(\delta\varphi(y))\rangle| \le o(1),$$

as $|x - y| \to \infty$. By establishing sufficiently rapid vanishing of free correlations, as a function of the scaled distance $d = \text{length} \times \kappa$, we can use this decay to establish independence and decay of correlations in the interacting problem. The paper [GJ5] in which we study the decoupling of different phase cells for this problem is a highlight of the development of constructive quantum field theory. It also represents the first incorporation of the Symanzik–Nelson approach to Euclidean quantum field theory [Sy, N] into the proof of renormalization estimates and the non-perturbative solution of an 'ultraviolet' problem in quantum field theory.

The constructive frontier

I recall an experience almost 30 years ago when explaining some early work in constructive quantum field theory at the Centre for Theoretical Physics in Trieste. Abdus Salam complimented me on the results. But then he went on to explain that he thought these questions were so difficult that by the time one could get to the physically interesting field theories, it would probably be past the year 2000! At that time, the turn of the century seemed too distant for a young man to contemplate or worry about.

However, now that we are practically at the millennium 2000, perhaps we should take a look at what has been achieved. These problems are truly difficult, and one cannot fault the rate of progress by suggesting that there is a shortage of new ideas or techniques. In fact I believe that the originality, the depth and the richness of the new methods, as well as the nature of the results that have been achieved, hold their own when compared with any other subfield either in mathematics or in theoretical physics. In fact, we do appear to have in hand all the tools needed to establish existence of a four-dimensional $SU(n)$ Yang–Mills theory (at least on a compact space-time). Balaban has extended earlier constructive methods to deal with lattice gauge theories, giving a powerful treatment of effective actions and gauge fixing [B]. A compact space-time avoids the question of how to establish 'confinement'.[3]

---

[3] Confinement (or the existence of a gap in the mass spectrum) is an important feature required to prove the existence of the infinite volume limit, using the cluster expansion methods for weakly coupled constructive quantum field theory. In gauge theories the problem is that the basic model has mass zero, and confinement is generated by a non-perturbative phenomenon. Feynman [F] gave a picture of confinement resulting from the curvature of the space of gauge potentials, and this general picture was pursued by Singer who interprets the curvature of the space of gauge potentials (connections) as giving rise to a mass. This can also be interpreted, as pointed out by Jackiw and others, as regarding the gap as arising from the A^4 gauge potential in a non-Abelian Yang–Mills theory. But this idea by itself is insufficient. The gauge potential has flat directions, and a detailed analysis of the competing effects is required to recover the expected mass which vanishes exponentially fast as the gauge coupling tends to zero. There is work in some simple quantum mechanical models by Simon [Si] which indicates that this idea may have some basis. However, at present, we are far from finding even a rough outline of an argument for a mathematical proof of this effect in a quantum gauge theory. In fact, we do not even have an existence proof for a multi-component, scalar quantum field theory with zero bare mass and with a potential with flat directions. This is certainly a good problem to work on, and would provide a good first step toward understanding these issues!

(And another approach by Feldman, Magnen, Sénéor, and Rivasseau has given partial results through a method of non-gauge covariant regularization.)

Leaving confinement aside, I hope that some careful, young and very strong mathematical physicist will make the Herculean effort necessary to look again at the existence of Yang–Mills theory. Certainly there is a great prize to be achieved to the single-minded person who ties this up![4]

Non-commutative geometry

One fascinating new direction for physics appears based on a revision of the nature of space itself. Traditionally we write down the equations of physics as a differential equation for a quantum object on a classical space. For example, the wave equation for an operator-valued quantum field $\varphi(x, t)$ (an operator-valued field) is posed on a classical space-time (x, t). With discovery of Connes' form of non-commutative geometry [C], and the discovery of a number of examples of quantum spaces, it becomes natural to build the quantum concepts into the space-time itself. For instance, the quantum unit disc is generated by commutation relations

$$[z, \bar{z}] = h(I - z\bar{z})(I - \bar{z}z),$$

where h is a quantization parameter [KL]. The quantum disc has the same $SU(1, 1)$ symmetry as the classical unit disc: if (a, b, c, d) denote the entries of an $SU(1, 1)$ matrix, then $w = (az + b)(cz + d)^{-1}$ and its adjoint $\bar{w}$ generate the same algebra as z and $\bar{z}$. In this case, the classical symmetry is preserved by quantization. This is a common situation, but sometimes extremely subtle to understand. Beginning efforts to describe gravity in a non-commutative space have also been proposed [DF].

Today there are many examples of quantum spaces, including spaces with quantum groups – as well as classical groups – as symmetries. Other quantum spaces have fractional dimension. It would be of great interest to define an interacting quantum field theory on a quantum space.

One classical field theory in this direction is Connes' picture of the standard model, based on a simple non-commutative space, namely two copies of an ordinary space-time separated by a discrete distance which has the physical interpretation of the Higgs particle mass. This is described beautifully in [C].

In non-commutative geometry, functions $f(x)$ on a space are generalized to operators a in an algebra $\mathfrak{U}$ of operators. The derivative da of a function a is given by the commutator $[Q, a] = Qa - aQ$ of a with a fundamental operator Q. Supersymmetry fits very naturally into this picture, for Q can be interpreted as the generator of supersymmetry. There is a natural operator γ with the property $\gamma = \gamma^* = \gamma^{-1}$ and such that $\gamma Q \gamma = -Q$. With $a^\gamma = \gamma a \gamma$, we have

$$(da)^\gamma = -da^\gamma.$$

The commutator here $[Q, a]$ is actually a *graded* commutator $da = Qa - a^\gamma Q$. Let us suppose that $U(g)$ is a unitary group of symmetries of Q, and that $\text{Tr}(e^{-Q^2}) < \infty$. Then for invariant a which is also an operator root of unity, $a^2 = I$,

---

[4] And such a person must be single-minded, not to be sidetracked by all the other interesting questions which one can now study!

the formula

$$3^Q(a;g) = \frac{1}{\sqrt{\pi}} \int_{-\infty}^{\infty} e^{-t^2} \mathrm{Tr}(\gamma U(g) a\, e^{-Q^2 + it\, da}) \, dt$$

defines a geometric invariant [J3]. The proof of this fact arises from studying the mathematical framework of equivariant, entire cyclic cohomology. In the case of $N = 2$ supersymmetry, this invariant displays an interesting *hidden SL(2, Z) symmetry*, a symmetry ordinarily associated only with conformally invariant structures. Consider the case of a two-dimensional Wess–Zumino model with a quasi-homogeneous, polynomial superpotential $V(z)$ with weights ω_j and satisfying certain elliptic estimates. This is a fundamentally non-conformal theory, where $U(g)$ is a symmetry of one component Q_1 of the space-time supercharge. In this case, taking $U(g) = e^{i\tau Q^2 + i\theta J}$, where J generates a rotation in the space of the two Q^2's, it turns out that for the case $a = I$, the invariant 3 can be evaluated by a homotopy argument, combined with a computation at a singular endpoint. The answer is $3 = \prod_j \vartheta(\tau, (1 - \omega_j)\theta) / \vartheta(\tau, \omega_j \theta)$. This result [Wit, KYY, J4] has a modular form, transforming under $SL(2, Z)$ with weight $c = \sum_j (1 - 2\omega_j)$. In fact this same structure appears in the exactly soluble models of rational conformal field theory with central charge c. Understanding such a hidden symmetry, which is not apparent in the underlying Lagrangian, is a challenge. In non-Abelian gauge theories, Vafa and Witten have identified fascinating potential duality symmetries [VW] which also relate to quantum geometry.

The future

It is difficult to predict the future, and at present the number of fruitful directions for research appears very large. However, I believe that major developments in the mathematical approach to quantum physics will take place in areas related to our discussion above – as well as in many others we are not now considering!

1 Quantum fields

I believe that the present methods contain the tools which make it possible to complete the construction of a nonlinear Yang–Mills field theory in four dimensions. Hopefully this will be carried out, at least for a Yang–Mills field theory on a compact space-time.

2 New mathematics

I believe that the influence of ideas from both classical and quantum physics will remain at the forefront of mathematical developments for a long time. We will witness fundamental advances in analysis, as well as in geometry where one will understand the new symmetries from physics, including mirror symmetry and duality, as well as gaining a deeper understanding of modular forms. Thus we should feel even more at home in the future with Wigner's converse.

3 New physics

There will be an increasing role for the geometric content of quantum physics, an area which has already revolutionized our thinking in the recent past. I believe that new variations on our picture of string physics will emerge, and that they will incorporate in a central way mathematical ideas of non-commutative geometry.

References

[B]   Tadeusz Balaban, Renormalization group approach to lattice gauge field theories. I. Generation of effective actions in a small field approximation and a coupling constant renormalization in 4D, *Comm. Math. Phys.* **109** (1987) 249–401.

[C]   Alain Connes, *Non-Commutative Geometry*, Academic Press 1994.

[CJ]   John Cannon and Arthur Jaffe, Lorentz convariance of the $\lambda(\varphi^4)_2$ quantum field theory, *Comm. Math. Phys.* **17** (1970) 261–321.

[D]   Freeman J. Dyson, Missed opportunities, *Bull. Am. Math. Soc.* **78** (1972) 635–52.

[DF]   Sergio Doplicher, Klaus Fredenhagen and John Roberts, The quantum structure of spacetime at the Planck scale and quantum fields, *Comm. Math. Phys.* **172** (1995) 187–220.

[F]   Richard Feynman, The quantitative behavior of Yang–Mills theory in $2+1$ dimensions, *Nucl. Phys. B* **188** (1981) 479–512.

[FO]   Joel Feldman and Konrad Osterwalder, The Wightman axioms and the mass gap for weakly coupled $(\Phi^4)_3$ quantum field theories, *Ann. Phys.* **97** (1976) 80–135.

[GJ]   James Glimm and Arthur Jaffe, *Quantum Physics*, Springer Verlag 1981, 1987.

[GJ1]   James Glimm and Arthur Jaffe, A $\lambda(\varphi^4)_2$ quantum field theory without cutoffs. I, *Phys. Rev.* **176** (1968) 1945–51.

[GJ3]   James Glimm and Arthur Jaffe, The $\lambda(\varphi^4)_2$ quantum field theory without cutoffs. III: The physical vacuum, *Acta Math.* **125** (1970) 204–67.

[GJ4]   James Glimm and Arthur Jaffe, The $\lambda(\varphi^4)_2$ quantum field theory without cutoffs. IV: Perturbations of the Hamiltonian, *J. Math. Phys.* **13** (1972) 1568–84.

[GJ5]   James Glimm and Arthur Jaffe, Positivity of the φ_3^4 Hamiltonian, *Fortschr. Phys.* **21** (1973) 327–76. Reprinted in *Constructive Quantum Field Theory, Selected Papers*, Birkhäuser, Boston 1985.

[GJS1] James Glimm. Arthur Jaffe and Thomas Spencer, The Wightman axioms and particle structure in the weakly coupled $P(\varphi)^2$ quantum field model, *Ann. Math.* **100** (1974) 585–632.

[GJS2] James Glimm, Arthur Jaffe and Thomas Spencer, Phase transitions for φ^4 quantum fields, *Comm. Math. Phys.* **45** (1975) 203–16.

[J1]   Arthur Jaffe, Existence theorems for a cut-off $\lambda\varphi^4$ field theory. In: *Conference on the Mathematical Theory of Elementary Particles*, R. Goodman and I. Segal, eds., MIT Press 1966.

[J2]   Arthur Jaffe, Divergence of perturbation theory for bosons, *Comm. Math. Phys.* **1** (1965) 127–49.

[J3]   Arthur Jaffe, Quantum harmonic analysis and geometric invariants, *Adv. Math.* to appear; and Quantum invariants, *Comm. Math. Phys.* to appear.

[J4]   Arthur Jaffe, Hidden symmetry; and On the elliptic genus, in preparation.

[JL1]   Arthur Jaffe and Andrzej Lesniewski, Supersymmetric field theory and infinite dimensional analysis. In: *Nonperturbative Quantum Field Theory, Proceedings of the 1987 Cargèse Summer School*, G. 't Hooft *et al.*, eds., Plenum Press 1988.

[JL2]   Arthur Jaffe and Andrzej Lesniewski, Geometry of supersymmetry. In: *Constructive Quantum Field Theory II*, A. Wightman and G. Velo, eds., Plenum Press, New York 1990.

[K]   Joseph Keller, *Comm. Pure Appl. Math.* **10** (1957) 523–30.

[KL]   Slawomir Klimek and Andrzej Lesniewski, Quantum Riemann surfaces, I. The unit disc, *Comm. Math. Phys.* **146** (1992) 103–22.

[KYY]   Toshiya Kawai, Yasuhiko Yamada and Sung-Kil Yang, Elliptic genera and $N=2$ superconformal field theory, *Nucl. Phys. B* **414** (1994) 191–212.

[N]   Edward Nelson, Quantum fields and Markoff fields. *Proc. Sympos. Pure Math.*,
        Univ. California, Berkeley, California 1971, Vol. XXIII, pp. 413–20, Am. Math.
        Soc. RI 1973.

[OS]   Konrad Osterwalder and Roland Sénéor, A nontrivial scattering matrix for weakly
        coupled $P(\varphi)_2$ models, *Helv. Phys. Acta* **49** (1976) 525–35.

[Si]   Barry Simon, Some quantum operators with discrete spectrum but classically
        continuous spectrum, *Ann. Phys.* **146** (1983) 209–20.

[Sy]   Kurt Symanzik, Euclidean quantum field theory. In: *Local Quantum Theory*,
        pp. 152–226, R. Jost, ed., Academic Press, New York 1969.

[VW]   Cumrun Vafa and Edward Witten, Dual string pairs with $N = 1$ and $N = 2$
        supersymmetry in four dimensions: S-duality and mirror symmetry, *Nucl. Phys. B
        Proc. Suppl.* **46** (1996) 225–47.

[Wig]   Eugene P. Wigner, The unreasonable effectiveness of mathematics in the natural
        sciences, *Comm. Pure Appl. Math.* **13** (1960) 1–14. Reprinted in *Mathematics and
        Science*, 291–306, World Sci. Publishing, Teaneck NJ 1990.

[Wit]   Edward Witten, On the Landau–Ginzburg description of $N = 2$ superconformal
        field theory, *Int. J. Mod. Phys.* **A9** (1994) 4783.

10. The unreasonable effectiveness of quantum field theory

Quantum field theory offers physicists a tremendously wide range of application: it is a language with which a vast variety of physical processes can be discussed and also provides a model for fundamental physics, the so-called 'standard model', which thus far has passed every experimental test. No other framework exists in which one can calculate so many phenomena with such ease and accuracy. Nevertheless, today some physicists have doubts about quantum field theory, and here I want to examine these reservations. So let me first review the successes.

Field theory has been applied over a remarkably broad energy range and whenever detailed calculations are feasible and justified, numerical agreement with experiment extends to many significant figures. Arising from a mathematical account of the propagation of fluids (both 'ponderable' and 'imponderable'), field theory emerged over a hundred years ago in the description within classical physics of electromagnetism and gravity. [1] Thus its first use was at macroscopic energies and distances, with notable successes in explaining pre-existing data (relationship between electricity and magnetism, planetary perihelion precession) and predicting new effects (electromagnetic waves, gravitational bending of light). Schrödinger's wave mechanics became a bridge between classical and quantum field theory: the quantum mechanical wave function is also a local field, which when 'second' quantized gives rise to a true quantum field theory, albeit a non-relativistic one. This theory for atomic and chemical processes works phenomenally well at electron-volt energy scales or at distances of $O(10^{-5}$ cm$)$. Its predictions, which do not include relativistic and radiative effects, are completely verified by experiment. For example the ground state energy of helium is computed to seven significant figures; the experimental value is determined with six figure accuracy; disagreement, which is seen in the last place, disappears once radiative and relativistic corrections are included. Precisely in order to incorporate relativity and radiation, quantum field theory of electromagnetism was developed, and successfully applied before World War II to absorption or emission of real photons. Calculation of virtual photon processes followed the war, after renormalization theory succeeded in hiding the infinities that appear in the formalism. [2] Here accuracy of calculation is achieved at the level of one part in 10^8 (as in the magnetic moment of the electron, whose measured value agrees completely with theory), where distances of $O(10^{-13}$ cm$)$ are being probed. Further development, culminating in the standard particle physics model, followed after further infinities that afflict theories that have massive vector mesons were tamed. Indeed the masses of the

* This work is supported in part by funds provided by the U.S. Department of Energy under contract #DE-FC02-94ER40818.

vector mesons mediating weak interactions (by now unified with electromagnetism) were accurately predicted, at a scale of 100 GeV.

I have summarized briefly quantum field theoretic successes within elementary particle physics. It is also necessary to mention the equally impressive analyses in condensed matter physics, where many fascinating effects (spontaneous symmetry breaking, both in the Goldstone–Nambu and Anderson–Higgs modes, quantum solitons, fractional charge, etc.) are described in a field theoretic language, which then also informs elementary particle theory, providing crucial mechanisms used in the latter's model building. This exchange of ideas demonstrates vividly the vitality and flexibility of field theory. Finally we note that quantum field theory has been extrapolated from its terrestrial origins to cosmic scales of distance and energy, where it fuels 'inflation' – a speculative, but completely physical analysis of early universe cosmology, which also appears to be consistent with available data.

With this record of accomplishment, why are there doubts about quantum field theory, and why is there vigorous movement to replace it with string theory, to take the most recent instance of a proposed alternative? Several reasons are discerned. Firstly, no model is complete – for example the standard particle physics model requires *ad hoc* inputs, and does not encompass gravitational interactions. Also, intermittently, calculational difficulties are encountered, and this is discouraging. But these shortcomings would not undermine faith in the ultimate efficacy of quantum field theory were it not for the weightiest obstacle: the occurrence of divergences when the formalism is developed into a computation of physical processes.

Quantum field theoretic divergences arise in several ways. First of all, there is the lack of convergence of the perturbation series, which at best is an asymptotic series. This phenomenon, already seen in quantum mechanical examples such as the anharmonic oscillator, is a shortcoming of an approximation method and I shall not consider it further.

More disturbing are the infinities that are present in every perturbative term beyond the first. These divergences occur after integrating or summing over intermediate states – a necessary calculational step in every non-trivial perturbative order. When this integration/summation is expressed in terms of an energy variable, an infinity can arise either from the infrared (low energy) domain or from the ultraviolet (high energy) domain, or both.

The former, infrared infinity afflicts theories with massless fields and is a consequence of various idealizations for the physical situation: taking the region of space-time which one is studying to be infinite, and supposing that massless particles can be detected with infinitely precise energy–momentum resolution, are physically unattainable goals and lead in consequent calculations to the aforementioned infrared divergences. In quantum electrodynamics one can show that physically realizable experimental situations are described within the theory by infrared-finite quantities. Admittedly, thus far we have not understood completely the infrared structure in the non-Abelian generalization of quantum electrodynamics – this generalization is an ingredient of the standard model – but we believe that no physical instabilities lurk there either. So the consensus is that infrared divergences do not arise from any intrinsic defect of the theory, but rather from illegitimate attempts at forcing the theory to address unphysical questions.

Finally, we must confront the high energy, ultraviolet infinities. These do appear to be intrinsic to quantum field theory, and no physical consideration can circumvent them: unlike the infrared divergences, ultraviolet ones cannot be excused away. But they can be 'renormalized'. This procedure allows the infinities to be sidestepped or hidden, and succeeds in unambiguously extracting numerical predictions from the standard model and from other 'physical' quantum field theories, with the exception of Einstein's gravity theory – general relativity – which thus far remains 'non-renormalizable'.

The apparently necessary presence of ultraviolet infinities has dismayed many who remain unimpressed by the pragmatism of renormalization: Dirac and Schwinger, who count among the creators of quantum field theory and renormalization theory, respectively, ultimately rejected their constructs because of the infinities. But even among those who accept renormalization, there is disagreement about its ultimate efficacy at defining a theory well. Some argue that sense can be made only of 'asymptotically free' renormalizable field theories (in these theories the interaction strength decreases with increasing energy). On the contrary, it is claimed that asymptotically non-free models, like electrodynamics and ϕ^4-theory, do not define quantum theories, even though they are renormalizable – it is said that they 'do not exist'. Yet electrodynamics is the most precisely verified quantum field theory, while the ϕ^4-model is a necessary component to the standard model, which thus far has met no experimental contradiction.

The ultraviolet infinities appear as a consequence of space-time localization of interactions, which occur at a point, rather than spread over a region. (Sometimes it is claimed that field theoretic infinities arise from the unhappy union of quantum theory with special relativity. But this does not describe all cases – later I shall discuss a non-relativistic, ultraviolet-divergent and renormalizable field theory.) Therefore choosing models with non-local interactions provides a way to avoid ultraviolet infinities. The first to take this route was Heisenberg, but his model was not phenomenologically viable. These days in string theory non-locality is built-in at the start, so that all quantum effects – including gravitational ones – are ultraviolet finite, but this has been achieved at the expense of calculability: unlike ultraviolet-divergent local quantum field theory, finite string theory has enormous difficulty in predicting definite physical effects, even though it has succeeded in reproducing previously mysterious results of quantum field theory – I have in mind the quantities associated with black-hole radiance.

My goal in this talk is to persuade you that the divergences of quantum field theory must not be viewed as unmitigated defects; on the contrary, they convey crucially important information about the physical situation, without which most of our theories would not be physically acceptable. The stage where such considerations play a role is that of symmetry, symmetry breaking and conserved quantum numbers, so next I have to explain these ideas.

Physicists are mostly agreed that the ultimate laws of Nature enjoy a high degree of symmetry. By this I mean that the formulation of these laws, be it in mathematical terms or perhaps in other accurate descriptions, is unchanged when various transformations are performed. Presence of symmetry implies absence of complicated and irrelevant structure, and our conviction that this is fundamentally true reflects an ancient aesthetic prejudice – physicists are happy in the belief that Nature in its fundamental workings is essentially simple. Moreover, there are practical consequences of the simplicity entailed by symmetry: it is easier to understand the predictions of physical laws. For example, working out the details of very-many-body motion is beyond the reach of actual calculations, even with the help of computers. But taking into account the symmetries that are present allows one to understand at least some aspects of the motion, and to chart regularities within it.

Symmetries bring with them conservation laws – an association that is precisely formulated by Noether's theorem. Thus time-translation symmetry, which states that physical laws do not change as time passes, ensures energy conservation; space-translation symmetry – the statement that physical laws take the same form at different spatial locations – ensures momentum conservation. For another example, we note that quantal description makes use of complex numbers (involving $\sqrt{-1}$). But physical quantities are real, so complex phases can be changed at will, without affecting physical content. This invariance against phase redefinition, called **gauge symmetry**, leads to charge conservation. The above exemplify a general fact: symmetries are linked to constants of motion. Identifying such constants, on the one hand, satisfies our urge to find regularity and permanence in natural phenomena, and on the other hand, provides us with a useful index for ordering physical data.

However, in spite of our preference that descriptions of Nature be enhanced by a large amount of symmetry and characterized by many conservation laws, it must be recognized that actual physical phenomena rarely exhibit overwhelming regularity. Therefore, at the very same time that we construct a physical theory with intrinsic symmetry, we must find a way to break the symmetry in physical consequences of the model. Progress in physics can be frequently seen as the resolution of this tension.

In classical physics, the principal mechanism for symmetry breaking, realized already within Newtonian mechanics, is through boundary and initial conditions on dynamical equations of motion – for example, radially symmetric dynamics for planetary motion allows radially non-symmetric, non-circular orbits with appropriate initial conditions. But this mode of symmetry breaking still permits symmetric configurations – circular orbits, which are rotationally symmetric, are allowed. In quantum mechanics, which anyway does not need initial conditions to make physical predictions, we must find mechanisms that prohibit symmetric configurations altogether.

In the simplest, most direct approach to symmetry breaking, we suppose that in fact dynamical laws are not symmetric, but that the asymmetric effects are 'small' and can be ignored 'in the first approximation'. Familiar examples are the breaking of rotational symmetry in atoms by an external electromagnetic field or of isospin symmetry by the small electromagnetic interaction. However, this explicit breaking of symmetry is without fundamental interest for the exact and complete theory; we need more intrinsic mechanisms that work for theories that actually are symmetric.

A more subtle idea is **spontaneous symmetry breaking**, where the dynamical laws are symmetric, but only asymmetric configurations are actually realized, because the symmetric ones are energetically unstable. This mechanism, urged upon us by Heisenberg, Anderson, Nambu and Goldstone, is readily illustrated by the potential energy profile possessing left–right symmetry and depicted in the Figure. The left–right symmetric value at the origin is a point of unstable equilibrium; stable equilibrium is attained at the reflection unsymmetric points $\pm a$. Once the system settles in one location or the other, left–right parity is absent. One says that the symmetry of the equations of motion is 'spontaneously' broken by the stable solution.

But here we come to the first instance where infinities play a crucial role. The above discussion of the asymmetric solution is appropriate to a classical physics description, where a physical state minimizes energy and is uniquely realized by one configuration

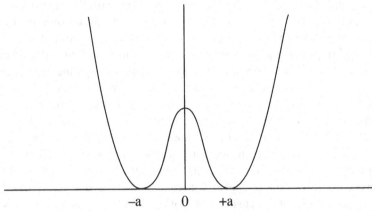

Fig. 1. Left–right symmetric particle energy or field theoretic energy density. The symmetric point at 0 is energetically unstable. Stable configurations are at $\pm a$, and a quantum mechanical particle can tunnel between them. In field theory, the energy barrier is infinite, tunneling is suppressed, the system settles into state $+a$ or $-a$ and left–right symmetry is spontaneously broken.

or the other at $\pm a$. However, quantum mechanically a physical state can comprise a superposition of classical states, where the necessity of superposing arises from quantum mechanical tunneling, which allows mixing between classical configurations. Therefore if the profile in the Figure describes the potential energy of a single quantum particle as a function of particle position, the barrier between the two minima carries finite energy. The particle can then tunnel between the two configurations of $\pm a$, and the lowest quantum state is a superposition, which in the end respects the left–right symmetry. Spontaneous symmetry breaking does not occur in quantum particle mechanics. However, in a field theory, the graph in the Figure describes spatial energy density as a function of the field, and the total energy barrier is the finite amount seen in the Figure, multiplied by the infinite spatial volume in which the field theory is defined. Therefore the total energy barrier is infinite, and tunneling is impossible. Thus spontaneous symmetry breaking can occur in quantum field theory, and Weinberg as well as Salam employed this mechanism for breaking unwanted symmetries in the standard model. But we see that this crucial ingredient of our present-day theory for fundamental processes is available to us precisely because of the infinite volume of space, which is also responsible for infrared divergences!

But infrared problems are not so significant, so let me focus on the ultraviolet infinities. These are important for a further, even more subtle mode of symmetry breaking, which also is crucial for the phenomenological success of our theories. This mode is called **anomalous** or **quantum mechanical** symmetry breaking, and in order to explain it, let me begin with a reminder that the quantum revolution did not erase our reliance on the earlier, classical physics. Indeed, when proposing a theory, we begin with classical concepts and construct models according to the rules of classical, pre-quantum physics. We know, however, such classical reasoning is not in accordance with quantum reality. Therefore, the classical model is reanalyzed by the rules of quantum physics, which comprise the true laws of Nature. This two-step procedure is called *quantization*.

Differences between the physical pictures drawn by a classical description and a quantum description are of course profound. To mention the most dramatic, we recall that dynamical quantities are described in quantum mechanics by operators, which need not commute. Nevertheless, one expects that some universal concepts transcend the classical/quantal dichotomy, and enjoy rather the same role in quantum physics as in classical physics.

For a long time it was believed that symmetries and conservation laws of a theory are not affected by the transition from classical to quantum rules. For example if a model possesses translation and gauge invariance on the classical level, and consequently energy/momentum and charge are conserved classically, it was believed that after quantization the quantum model is still translation- and gauge-invariant so that the energy/momentum and charge operators are conserved within quantum mechanics, that is, they commute with the quantum Hamiltonian operator. But now we know that in general this need not be so. Upon quantization, some symmetries of classical physics may disappear when the quantum theory is properly defined in the presence of its infinities. Such tenuous symmetries are said to be **anomalously** broken; although present classically, they are absent from the quantum version of the theory, unless the model is carefully arranged to avoid this effect.

The nomenclature is misleading. At its discovery, the phenomenon was unexpected and dubbed 'anomalous'. By now the surprise has worn off, and the better name today is 'quantum mechanical' symmetry breaking.

Anomalously or quantum mechanically broken symmetries play several and crucial roles in our present-day physical theories. In some instances they save a model from possessing too much symmetry, which would not be in accord with experiment. In other instances, the desire to preserve a symmetry in the quantum theory places strong constraints on model building and gives experimentally verifiable predictions; more about this later. [3]

Now I shall describe two specific examples of the anomaly phenomenon. Consider first massless fermions moving in an electromagnetic field background. Massive, spin-$\frac{1}{2}$ fermions possess two spin states – up and down – but massless fermions can exist with only one spin state, called a **helicity** state, in which spin is projected along the direction of motion. So the massless fermions with which we are here concerned carry only one helicity and these are an ingredient in present-day theories of quarks and leptons. Moreover, they also arise in condensed matter physics, not because one is dealing with massless, single-helicity particles, but because a well-formulated approximation to various many-body Hamiltonians can result in a first order matrix equation that is identical to the equation for single-helicity massless fermions, i.e., a massless Dirac–Weyl equation for a spinor Ψ.

If we view the spinor field Ψ as an ordinary mathematical function, we recognize that it possesses a complex phase, which can be altered without changing the physical content of the equation that Ψ obeys. We expect therefore that this instance of gauge invariance implies charge conservation. However, in a quantum field theory Ψ is a quantized field operator, and one finds that in fact the charge operator Q is not conserved; rather

$$\frac{\mathrm{d}Q}{\mathrm{d}t} = \frac{i}{\hbar}[H, Q] \propto \int_{\text{volume}} \mathbf{E} \cdot \mathbf{B} \tag{1}$$

where **E** and **B** are the background electric and magnetic fields in which our massless fermion is moving; gauge invariance is lost!

One way to understand this breaking of symmetry is to observe that our model deals with massless fermions, and conservation of charge for **single-helicity** fermions makes sense only if there are no fermion masses. But quantum field theory is beset by its ultraviolet infinities that must be controlled in order to do a computation. This is accomplished by regularization and renormalization, which introduces mass scales for the fermions, and we see that the symmetry is anomalously broken by the ultra-violet infinities of the theory.

The phase invariance of single-helicity fermions is called **chiral symmetry**, and this has many important roles in the standard model, which involves many kinds of fermion fields, corresponding to the various quarks and leptons. In those channels where a gauge vector meson couples to the fermions, chiral symmetry must be main-tained to ensure gauge invariance. Consequently fermion content must be carefully adjusted so that the anomaly disappears. This is achieved, because the proportionality constant in the failed conservation law (1) involves a sum over all the fermion charges, $\sum_n q_n$, so if that quantity vanishes the anomaly is absent. In the standard model the sum indeed vanishes, separately for each of the three fermion families. For a single family this works out as follows:

$$
\begin{array}{lll}
\text{three quarks} & q_n = \tfrac{2}{3} \Rightarrow & 2 \\
\text{three quarks} & q_n = -\tfrac{1}{3} \Rightarrow & -1 \\
\text{one lepton} & q_n = -1 \Rightarrow & -1 \\
\text{one lepton} & q_n = 0 \Rightarrow & 0 \\
& \sum_n q_n = & 0
\end{array}
$$

In channels to which no gauge vector meson couples, there is no requirement that the anomaly vanish, and this is fortunate: a theoretical analysis shows that gauge invariance in the up–down quark channel prohibits the two-photon decay of the neutral pion (which is composed of up and down quarks). But the decay does occur with the invariant decay amplitude of $(0.025 \pm 0.001)(\text{GeV})^{-1}$. Before anoma-lous symmetry breaking was understood, this decay could not be fitted into the standard model, which seemed to possess the decay-forbidding chiral symmetry. Once it was realized that the relevant chiral symmetry is anomalously broken, this obstacle to phenomenological viability of the standard model was removed. Indeed, since the anomaly is completely known, the decay amplitude can be comple-tely calculated (in the approximation that the pion is massless) and one finds $(0.025)(\text{GeV})^{-1}$, in excellent agreement with experiment.

We must conclude that Nature knows about and makes use of the anomaly mechanism: fermions are arranged into gauge-anomaly-free representations, and the requirement that anomalies disappear 'explains' the charges of elementary fermions; the pion decays into two photons because of an anomaly in an ungauged channel. It seems therefore that in local quantum field theory these phenomenologi-cally desirable results are facilitated by ultraviolet divergences, which give rise to symmetry anomalies.

The observation that infinities of quantum field theory lead to anomalous symme-try breaking allows comprehension of a second example of quantum mechanical breaking of yet another symmetry – that of scale invariance. Like the space-time translations mentioned earlier, which led to energy–momentum conservation, scale transformations also act on space-time coordinates, but in a different manner: they dilate the coordinates, thereby changing the units of space and time measurements. Such transformations will be symmetry operations in models that possess no fundamental parameters with time or space dimensionality, and therefore do not contain an absolute scale for units of space and time. Our quantum chromodynamical model (QCD) for quarks is free of such dimensional parameters, and it would appear that this theory is scale-invariant – but Nature certainly is not! The observed variety of different objects with different sizes and masses exhibits many different and inequivalent scales. Thus if scale symmetry of the *classical* field theory, which underlies the *quantum* field theory of QCD, were to survive quantization, experiment would have grossly contradicted the model, which therefore would have to be rejected. Fortunately, scale symmetry is quantum mechanically broken, owing to the scales that are introduced in the regularization and renormalization of ultraviolet singularities. Once again a quantum field theoretic pathology has a physical effect, a beneficial one: an unwanted symmetry is anomalously broken, and removed from the theory.

Another application of anomalously broken scale invariance, especially as realized in the renormalization group program, concerns high energy behavior in particle physics and critical phenomena in condensed matter physics. A scale-invariant quantum theory could not describe the rich variety of observed effects, so it is fortunate that the symmetry is quantum mechanically broken. [4]

A different perspective on the anomaly phenomenon comes from the path integral formulation of quantum theory, where one integrates over classical paths the phase exponential of the classical action.

$$\text{Quantum mechanics} \iff \int_{\text{(measure on paths)}} e^{i(\text{classical action})/\hbar}. \qquad (2)$$

When the classical action possesses a symmetry, the quantum theory will respect that symmetry if the measure on paths is unchanged by the relevant transformation. In the known examples (chiral symmetry, scale symmetry) anomalies arise precisely because the measure fails to be invariant and this failure is once again related to infinities: the measure is an **infinite** product of measure elements for each point in the space-time where the quantum (field) theory is defined; regulating this infinite product destroys its apparent invariance.

Yet another approach to chiral anomalies, which arise in (massless) fermion theories makes reference to the first instance of regularization/renormalization, used by Dirac to remove the negative-energy solutions to his equation. Recall that to define a quantum field theory of fermions, it is necessary to fill the negative-energy sea and to renormalize the infinite mass and charge of the filled states to zero. In modern formulations this is achieved by 'normal ordering' but for our purposes it is better to remain with the more explicit procedure of subtracting the infinities, i.e. renormalizing them.

It can then be shown that in the presence of a gauge field, the distinction between 'empty' positive-energy states and 'filled' negative-energy states cannot be drawn in a gauge-invariant manner, for massless, single-helicity fermions. Within this framework, the chiral anomaly comes from the gauge non-invariance of the infinite negative-energy sea. Since anomalies have physical consequences, we must assign physical reality to this infinite negative-energy sea. [5]

Actually, in condensed matter physics, where a Dirac-type equation governs electrons, owing to a linearization of dynamical equations near the Fermi surface, the negative-energy states **do** have physical reality: they correspond to filled, bound states, while the positive energy states describe electrons in the conduction band. Consequently, chiral anomalies also have a role in condensed matter physics, when the system is idealized so that the negative-energy sea is taken to be infinite. [6]

In this condensed matter context another curious, physically realized, phenomenon has been identified. When the charge of the filled negative states is renormalized to zero, one is subtracting an infinite quantity and rules have to be agreed upon so that no ambiguities arise when infinite quantities are manipulated. With this agreed subtraction procedure, the charge of the vacuum is zero, and filled states of positive energy carry integer units of charge. Into the system one can insert a soliton – a localized structure that distinguishes between different domains of the condensed matter. In the presence of such a soliton, one needs to recalculate charges using the agreed rules for handling infinities and one finds, surprisingly, a non-integer result, typically half-integer: the negative-energy sea is distorted by the soliton to yield half a unit of charge. The existence of fractionally charged states in the presence of solitons has been experimentally identified in polyacetylene. [7] We thus have another example of a physical effect emerging from infinities of quantum field theory.

Let me conclude my qualitative discussion of anomalies with an explicit example from quantum mechanics, whose wave functions provide a link between particle and field theoretic dynamics. My example also dispels any suspicion that ultraviolet divergences and the consequent anomalies are tied to the complexities of relativistic quantum field theory: the non-relativistic example shows that locality is what matters.

Recall first the basic dynamical equation of quantum mechanics: the time-independent Schrödinger equation for a particle of mass m moving in a potential $V(\mathbf{r})$ with energy E.

$$\left(-\nabla^2 + \frac{2m}{\hbar^2} V(\mathbf{r})\right)\psi(\mathbf{r}) = \frac{2m}{\hbar^2} E\psi(\mathbf{r}). \tag{3}$$

In its most importance physical applications, this equation is taken in three spatial dimensions and $V(\mathbf{r})$ is proportional to $1/r$ for the Coulomb force relevant in atoms. Here we want to take a different model with potential that is proportional to the inverse square, so that the Schrödinger equation is presented as

$$\left(-\nabla^2 + \frac{\lambda}{r^2}\right)\psi(\mathbf{r}) = k^2\psi(\mathbf{r}), \qquad k^2 \equiv \frac{2m}{\hbar^2} E. \tag{4}$$

In this model, transforming the length scale is a symmetry: because the Laplacian scales as r^{-2}, λ is dimensionless and in (4) there is no intrinsic unit of length. A consequence of scale invariance is that the scattering phase shifts and the S-matrix, which in general depend on energy, i.e. on k, are energy-independent in scale-invariant models, and indeed when the above Schrödinger equation is solved, one verifies this prediction of the symmetry by finding an energy-independent S-matrix. Thus scale invariance is maintained in this example – there are no surprises.

Let us now look to a similar model, but in two dimensions with a δ-function potential, which localizes the interaction at a point.

$$(-\nabla^2 + \lambda\delta^2(\mathbf{r}))\psi(\mathbf{r}) = k^2\psi(\mathbf{r}). \tag{5}$$

Since in two dimensions the two-dimensional δ-function scales as $1/r^2$, the above model also appears scale-invariant; λ is dimensionless. But in spite of the simplicity of the local contact interaction, the Schrödinger equation suffers a short-distance, ultraviolet singularity at $\mathbf{r} = 0$, which must be renormalized. Here is not the place for a detailed analysis, but the result is that only the s-wave possesses a non-vanishing phase shift δ_0, which shows a logarithmic dependence on energy.

$$\operatorname{ctn} \delta_0 = \frac{2}{\pi} \ln kR + \frac{1}{\lambda} \tag{6}$$

R is a scale that arises in the renormalization, and scale symmetry is decisively and quantum mechanically broken. [8] Moreover, the scattering is non-trivial solely as a consequence of broken scale invariance. It is easily verified that the two-dimensional δ-function in classical theory, where it is scale-invariant, produces no scattering.

Furthermore the δ-function model may be second quantized, by promoting the wavefunction to a field operator $\hat{\psi}$, and positing a field theoretic Hamiltonian density operator of the form

$$\mathcal{H} = \frac{\hbar^2}{2m} \nabla \hat{\psi}^* \cdot \nabla \hat{\psi} + \frac{\lambda}{2} (\hat{\psi}^* \hat{\psi})^2. \tag{7}$$

The physics of a second quantized non-relativistic field theory is the same as that of many-body quantum particle mechanics, and the two-body problem (with center-of-mass motion removed) is governed by the above-mentioned Schrödinger equation: the δ-function interaction is encoded in the $(\lambda/2)(\hat{\psi}^* \hat{\psi})^2$ interaction term, which is local.

Analysis of the field theory confirms its apparent scale invariance, but the local interaction produces ultraviolet divergences, which must be regulated and renormalized, thereby effecting an anomalous, quantum mechanical breaking of the scale symmetry. [9]

The above list of examples persuades me that the infinities of local quantum field theory – be they ultraviolet divergences, or an infinite functional measure, or the infinite negative-energy sea – are not merely undesirable blemishes on the theory, which should be hidden or, even better, not present in the first place. On the contrary the successes of various field theories in describing physical phenomena depend on the occurrence of these infinities. One cannot escape the conclusion that Nature makes use of anomalous symmetry breaking, which occurs in local field theory owing to underlying infinities in the mathematical description.

The title of my talk expresses the surprise at this state of affairs: surely it is unreasonable that some of the effectiveness of quantum field theory derives from its infinities. Of course my title is also a riff on Wigner's well-known aphorism about the unreasonable effectiveness of mathematics in physics. [10] We can understand the effectiveness of mathematics: it is the language of physics and **any** language is effective in expressing the ideas of its subject. Field theory, in my opinion, is also a language, a more specialized language that we have invented for describing fundamental systems with many degrees of freedom. It follows that all relevant phenomena will be expressible in the chosen language, but it may be that some features can be expressed only awkwardly. Thus for me chiral and scale symmetry breaking are completely natural effects, but their description in our present language – quantum field theory – is awkward and leads us to extreme formulations, which make use of infinities. One hopes that there is a more felicitous description, in an as yet undiscovered language. It is striking that anomalies afflict precisely those symmetries that depend on absence of mass: chiral symmetry, scale symmetry. Perhaps when we have a natural language for anomalous symmetry breaking we shall also be able to speak in a comprehensible way about mass, which today remains a mystery.

Some evidence for a description of anomalies without the paradox of field theoretic infinities comes from the fact that they have a very natural mathematical expression. For example the $\mathbf{E} \cdot \mathbf{B}$ in our anomalous non-conservation of chiral charge is an example of a Chern–Pontryagin density, whose integral measures the topological index of gauge field configurations and enters in the Atiyah–Singer index theorem. Also the fractional charge phenomenon, which physicists found in Dirac's infinite negative-energy sea, can alternatively be related to the mathematicians' Atiyah–Patodi–Singer spectral flow. [11]

The relation between mathematical entities and field theoretical anomalies was realized 20 years ago and has led to a flourishing interaction between physics and mathematics, which today culminates in the string program. However, it seems to me that now mathematical ideas have taken the lead, while physical understanding lags behind. In particular, I wonder where within the completely finite and non-local dynamics of string theory we are to find the mechanisms for symmetry breaking that are needed to describe the world around us. If string theory is valid, these mechanisms are presumably contained within it, but at the present time they are obscured by the overwhelming formalism. Perhaps that thicket of ideas can be penetrated to find the answer to my question, just as the topics of black-hole entropy and radiance became illuminated after the relevant string degrees of freedom were isolated.

Notes

1. For an account of the origins of classical field theory, see L.P. Williams, *The Origins of Field Theory* (Random House, New York, NY 1966).
2. For an account of the origins of quantum field theory, specifically quantum electrodynamics, see S.S. Schweber, *QED and the Men Who Made It* (Princeton University Press, Princeton, NJ 1994).
3. For an account of quantum anomalies in conservation laws, see R. Jackiw, 'Field theoretic investigations in current algebra' and 'Topological investigation of quantized gauge theories' in *Current Algebra and Anomalies*, S. Treiman, R. Jackiw, B. Zumino and E. Witten eds. (Princeton University Press/World Scientific, Princeton, NJ/Singapore 1985); or S. Adler, 'Perturbation theory anomalies' in *Lectures on Elementary Particles and Quantum Field Theory*, S. Deser, M. Grisaru and H. Pendleton eds. (MIT Press, Cambridge, MA 1970).
4. For a review, see K. Wilson, 'The renormalization group, critical phenomena and the Kondo problem', *Rev. Mod. Phys.* **47**, 773 (1975).
5. See R. Jackiw, 'Effect of Dirac's negative energy sea on quantum numbers', *Helv. Phys. Acta* **59**, 835 (1986).
6. For a selection of applications, see H.B. Nielsen and M. Ninomiya, 'The Adler–Bell–Jackiw anomaly and Weyl fermions in a crystal', *Phys. Lett.* **130B**, 389 (1983); I. Krive and A. Rozhavsky, 'Evidence for a chiral anomaly in solid state physics', *Phys. Lett.* **113a**, 313 (1983); M. Stone and F. Gaitan, 'Topological charge and chiral anomalies in Fermi superfluids', *Ann. Phys. (NY)* **178**, 89 (1987).
7. See R. Jackiw and J.R. Schrieffer, 'Solitons with fermion number 1/2 in condensed matter and relativistic field theories', *Nucl. Phys.* **B190**, [FS3], 253 (1981) and R. Jackiw, 'Fractional fermions', *Comments Nucl. Part. Phys.* **13**, 15 (1984).

8. See R. Jackiw, 'Introducing scale symmetry', *Physics Today* **25**, No. 1, 23 (1972) and 'Delta-function potentials in two- and three-dimensional quantum mechanics' in *M.A.B. Bég Memorial Volume*, A. Ali and P. Hoodbhoy eds. (World Scientific, Singapore 1991).

9. O. Bergman, 'Non-relativistic field theoretic scale anomaly', *Phys. Rev. D* **46**, 5474 (1992); O. Bergman and G. Lozano, 'Aharonov–Bohm scattering, contact interactions and scale invariance', *Ann. Phys. (NY)* **229**, 416 (1994); B. Holstein, 'Anomalies for pedestrians', *Am. J. Phys.* **61**, 142 (1993).

10. E.P. Wigner, 'The Unreasonable Effectiveness of Mathematics in the Natural Sciences'; see also his 'Symmetry and Conservation Laws' and 'Invariance in Physical Theory'. All three articles are reprinted in *The Collected Works of Eugene Paul Wigner*, Vol. VI, Part B, J. Mehra ed. (Springer Verlag, Berlin, 1995).

11. For a review, see A.J. Niemi and G. Semenoff, 'Fermion Number Fractionization in Quantum Field Theory', *Physics Rep.* **135**, 99 (1986).

Discussions

Gross: I want to take exception to some of the remarks that you (Jackiw) made in a very beautiful presentation. I agree with you completely about the importance of anomalies. But I think it's importance to distinguish one infinity from another —

Jackiw: Good infinities from bad infinities.

Gross: No, no, no, no. Different kinds of infinity. There are infinities in quantum field theory, as you point out, having to do with the fact that we often work in the idealized case of an infinite volume. There are infinities which I believe are absolutely intrinsic to the subject and to any theory that supersedes it like string theory, which have to do with the fact that we are treating a system with an infinite number of degrees of freedom. And finally, there are the infinities that occur in calculation and have plagued the theory from the beginning, the ultraviolet infinities. Now the statement I made about QCD, which I will stick to and we can debate, is that QCD, which surely has infrared problems when we consider infinite volume systems, and surely has an infinite number of degrees of freedom, has no ultraviolet infinities, period. As you pointed out in your long list of how to derive the anomaly as a very deep feature of quantum field theory, it is only the original way of discovering the anomaly that seemed to indicate that it had something to do with ultraviolet infinities. As you know very well, it really doesn't. And the proof of that, if you want, is string theory. A theory, which as you know, being a gauge theory, for its consistency relies on the absence of anomalies. These anomalies can occur in string theory although there are absolutely no ultraviolet infinities, or singularities, or divergences at all. However, of course, string theory has the feature which leads to anomalies, which is not infinite ultraviolet singularities, but an infinite number of degrees of freedom. Similarly, QCD has an infinite volume, and an infinite number of degrees of freedom, but no ultraviolet infinities, period. The anomaly that arises in QCD is due to the infinite number of degrees of freedom, as in string theory.

Jackiw: Well I don't think there's any point in arguing whether one should say that infinite number of degrees of freedom cause divergences, or whether one should say that the energy associated with this infinite number of degrees of freedom, which grows and becomes infinite, causes the divergences. These distinctions are just worthless. If I look at the QCD theory written on the blackboard, on the transparencies, I see chiral symmetry, I see scale symmetry. If those symmetries are absent then there is something wrong with my reasoning, and what is wrong with my reasoning is traceable to the occurrence of infinities, which may be massaged in such a way that they don't appear in subsequent discussions. But they have to be there, otherwise the theory would possess all its superficial features. When you look at the Hamiltonian of a classical system and study its properties, you are not misled. When you look at that same expression viewed quantum mechanically, you are misled. Why are you misled? Because of infinities, or infinite number of degrees of freedom, that could be the infinity.

Gross: An infinite number of degrees of freedom has nothing to do with ultraviolet infinities. There are theories that are completely finite theories, even more finite than QCD, in which there are no renormalization effects at all. These are highly symmetric and special theories. The axial anomalies and the other anomalies, in my opinion, arise from an infinite number of degrees of freedom, and not from singularities of physical observables at short distances, which is what ultraviolet singularities mean.

One can of course encounter such infinities in approximate methods of calculation. That is simply, physically, for a very simple reason: one is dividing finite quantities by zero, getting infinity. If you try to express, in an asymptotically free theory, finite quantities in terms of zero bare couplings, which is what perturbative methods tell you to do, then you will encounter on the way infinities. But it's very easy nowadays to avoid these infinities. We do so when we turn QCD into a well-defined mathematical question, the way Arthur was describing, or put it on a lattice and start doing calculations. Never in any stage of that game do we have to encounter infinities.

Jackiw: Right. And also on the lattice you don't encounter anomalies.

Gross: You do encounter them in the continuum limit. You get the pion and you would get its decay mode correctly, thus you can get anomalies in lattice QCD.

Jackiw: The quantum mechanical example in my lecture may be viewed as involving infinite number of degrees of freedom, but maybe different from the way you're counting: here the infinity is in the infinite dimensionality of the quantum mechanical Hilbert space. Alternatively one sees that the scale anomaly in the quantum mechanical example comes from the short-distance behavior, whose effect is enhanced by the quantum mechanical infinity of degrees of freedom. Classically the short-distance behavior in the example does not even affect equations of motion.

Gross: I don't think that that particular Schrödinger equation example captures the essence of either the singularities or the anomalies in quantum field theory.

11. Comments: The quantum field theory of physics and of mathematics

One recognizes that there has been, and continues to be, a great deal of common ground between statistical mechanics and quantum field theory (QFT). Many of the effects and methods of statistical physics find parallels in QFT, particularly in the application of the latter to particle physics. One encounters spontaneous symmetry breaking, renormalization group, solitons, effective field theories, fractional charge, and many other shared phenomena.

Professor Fisher [1] has given us a wonderful overview of the discovery and role of the renormalization group (RG) in statistical physics. He also touched on some of the similarities and differences in the foundations of the RG in condensed matter and high-energy physics, which were amplified in the discussion. In the latter subject, in addition to the formulation requiring cutoff-independence, we have the very fruitful Callan–Symanzik equations. That is, in the process of renormalizing the divergences of QFT, arbitrary, finite mass-scales appear in the renormalized amplitudes. The Callan–Symanzik equations are the consequence of the requirement that the renormalized amplitudes in fact be independent of these arbitrary masses. This point of view is particularly useful in particle physics, although it does make its appearance in condensed matter physics as well.

The very beautiful subject of conformal field theory spans all three topics we are considering: critical phenomena, quantum field theory, and mathematics. The relationship between conformal field theory and two-dimensional critical phenomena has become particularly fruitful in recent years. Conformal field theory in its own right, as well as being an aspect of string theory, has also played an important role in the recent vigorous interchanges between physics and mathematics. A wide variety of mathematical techniques have become commonplace tools to physicists working in these areas. Similarly, questions raised by conformal field theorists, topological field theorists, and string theorists have presented many new directions and opportunities for mathematics, particularly in geometry and topology. The relationship has been rich and symbiotic.

By contrast, it is my view that the more traditional mathematical physics of constructive and axiomatic field theory, and related endeavors, have not come any closer to the physicists' view of QFT. In fact, with the increased appreciation and understanding of effective field theories, the gap between the two communities might even be widening, if I understand correctly. It is this issue that I wish to address.

---

[1] Supported in part by the DOE under grant DE-FGO2-92ER40706.

It should be clear that the quantum field theory of mathematics is very different from that of physics. In fact, I have long had the opinion that these may not even be the same theories at all. That is, there are (at least) two classes of quantum field theories, which for historical reasons go by the same name, despite being very different. The quantum field theory considered by the mathematics community is built on an axiomatic structure, and requires that the infinite volume system be consistent at *all* distance scales, infrared as well as ultraviolet. By contrast what physicists mean by a field theory is, in contemporary language, an effective field theory, which is applicable within a well-defined domain of validity, usually below some energy scale. Consistency is not required at short distances, i.e., at energies above a specified energy scale.

Does any four-dimensional QFT exist, in the mathematical sense? The status of this topic was reviewed by Jaffe [2]. As yet, no four-dimensional QFT has been demonstrated to meet all the necessary requirements, although it is believed that pure Yang–Mills theory will eventually attain the status of a consistent theory. What about the standard model? There are two possibilities. Either it is just a matter of time for the necessary ingredients to be assembled to produce a mathematically consistent four-dimensional QFT describing the standard model, or no such theory exists. Suppose a given candidate field theory of interest is shown in fact not to be consistent. One possible response is to embed the candidate theory in a larger system with more degrees of freedom, i.e., additional fields, and re-examine the consistency of the enlarged system. The hope is that eventually this sequence stops, and there is a consistent QFT. However, to repeat, the logical possibility exists that this procedure does not lead to a consistent four-dimensional local QFT.

If no such consistent local QFT existed this would not have grave consequences for physical theory as we know it. From the physicists' point of view, the nesting of a sequence of theories is a familiar strategy carrying a physical description to higher energies. Embedding QED in the standard model, and then into possible grand unified QFT, extends the description of fundamental interactions to at least 100 GeV, and hopefully to still higher energies. However, the last *field theory* in this sequence may not be consistent at the shortest distance scales in the mathematical sense. In any case, eventually in this regression to shorter distances, quantum effects of gravity are encountered. The question of consistency then must change. Then it becomes plausible that to proceed to even shorter distances, something other than local QFT or *local* quantum gravity will be required. String theory provides one possibility for an extension to distances where quantum gravity is relevant. The standard model is then just an effective low-energy representation of string theory. It has even been speculated, on the basis of string strong–weak duality, that string theory itself may only be an effective theory of some other theory (for example the as yet unknown M-theory) [3]. Therefore, the physicist does not (need to) care if there is any completely consistent local QFT, valid at the shortest distances. The appropriate theories to be considered are effective theories; usually, but not always effective *field* theories.

Jackiw [4] argues elegantly and persuasively for physical information carried by certain infinities of QFT, making their presence known in anomalies and spontaneous symmetry breaking. In QFT, the ultraviolet divergences should not just be regarded as awkward flaws, but rather as a feature which can lead to physical consequences. However, it should not be concluded from his analysis that a theory without ultraviolet infinities cannot describe the physical phenomena appropriate to anomalies and spontaneous symmetry breaking. Such phenomena can be accommodated without ultraviolet divergences; string theory again providing one example, although the language to describe the phenomena will differ. Though ultraviolet divergences are an essential feature of local QFT, or effective field theory, they are not necessary for a description of the physics. Jackiw asks whether the string theory program has illuminated any physical questions. I should like to respond briefly in the affirmative. String theory has provided us with finite, well-defined examples of quantum gravity; the only class of such theories known at present, with an essential aspect being non-locality. It had long been conjectured that one could not construct a non-local theory which was Lorentz invariant, positive definite, causal and unitary. One wondered whether there was any finite quantum gravity. Certainly string theory sets these earlier prejudices aside and allows us to consider a much broader class of theories in confronting the physical world. One should acknowledge that these are issues that have long been on the physicists' agenda, and are not solely of mathematical interest. In this evolution of our understanding one still retains the basic assumptions of quantum mechanics, even though locality is no longer a sacred physical principle.

Mathematicians speak of field theories such as quantum electrodynamics (QED) as heuristic field theories, or even as models, since no proof of mathematical consistency, at all distances, exists in the sense mentioned earlier. I feel that this description of QED is pejorative, albeit unintended. In fact QED is the most precise physical theory ever constructed, with well-defined calculational rules for a very broad range of physical phenomena, and extraordinary experimental verification. There is even a plausible explanation of why the fine-structure constant is small, based on a renormalization group extrapolation in grand unified theories. Of course, we understand that QED is an effective field theory, but it is a well-defined theory in the sense of physical science. We know that to extend its domain of validity one may embed it in the so-called standard model (itself an effective field theory), the electro-weak sector of which has been tested over an enormous energy range (eV to 100 GeV; 11 decades), although not with the precision of QED. Thus both QED and the standard model are full-fledged theories of physical phenomena in every sense of the word!

The investigation of the mathematical consistency of four-dimensional local QFT is an interesting question in its own right. No matter what the outcome, we will gain important insights into the structure of QFT. However, the answers to such questions are not likely to change the way we do particle physics. Then how can mathematical physics make contact with issues of concern to particle physics? What are the right questions? Some suggestions immediately come to mind. What is a mathematically consistent *effective* field theory? Is this even a well-posed problem? If so, what restrictions does it place on effective theories? Can any candidates be discarded? To begin with, one should not expect that effective field theories are local field theories, as they involve infinite polynomials in fields and their derivatives. Nor do they have to be consistent at the shortest distances. An approach to some of these issues has been made by Gomis and Weinberg [5]. They require that the infinities from loop graphs be constrained by the symmetries of the bare action, such that there is a counterterm available to absorb every infinity. This is a necessary requirement for a theory to make sense perturbatively. Their criteria are automatically satisfied if the bare action arises from global, linearly realized symmetries. However, this becomes a non-trivial requirement if either there are non-linearly realized symmetries or gauge symmetries in the bare action. In constructive field theory one encounters a cutoff version of the local field theory being studied at intermediate stages of the analysis. These can be regarded as effective field theories, but to be relevant they must be Lorentz invariant. However, one needs to consider a wider class of effective theories than is at present considered by constructive field theorists if the work is to have an impact on the concerns of particle physicists. In any case, there is certainly a great deal more to do in making effective field theories more precise.

In summary, I have argued that the QFT of mathematicians and of physicists are quite different, although both go by the same name. To bridge the gap, one recognizes that there are many important mathematical problems posed by effective field theories, but these have not received the attention they deserve. Further, the existence of consistent string theories challenges the idea that locality is essential in the description of particle physics at the shortest distances.

References

[1] Michael E. Fisher, 'Renormalization group theory: its basis and formulation in statistical physics,' this volume.
[2] Arthur Jaffe, 'Where does quantum field theory fit into the big picture?,' this volume.
[3] Edward Witten, Loeb Lectures, Harvard University, Fall 1995.
[4] Roman Jackiw, 'The unreasonable effectiveness of quantum field theory,' MIT-CTP-2500, hep-th/9602122, and in this volume.
[5] Joaquim Gomis and Steven Weinberg, 'Are nonrenormalizable gauge theories renormalizable in the modern sense?' RIMS-1036, UTTG-18-95, hep-th/9510087.

Session discussions

Jaffe: I'd just like to mention, I didn't say it in my talk that we look at effective field theories, but we do. And the methods of proof involve the study of effective field theories, and it's the study of these effective field theories in the case of four-dimensional Yang–Mills fields that suggests that there may be existence in that case. But that's only a suggestion, and it will take a lot of work to prove a mathematical theorem.

Gross: I would like to object to the suggestion that perhaps that QCD might not exist as a mathematically sound theory. I think that if one really suspected that that was the case, that there was some flaw in QCD, presumably at short distances, then there would be many people continuing Balabon's work. Because, since the energy scale of QCD is a GeV and the couplings are strong, that flaw would be of immediate, practical importance. The reason, I believe, that people aren't carrying this program, to Arthur's regret, is that everybody believes that the answer is yes, that there is a mathematically consistent QCD, and that one day it will be proven rigorously.

Jaffe: Well, everybody believed in Fermat's theorem, but it took a lot of work to actually prove it. [laughter]

Gross: Oh, I think it's an important issue. I support you. I think somebody should do this. Not me, not you, but somebody should do this.

Jaffe: It's actually a very important piece of work to do.

Gross: Absolutely.

Jaffe: I think the reason people are not working on it is that it's hard. And people don't like to work on hard problems.

Gross: Believe me, if somebody believed that there was a 50% chance that they could prove that QCD doesn't exist, there'd be a lot of people working on it.

Jaffe: It's much harder to prove things don't exist than do exist. By elementary scaling arguments we know that nonlinear, scalar, quartic interactions do not exist in five or more dimensions. However, most theoretical physicists believe that QED is inconsistent, yet it gives agreement with experiment to phenomenal accuracy.

Schnitzer: I had not intended to suggest that QCD would not end up consistent. What I had in mind was the standard model, where we have to live with scalar fields. In that case I think there's considerably more doubt that can be raised. Then for that example we again reopen the issue of what we mean by an effective theory.

Fisher: In support of Arthur's viewpoint, it must be said that the true facts are not much advertised, and, generally, there is not much reward for proving things rigorously. But I would like to stress that there are a number of important problems in statistical mechanics, even just at the classical level, that have never been addressed rigorously. One of the most outrageous, if you like, is the situation with regard to dipolar forces in the presence of external (electric or magnetic) fields. For example, the proof of the equivalence of the correlation functions between the different ensembles is lacking. Many of these problems are, in fact, very difficult: but the rewards for solving any of them are usually small. Nevertheless, some of these questions entail conceptual problems that I do not believe are properly understood even at a heuristic level. And, nowadays, such issues often come to light when one attempts serious, large-scale simulations and calculations.

Part Five. Quantum field theory and space-time

Introduction

JOHN STACHEL

One of the purposes of this conference is consideration of the historical background
to the topics under discussion, so I shall open with a historical introduction to
'quantum gravity'.[1]

Once he completed work on the general theory of relativity at the end of 1915,
Einstein began to concern himself with its implications for microphysics. In his first
(1916) paper on gravitational radiation,[2] Einstein argued that quantum effects
must modify the general theory:

> Nevertheless, due to the inneratomic movement of electrons, atoms would have to radiate not
> only electromagnetic but also gravitational energy, if only in tiny amounts. As this is hardly
> true in nature, it appears that quantum theory would have to modify not only Maxwellian elec-
> trodynamics, but also the new theory of gravitation.
>
> (p. 209)

He reiterated this conclusion in his second (1918) paper:[3]

> As already emphasized in my previous paper, the final result of this argument, which demands a
> [gravitational] energy loss by a body due to its thermal agitation, must arouse doubts about the
> universal validity of the theory. It appears that a fully developed quantum theory must also
> bring about a modification of the theory of gravitation.
>
> (p. 164)

He soon began to speculate about whether gravitation plays a role in the atomistic
structure of matter:[4]

---

[1] I have not made an exhaustive search of the literature, so these comments are far from definitive. I
am indebted to Gennady Gorelik, 'The first steps of quantum gravity and the Planck values,' in *Studies
in the History of General Relativity (Einstein Studies, vol. 3)*, Jean Eisenstaedt and A. J. Kox, eds.
(Birkhäuser, Boston/Basel/Berlin, 1992) (hereafter referred to as Gorelik 1992), pp. 366–82; and
Gennady E. Gorelik and Victor Ya. Frenkel, *Matvei Petrovich Bronstein and Soviet Theoretical Physics
in the Thirties* (Birkhäuser, Basel/Boston/Berlin, 1994) (hereafter Gorelik and Frenkel 1994) for a number
of references to the early literature.

[2] Albert Einstein, 'Näherungsweise Integration der Feldgleichungen der Gravitation,' *Preussische Akademie der Wissenschaften (Berlin). Sitzungsberichte* (1916): 688–96, translated as 'Approximate integration of the field equations of gravitation,' in *The Collected Papers of Albert Einstein*, vol. 6, *The Berlin
Years: Writings 1914–1917, English Translation of Selected Texts* (Princeton University Press, 1997),
Alfred Engel, transl., pp. 201–10.

[3] Albert Einstein, 'Über Gravitationswellen,' *Preussische Akademie der Wissenschaften (Berlin). Sitzungsberichte* (1918): 154–67.

[4] Albert Einstein, 'Spielen Gravitationsfelder im Aufbau der materiellen Elementarteilchen eine wesent-
liche Rolle?,' *Preussische Akademie der Wissenschaften (Berlin). Sitzungberichte* (1919): 349–56; translated as 'Do gravitational fields play an essential part in the structure of the elementary particles of
matter?' in *The Principle of Relativity*, Otto Blumenthal, ed., W. Perret and J. B. Jeffery, transls.
(Methuen, London, 1923), reprint 1952 (Dover, New York), pp. 191–98.

[T]here are reasons for thinking that the elementary formations which go to make up the atom are held together by gravitational forces.... The above reflexions show the possibility of a theoretical construction of matter out of the gravitational field and the electromagnetic field alone...

In order to construct such a model of an 'elementary particle', Einstein shows, it is necessary to modify the original gravitational field equations. The details of his model are unimportant; the main point of interest is that his attention now shifted to the search for a unified theory of the electromagnetic and gravitational fields, out of which he hoped to be able to explain the structure of matter. Quantum effects are to be derived from such a theory, rather than postulated *ad hoc*. This remained his approach for the rest of his life: the search for a 'natural' mathematical extension of the general theory, with the hope that it would somehow explain the quantization of matter and energy. Einstein lost interest in the possible application of quantization procedures – either of the old quantum theory or the new quantum mechanics – to the gravitational field proper.[5]

Others took up this challenge. In 1927 Oskar Klein argued that quantum theory must ultimately modify the role of spatio-temporal concepts in fundamental physics:[6]

As is well known, the concepts of space and time have lost their immediate sense as a result of the development of quantum theory. This is connected, as Bohr in particular has emphasized, with the fact that, according to the quantum theory, our measuring tools (light and free material particles) manifest a behavior that is characterized by the dilemma particle–wave, as a result of which a sharp definition of a spatio-temporal coincidence, which lies at the foundation of the usual concepts of space and time, becomes impossible. From this standpoint it is to be expected that the general theory of relativity stands in need of a revision in the sense of the quantum postulate, as also results from the fact that various of its consequences stand in contradiction to the requirements of the quantum theory.

Then follows a footnote:

For example, Einstein light deflection by a collection of moving material bodies in a thermal equilibrium state would destroy the equilibrium of the radiation, which allows one to suspect a sort of gravitational Compton effect. One would expect a corresponding result in the case of statistical equilibrium between gravitational waves and light waves according to the general theory of relativity, due to the non-occurrence of Planck's constant in the relevant equations.

After discussing the complementarity between the space-time description and the conservation laws for energy and momentum, Klein goes on:

When one considers the apparently rather accidental role of the conservation laws in analytical mechanics as four of the many possible integrals of the equations of motion, then it could appear, in accord with the viewpoint just mentioned, as if the spatio-temporal magnitudes, which are of such central significance for our immediate experience, will retreat to the background in the mathematical structure of the future quantum theory. In spite of the great

[5] There is an intriguing comment by Y. I. Frenkel, in a paper written for the Schilpp volume *Albert Einstein: Philosopher-Scientist*, but not submitted: 'Einstein was probably the first to point out the connection of gravitational waves and the corresponding particles in a conversation with the author back in 1925' (cited in Gorelik and Frenkel 1994, p. 85; translation kindly corrected by Dr Gorelik).

[6] Oskar Klein, 'Zur fünfdimensionalen Darstellung der Relativitätstheorie,' *Zeitschrift für Physik* **46** (1927): 188–208. Klein was working with Bohr in Copenhagen at this time, and his comments may well reflect Bohr's views.

fruitfulness of quantum mechanics, many things speak for the view that the starting point for a general mathematical formulation of quantum theory is to be sought in the equations of classical field theory rather than in the equations of motion of mechanics. Thereby, the question of the role of spatio-temporal quantities (or rather of their quantum-mechanical analogues) appears in different light, for a large part of the state variables characterizing field theory (the Einstein gravitational potentials) are indeed formally immediately tied to the space-time concepts.

(pp. 188–9)

Others were less cautious in their approach to quantum gravity. In their first paper on quantum electrodynamics, Heisenberg and Pauli asserted that:[7]

quantization of the gravitational field, which appears to be necessary for physical reasons [in a footnote, they refer to the works of Einstein and Klein cited above], may be carried out without any new difficulties by means of a formalism fully analogous to that applied here.

(p. 3)

Almost seventy years have elapsed since this casual prediction, and we are still without a quantum theory of gravity!

But certainly not for lack of attempts. Leon Rosenfeld was the first to try.[8] He applied a method developed by Pauli, with whom he was working, for quantization of fields with gauge groups (whose Hamiltonian formulation therefore involves constraints) to the linearized Einstein field equations. After setting up the formalism to deal with both the electromagnetic and linearized gravitational fields, he showed that the gravitational field, like the electromagnetic, can only be quantized using commutators (Bose quantization). In the second paper mentioned in Footnote 8, he calculated the gravitational self-energy of a light quantum, which diverges. He attributed this infinity to the fact 'that one cannot attribute a finite radius to a light quantum. The analogy with the case of the electron need hardly be emphasized' (p. 596).[9] He then proceeded to discuss the various possible transition processes between light and gravitational quanta, and concluded: 'If we imagine a cavity initially filled with radiation (without dust particles!), then the first-approximation gravitational interactions between light quanta suffice to establish Planckian [i.e., thermal] equilibrium (with a speed proportional to $1/\kappa$ [$\kappa = 8\pi G/c^4$])' (p. 598).[10]

During the thirties, the idea of a gravitational quantum came to be generally (but as we shall see, not universally) accepted by theoretical physicists, apart from the small relativity community that followed Einstein in his continued search for a (classical) unified theory. In the early thirties, there was some discussion of Bohr's idea that

---

[7] Werner Heisenberg and Wolfgang Pauli, 'Zur Quantenelektrodynamik der Wellenfelder,' *Zeitschrift für Physik* **56** (1929): 1–61.

[8] Leon Rosenfeld, 'Zur Quantelung der Wellenfelder,' *Annalen der Physik* **5** (1930): 1113–52; 'Über die Gravitationswirkungen des Lichtes,' *Zeitschrift für Physik* **65** (1930): 589–99.

[9] In response to a query from Pauli (see Pauli to Rosenfeld, 12 April 1931, in Wolfgang Pauli, *Wissenschaftliche Briefwechsel*, vol. 3, *1940–1949*, Karl von Meyenn, ed. [Springer, Berlin/Heidelberg/New York/Tokyo, 1993], p. 746) Rosenfeld added a supplement to his paper showing that the gravitational self-energy of any one-photon state is infinite. Solomon soon showed that this divergence was not due to the zero-point energy of the field. See Jacques Solomon, 'Nullpunktsenergie der Strahlung und Quantentheorie der Gravitation,' *Zeitschrift für Physik* **71** (1931): 162–70.

[10] Rosenfeld himself later came to question the necessity of quantizing the gravitational field, and did not include his work on this topic in a collection of papers that he selected; instead he reprinted critical comments on field quantization, including arguments against the need for a quantum gravity. See Leon Rosenfeld, 'On quantization of fields,' in *Selected Papers of Leon Rosenfeld*, Robert S. Cohen and John Stachel, eds. (Reidel, Dordrecht/Boston/London, 1979) (hereafter Rosenfeld 1979), pp. 442–5, and 'Quantum theory and gravitation,' ibid., pp. 598–608.

the neutrino might be related to the gravitational quantum.[11] The most elaborate discussion I have found is in a paper by Blokhintsev and Gal'perin,[12] cited at length in Gorelik and Frenkel 1994 (pp. 96–97; see Footnote 1), who claim this is the first time the term 'graviton' (for the gravitational quantum) appears in print.

In the late thirties, Fierz developed a formalism for free quantum fields of arbitrary spin and mass,[13] which was soon generalized by Pauli and Fierz to include electromagnetic interactions.[14] In particular, they noted that the massless spin-2 field obeyed equations identical to the linearized Einstein field equations. Pauli prepared a 'Report on the general properties of elementary particles' for the projected 1939 Solvay Congress, which, owing to the outbreak of World War II, was never held. At the urging of Heisenberg,[15] he included a section entitled 'remarks on gravitational waves and gravitational quanta (spin 2).' This report was only published recently;[16] while most of its contents were published elsewhere by Pauli, the section on gravitation was not. Nevertheless, Heisenberg – as, no doubt, were many other leading physicists – was aware of its contents.[17] In this section Pauli develops a Lagrangian for the massless spin-2 field formally identical with the one for the linearized Einstein equations. He notes that the invariance of the theory under a group of what he calls 'gauge transformations of the second kind' (by analogy with the gauge transformations of Maxwell's theory) corresponds to the invariance under infinitesimal coordinate transformations of the linearized Einstein theory. He quantizes the

[11] See Wolfgang Pauli, *Wissenschaftlicher Briefwechsel*, vol. 2, *1930–1939*, Karl von Meyenn, ed. (Springer Verlag, Berlin/Heidelberg/New York/Tokyo, 1985), Bohr to Pauli, 15 March 1934, p. 308: 'The idea was that the neutrino, for which one assumes a zero rest mass, could hardly be anything else than a gravitational wave with appropriate quantization' (transl. from Niels Bohr, *Collected Works*. vol. 7, *Foundations of Quantum Physics II (1933–1958)*, J. Kalcar, ed. (Elsevier, Amsterdam/Lausanne/New York/Oxford/Shannon/Tokyo, 1996), p. 479). Fermi had evidently had a similar idea, but was aware of the problem of the different spins. See ibid., Pauli to Heisenberg, 6 February 1934, p. 277: 'Fermi would prefer to make a connection between neutrinos and half gravitational quanta.' As late as November 1934, Pauli cautiously stated: 'While up to now it has been held almost certain that gravitational phenomena play practically no role in nuclear physics, it now seems that the possibility cannot be immediately rejected, that the phenomena of beta-radiation could be connected with the square root of kappa [the gravitational constant]' ('Raum, Zeit, und Kausalität in der modernen Physik,' *Scientia* **59** (1936): 65–76, p. 76). This suggests that Pauli may have had in mind construction of a graviton from two neutrinos, along the lines of de Broglie's neutrino theory of light.

[12] Dmitri Ivanovich Blokhintsev and F. M. Gal'perin, 'Gipoteza neitrino i zakon sokhraneniya energii,' *Pod znamenem marxisma* (1934), no. 6, pp. 147–57. As cited by Gorelik and Frenkel, they wrote: 'The comparison displayed above indicates that the graviton and the neutrino have much in common. This probably testifies that in general the highly improbable process of graviton radiation becomes practically observable in beta-decay. If the neutrino turns out to be the graviton this would mean that contemporary physics had approached the limits beyond which there would be no present insurmountable barrier between gravitation and electromagnetism. Due to theoretical considerations it is hard to identify gravitons with the neutrino since it is hard to admit that they have the same spin 1/2 as the neutrino. In this respect gravitons have much more in common with light quanta. It is impossible, however, to totally rule out a theoretical possibility of their identification. So far it is much more correct to regard the neutrino as an independent type of particle' (Gorelik and Frenkel 1994, p. 97).

[13] Marcus Fierz, 'Über die relativistische Theorie kräftefreier Teilchen mit beliebigem Spin,' *Helvetica Physica Acta* **12** (1939): 3–37.

[14] Wolfgang Pauli and Marcus Fierz, 'Über relativistische Wellengleichungen von Teilchen mit beliebigem Spin im elektromagnetischen Feld,' *Helvetica Physica Acta* **12** (1939): 297–300; ibid., 'On relativistic wave equations for particles of arbitrary spin in an electromagnetic field,' *Royal Society (London). Proceedings* **A173** (1939): 211–32.

[15] See Pauli to Heisenberg, 10 June 1939, in Wolfgang Pauli, *Wissenschaftlicher Briefwechsel*, vol. 2, *1930–1939*, Karl von Meyenn, ed. (Springer, Berlin/Heidelberg/New York/Tokyo, 1985), p. 662; and Heisenberg to Pauli, 12 June 1939, ibid., p. 665.

[16] See Wolfgang Pauli, *Wissenschaftlicher Briefwechsel*, vol. 2, *1930–1939*, Karl von Meyenn, ed. (Springer, Berlin/Heidelberg/New York/Tokyo, 1985), pp. 833–901; the section on gravitation is on pp. 897–901.

[17] See, for example, Pauli to Schrödinger, 5 November 1939, ibid., pp. 823–25.

theory in analogy with Fermi's method for electrodynamics, requiring the linearized harmonic coordinate condition operators to annihilate the wave function. This allows the imposition of commutation relations on all components of the linearized metric using the invariant D-function, with which the harmonic constraints are compatible. He concludes:

The so-defined gravitational quanta or 'gravitons' have the...spin 2...It is certainly a limitation of the quantum-mechanical side of this treatment, that one leaves it at that approximation, in which the general-relativistic field equations are linear. This limitation is most closely connected with the well-known divergence difficulties of field theory.

(See Footnote 16, p. 901)

Apart from these divergence difficulties common to all quantum field theories at that time, Pauli, like Rosenfeld, seemed satisfied with linearized quantum gravity. But in 1935, well before Pauli's work and unnoted by him and most others in the west, a Russian physicist had already applied Fermi's technique to the linearized Einstein equations, but come to much more sceptical conclusions about the physical significance of the resulting theory. This was Matvei Petrovich Bronstein, working at the Leningrad Physico-Technical Institute.

Bronstein was no stranger to the problem of quantum gravity. He had been pondering the issue since 1930, when he opined: 'It is a task of the nearest future to identify the ties between quantum mechanics and the theory of gravitation' (see Gorelik and Frenkel, p. 88). In 1931, he summed up an article on unified field theories by stating: 'It seems that Einstein's program of unified field theory will prove impossible... what will be needed is a sort of marriage between relativity and quantum theory' (see Gorelik and Frenkel, p. 88). By 1933 he was stressing the crucial significance of the three dimensional constants c, G, and h in defining the past and future development of physics, offering the chart shown in the Figure, in which 'solid lines correspond to existing theories, dotted ones to problems not yet solved' (Bronstein 1933, p. 15, trans. in Gorelik 1992, p. 372).

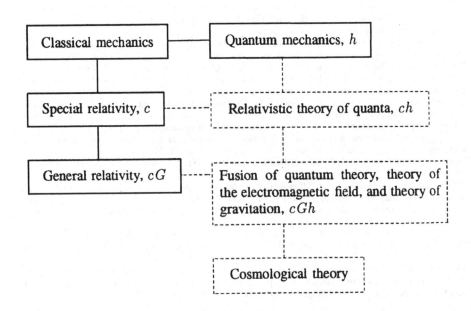

After relativistic quantum theory is formulated, the next task would be to realize the third step... namely, to merge quantum theory (*h* constant) with special relativity (*c* constant) and the theory of gravitation (*G* constant) into a single whole.

(Cited from Gorelik and Frenkel 1994, p. 90.)[18]

In 1935 Bronstein presented a Doctoral Thesis on 'Quantizing Gravitational Waves' to a committee that included Vladimir Fock and Igor Tamm.[19] The work was published in Russian in 1936,[20] as well as in a condensed German version.[21] After carrying out the quantization, developing the quantum analogue of Einstein's quadrupole radiation formula, and deducing the Newtonian law of attraction from the interchange of longitudinal gravitational quanta, Bronstein proceeds to more critical reflections on the physical significance of his results.

He carries out an analysis of the measurability of the linearized Christoffel symbols, which he takes to be the components of the gravitational field. By analogy with the Bohr–Rosenfeld analysis of the measurability of the electromagnetic field components, he shows that there are limitations on the measurability of gravitational field components implied by the uncertainty relations between position and momentum of the test bodies whose acceleration is used to measure the field. But he asserts that there is an additional limitation without an electromagnetic analogue: for measurement of the components of the electromagnetic field, it is permissible to use electrically neutral test bodies, which have no effect on the field being measured. But in the gravitational case, owing to the universality of the gravitational interaction, the effect of the energy–momentum of the test bodies on the gravitational field cannot be neglected even in linear approximation. Bronstein derives an expression for the minimum uncertainty in a measurement of the components of the affine connection that depends inversely on the mass density of the test body, just as Bohr–Rosenfeld's result did on the charge density. He now comes to what he sees as the crucial difference between the two cases.

Here we should take into account a circumstance that reveals the fundamental distinction between quantum electrodynamics and the quantum theory of the gravitational field. Formal quantum electrodynamics that ignores the structure of the elementary, charge does not, in principle, limit the density of ρ. When it is large enough we can measure the electric field's components with arbitrary precision. In nature, there are probably limits to the density of electric charge... but formal quantum electrodynamics does not take these into account... The quantum theory of gravitation represents a quite different case: it has to take into account the fact that the gravitational radius of the test body ($\kappa\rho V$) must be less than its linear dimensions $\kappa\rho V < V^{1/3}$.

(Bronstein 1936b, p. 217, transl. from Gorelik and Frenkel 1994,
p. 105 and Gorelik 1992, pp. 376–77)

He acknowledges that 'this was a rough approximation, because with the measuring device's relatively large mass departures from the superposition principle [i.e., the linearized approximation] will, probably be noticeable;' but thinks 'a similar result

---

[18] In 1934 Pauli also discussed 'the three fundamental natural constants', but added: 'for the sake of simplicity we ignore gravitational phenomena for the present' (the article was not published until 1936 in *Scientia*; see the reference in note 11).

[19] For Bronstein's life and work, see Gorelik and Frenkel 1994.

[20] Matvei Petrovich Bronstein, 'Kvantovanie gravitatsionnykh voln [Quantization of gravitational waves],' *Zhurnal Eksperimentalnoy i Teoreticheskoy Fiziki* **6** (1936): 195–236.

[21] Matvei Petrovich Bronstein, 'Quantentheorie schwacher Gravitationsfelder,' *Physikalische Zeitschrift der Sowjetunion* **9** (1936): 140–57.

will survive in a more exact theory since it does not follow from the superposition principle. It merely corresponds to the fact that in General Relativity there cannot be bodies of arbitrarily large mass with the given volume.' (Gorelik and Frenkel, pp. 105–106.) He concludes:

The elimination of the logical inconsistencies connected with this result requires a radical reconstruction of the theory, and in particular, the rejection of a Riemannian geometry dealing, as we have seen here, with values unobservable in principle, and perhaps also rejection of our ordinary concepts of space and time, modifying them by some much deeper and nonevident concepts. *Wer's nicht glaubt, bezahlt einen Taler.*[22]

(Bronstein 1936b, transl. from Gorelik 1992, p. 377)

At least one physicist outside the Soviet Union referred to Bronstein's work in print. In 1938 Jacques Solomon resumed Bronstein's argument and concluded:[23]

... [I]n the case when the gravitational field is not weak, the very method of quantization based on the superposition principle fails, so that it is no longer possible to apply a relation such as [the equation setting a lower limit on the measurability of the linearized field strength] in an unambiguous way ... Such considerations are of a sort to put seriously in doubt the possibility of reconciling the *present* formalism of field quantization with the *non-linear* theory of gravitation.

(p. 484)

In one of the many tragic ironies of history, both of the pre-war advocates of the need for a radically different approach to quantum gravity perished prematurely. Bronstein was arrested by the Soviet State Security (NKVD) on August 6, 1937; only twenty years later did his widow, the writer Lydia Chukovskaya, learn the exact date of his death: February 18, 1938.[24] Solomon, a Communist militant active in the underground resistance to the German occupation of France, was arrested with his wife, Helene Langevin, in March 1942. He was killed by the Germans on May 23, 1942; she was sent to Auschwitz, but survived the war.[25] Between them, Stalin and Hitler saw to it that the post-World War II discussion of quantum gravity would take place without two important voices.

Bronstein was a student of Yakov Frenkel, who was quite skeptical about the whole project of quantizing the gravitational field equations. In an article prepared for the Schilpp volume (see Footnote 5), he argued against the analogy between the gravitational and electromagnetic fields that is the basis of the graviton approach. Since the article is unavailable, I shall quote a summary with citations from Gorelik and Frenkel 1994:

He argued that 'the electromagnetic field [was] *matter* while the gravitational field merely [determined] the metrical properties of the space-time continuum.' He insisted that 'strictly speaking there [were] no such things as gravitational energy or gravitational momentum since the corresponding values [did] not form a true tensor and [were] nothing more than a *pseudotensor*.' It was his conviction that the attempts to quantize gravitation were senseless since 'the gravitational field [had] macroscopic, rather than microscopic meaning; it merely

---

[22] The German phrase – 'Let him who doubts it pay a Thaler' – comes from the Grimm brothers' tale, 'Der tapfere Schneider.'

[23] Jacques Solomon, 'Gravitation et quanta,' *Journal de Physique et de Radium* **9** (1938): 479–85.

[24] See Gorelik and Frenkel 1994, pp. 144–47; and Lydia Chukovskaya, *The Akhmatova Journals/Volume I 1938–41* (Farrar, Strauss and Giroux, New York, 1994).

[25] See Leon Rosenfeld, 'Jacques Solomon,' in Rosenfeld 1979, pp. 297–301; Martha Cecilia Bustamente, 'Jacques Solomon (1908–1942): Profil d'un physicien theoricien dans la France des années trente,' *Revue d'histoire des sciences* **50**, 49–87, 1997.

[supplied] a *framework* for physical events occurring in space and time while quantizing [could] be applied only to microscopic processes in *material fields*.' These considerations were little affected by the developments in physics that occurred after November 1935 – this and his remark during Bronstein's defense [173, p. 319] allows us to surmise that his position was the same in 1935.

(p. 85)

Another physicist who expressed doubts about the need to quantize general relativity was van Dantzig, who considered general relativity to be a sort of thermodynamic limit of a deeper-level theory of interactions between particles.[26]

Thus, by the mid-thirties the three positions that were to dominate post-World War II discussions of quantum gravity (at least among physicists not wedded to Einstein's unified field theory program) had already been formulated.

(1) Quantum gravity should be formulated by analogy with quantum electrodynamics. In particular, one should start from quantization of the linearized gravitational field equations. Technical problems that arise will be similar to those arising in quantum electrodynamics and will presumably be solved *pari passu* with the problems of the latter (Pauli, Rosenfeld, Fierz).[27]

(2) The unique features of gravitation will require special treatment. One possibility is that the full theory, with its non-linear field equations, must be quantized. This implies that, while the general techniques of quantum field theory may be relevant, they must be generalized in such a way as to be applicable in the absence of a background metric.

It is also possible that one must search for 'some much deeper and non-evident concepts' than the metric field (Bronstein) to quantize. This would suggest the validity of the third position:

(3) General relativity is essentially a macroscopic theory, to which the techniques of quantum field theory should not be (mis)applied (Frenkel, van Dantzig).

It is sobering to realize that these alternatives were clearly posed sixty years ago, particularly when one recalls that just sixty years separate Maxwell's treatise from the mid-thirties! After a hiatus during the World War II years,[28] work on the first line of research continued to dominate work on quantum gravity by quantum field theorists, at least until the discovery of the perturbative non-renormalizability of the Einstein equations. I shall not say more about this approach, since a fairly good picture of the post-war work is found in the 'Foreword' and introduction on 'Quantum Gravity' to the Feynman lectures on gravitation – to say nothing of the lectures themselves![29]

---

[26] See D. van Dantzig, 'Some possibilities of the future development of the notions of space and time,' *Erkenntnis* **7** (1938): 142–46; 'On the relation between geometry and physics and the concept of space-time,' in *Fünfzig Jahre Relativitätstheorie*, P. Kervaire, ed., *Helvetic Physica Acta Supplementum IV* (Birkhäuser, Basel 1956), pp. 48–53.

[27] Indeed, in 1949 Bryce DeWitt, using Schwinger's covariant technique, recalculated the gravitational self-energy of the photon and showed that it vanished identically. See Carl Bryce Seligman [DeWitt], 'I. The theory of gravitational interactions. II. The interactions of gravity with light,' Ph.D. Thesis, Harvard University, December 1949. Note that DeWitt emphasized the need to quantize the full, non-linear theory, and never regarded quantization of the linearized equations as more than a preliminary exercise.

[28] The only papers I have found on quantum gravity during the war years are a couple by Marie-Antoinette Tonnelat.

[29] Richard P. Feynman, Fernando B. Morinigo and William G. Wagner, *Feynman Lectures on Quantum Gravitation*, Bryan Hatfield, ed. (Addison-Wesley, Reading, MA, 1995), 'Foreword' by John Preskill and Kip S. Thorne, pp. vii–xxx; 'Quantum Gravity,' by Brian Hatfield, pp. xxxi–xl.

Rosenfeld himself became an advocate of the third viewpoint. He expressed himself most clearly on this issue during a discussion at the 1962 Warsaw–Jablonna relativity meeting:[30]

It is thinkable, however, that Einstein's equation has the character of a thermodynamic equation which breaks down and is not valid in case of extreme fluctuations... [S]ince we know the validity of the gravitational equations only for macroscopic systems, it is perhaps not so impossible to look at them as a kind of thermodynamic equations.

(pp. 221–22)

If we take the essence of the third viewpoint to be that general relativity should emerge as some sort of limiting case of a more fundamental theory, which is the one to quantize, then its main contemporary advocates are, of course, the string theorists. Penrose's work on twistor theory also falls within this category, about which I shall say no more; but I will conclude with some remarks on the post-war history of the second approach.

Work on this approach to quantum gravity started in the U.S. in the late 1940s, in complete ignorance of Bronstein's work and indeed, sometimes even of Rosenfeld's.[31] Four major groups soon formed: the Bergmann group, first at Brooklyn Polytechnic and then Syracuse; the DeWitt group at the University of North Carolina; the Wheeler group at Princeton; and Arnowitt, Deser and Misner, who constituted a group already (three people made a large relativity center in those days). The Syracuse group and Arnold, Deser and Misner worked primarily on the canonical formulation of general relativity; the North Carolina group on covariant quantization methods; and the Princeton group on both.

Starting with these groups, we may say, broadly speaking, that there have been three generations of workers on quantum gravity. And as everyone acquainted with quarks and leptons knows, three generations can accomplish things that two cannot! Fortunately, all three generations are represented here.

The first generation, that of the post-World War II founding fathers, is represented by Bryce DeWitt, who came to the subject from the side of quantum field theory. At the time, those of his colleagues who bothered to take any interest in the subject at all were treating general relativity as a Poincaré-invariant theory that happened to be non-linear and to have a very nasty gauge group, and trying to force it into the bed of Procrustes of special-relativistic quantum field theory. Bryce resisted that trend, and as he very gently indicated in his remarks yesterday, was not greatly encouraged in his efforts by his field-theory colleagues. Nevertheless, he has persisted with devotion to the subject to this day, uncovering many important features – and problems – of the elusive quantum gravity.

The next generation is the one that had to come to terms with the stubborn resistance of general relativity to all attempts at conventional quantization, and draw the lessons by identifying a series of issues that highlighted the reasons for these persistent failures. That generation is represented by Abhay Ashtekar, who

---

[30] See *Proceedings on the Theory of Gravitation/Conference in Warszawa and Jablonna 25–31 July 1962*, L. Infeld, ed. (Gauthier-Villars, Paris & PWN, Warszawa, 1964). Rosenfeld's comments are on pp. 144–45 and 219–22. See also the references in note 10. The views expressed here are very close to those of van Dantzig (see note 26).

[31] Indeed, his work created considerable consternation when it was discovered by the Syracuse group several years after Bergmann had started work on his program, which in part reduplicated some of Rosenfeld's (Ralph Schiller, personal communication).

has so well summarized the problems of, and done so much to blaze new paths in quantum gravity.

The contemporary generation takes for granted the need for new concepts and new techniques in the ongoing attempt to somehow combine general relativity and quantum field theory into some sort of coherent whole. Carlo Rovelli has contributed much toward the new efforts, while never forgetting the debt owed to, and while continuing to learn from, the work of the preceding generations.

Acknowledgement

I thank Dr Gennady Gorelik for a number of helpful comments.

12. Quantum field theory and space-time – formalism and reality

BRYCE DEWITT

Although thinkers had been groping for two hundred years for the idea of a field – something that would transmit forces across the vastness of interstellar space – it was not until Faraday and Maxwell that field theory in the modern sense was born. Not really in the modern sense, of course, because it was not initially decided whether fields are merely mathematical descriptions of a deeper reality or are themselves physical entities that occupy the absolute space of Newton. Maxwell certainly regarded the electromagnetic field as something physical, but he was curiously unwilling to ascribe to it an independent existence as a 'nonmechanical' entity in its own right.

Even Einstein, in the early years of this century, was unwilling to ascribe an independent existence to the gravitational field. Instead he made the bold advance of expressing it, not as something that occupies space-time, but as a property, *curvature*, of space-time itself, and he tried to develop a geometrical theory of the electromagnetic field as well.

The question 'Are fields merely mathematical descriptions of a deeper reality, or are they independent physical entities?' and questions like it, recur over and over in different guises. As titillating as such questions may be to some philosophers, the modern response to them is surely that they are irrelevant. The mathematical concept of a field is used in a host of different contexts, and one of the marvelous aspects of theoretical physics is that familiarity with a relatively small arsenal of such basic concepts suffices for the physicist to feel equally at home in a seminar on particle physics, in condensed matter physics, in statistical mechanics, or even in string theory. These concepts provide our only window on reality.

The advent of quantum mechanics did not change the picture. It gave rise to a rich new supply of fields. The first was Schrödinger's wave function itself: a complex-valued scalar field on space-time. Schrödinger hoped to use the wave function to get rid of 'quantum jumps'. But it can also be used for precisely the opposite purpose by noting that it can be regarded as a dynamical system like any other field. What is more natural than to quantize *it*, a process known as *second* quantization. The result is like hitting a jackpot. The unquantized wave function describes only a single particle; the quantized one describes a potential infinity of indistinguishable particles and serves as a creation or annihilation operator. (Thus, discreteness comes back with a vengeance.) Moreover, the quantization can be performed in two ways, one leading to Bose statistics, the other to Fermi statistics. Most important of all, the differential equation satisfied by the field can be modified so that it is no longer linear, like the original Schrödinger equation, but nonlinear instead. The nonlinearities describe

176

interactions between the particles, and the resulting theory is used to this day in condensed matter physics.

The original Schrödinger wave function was a field in ordinary $(3 + 1)$-dimensional space-time. It was understood very early that if one restricts oneself to that sector of the nonlinear second-quantized theory which corresponds to a fixed number N of particles, the resulting formalism can equally well be expressed in terms of a wave function satisfying a *linear* equation in a space of $3N$ dimensions plus one time dimension. Schrödinger had hoped to confine the discussion of quantum problems to ordinary $(3 + 1)$-dimensional space-time, but much more important, in fact, is the wave function in the $3N$ dimension *configuration space* of each problem. It is the potential complexity of the wave function in a space of a large number of dimensions that allows the simultaneous emergence of both classical behavior and quantum jumps, as was shown in a beautiful but neglected paper by Nevill Mott in 1929, entitled 'The wave mechanics of α-ray tracks.'[1]

The method of second quantization not only gives an explanation of the absolute indistinguishability of particles of a given species but also opens up the possibility of characterizing different species by associating them with different kinds of fields on space-time. Moreover, these fields can be chosen so as to respect the principles of relativity. Thus we have scalar fields satisfying the Klein–Gordon equation and vector fields satisfying the Proca equation, the Maxwell equation, or the Yang–Mills equation. Of course, since physicists are not mathematicians, the last named equation, which really belongs in the framework of the theory of connections on principal fiber bundles, did not come till rather later. Another famous case, in which the mathematics was reinvented in a physicist's typical clumsy way, was Dirac's discovery of the spinor field satisfying his epochal equation. Nowadays, physicists classify particles by their 'spin' and 'parity', referring to particular representations of the covering group of the full Lorentz group, and they know that the constraints of relativity theory limit the freedom they have in their choice of statistics when dealing with the old nonrelativistic Schrödinger equation.

Just as nonlinearities can be introduced into the old Schrödinger equation to describe interacting particles, so can they be introduced into relativistic quantum field theories. The paradigm of a nonlinear relativistic theory is the fabulously successful quantum electrodynamics, in which the Dirac spinor field is coupled to the Maxwell field to describe the interactions between electrons and photons. In recent decades increased attention has been devoted to fields that are intrinsically nonlinear, such as the Yang–Mills and gravitational fields. These fields, which are known as *gauge fields*, cannot even be described except in terms of nonlinear equations. The Yang–Mills field has played a fundamental role in the strikingly successful Standard Model, which unites electrodynamics with the theory of weak interactions. The gravitational field has so far played no role in quantum theory – *except* for throwing down the great challenge: how does it fit with quantum theory? What *is* quantum gravity?

I do not wish to discuss the technical obstacles that have impeded the development of a complete and consistent quantum theory of gravity. Rather, let me step back and ask, 'Where are we heading?' Other fields have been previously described as being *in* or *on* space-time. The gravitational field is supposed to be a geometrical property *of* space-time. From a mathematical point of view there really is no distinction here. The

---

[1] N.F. Mott, *Proc. Roy. Soc.* (London) **126**, 79 (1929).

metric field, which defines distance scales, is the basic descriptor of space-time geometry, and it satisfies dynamical equations, Einstein's equations, like any other field. That it can carry energy from one place to another is so well confirmed by the behavior of one particular binary pulsar, that very expensive antennae are being constructed to detect the energy flow. There is certainly no reason for the theorist not to be carried along by the mathematics and by very basic arguments concerning the consistency of quantum mechanics. Since it interacts with all other dynamical systems, the gravitational field simply *must* be quantized, whether by string theory or through some as yet unsuspected approach.

Yet the very idea of quantizing gravity raises a host of unsettling questions. Let us begin with the simplest. From a classical point of view, a pure gravitational field is a hard vacuum. To be sure, space-time is curved, but nothing else is there. So what *is* a vacuum? In a lecture given in Leyden in 1920, Einstein appealed to his audience not to abandon the old idea of the ether. (Yes, Einstein made such an appeal.) He was not proposing to deny the principles of relativity but rather to refurbish the idea of the ether as an entity with *texture*. Curvature is a part of that texture and, in Einstein's view, the ether undoubtedly possesses other textures as well. This view, in fact, is rather modern. In quantum field theory one regards the ether, that is, the vacuum, as being exceedingly complex, a cauldron of interacting particles, each having only a fleeting, or *virtual*, existence. The vacuum, in fact, carries a blueprint for the whole field dynamics, the expression of which was first provided by Julian Schwinger in his introduction of the *vacuum persistence amplitude*. For every briefly time independent classical *background field* there is a corresponding quantum vacuum state. The vacuum persistence amplitude is the probability amplitude that the fields will remain in this state if they start out in this state. By noting the response of this amplitude to variations in the background field, Schwinger was able to point the way to a complete description of the quantum field dynamics.

One immediate consequence of the complexity of the vacuum is that one can no longer say that the vacuum is the state in which nothing else is present. A good example is provided by the Casimir effect: two plane parallel conductors are placed near each other in a standard laboratory vacuum. Neither carries an electric charge and there is no potential difference between them; yet they exert an attractive force on each other. Moreover, the energy density of the vacuum between them is not zero, but *negative*, and, if the law of conservation of energy is not to be violated, must act as a source for the gravitational field, like any other source of energy, including gravitational. Curvature produces a similar effect. The stress-energy density in the vacuum in a region where space-time is curved is not zero, but nonvanishing in an exceedingly complex way. In a proper quantum theory of gravity the effect of this nonvanishing stress-energy on the gravitational field itself (that is, on the curvature) would have to be taken into account, leading to a modification of Einstein's equations.

These examples are akin to the well known phenomenon of vacuum polarization: a nonvanishing electric field (produced by a proton, for example) is introduced into the laboratory vacuum. The space occupied by this field is no longer electrically neutral but carries an induced charge density. This charge density in turn modifies the electric field, leading to corrections in Maxwell's equations. In our case a pair of neutral conductors or a gravitational field (curvature) is introduced into the vacuum. A nonvanishing stress-energy density is induced, leading to corrections in Einstein's equations.

Vacuum polarization is usually regarded as a phenomenon of virtual particles, arising when stationary fields (electrical, gravitational, etc.) are introduced as backgrounds to a laboratory vacuum. 'Real' particles can be produced by *non*stationary background fields. Examples are electron-positron pairs produced when the oscillating electromagnetic field carried by a photon passes close to an atomic nucleus, another is the Hawking radiation produced by the rapidly changing curvature arising from gravitational collapse to a black hole.

The alert listener will notice that I have been using the words 'particle' and 'vacuum' here as if I knew what I were talking about. The word 'particle' usually elicits mental pictures of tracks in bubble chambers or sudden changes in velocity of Millikan oil drops. In fact the theorist uses these words in a narrow technical sense. For each stationary classical background field there is a ground state of the associated quantized field. This is the vacuum *for that background.* 'Particles' are associated with excited states of the field, these 'states' being points or rays in a space known as *Fock space.* As a general rule these technical definitions correspond well with our common sense notions of 'vacuum' and 'particle', but not always. One of the most important results of the development of the quantum theory of fields in exotic backgrounds (notably in curved space-time) is the recognition that how a given so-called 'particle detector' responds in a given Fock state depends both on the nature of the detector and on its state of motion, as well as, of course, on the state, the Fock state. Thus a standard detector will become excited (that is, detect something) when it is accelerated in a Minkowski vacuum, but will detect nothing when it is in uniform motion. This is the Unruh effect. The same detector will detect nothing when it is in free fall in the so-called Hartle–Hawking vacuum outside a black hole but will become excited when it is at rest with respect to the hole. There is another vacuum, the so-called Boulware vacuum, in which its behavior is exactly opposite.

The blurring of the concepts of vacuum and particle carries over to a blurring of the distinction between real and virtual particles. Both real and virtual particles carry electric (and other) charges, as well as stress-energy, which modify the fields of which these quantities are sources, and produce quantum corrections to classical field equations. All of these effects can be regarded as part of the general *back reaction problem.* Back reaction effects are best analyzed, from a conceptual point of view, within the framework of the so-called *quantum effective action*, which is essentially $-i$ times the logarithm of Schwinger's vacuum persistence amplitude. The quantum effective action is a function*al* of the background fields, that is, a function on the *space of field histories* – not the classical space of field histories but its complex extension. Solutions of the quantum corrected field equations are stationary points of the quantum effective action, just as solutions of the classical field equations are stationary points of the classical action.

The quantum effective action is most easily defined (and computed) in terms of Feynman's functional integral for the vacuum persistence amplitude, which is essentially a formal integral of Schwinger's functional differential (or variational) equations for the amplitude, and which nowadays is usually taken as the point of departure for what we *mean* by the quantum theory.

An analysis of the quantum effective action that goes beyond the merely superficial leads one to study, in a nontrivial way, the structure and geometry of the space of field histories. For gauge theories including gravity, the space of field histories is a principal fiber bundle having Killing vector fields as well as a natural metric and a natural

connection. The frame bundle of the space of field histories also has a special connection, with a number of remarkable properties that were first noted by Grigori
Vilkovisky. Judicious use of these properties in the case of Yang–Mills fields has
recently led to a new way of computing physical amplitudes and carrying out renormalization procedures, which dispenses entirely with the famous 'ghosts' originally
introduced by Feynman to preserve the unitarity of quantum theory. A careful
study of the functional integral itself, which is an integral *over* the space of field
histories, has led to the identification of a natural *measure* for the integral, which
turns out to be intimately related to the Wick rotation procedure used in evaluating
the integral in the loop expansion.

What are we doing here? We started out with fields – scalar, vector, spinor – on
ordinary space-time. We then passed to fields (the Schrödinger fields) on space-
times of $3N + 1$ dimensions. And now we are studying fields (Killing vectors, connections and so on) on an infinite dimensional space, the space of field histories. I could
have called attention, in passing, to other structurally nontrivial infinite dimensional
spaces: Hilbert space (that's the space of Schrödinger wave functions at given instants
of time), Fock space (the space of arbitrary fields on chosen sections of space-time). I
have not mentioned at all the higher finite-dimensional spaces that one meets in
Kaluza–Klein theory, in supergravity, or in string theory.

In which of these lies reality? In the remainder of this talk I shall try to suggest not
that reality lies in the whole mathematical structure (because we do not yet have a
theory of everything) but that our *grasp* of reality, however imperfect it is, requires
the whole mathematical structure. I shall do this by discussing the most unsettling
question of all that a quantum theory of gravity raises: how is the quantum theory
in the end to be interpreted?

Since gravity plays no role at all at the quantum level except under the most
extreme – so-called Planckian – conditions, the natural setting for quantum gravity
is quantum cosmology, the theory of the Big Bang and its after-effects. In papers on
quantum cosmology one often introduces the so-called Wheeler–DeWitt equation. I
discuss this equation here not because my name is associated with it (in fact I think it
is a very bad equation, based on the relativity-violating procedure of cutting space-
time into 'time slices') but because it suffices for me to make my point. It is a
second-order functional differential equation satisfied by a complex-valued functional
Ψ. This functional Ψ is a scalar field on the Cartesian product of the space of 3-
dimensional geometries that a given section (or 'time slice') of space-time may
assume and the space of values that all nongravitational physical fields may assume
over that section. Here we have yet another field, Ψ, and another space, call it Σ.
For each 3-geometry *cum* set of nongravitational fields, that is, at each point of Σ,
Ψ has a certain complex value. This value is something like a probability amplitude
for that 3-geometry *cum* field distribution. But not quite, because the Wheeler–
DeWitt equation does not yield a simple probability conservation law with a positive
probability. It also does not yield easily a time with respect to which one may view the
evolution of probability, which is not surprising because the very last thing that a
general relativistic theory is going to give is an absolute time. But these are technical
problems, which are not our primary concern here. The important point is that Ψ is
something like a wave function for the whole universe!

Some workers have shied away from this concept, and from the technical difficulties as well, by proposing that one *quantize* the Wheeler–DeWitt equation. That

would be *third* quantization and conjures up creation and annihilation operators for whole universes! Fortunately one does not have to go to this extreme.

What is certain is that gravity leads unavoidably to a consideration of the whole universe as a quantum object. What can that mean? There is nothing outside the universe with respect to which one can view its quantum behavior. It is an *isolated* system. And in my view, we are not going to understand quantum gravity until we are able to understand the quantum mechanics of isolated systems.

Fortunately, in the last few years great progress has been made on this problem. The key word here is *decoherence*, the development of random phase relationships between parts of the wave function. We now know that there are two major kinds of decoherence that play dominant roles in nature: the decoherence of the state of a light body (typically subatomic) brought about by careful coupling to a massive body (known as an apparatus) and the decoherence of the state of a massive body that results from collisions that it undergoes with light bodies in the ambient medium around it. The first is known as a *quantum measurement* and was first fully understood in the modern sense by Everett in 1957. The second, which was understood in its essentials already by Mott, is ubiquitous, and is what gives rise to the emergence of classical behavior in the world around us, and makes it possible for us to carefully prepare an apparatus in a state that will yield a good measurement on a 'quantum' system.

It is important to note that the apparatus is part of the wave function and that even we ourselves can be part of it. As for the mathematics that describes an isolated quantum system it plainly paints a picture consisting of *many worlds*; all the possible outcomes of a quantum measurement are there. Each world is unaware of the others because of decoherence. The concept of decoherence can be made precise by introducing so-called decoherence functions, which allow one to describe in a uniform manner both types of decoherence previously mentioned. People such as Gell-Mann and Hartle who, along with Omnès and many others, have developed the decoherence functional formalism, often (and correctly) refer to their work as *post-Everett*. Their mathematics again paints a picture consisting of many worlds or, as they prefer to call them, *many histories* or independent *consistent* histories.

An important feature of their work concerns the notion of probability, in the quantum sense. In standard nonrelativistic quantum mechanics (despite what some of my colleagues have tried to tell me) probability – that is, the standard language of the probability calculus – emerges willy nilly from the formalism. The probability interpretation of quantum mechanics does *not* have to be grafted onto the theory as an independent postulate. This is true both in the Everett and post-Everett formalisms. But whether the probability concept is a useful one for the subunits (or observers) of an isolated system depends on the quantum state of that system. (There is no outside system with respect to which an absolute notion of probability can be defined.) Whether probability is a useful concept depends on the nature of the decoherence function to which, for each choice of coarse graining, the state gives rise. The post-Everett workers have identified those mathematical properties of the decoherence function that translate directly into probability statements with some reasonable degree of precision. Their work is of great importance for a general relativistic quantum theory (that is, quantum gravity) because of the intimate relation between the concepts of probability and *time*. In a general relativistic theory, time is a phenomenological concept, not an absolute one. It is what a clock measures and can never be infinitely precise. Probability too, in the end, is a phenomenological concept. It can never be infinitely precise.

What does all this teach us? In my view it teaches us the following. We should always push the mathematical formalism of physics, and its internal logic, to their ultimate conclusions. We should let the logic carry us along, at every level, subject to corrections from experimental physics. In the end we shall not have arrived at reality, but we shall have obtained the best grasp of reality that we can get.

Discussions

DeWitt: May I abuse my position on the podium since I didn't spend a full 50 minutes. I thought somebody might like to ask me the question: how does the probability interpretation automatically emerge from the formalism? The argument is due to my old friend David Deutsch,[1] and I'd like to show it to you because it's not generally known.

Let $|\Psi\rangle$ be the state vector of a system S with observable **s** having eigenvalues s, which for simplicity will be assumed to come from a discrete set. The problem is to find an interpretation for the coefficients

$$c_s = \langle s|\Psi\rangle \qquad (1)$$

in the superposition

$$|\Psi\rangle = \sum_s c_s|s\rangle, \qquad (2)$$

$|s\rangle$ being the eigenvectors of **s**. (We suppress other labels and assume all vectors to be normalized to unity.) Suppose, in the superposition (2), that a particular eigenvalue of **s** fails to appear, i.e., that $c_s = 0$ for that eigenvalue. Then it is natural to say that if an apparatus designed to measure **s** were coupled to s, it would have *zero likelihood* of observing the value. But what can one say about the nonvanishing coefficients?

Denote by S the set of eigenvalues of **s**. Let F be a finite subset of S contained in the support of the function c_s. Let P be the group of permutations of F and, for each $\Pi \in P$ let $U(\Pi)$ be the operator defined by

$$U(\Pi)|s\rangle = |\Pi(s)\rangle. \qquad (3)$$

It is easy to show that $U(\Pi)$ is a unitary operator for each Π and that

$$U(\Pi)^* = U(\Pi^{-1}). \qquad (4)$$

Let T be an external agent that can act on S in such a way that the state vector $|\Psi\rangle$, whatever it is, undergoes the change

$$|\Psi\rangle \rightarrow U(\Pi)|\Psi\rangle, \qquad (5)$$

and suppose that T can choose to do this for any Π in P. Let R be a second agent that can make arbitrary measurements on S after T has acted on it. One may conveniently speak of T as a *transmitter* and R as a *receiver*. Thus if R knows that the state vector of S, before T acts on it, is $|\Psi\rangle$, and if T sends a *signal* to R, by choosing a particular permutation in (5), then R may try to determine whether there has been

---

[1] Oxford preprint (1989), unpublished.

a change in $|\Psi\rangle$. His best strategy for determining, first of all, whether anything other than a null signal has been sent is to measure the projection operator $|\Psi\rangle\langle\Psi|$. If he obtains the value 1 for this observable he cannot tell whether $|\Psi\rangle$ has been changed, but if he obtained the value 0 then he knows that it has.

Suppose $|\Psi\rangle$ has the property that, no matter which permutation T chooses to perform, R cannot detect the change in $|\Psi\rangle$ by any measurement whatever. That is, the permutations themselves are undetectable and hence leave the physics of S unchanged. It is natural, under these circumstances, to say that if a measurement of **s** were to be made then all the outcomes s that lie in F are *equally likely*. Note that although this terminology is probabilistic the statement itself is purely factual, concerning the impossibility of using permutations to send signals, and has no probabilistic antecedents. As to the appropriateness of the terminology, just ask yourselves what odds you would place on the relative likelihood of the various outcomes in F.

When all the outcomes in F are equally likely it is impossible for R to obtain the value 0 when he measures $|\Psi\rangle\langle\Psi|$. But this means

$$\mathbf{U}(\Pi)|\Psi\rangle = e^{i\chi(\pi)}|\Psi\rangle \quad \text{for all } \Pi \in \mathbf{P} \tag{6}$$

for some $\chi: \mathbf{P} \rightarrow \mathbf{R}$. From (3) and (4) it follows that

$$\langle s|\Psi\rangle = e^{-i\chi(\pi)}\langle s|\mathbf{U}(\Pi)|\Psi\rangle = e^{-i\chi(\pi)}\langle\Psi|\mathbf{U}(\Pi)^*|s\rangle^*$$
$$= e^{-i\chi(\pi)}\langle\Psi|\Pi^{-1}(s)\rangle^* = e^{-i\chi(\pi)}\langle\Pi^{-1}(s)|\Psi\rangle, \tag{7}$$

and hence

$$|\langle s|\Psi\rangle| = |\langle\Pi^{-1}(s)|\Psi\rangle| \quad \text{for all } \Pi \in P. \tag{8}$$

Conversely it is easy to show that if (8) holds, then an operator $U(\Pi)$ satisfying (3) exists for all Π.

It follows that all the members of a finite subset F of eigenvalues of **s** are equally likely to be observed in a measurement if and only if $|\langle s|\Psi\rangle|$ is constant over F. Suppose F has n members and happens to coincide with the support of the function c_s. Then no eigenvalues will be observed other than those contained in F, and it is natural to push the probabilistic terminology one step further by saying that each of the eigenvalues in F has *probability* $1/n$ of being observed. When this happens one has

$$1 = \langle\Psi|\Psi\rangle = \sum_{s'}\langle\Psi|s'\rangle\langle s'|\Psi\rangle = n|\langle s|\Psi\rangle|^2, \quad s \in F, \tag{9}$$

so this probability can be expressed in the form

$$1/n = |\langle s|\Psi\rangle|^2 = |c_s|^2, \quad s \in F. \tag{10}$$

I stress again that although the terminology of probability theory is now being used, the words themselves have no probabilistic antecedents. They are defined neither in terms of an *a priori* metaphysics nor in terms of the mathematical properties of the state vector, but in terms of factual physical properties of the system S in the state that the state vector represents. However, once the terminology of probability theory has been introduced there need be no hesitation in using it in exactly the same way as it is used in the standard probability calculus. That is, the probability calculus, in particular the calculus of *joint* probabilities, may be freely used to motivate further definitions.

With this in mind I turn to the interpretation of $|\langle s|\Psi\rangle|^2$ when it is not constant over the support of the function c_s. In this case the elucidation of the meaning $|\langle s|\Psi\rangle|^2$ requires a more elaborate, nonetheless firmly physically based, argument. Introduce two auxiliary physical systems, Q and R, in addition to S, together with their state vectors $|\varphi\rangle$ and $|\chi\rangle$. Let A be a subset of m distinct eigenvalues q of an observable $\mathbf{q}$ of Q, and let B be a subset of $n - m$ distinct eigenvalues r of an observable $\mathbf{r}$ in R. Let the r's in B be all different from the q's in A so that the set $A \cup B$ has n distinct elements.

Suppose $|\varphi\rangle$ and $|\chi\rangle$ are such that all the q's in A are equally likely and all the r's in B are equally likely. Suppose, furthermore, that $\langle q|\varphi\rangle = 0$ when $q \notin A$ and $\langle r|\chi\rangle = 0$ when $r \notin B$. Then each q in A has probability $1/m$ of being observed and each r in B has probability $1/(n - m)$ of being observed. In mathematical language

$$|\langle q|\varphi\rangle|^2 = 1/m, \quad q \in A, \tag{11}$$

$$|\langle r|\chi\rangle|^2 = 1/(n - m), \quad r \in B. \tag{12}$$

In the combined state vector space of the systems Q, R and S consider the operator

$$\mathbf{u} = \mathbf{q} \otimes \mathbf{1} \otimes |s\rangle\langle s| + \mathbf{1} \otimes \mathbf{r} \otimes (\mathbf{1} - |s\rangle\langle s|). \tag{13}$$

This operator, which is an observable of the combined system, can be measured as follows: First measure $\mathbf{s}$. If s is obtained then measure $\mathbf{q}$. If s is not obtained measure $\mathbf{r}$ instead. The final outcome in either case is the measured value of $\mathbf{u}$. Note that $\mathbf{u}$ has n distinct eigenvalues u lying in the set $A \cup B$, and these are the only eigenvalues that can turn up if the state vector of the combined system is

$$|\Psi\rangle = |\varphi\rangle|\chi\rangle|\psi\rangle. \tag{14}$$

Now suppose the state of S (state vector $|\psi\rangle$) is such that, when $\mathbf{u}$ is measured, all the n outcomes u lying in $A \cup B$ are equally likely, as defined by the permutation-signaling prescription. Then each u in $A \cup B$ has probability $1/n$ of being observed. Given the prescription for measuring u, it follows from the calculus of joint probabilities that

$$\frac{1}{n} = p \times \frac{1}{m} \quad \text{or} \quad (1 - p)\frac{1}{n - m} \tag{15}$$

where p is *the probability that* s *will be observed when* $\mathbf{s}$ *is measured*. Note that p is defined via the calculus of joint probabilities and that both possibilities in (15) lead to

$$p = \frac{m}{n}. \tag{16}$$

To relate p to $|\langle s|\psi\rangle|^2$ consider the projection operator on the eigenvalue u of $\mathbf{u}$:

$$\mathbf{P_u} = \sum_q \delta_{uq}|q\rangle\langle q| \otimes \mathbf{1} \otimes |s\rangle\langle s| + \sum_r \delta_{ur}\mathbf{1} \otimes |r\rangle\langle r| \otimes (\mathbf{1} - |s\rangle\langle s|). \tag{17}$$

If $u \in A \cup B$ then

$$\frac{1}{n} = \langle \Psi | \mathbf{P}_u | \Psi \rangle$$

$$= \sum_{q \in A} \delta_{uq} |\langle q | \varphi \rangle|^2 |\langle s | \psi \rangle|^2 + \sum_{r \in B} \delta_{ur} |\langle r | \chi \rangle|^2 (1 - |\langle s | \psi \rangle|^2)$$

$$= \frac{1}{m} |\langle s | \psi \rangle|^2 \quad \text{or} \quad \frac{1}{n-m} (1 - |\langle s | \psi \rangle|^2), \tag{18}$$

in which (11) and (12) are used in passing to the last line. Comparing (18) with (15) one sees immediately that

$$p = |\langle s | \psi \rangle|^2 = |c_s|^2. \tag{19}$$

Note that this result does not require the support of the function c_s to be a finite subset of S.

The above analysis gives a factual meaning to all rational probabilities. The extension to irrational probabilities is achieved by introducing the notion of 'at least as likely' and making a kind of Dedekind cut in the space of state vectors. But such an extension is, at best, unnecessarily pedantic and, at worst, unphysical, for the following reasons. Firstly, probabilities themselves are physically measurable only when viewed as frequencies, and physically measured numbers are always rational. Secondly, the set of comparison states in the Dedekind cut definition is infinite in number, so the definition is a non-operational one. It cannot be checked by any physical process. Setting these quibbles aside, however, it is clear that the probability interpretation of quantum mechanics emerges from the standard operator–state-vector formalism and does not have to be grafted onto the theory as an independent postulate.

Wightman: Where did the probability distribution, the equal probability distribution come from? The one you started from.

DeWitt: I said: Suppose $|\psi\rangle$ is such that you cannot detect any change in it by the action of the permutation operator. Then how are you going to view $|\psi\rangle$? No value of s is preferred over any other. What odds will you inevitably give? David Deutsch even discusses a calculus of values. Are you going to bet that one value of s would be more likely than another?

Wightman: Depends on the state.

DeWitt: The state is one in which you cannot detect any change under permutation.

Coleman: It seems to me that this is an argument of the structure: if we are to assign a probability interpretation, this is how we must assign it. We can talk about it at length, but clearly the pressure of time keeps us from that. But even within that, it seems to be an argument only for these special states. If you don't have a state of the special form, it's certainly 'natural' to generalize the calculus you've developed for these special states to the general case. But I don't see that it's logically inevitable.

DeWitt: No. Suppose we have a state that has varying values for c_s. We need an interpretation for it. Suppose there exist these two other auxiliary systems, Q and R, with the given properties. Then we are driven to the usual probability interpretation. It's true, one has to introduce the auxiliary systems.

Coleman: I have to think about this more, I should have kept my mouth shut. Let's talk about it privately.

Weingard: The question I'd like to ask is, there is an interpretation of quantum mechanics, Bohm's theory, according to which it's possible to have a consistent dynamics for actual particle trajectories, that's consistent with Schrödinger's evolution, without obeying the probability rule. How can you possibly derive the probability interpretation from regular quantum mechanics, because there psi squared does not give the probability?

DeWitt: Well, Bohm's theory is not quantum mechanics, I'm talking about conventional quantum mechanics.

Weingard: But still you're saying you can derive the probability interpretation from conventional quantum mechanics. How can that be if there's an interpretation where that probability interpretation, that probability rule, doesn't hold, that you still have the same formalism. The point is that the actual probability distribution is not that given by psi squared. You need special initial conditions on Bohm theory. That's why Bohm has to introduce a postulate that psi squared equals the probability distribution. That's why the interpretation can't follow just from the quantum mechanics.

DeWitt: Well, I would say that's a weakness in Bohm's theory then.

Weingard: No, this is about logic.

13. Quantum field theory of geometry[1]

ABHAY ASHTEKAR AND JERZY LEWANDOWSKI

1 Introduction

Several speakers at this conference have emphasized the conceptual difficulties of quantum gravity (see particularly [1–3]). As they pointed out, when we bring in gravity, some of the basic premises of quantum field theory have to undergo radical changes: *we must learn to do physics in the absence of a background space-time geometry*. This immediately leads to a host of technical difficulties as well, for the familiar mathematical methods of quantum field theory are deeply rooted in the availability of a fixed space-time metric, which, furthermore, is generally taken to be flat. The purpose of this contribution is to illustrate how these conceptual and technical difficulties can be overcome.

For concreteness, we will use a specific non-perturbative approach and, furthermore, limit ourselves to just one set of issues: exploration of the nature of quantum geometry. Nonetheless, the final results have a certain degree of robustness and the constructions involved provide concrete examples of ways in which one can analyze genuine field theories, with an infinite number of degrees of freedom, in absence of a background metric. As we will see, the underlying diffeomorphism invariance is both a curse and a blessing. On the one hand, since there is so little background structure, concrete calculations are harder and one is forced to invent new regularization methods. On the other hand, when one does succeed, the final form of results is often remarkably simple since the requirement of diffeomorphism invariance tends to restrict the answers severely. The final results are often unexpected and qualitatively different from the familiar ones from standard quantum field theories.

Let us begin with a brief discussion of the issue on which we wish to focus. In his celebrated inaugural address, Riemann suggested [4] that geometry of space may be more than just a fiducial, mathematical entity serving as a passive stage for physical phenomena, and may in fact have direct physical meaning in its own right. As we know, general relativity provided a brilliant confirmation of this vision: Einstein's equations put geometry on the same footing as matter. Now, the physics of this century has shown us that matter has constituents and the 3-dimensional objects we perceive as solids in fact have a discrete underlying structure. The continuum description of matter is an approximation which succeeds brilliantly in the macroscopic regime but fails hopelessly at the atomic scale. It is therefore natural to ask if the same is true of geometry. Does geometry also have constituents at the Planck scale? What are its atoms? Its elementary excitations? Is the space-time continuum

---

[1] Talk presented by A.A.

only a 'coarse-grained' approximation? If so, what is the nature of the underlying quantum geometry?

To probe such issues, it is natural to look for hints in the procedures that have been successful in describing matter. Let us begin by asking what we mean by quantization of physical quantities. Take a simple example – the hydrogen atom. In this case, the answer is clear: whereas the basic observables – energy and angular momentum – take on a continuous range of values classically, in quantum mechanics their spectra are discrete. So, we can ask if the same is true of geometry. Classical geometrical observables such as areas of surfaces and volumes of regions can take on continuous values on the phase space of general relativity. Are the spectra of corresponding quantum operators discrete? If so, we would say that geometry is quantized.

Thus, it is rather easy to pose the basic questions in a precise fashion. Indeed, they could have been formulated soon after the advent of quantum mechanics. Answering them, on the other hand, has proved to be surprisingly difficult. The main reason, we believe, is the inadequacy of the standard techniques. More precisely, in the traditional approaches to quantum field theory, one *begins* with a continuum, background geometry. To probe the nature of quantum geometry, on the other hand, we should not begin by assuming the validity of this model. We must let quantum gravity decide whether this picture is adequate at the Planck scale; the theory itself should lead us to the correct microscopic model of geometry.

With this general philosophy, in this talk we will use a non-perturbative, canonical approach to quantum gravity to probe the nature of quantum geometry. In this approach, one uses $SU(2)$ connections on a 3-manifold as configuration variables; 3-metrics are constructed from 'electric fields' which serve as the conjugate momenta. These are all dynamical variables; to begin with, we are given only a 3-manifold without *any* fields. Over the past three years, this approach has been put on a firm mathematical footing through the development of a new functional calculus on the space of gauge equivalent connections [4–12]. This calculus does not use any background fields (such as a metric) and is therefore well-suited for a fully non-perturbative exploration of the nature of quantum geometry.

In section 2, we will introduce the basic tools from this functional calculus and outline our general strategy. This material is then used in section 3 to discuss the main results. In particular, operators corresponding to areas of 2-surfaces and volumes of 3-dimensional regions are regulated in a fashion that respects the underlying diffeomorphism invariance. They turn out to be self-adjoint on the underlying (kinematical) Hilbert space of states. A striking property is that their spectra are *purely* discrete. This indicates that the underlying quantum geometry is far from what the continuum picture might suggest. Indeed, the fundamental excitations of quantum geometry are 1-dimensional, rather like polymers, and the 3-dimensional continuum geometry emerges only on coarse graining [13, 32]. In the case of the area operators, the spectrum is explicitly known. This detailed result should have implications for the statistical mechanical origin of the black hole entropy [15, 16] and the issue is being investigated. Section 4 discusses a few ramifications of the main results.

Our framework belongs to what Carlo Rovelli referred to in his talk as 'general quantum field theory'. Thus, our constructions do not directly fall in the category of axiomatic or constructive quantum field theory and, by and large, our calculations do not use the standard methods of perturbative quantum field theory. Nonetheless, we *do* discuss the quantum theory of a system with an infinite number of degrees of freedom (which, moreover, is diffeomorphism covariant) and face the issues of regularization squarely. For this, we begin '*ab initio*', construct the Hilbert space of states, introduce on it well-defined operators which represent (regulated) geometric observables and examine their properties.

Details of the results discussed here can be found in [17–19]. At a conceptual level, there is a close similarity between the basic ideas used here and those used in discussions based on the 'loop representation' [13, 20–22]. (For a comparison, see [17, 19].) Indeed, the development of the functional calculus which underlies this analysis was itself motivated, in a large measure, by the pioneering work on loop representation by Rovelli and Smolin [23].

Finally, we emphasize that this is *not* a comprehensive survey of non-perturbative quantum gravity; our main purpose, as mentioned already, is to illustrate how one can do quantum field theory in absence of a space-time background and to point out that results can be unexpected. Indeed, even the use of general relativity as the point of departure is only for concreteness; the main results do not depend on the details of Einstein's equations.[2]

2 Tools

This section is divided into four parts. The first summarizes the formulation of general relativity based on connections; the second introduces the quantum configuration space; the third presents an intuitive picture of the *non-perturbative* quantum states and the fourth outlines our strategy to probe quantum geometry.

2.1 From metrics to connections

The non-perturbative approach we wish to use here has its roots in canonical quantization. The canonical formulation of general relativity was developed in the late fifties and early sixties in a series of papers by Bergmann, Dirac, and Arnowitt, Deser and Misner. In this formulation, general relativity arises as a dynamical theory of 3-metrics. The framework was therefore named *geometrodynamics* by Wheeler and used as a basis for canonical quantization both by him and his associates and by Bergmann and his collaborators. The framework of geometrodynamics has the advantage that classical relativists have a great deal of geometrical intuition and physical insight into the nature of the basic variables: 3-metrics g_{ab} and extrinsic curvatures K_{ab}. For these reasons, the framework has played a dominant role, e.g., in numerical relativity. However, it also has two important drawbacks. First, it sets the mathematical treatment of general relativity quite far from that of theories of other interactions where the basic dynamical variables are connections rather than metrics. Second, the equations of the theory are rather complicated in terms of metrics and extrinsic curvatures; being non-polynomial, they are difficult to carry over to quantum theory with a reasonable degree of mathematical precision.

---

[2] Nonetheless, since there were several remarks in this conference on the viability of quantum general relativity, it is appropriate to make a small digression to clarify the situation. It is well known that general relativity is perturbatively non-renormalizable. However, there *do* exist quantum field theories which share this feature with general relativity but are *exactly soluble*. A striking example is $(GN)_3$, the Gross–Neveau model in 3 dimensions. Furthermore, in the case of general relativity, there are *physical* reasons which make perturbative methods especially unsuitable. Whether quantum general relativity can exist non-perturbatively is, however, an open question. For further details and current status, see e.g. [24].

For example, consider the standard Wheeler–DeWitt equation:

$$\left[\frac{G\hbar}{\sqrt{g}}\left(g^{ab}g^{cd}-\frac{1}{2}g^{ac}g^{bd}\right)\frac{\delta}{\delta g_{ac}}\frac{\delta}{\delta g_{bd}}-\frac{\sqrt{g}}{G\hbar}R(g)\right]\circ\Psi(g)=0 \qquad (2.1)$$

where g is the determinant of the 3-metric g_{ab} and R its scalar curvature. As is often emphasized, since the kinetic term involves products of functional derivatives evaluated at the same point, it is ill-defined. However, there are also other, deeper problems. These arise because, in field theory, the quantum configuration space – the domain space of wave functions Ψ – is larger than the classical configuration space. While we can restrict ourselves to suitably smooth fields in the classical theory, in quantum field theory we are forced to allow distributional field configurations. Indeed, even in the free field theories in Minkowski space, the Gaussian measure that provides the inner product is concentrated on genuine distributions. This is the reason why in quantum theory fields arise as operator-valued distributions. One would expect that the situation would be at least as bad in quantum gravity. If so, even the products of the 3-metrics that appear in front of the momenta, as well as the scalar curvature in the potential term, would fail to be meaningful. Thus, the left hand side of the Wheeler–DeWitt equation is seriously ill-defined and must be regularized appropriately.

However, as we just said, the problem of distributional configurations arises already in the free field theory in Minkowski space-time. There, we do know how to regularize physically interesting operators. So, why can we not just apply those techniques in the present context? The problem is that those techniques are tied to the presence of a background Minkowski metric. The covariance of the Gaussian measure, for example, is constructed from the Laplacian operator on a space-like plane defined by the induced metric, and normal ordering and point-splitting regularizations also make use of the background geometry. In the present case, we do *not* have background fields at our disposal. We therefore need to find another avenue. What is needed is a suitable functional calculus – integral and differential – that respects the diffeomorphism invariance of the theory.

What space are we to develop this functional calculus on? Recall first that, in the canonical approach to diffeomorphism invariant theories such as general relativity or supergravity, the key mathematical problem is that of formulating and solving the quantum constraints. (In Minkowskian quantum field theories, the analogous problem is that of defining the regularized quantum Hamiltonian operator.) It is therefore natural to work with variables which, in the classical theory, simplify the form of the constraints. It turns out that, from this perspective, connections are better suited than metrics [25, 27].

We will conclude by providing explicit expressions of these connections. Recall first that in geometrodynamics we can choose, as our canonical pair, the fields (E_i^a, K_a^i) where E_i^a is a triad (with density weight one) and K_a^i, the extrinsic curvature. Here a refers to the tangent space of the 3-manifold and i is the internal $SO(3)$ – or, $SU(2)$, if we wish to consider spinorial matter – index. The triad is the square-root of the metric in the sense that $E_i^a E^{bi} =: g g^{ab}$, where g is the determinant of the covariant 3-metric g_{ab}, and K_a^i is related to the extrinsic curvature K_{ab} via: $K_a^i = (1/\sqrt{g})K_{ab}E^{bi}$. Let us make a change of variables:

$$(E_i^a, K_a^i) \mapsto (A_a^i := \Gamma_a^i - K_a^i, E_i^a), \qquad (2.2)$$

where Γ_a^i is the spin connection determined by the triad. It is not difficult to check that this is a canonical transformation on the real phase space [25, 26]. It will be convenient to regard A_a^i as the configuration variable and E_i^a as the conjugate momentum so that the phase space has the same structure as in the $SU(2)$ Yang–Mills theory. The basic Poisson bracket relations are:

$$\{A_a^i(x), E_j^b(y)\} = G\delta_a^b\delta_j^i\delta^3(x, y), \tag{2.3}$$

where the gravitational constant, G, features in the Poisson bracket relations because $E_i^a(x)$ now has the physical dimensions of a triad rather than that of Yang–Mills electric field. In terms of these variables, general relativity has the same kinematics as Yang–Mills theory. Indeed, one of the constraints of general relativity is precisely the Gauss constraint of Yang–Mills theory. Thus, the phase spaces of the two theories are the same and the constraint surface of general relativity is embedded in that of Yang–Mills theory. Furthermore, in terms of these variables, the remaining constraints of general relativity simplify considerably. Indeed, there is a precise sense in which they are the simplest non-trivial equations one can write down in terms of A_a^i and E_i^a without reference to any background field [28, 29]. Finally (in the spatially compact context) the Hamiltonian of general relativity is just a linear combination of constraints.

To summarize, one can regard the space $\mathcal{A}/\mathcal{G}$ of $SU(2)$ connections modulo gauge transformations on a ('spatial') 3-manifold Σ as the classical configuration space of general relativity.

2.2 Quantum configuration space

As we have already indicated, in Minkowskian quantum field theories, the quantum configuration space includes distributional fields which are absent in the classical theory, and physically interesting measures are typically concentrated on these 'genuinely quantum' configurations. The overall situation is the same in general relativity.

Thus, the quantum configuration space $\overline{\mathcal{A}/\mathcal{G}}$ is a certain completion of $\mathcal{A}/\mathcal{G}$ [5, 6]. It inherits the quotient structure of $\mathcal{A}/\mathcal{G}$, i.e., $\overline{\mathcal{A}/\mathcal{G}}$ is the quotient of the space $\overline{\mathcal{A}}$ of generalized connections by the space $\overline{\mathcal{G}}$ of generalized gauge transformations. To see the nature of the generalization involved, recall first that each smooth connection defines a holonomy along paths[3] in Σ: $h_p(A) := \mathcal{P}\exp - \int_p A$. Generalized connections capture this notion. That is, each $\bar{A}$ in $\overline{\mathcal{A}}$ can be defined [7, 9] as a map which assigns to each oriented path p in Σ an element of $\bar{A}(p)$ of $SU(2)$ such that: i) $\bar{A}(p^{-1}) = (\bar{A}(p))^{-1}$; and ii) $\bar{A}(p_2 \circ p_1) = A(p_2) \cdot \bar{A}(p_1)$, where p^{-1} is obtained from p by simply reversing the orientation, $p_2 \circ p_1$ denotes the composition of the two paths (obtained by connecting the end of p_1 with the beginning of p_2) and $\bar{A}(p_2) \cdot \bar{A}(p_1)$ is the composition in $SU(2)$. A generalized gauge transformation is a map g which assigns to each point v of Σ an $SU(2)$ element $g(x)$ (in an arbitrary, possibly discontinuous fashion). It acts on $\bar{A}$ in the expected manner, at the end points of paths: $\bar{A}(p) \rightarrow g(v_+)^{-1} \cdot \bar{A}(p) \cdot g(v_-)$, where v_- and v_+ are respectively the beginning and the end point of p. If $\bar{A}$ happens to be a smooth connection, say

---

[3] For technical reasons, we will assume that all paths are analytic. An extension of the framework to allow for smooth paths is being carried out [31]. The general exception is that the main results will admit natural generalizations to the smooth category.

A, we have $\bar{A}(p) = h_p(A)$. However, in general, $\bar{A}(p)$ cannot be expressed as a path ordered exponential of a smooth 1-form with values in the Lie algebra of $SU(2)$ [6]. Similarly, in general, a generalized gauge transformation cannot be represented by a smooth group valued function on Σ.

At first sight the spaces $\bar{A}$, $\bar{G}$ and $\overline{A/G}$ seem too large to be mathematically controllable. However, they admit three characterizations which enables one to introduce differential and integral calculus on them [5, 6, 8]. We will conclude this sub-section by summarizing the characterization – as suitable limits of the corresponding spaces in lattice gauge theory – which will be most useful for the main body of this paper.

Fix a graph γ in the 3-manifold Σ. In physics terminology, one can think of a graph as a 'floating lattice', i.e., a lattice whose edges are not required to be rectangular. (Indeed, they may even be non-trivially knotted!) Using the standard ideas from lattice gauge theory, we can construct the configuration space associated with the graph γ. Thus, we have the space A_γ, each element A_γ of which assigns to every edge in γ an element of $SU(2)$, and the space G_γ, each element g_γ of which assigns to each vertex in γ an element of $SU(2)$. (Thus, if N is the number of edges in γ and V the number of vertices, A_γ is isomorphic with $[SU(2)]^N$ and G_γ with $[SU(2)]^V$.) G_γ has the obvious action on A_γ: $A_\gamma(e) \rightarrow g(v_+)^{-1} \cdot A_\gamma(e) \cdot g(v_-)$. The (gauge invariant) configuration space associated with the floating lattice γ is just A_γ/G_γ. The spaces $\bar{A}$, $\bar{G}$ and $\overline{A/G}$ can be obtained as well-defined (projective) limits of the spaces A_γ, G_γ and A_γ/G_γ [8, 6]. Note however that this limit is *not* the usual 'continuum limit' of a lattice gauge theory in which one lets the edge length go to zero. Here, we are already in the continuum and have available to us *all possible* floating lattices from the beginning. We are just expressing the quantum configuration space of the continuum theory as a suitable limit of the configuration spaces of theories associated with all these lattices.

To summarize, the quantum configuration space $\overline{A/G}$ is a specific extension of the classical configuration space A/G. Quantum states can be expressed as complex-valued, square-integrable functions on $\overline{A/G}$, or, equivalently, as $\bar{G}$-invariant square-integrable functions on $\bar{A}$. As in Minkowskian field theories, while A/G is dense in $\overline{A/G}$ topologically, measured theoretically it is generally sparse; typically, A/G is contained in a subset of zero measure of $\overline{A/G}$ [8]. Consequently, what matters is the value of wave functions on 'genuinely' generalized connections. In contrast with the usual Minkowskian situation, however, $\bar{A}$, $\bar{G}$ and $\overline{A/G}$ are all *compact* spaces in their natural (Gel'fand) topologies [4–8]. This fact simplifies a number of technical issues.

2.3 Hilbert space

Since $\overline{A/G}$ is compact, it admits regular (Borel, normalized) measures and for every such measure we can construct a Hilbert space of square-integrable functions. Thus, to construct the Hilbert space of quantum states, we need to select a specific measure on $\overline{A/G}$.

It turns out that $\bar{A}$ admits a measure μ° that is preferred by both mathematical and physical considerations [6, 7]. Mathematically, the measure μ° is natural because its definition does not involve introduction of any additional structure: it is induced on $\bar{A}$ by the Haar measure on $SU(2)$. More precisely, since A_γ is isomorphic to $[SU(2)]^N$, the Haar measure on $SU(2)$ induces on it a measure μ_γ° in the obvious fashion. As we vary γ, we obtain a family of measures which turn out to be consistent in an appropriate sense and therefore induce a measure μ^o on $\overline{\mathcal{A}}$. This measure has the following attractive properties [6]: i) it is faithful: i.e., for any continuous, non-negative function f on $\overline{\mathcal{A}}$, $\int d\mu^o f \geq 0$, equality holding if and only if f is identically zero; and ii) it is invariant under the (induced) action of $\mathrm{Diff}[\Sigma]$, the diffeomorphism group of Σ. Finally, μ^o induces a natural measure $\tilde{\mu}^o$ on $\overline{\mathcal{A}/\mathcal{G}}$: $\tilde{\mu}^o$ is simply the push-forward of μ^o under the projection map that sends $\overline{\mathcal{A}}$ to $\overline{\mathcal{A}/\mathcal{G}}$. Physically, the measure $\tilde{\mu}^o$ is selected by the so-called 'reality conditions'. More precisely, the classical phase space admits an (over)complete set of naturally defined configuration and momentum variables which are real, and the requirement that the corresponding operators on the quantum Hilbert space be self-adjoint selects for us the measure $\tilde{\mu}^o$ [11].

Thus, it is natural to use $\tilde{\mathcal{H}}^o := L^2(\overline{\mathcal{A}/\mathcal{G}}, d\tilde{\mu}^o)$ as our Hilbert space. Elements of $\tilde{\mathcal{H}}^o$ are the kinematic states; we are yet to impose quantum constraints. Thus, $\tilde{\mathcal{H}}^o$ is the classical analog of the *full* phase-space of quantum gravity (prior to the introduction of the constraint sub-manifold). Note that these quantum states can be regarded also as *gauge invariant* functions on $\overline{\mathcal{A}}$. In fact, since the spaces under consideration are compact and measures normalized, we can regard $\tilde{\mathcal{H}}^o$ as the gauge invariant *subspace* of the Hilbert space $\tilde{\mathcal{H}}^o := L^2(\overline{\mathcal{A}}, d\mu^o)$ of square-integrable functions on $\overline{\mathcal{A}}$ [7, 8]. *In what follows, we will often do so.*

What do 'typical' quantum states look like? To provide an intuitive picture, we can proceed as follows. Fix a graph γ with N edges and consider functions Ψ_γ of generalized connections of the form $\Psi_\gamma(\bar{A}) = \psi(\bar{A}(e_1), \dots, \bar{A}(e_N))$ for *some* smooth function ψ on $[SU(2)]^N$, where $e_1, \dots, e_N$ are the edges of the graph γ. Thus, the functions Ψ_γ know about what the generalized connections do only to those paths which constitute the edges of the graph γ; they are precisely the quantum states of the gauge theory associated with the 'floating lattice' γ. This space of states, although infinite-dimensional, is quite 'small' in the sense that it corresponds to the Hilbert space associated with a system with only a *finite* number of degrees of freedom. However, if we vary γ through all possible graphs, the collection of all states that results is very large. Indeed, one can show that it is *dense* in the Hilbert space $\mathcal{H}^o$. (If we restrict ourselves to Ψ_γ which are gauge invariant, we obtain a dense sub-space in $\tilde{\mathcal{H}}^o$.)

Since each of these states Ψ_γ depends only on a finite number of variables, borrowing the terminology from the quantum theory of free fields in Minkowski space, they are called *cylindrical functions* and denoted by Cyl. Gauge invariant cylindrical functions represent the 'typical' kinematic states. In many ways, Cyl is analogous to the space $C_0^\infty(R^3)$ of smooth functions of compact support of R^3 which is dense in the Hilbert space $L^2(R^3, d^3x)$ of quantum mechanics. Just as one often defines quantum operators – e.g., the position, the momentum and the Hamiltonians – on C_0^∞ first and then extends them to an appropriately larger domain in the Hilbert space $L^2(R^3, d^3x)$, we will define our operators first on Cyl and then extend them appropriately.

Cylindrical functions provide considerable intuition about the nature of quantum states we are led to consider. These states represent 1-dimensional polymer-like excitations of geometry/gravity rather than 3-dimensional wavy undulations on flat space. Just as a polymer, although intrinsically 1-dimensional, exhibits 3-dimensional properties in sufficiently complex and densely packed configurations, the fundamental 1-dimensional excitations of geometry can be packed appropriately to provide a geometry which, when coarse-grained on scales much larger than the Planck length, lead us to continuum geometries [13, 14]. Thus, in this description, gravitons can arise only as approximate notions in the low energy regime [32]. At the basic level, states in $\tilde{\mathcal{H}}^o$ are fundamentally different from the Fock states of Minkowskian quantum field theories. The main reason is the underlying diffeomorphism invariance: in absence of a background geometry, it is not possible to introduce the familiar Gaussian measures and associated Fock spaces.

2.4 Statement of the problem

We can now outline the general strategy that will be followed in section 3.

Recall that the classical configuration variable is an $SU(2)$ connection[4] A_a^i on a 3-manifold Σ, where i is the $su(2)$-internal index with respect to a basis τ_i. Its conjugate momentum E_j^b has the geometrical interpretation of an orthonormal triad with density weight one [25]. Therefore, geometrical observables – functionals of the 3-metric – can be expressed in terms of this field E_i^a. Fix within the 3-manifold Σ any analytic, finite 2-surface. The area A_S of S is a well-defined, real-valued function on the *full* phase space of general relativity (which happens to depend only on E_i^a). It is easy to verify that these kinematical observables can be expressed as:

$$A_S := \int_S dx^1 \wedge dx^2 [E_i^3 E^{3i}]^{\frac{1}{2}}, \qquad (2.4)$$

where, for simplicity, we have used adapted coordinates such that S is given by $x^3 = 0$, and x^1, x^2 parameterize S, and where the internal index i is raised by the inner product we use on $su(2)$, $k(\tau_i, \tau_j) = -2\text{Tr}(\tau_i \tau_j)$. Similarly, if R is any 3-dimensional open region within Σ, the associated volume is a function on the phase space given by:

$$V_R := \int_R dx^1 \wedge dx^2 \wedge dx^3 \left[\frac{1}{3} \eta_{abc} \epsilon^{ijk} E_i^a E_j^b E_k^c \right]^{\frac{1}{2}}, \qquad (2.5)$$

where η_{abc} is the (metric independent, natural) Levi–Civita density of weight -1 on Σ. Our task is to find the corresponding operators on the kinematical Hilbert space $\tilde{\mathcal{H}}^o$ and investigate their properties.

There are several factors that make this task difficult. Intuitively, one would expect $E_i^a(x)$ to be replaced by the 'operator-valued distribution' $-i\hbar G \delta/\delta A_a^i(x)$. (See the basic Poisson bracket relation (2.3).) Unfortunately, the classical expression of A_S involves *square-roots of products* of E's and hence the formal expression of the corresponding operator is badly divergent. One must introduce a suitable regularization scheme. However, we do not have at our disposal the usual machinery of Minkowskian field theories and even the precise rules that are to underlie such a regularization are not *a priori* clear.

There are, however, certain basic expectations that we can use as guidelines: i) the resulting operators should be well-defined on a dense sub-space of $\tilde{\mathcal{H}}^o$; ii) their final expressions should be diffeomorphism covariant, and hence, in particular, independent of any background fields that may be used in the intermediate steps of the regularization procedure; and iii) since the classical observables are real-valued, the operators should be self-adjoint. These expectations seem to be formidable at first.

---

[4] We assume that the underlying 3-manifold Σ is orientable. Hence, principal $SU(2)$ bundles over Σ are all topologically trivial. Therefore, we can represent the $SU(2)$ connections on the bundle by a $su(2)$-valued 1-form on Σ. The matrices τ_i are anti-Hermitian, given, for example, by $(-i/2)$-times the Pauli matrices.

Indeed, these demands are rarely met even in Minkowskian field theories, in presence of interactions, it is extremely difficult to establish rigorously that physically interesting operators are well-defined and self-adjoint. As we will see, the reason why one can succeed in the present case is two-fold. First, the requirement of diffeomorphism covariance is a powerful restriction that severely limits the possibilities. Second, the background independent functional calculus is extremely well-suited for the problem and enables one to circumvent the various road blocks in subtle ways.

Our general strategy will be following. We will define the regulated versions of area and volume operators on the dense sub-space Cyl of cylindrical functions and show that they are essentially self-adjoint (i.e., admit unique self-adjoint extensions to $\tilde{\mathcal{H}}^o$). This task is further simplified because the operators leave each sub-space $\mathcal{H}_\gamma$ spanned by cylindrical functions associated with any one graph γ invariant. This in effect reduces the field theory problem (i.e., one with an infinite number of degrees of freedom) to a quantum mechanics problem (in which there are only a finite number of degrees of freedom). Finally, the operators in fact leave invariant certain *finite* dimensional sub-space of $\mathcal{H}^o$ (associated with 'extended spin networks' [17]). This powerful simplification further reduces the task of investigating the properties of these operators; in effect, the quantum mechanical problem (in which the Hilbert space is still infinite-dimensional) is further simplified to a problem involving spin systems (where the Hilbert space is finite-dimensional). It is because of these simplifications that a detailed analysis becomes possible.

3 Quantum geometry

Our task is to construct a well-defined operator $\hat{A}_S$ and $\hat{V}_R$ starting from the classical expression (2.4, 2.5). As is usual in quantum field theory, we will begin with the formal expression obtained by replacing E_i^a in (2.4, 2.5) by the corresponding operator valued distribution $\hat{E}_i^a$ and then regulate it to obtain the required operators. (For an early discussion of non-perturbative regularization, see, in particular, [33].) For brevity, we will discuss the area operators in some detail and then give the final result for the volume operators. Furthermore, to simplify the presentation, we will assume that S is covered by a single chart of adapted coordinates. Extension to the general case is straightforward: one mimics the procedure used to define the integral of a differential form over a manifold. That is, one takes advantage of the coordinate-invariance of the resulting 'local' operator and uses a partition of unity.

3.1 Regularization

The first step in the regularization procedure is to smear (the operator analog of) $E_i^3(x)$ and point-split the integrand in (2.4). Since in this integrand the point x lies on the 2-surface S, let us try to use a 2-dimensional smearing function. Let $f_\epsilon(x, y)$ be a 1-parameter family of fields on S which tend to the $\delta(x, y)$ as ϵ tends to zero; i.e., such that

$$\lim_{\epsilon \to 0} \int_S \mathrm{d}^2 y \, f_\epsilon(x^1, x^2; y^1, y^2) g(y^1, y^2) = g(x^1, x^2), \qquad (3.1)$$

for all smooth densities g of weight 1 and of compact support on S. (Thus, $f_\epsilon(x, y)$ is a density of weight 1 in x and a function in y.) The smeared version of $E_i^3(x)$ will be defined to be:

$$[E_i^3]_f(x) := \int_S d^2y \, f_\epsilon(x, y) E_i^3(y), \qquad (3.2)$$

so that, as ϵ tends to zero, $[E_i^3]_f$ tends to $E_i^3(x)$. The point-splitting strategy now provides a 'regularized expression' of area:

$$[A_S]_f := \int_S d^2x \left[\int_S d^2y, f_\epsilon(x, y) E_i^3(y) \int_S d^2z \, f_\epsilon(x, z) E^{3i}(z) \right]^{\frac{1}{2}}$$

$$= \int_S d^2x [[E_i^3]_f(x)[E^{3i}]_f(x)]^{\frac{1}{2}}, \qquad (3.3)$$

which will serve as the point of departure in the subsequent discussion. To simplify technicalities, we will assume that the smearing field $f_\epsilon(x, y)$ has the following additional properties for sufficiently small $\epsilon > 0$: i) for any given y, $f_\epsilon(x, y)$ has compact support in x which shrinks uniformly to y; and, ii) $f_\epsilon(x, y)$ is non-negative. These conditions are very mild and we are thus left with a large class of regulators.[5]

First, let us fix a graph γ and consider a cylindrical function Ψ_γ on $\bar{\mathcal{A}}$,

$$\Psi_\gamma(\bar{A}) = \psi(\bar{A}(e_1), \ldots, \bar{A}(e_N)) \equiv \psi(g_1, \ldots, g_n) \qquad (3.4)$$

where, as before, N is the total number of edges of γ, $g_k = \bar{A}(e_k)$ and ψ is a smooth function on $[SU(2)]^N$. One can show [17] that the action of the regulated triad operator on such a state is given by:

$$[\hat{E}_i^3]_f(x) \cdot \Psi_\gamma = \frac{i\ell_P^2}{2} \left[\sum_{I=1}^{n} \kappa_I f_\epsilon(x, v_{\alpha_I}) X_I^i \right] \cdot \psi(g_1, \ldots, g_N). \qquad (3.5)$$

Here, X_I^i are the left/right invariant vector fields on the Ith group copy in the argument of ψ in the ith internal direction, i.e., they are operators assigned to the edge e_I by the following formula

$$X_I^i \cdot \psi(g_1, \ldots, g_N) = \begin{cases} (g_I \tau^i)_B^A \dfrac{\partial \psi}{\partial (g_I)_B^A}, & \text{when } e_I \text{ is outgoing} \\[2ex] -(\tau^i g_I)_B^A \dfrac{\partial \psi}{\partial (g_I)_B^A}, & \text{when } e_I \text{ is incoming} \end{cases} \qquad (3.6)$$

and κ_I are real numbers given by:

$$\kappa_I = \begin{cases} 0, & \text{if } e_I \text{ is contained in } S \text{ or does not intersect } S, \\ +1, & \text{if } e_I \text{ has an isolated intersection with } S \text{ and lies above } S, \\ -1, & \text{if } e_I \text{ has an isolated intersection with } S \text{ and lies below } S. \end{cases} \qquad (3.7)$$

The right side again defines a cylindrical function based on the (same) graph γ. Denote by $\mathcal{H}_\gamma^o$ the Hilbert space $L^2(\mathcal{A}_\gamma, d\mu_\gamma^o)$ of square-integrable cylindrical functions associated with a fixed graph γ. Since μ_γ^o is the induced Haar measure on $\mathcal{A}_\gamma$ and since the operator is just a sum of right/left invariant vector fields, standard results in analysis imply that, with domain Cyl_γ^1 of all C^1 cylindrical functions based on γ, it is an essentially self-adjoint on $\mathcal{H}_\gamma^o$. Now, it is straightforward to

---

[5] For example, $f_\epsilon(x, y)$ can be constructed as follows. Take *any* non-negative function f of compact support on S such that $\int d^2x f(x) = 1$ and set $f_\epsilon(x, y) = (1/\epsilon^2)f((x - y)/\epsilon)$. Here, we have used the given chart to write $x - y$ and give $f_\epsilon(x, y)$ a density weight in x.

verify that the operators on $\mathcal{H}_\gamma^o$ obtained by varying γ are all consistent[6] in the appropriate sense. Hence, it follows from the general results in [9] that $[\hat{E}_i^3]_f(x)$, with domain Cyl^1 (the space of all C^1 cylindrical functions), is an essentially self-adjoint operator on $\mathcal{H}^o$. For notational simplicity, we will denote its self-adjoint extension also by $[\hat{E}_i^3]_f(x)$. (The context should make it clear whether we are referring to the essentially self-adjoint operator or its extension.)

Let us now turn to the integrand of the smeared area operator (corresponding to (3.3)). Denoting the determinant of the intrinsic metric on S by g_S, we have:

$$[\hat{g}_S]_f(x)\cdot\Psi_\gamma := [E_i^3]_f(x)[E^{3i}]_f(x)\cdot\Psi_\gamma$$

$$= -\frac{\ell_P^4}{4}\left[\sum_{I,J}\kappa(I,J)f_\epsilon(x,v_{\alpha_I})f_\epsilon(x,v_{\alpha_J})X_I^iX_J^i\right]\cdot\Psi_\gamma,\qquad(3.8)$$

where the summation goes over all the oriented pairs (I,J); v_{α_I} and v_{α_J} are the vertices at which edges e_I and e_J intersect S; $\kappa(I,J)=\kappa_I\kappa_J$ equals 0 if either of the two edges e_I and e_J fails to intersect S or lies entirely in S, $+1$ if they lie on the same side of S, and -1 if they lie on the opposite sides. (For notational simplicity, from now on we shall not keep track of the position of the internal indices i; as noted in Sec. 2.3, they are contracted using the invariant metric on the Lie algebra $su(2)$.) The next step is to consider vertices v_α at which γ intersects S and simply rewrite the above sum by regrouping terms by vertices. The result simplifies if we choose ϵ sufficiently small so that $f_\epsilon(x,v_{\alpha_I})f_\epsilon(x,v_{\alpha_J})$ is zero unless $v_{\alpha_I}=v_{\alpha_J}$. We then have:

$$[\hat{g}_S]_f(x)\cdot\Psi_\gamma = -\frac{\ell_P^4}{4}\left[\sum_\alpha(f_\epsilon(x,v_\alpha))^2\sum_{I_\alpha,J_\alpha}\kappa(I_\alpha,J_\alpha)X_{I_\alpha}^iX_{J_\alpha}^i\right]\cdot\Psi_\gamma,\qquad(3.9)$$

where the index α labels the vertices on S, and I_α and J_α label the edges at the vertex α.

The next step is to take the square-root of this expression. The same reasoning that established the self-adjointness of $[\hat{E}_i^3]_f(x)$ now implies the $[\hat{g}_S]_f(x)$ is a non-negative self-adjoint operator and hence has a well-defined square-root which is also a positive definite self-adjoint operator. Since we have chosen ϵ to be sufficiently small, for any given point x in S, $f_\epsilon(x,v_\alpha)$ is non-zero for at most one vertex v_α. We can therefore take the sum over α outside the square-root. One then obtains

$$([\hat{g}_S]_f)^{\frac{1}{2}}(x)\cdot\Psi_\gamma = \frac{\ell_P^2}{2}\sum_\alpha f_\epsilon(x,v_\alpha)\left[\sum_{I_\alpha,J_\alpha}\kappa(I_\alpha,J_\alpha)X_{I_\alpha}^iX_{J_\alpha}^i\right]^{\frac{1}{2}}\cdot\Psi_\gamma.\qquad(3.10)$$

Note that the operator is neatly split; the x-dependence all resides in f_ϵ and the operator within the square-root is 'internal' in the sense that it acts only on copies of $SU(2)$.

Finally, we can remove the regulator, i.e., take the limit as ϵ tends to zero. By integrating both sides against test functions on S and then taking the limit, we conclude that the following equality holds in the distributional sense:

$$\widehat{\sqrt{g_S}}(x)\cdot\Psi_\gamma = \frac{\ell_P^2}{2}\sum_\alpha\delta^{(2)}(x,v_\alpha)\left[\sum_{I_\alpha,J_\alpha}\kappa(I_\alpha,J_\alpha)X_{I_\alpha}^iX_{J_\alpha}^i\right]^{\frac{1}{2}}\cdot\Psi_\gamma.\qquad(3.11)$$

---

[6] Given two graphs, γ and γ', we say that $\gamma\geq\gamma'$ if and only if every edge of γ' can be written as a composition of edges of γ. Given two such graphs, there is a projection map from $\mathcal{A}_\gamma$ to $\mathcal{A}_{\gamma'}$, which, via pull-back, provides a unitary embedding $U_{\gamma,\gamma'}$ of $\mathcal{H}_{\gamma'}^o$ into $\mathcal{H}_\gamma^o$. A family of operators $\mathcal{O}_\gamma$ on the Hilbert spaces H_{o} is said to be consistent if $U_{\gamma,\gamma'}\mathcal{O}_{\gamma'}=\mathcal{O}_\gamma U_{\gamma,\gamma'}$ and $U_{\gamma,\gamma'}D\gamma'\subset D_\gamma$ for all $g\geq g'$.

Hence, the regularized area operator is given by:

$$\hat{A}_S \cdot \Psi_\gamma = \frac{\ell_P^2}{2} \sum_\alpha \left[\sum_{I_\alpha, J_\alpha} \kappa(I_\alpha, J_\alpha) X_{I_\alpha}^i X_{J_\alpha}^i \right]^{\frac{1}{2}} \cdot \Psi_\gamma. \qquad (3.12)$$

(Here, as before, α labels the vertices at which γ intersects S, and I_α labels the edges of γ at the vertex v_α.) With Cyl^2 as it domain, $\hat{A}_S$ is essentially self-adjoint on the Hilbert space $\mathcal{H}^\circ$.

The classical expression A_S of (2.4) is a rather complicated. It is therefore somewhat surprising that the corresponding quantum operators can be constructed rigorously and have quite manageable expressions. The essential reason is the underlying diffeomorphism invariance which severely restricts the possible operators. Given a surface and a graph, the only diffeomorphism invariant entities are the intersection vertices. Thus, a diffeomorphism covariant operator can only involve structure at these vertices. In our case, it just acts on the copies of $SU(2)$ associated with various edges at these vertices.

We will close this discussion by simply writing the final expression of the volume operator:

$$\hat{V}_R \cdot \Psi_\gamma := \frac{\ell_P^3}{4\sqrt{3}} \sum_\alpha \left| \sum_{I_\alpha, J_\alpha, K_\alpha} i\epsilon^{ijk} \epsilon(I_\alpha, J_\alpha, K_\alpha) X_{I_\alpha}^i X_{J_\alpha}^j X_{K_\alpha}^k \right|^{\frac{1}{2}} \cdot \Psi_\gamma, \qquad (3.13)$$

where the first sum now is over vertices which lie in the region R and $\epsilon(I_\alpha, J_\alpha, K_\alpha)$ is 0 if the three edges are linearly dependent at the vertex v_α and otherwise ± 1 depending on the orientation they define. With Cyl^3 as its domain, $\hat{V}_R$ is essentially self-adjoint on $\mathcal{H}^\circ$.

To summarize, the diffeomorphism covariant functional calculus has enabled us to regulate the area and volume operators. While in the intermediate steps we have used additional structures – such as charts – the final results make no reference to these structures; the final expressions of the quantum operators have the same covariance properties as those of their classical counterparts.

3.2 General properties of geometric operators

We will now discuss the key properties of these geometric operators and point out a few subtleties. As in the previous subsection, for definiteness, the detailed comments will refer to the area operators. It should be clear from the discussion that analogous remarks hold for the volume operators as well.

1. Discreteness of the spectrum: By inspection, it follows that the total area operator $\hat{A}_S$ leaves the sub-space of Cyl_γ^2 which is associated with any one graph γ invariant and is a self-adjoint operator on the sub-space $\mathcal{H}_\gamma^\circ$ of $\mathcal{H}^\circ$ corresponding to γ. Next, recall that $\mathcal{H}_\gamma^\circ = L^2(\mathcal{A}_\gamma, d\mu^\circ)$, where $\mathcal{A}_\gamma$ is a compact manifold, isomorphic with $(SU(2))^N$ where N is the total number of edges in γ. The restriction of $\hat{A}_S$ to $\mathcal{H}_\gamma^\circ$ is given by certain commuting elliptic differential operators on this compact manifold. Therefore, all its eigenvalues are discrete. Now suppose that the complete spectrum of $\hat{A}_S$ on $\mathcal{H}^\circ$ has a continuous part. Denote by P_c the associated projector. Then, given any Ψ in $\mathcal{H}^\circ$, $P_c \cdot \Psi$ is orthogonal to $\mathcal{H}_\gamma^\circ$ for any graph γ, and hence to the space Cyl of cylindrical functions. Since Cyl^2 is dense in $\mathcal{H}^\circ$, $P_c \cdot \Psi$ must vanish for all Ψ in $\mathcal{H}^\circ$. Hence, the spectrum of $\hat{A}_S$ has no continuous part.

Note that this method is rather general: it can be used to show that *any* self-adjoint operator on $\mathcal{H}^o$ which maps (the intersection of its domain with) $\mathcal{H}^o_\gamma$ to $\mathcal{H}^o_\gamma$, and whose action on $\mathcal{H}^o_\gamma$ is given by elliptic differential operators, has a purely discrete spectrum on $\mathcal{H}^o$. Geometrical operators constructed purely from the triad field tend to satisfy these properties.

In the case of area operators, one can do more: the complete spectrum has been calculated. The eigenvalues are given by [17]:

$$a_S = \frac{\ell_P^2}{2} \sum_\alpha [2j_\alpha^{(d)}(j_\alpha^{(d)} + 1) + 2j_\alpha^{(u)}(j_\alpha^{(u)} + 1) - j_\alpha^{(d+u)}(j_\alpha^{(d+u)} + 1)]^{\frac{1}{2}} \qquad (3.14)$$

where α labels a finite set of points in S and the non-negative half-integers assigned to each α are subject to the inequality

$$j^{(d)} + j^{(u)} \geq j^{(d+u)} \geq |j^{(d)} - j^{(u)}|. \qquad (3.15)$$

There is, in particular, the smallest, non-zero eigenvalue, the 'elementary quantum of area': $a_S^0 = (\sqrt{3/4})\ell_P^2$. Note, however, that the level spacing between eigenvalues is *not* regular. For large a_S, the difference between consecutive eigenvalues in fact goes to zero as $1/\sqrt{a_S}$. (For comparison with other results [13, 20], see [17].)

2. Area element: Note that not only is the total area operator well-defined, but in fact it arises from a local area element, $\widehat{\sqrt{g_S}}$, which is an operator-valued distribution in the usual sense. Thus, if we integrate it against test functions, the operator is densely defined on $\mathcal{H}^o$ (with C^2 cylindrical functions as domain) and the matrix elements

$$\langle \Psi'_{\gamma'}, \widehat{\sqrt{g_S}}(x) \cdot \Psi_\gamma \rangle \qquad (3.16)$$

are 2-dimensional distributions on S. Furthermore, since we did not have to renormalize the regularized operator (3.10) before removing the regulator, there are *no* free renormalization constants involved. The local operator is completely unambiguous.

3. $[\hat{g}_S]_f$ versus its square-root: Although the regulated operator $[\hat{g}_S]_f$ is well-defined, if we let ϵ to go zero, the resulting operator is in fact divergent: roughly, it would lead to the square of the 2-dimensional δ distribution. Thus, the determinant of the 2-metric is not well-defined in the quantum theory. As we saw, however, the square-root of the determinant *is* well-defined: we have to first take the square-root of the *regulated* expression and *then* remove the regulator. This, in effect, is the essence of the regularization procedure.

To get around this divergence of $\hat{g}_S$, as is common in Minkowskian field theories, we could have first rescaled $[\hat{g}_S]_f$ by an appropriate factor and then taken the limit. The result can be a well-defined operator, but it will depend on the choice of the regulator, i.e., additional structure introduced in the procedure. Indeed, if the resulting operator is to have the same density character as its classical analog $g_S(x)$ – which is scalar density of weight two – then the operator cannot respect the underlying diffeomorphism invariance, for there is no metric/chart independent distribution on S of density weight two. Hence, such a 'renormalized' operator is not useful to a fully non-perturbative approach. For the square-root, on the other hand, we need a local density of weight *one*. And the 2-dimensional Dirac distribution provides this; now there is no *a priori* obstruction for a satisfactory operator corresponding to the area element to exist. This is an illustration of what appears to be typical in non-perturbative approaches to quantum gravity: either the limit of the operator exists as the regulator is removed without the need of renormalization or it inherits background dependent renormalization fields (rather than constants).

4. Gauge invariance: The classical area element $\sqrt{g_S}$ is invariant under the internal rotations of triads E_i^a; its Poisson bracket with the Gauss constraint functional vanishes. This symmetry is preserved in the quantum theory: the quantum operator $\widehat{\sqrt{g_S}}$ commutes with the induced action of $\bar{\mathcal{G}}$ on the Hilbert space $\mathcal{H}^o$. Thus, $\widehat{\sqrt{g_S}}$ and the total area operator $\hat{A}_S$ map the space of gauge invariant states to itself: they project down to the Hilbert space $\tilde{\mathcal{H}}^o$ of kinematic states. In the classical theory, the allowed values of the area operators on the full phase space are the same as those on the constraint surface. That is, the passage from all kinematical states to the dynamically permissible ones does not give rise to restrictions on the 'classical spectrum' of these operators. The same is true in the quantum theory. The spectrum of $\hat{A}_S$ on $\tilde{\mathcal{H}}^o$ is the same as that on $\mathcal{H}^o$. (Only the degeneracies of eigenvectors change.)

4 Discussion

In section 1, we began by formulating what we mean by quantization of geometry. Are there geometrical observables which assume continuous values on the classical phase space but whose quantum analogs have discrete spectra? In order to explore these issues, we had to use a fully non-perturbative framework which does not use a background geometry. In the last two sections, we answered the question in the affirmative in the case of area and volume operators. The discreteness came about because, at the microscopic level, geometry has a distributional character with 1-dimensional excitations. This is the case even in semi-classical states which approximate classical geometries macroscopically [13, 14]. We wish to emphasize that these results have been obtained in the framework of a (non-traditional but) rigorous quantum field theory. In particular, the issues of regularization have been addressed squarely and the calculations are free of hidden infinities.

We will conclude by examining the main results from various angles.

1. Inputs: The picture of quantum geometry that has emerged here is strikingly different from the one in perturbative, Fock quantization. Let us begin by recalling the essential ingredients that led us to the new picture.

This task is made simpler by the fact that the new functional calculus provides the degree of control necessary to distill the key assumptions. There are only two essential inputs. The first assumption is that the Wilson loop variables, $T_\alpha = \mathrm{Tr}\,\mathcal{P}\exp\int_\alpha -A$, should serve as the configuration variables of the theory, i.e., that the Hilbert space of (kinematic) quantum states should carry a representation of the C^*-algebra generated by the Wilson loop functional on the classical configuration space $\mathcal{A}/\mathcal{G}$. The second assumption singles out the measure $\tilde{\mu}^o$. In essence, if we assume that $\hat{E}_i^a$ be represented by $-i\hbar\delta/\delta A_a^i$, the 'reality conditions' lead us to the measure $\tilde{\mu}^o$ [11]. Both these assumptions seem natural from a mathematical physics perspective.

However, a deeper understanding of their *physical* meaning is still needed for a better understanding of the overall situation.[7]

2. Kinematics versus dynamics: As we emphasized in the main text, in the classical theory, geometrical observables are defined as functionals on the *full* phase space; these are kinematical quantities whose definitions are quite insensitive to the precise nature of dynamics, presence of matter fields, etc. Thus, in the connection dynamics description, all one needs is the presence of a canonically conjugate pair consisting of a connection and a (density weighted) triad. Therefore, one would expect the result on the area operator presented here to be quite robust. In particular, they should continue to hold if we bring in matter fields or extend the theory to supergravity.

There is, however, a subtle caveat: in field theory, one cannot completely separate kinematics and dynamics. For instance, in Minkowskian field theories, the kinematic field algebra typically admits an infinite number of *inequivalent* representations and a given Hamiltonian may not be meaningful on a given representation. Therefore, whether the kinematical results obtained in any one representation actually hold in the physical theory depends on whether that representation supports the Hamiltonian of the model. In the present case, therefore, a key question is whether the quantum constraints of the theory can be imposed meaningfully on $\tilde{H}^\circ$.[8] Results to date indicate (but do not yet conclusively prove) that this is likely to be the case for general relativity. The general expectation is that this would be the case also for a class of theories such as supergravity, which are 'near' general relativity. The results obtained here would continue to be applicable for this class of theories.

3. Dirac observable: Note that $\hat{A}_S$ has been defined for *any* surface S. Therefore, these operators will not commute with constraints: they are not Dirac observables. To obtain a Dirac observable, one would have to specify S *intrinsically*, using, for example, matter fields. In view of the Hamiltonian constraint, the problem of providing an explicit specification is extremely difficult. However, this is true already in the classical theory. In spite of this, *in practice* we do manage to specify surfaces (or 3-dimensional regions) and furthermore compute their areas (volumes) using the standard formula from Riemannian geometry which is quite insensitive to the details of how the surface (region) was actually defined. Similarly, in the quantum theory, *if* we could specify a surface S (region R) intrinsically, we could compute the spectrum of $\hat{A}_S$ and $\hat{V}_R$ using results obtained in this paper.

4. Manifold versus geometry: In this paper, we began with an orientable, analytic, 3-manifold Σ and this structure survives in the final description. As noted in footnote 1, we believe that the assumption of analyticity can be weakened without changing the qualitative results. Nonetheless, a smoothness structure of the underlying manifold will persist. What is quantized is 'geometry' and not smoothness. Now, in $2+1$ dimensions, using the loop representation one can recast the final description in a

---

[7] In particular, in the standard spin-2 Fock representation, one uses quite a different algebra of configuration variables and uses the flat background metric to represent it. It then turns out that the Wilson loops are *not* represented by well-defined operators; our first assumption is violated. One can argue that in a fully non-perturbative context, one cannot mimic the Fock space strategy. Further work is needed, however, to make this argument water-tight.

[8] Note that this issue arises in *any* representation once a sufficient degree of precision is reached. In geometrodynamics, this issue is not discussed simply because generally the discussion is rather formal.

purely combinatorial fashion (at least in the so-called 'time-like sector' of the theory). In this description, at a fundamental level, one can avoid all references to the underlying manifold and work with certain abstract groups which, later on, turn out to be the homotopy groups of the 'reconstructed/derived' 2-manifold (see, e.g., section 3 in [30]). One might imagine that if and when our understanding of knot theory becomes sufficiently mature, one would also be able to get rid of the underlying manifold in the $3+1$ theory and introduce it later as a secondary/derived concept. At present, however, we are quite far from achieving this.

In the context of geometry, however, a detailed combinatorial picture *is* emerging. Geometrical quantities are being computed by counting; integrals for areas and volumes are being reduced to genuine sums. (However, the sums are *not* the 'obvious' ones, often used in approaches that *begin* by postulating underlying discrete structures. In the computation of area, for example, one does not just count the number of intersections; there are precise and rather intricate algebraic factors that depend on the representations of $SU(2)$ associated with the edges of each intersection.) It is striking to note that, in the same address [4] in which Riemann first raised the possibility that geometry of space may be a physical entity, he also introduced ideas on discrete geometry. The current program comes surprisingly close to providing us with a concrete realization of these ideas.

To summarize, it *is* possible to do physics in absence of a background space-time geometry. It does require the use of new mathematical methods, such as diffeomorphism covariant functional calculus. However, one can obtain concrete, physically motivated results which are quite surprising from the viewpoint of Minkowskian field theories.

Acknowledgments

We would like to thank John Baez, Bernd Bruegman, Don Marolf, Jose Mourao, Thomas Thiemann, Lee Smolin, John Stachel and especially Carlo Rovelli for discussions. This work was supported in part by the NSF grants PHY93-96246 and PHY95-14240, the KBN grant 2-P302 11207 and the Eberly fund of the Pennsylvania State University. JL thanks the members of the Max Planck Institute for the hospitality.

References

[1] A.S. Wightman, The usefulness of a general theory of quantized fields, this volume.

[2] B.S. DeWitt, Quantum field theory and space-time – formalism and reality, this volume.

[3] C. Rovelli, 'Localization' in quantum field theory: how much of QFT is compatible with what we know about space-time?, this volume.

[4] B. Riemann, Über die Hypothesen, welche der Geometrie zugrunde liegen (University of Göttingen, 1854).

[5] A. Ashtekar and C.J. Isham, *Classical Quantum Grav.* **9**, 1433 (1992).

[6] A. Ashtekar and J. Lewandowski, 'Representation theory of analytic holonomy C^* algebras', in *Knots and Quantum Gravity*, J. Baez (ed.), (Oxford University Press, Oxford 1994); *J. Math. Phys.* **36**, 2170 (1995).

[7] J. Baez, *Lett. Math. Phys.* **31**, 213 (1994); 'Diffeomorphism invariant generalized measures on the space of connections modulo gauge transformations', hep-th/9305045, in the *Proceedings of the Conference on Quantum Topology*, D. Yetter (ed.) (World Scientific, Singapore, 1994).

[8] D. Marolf and J.M. Mourão, *Commun. Math. Phys.* **170**, 583 (1995).

[9] A. Ashtekar and J. Lewandowski, *J. Geom. Phys.* **17**, 191 (1995).

[10] J. Baez, 'Spin network states in gauge theory', *Adv. Math.* **177**, 253 (1996); 'Spin networks in non-perturbative quantum gravity', in *The Interface of Knots and Physics*, L. Kauffman (ed.) (American Mathematical Society, 1996).

[11] A. Ashtekar, J. Lewandowski, D. Marolf, J. Mourão and T. Thiemann, *J. Math. Phys.* **36**, 6456 (1995).

[12] A. Ashtekar, J. Lewandowski, D. Marolf, J. Mourão and T. Thiemann, *J. Funct. Analysis* **135**, 519 (1996).

[13] A. Ashtekar, C. Rovelli and L. Smolin, *Phys. Rev. Lett.* **69**, 237 (1992).

[14] A. Ashtekar and L. Bombelli (in preparation).

[15] J. Beckenstein and V.F. Mukhanov, 'Spectroscopy of quantum black holes', *Phys. Lett.* **B360**, 7 (1995).

[16] S. Carlip, *Classical Quantum Grav.* **12**, 2853 (1995); 'Statistical mechanics and black-hole entropy', preprint gr-qc/9509024 (1995).

[17] A. Ashtekar and J. Lewandowski, 'Quantum theory of geometry I: Area operators', gr-qc 9602046; *Classical Quantum Grav.* **A14**, 55 (1997).

[18] A. Ashtekar and J. Lewandowski, 'Quantum theory of geometry II: volume operators', *Adv. Theo. Math. Phys.* **1**, 388 (1997).

[19] J. Lewandowski, 'Volume and quantizations', *Classical Quantum Grav.* **14**, 71 (1997).

[20] C. Rovelli and L. Smolin, *Nucl. Phys.* **B442**, 593 (1995).

[21] R. Loll, *Phys. Rev. Lett.* **75**, 3084 (1995).

[22] R. Loll, 'Spectrum of the volume operator in quantum gravity', *Nucl. Phys.* B (to appear).

[23] C. Rovelli and L. Smolin, *Nucl. Phys.* **B331**, 80 (1990).

[24] A. Ashtekar, 'Polymer geometry at Planck scale and quantum Einstein's equations', *Acta Cosmologica* **XXI**, 85–110 (1996); CGPG-95/11-5, hep-th/9601054.

[25] A. Ashtekar, *Phys. Rev. Lett.* **57**, 2244 (1986); *Phys. Rev.* **D36**, 1587 (1987).

[26] A. Ashtekar, in *Mathematics and General Relativity*, J. Isenberg (ed.) (American Mathematical Society, Providence, 1987).

[27] A. Ashtekar, *Lectures on Non-Perturbative Canonical Gravity*. Notes prepared in collaboration with R.S. Tate (World Scientific, Singapore, 1991).

[28] T. Thiemann, 'Reality conditions inducing transforms for quantum gauge field theory and quantum gravity', *Classical Quantum Grav.* **13**, 1383 (1996).

[29] A. Ashtekar, 'A generalized Wick transform for gravity', *Phys. Rev.* **D53**, R2865 (1996).

[30] A. Ashtekar, in *Gravitation and Quantizations*, B. Julia and J. Zinn-Justin (eds.) (Elsevier, Amsterdam, 1995).

[31] J. Baez and S. Sawin, 'Functional integration on spaces of connections', q-alg/9507023; J. Lewandowski and T. Thiemann (in preparation)

[32] J. Iwasaki and C. Rovelli, *Int. J. Modern. Phys.* **D1**, 533 (1993); *Classical Quantum Grav.* **11**, 2899 (1994).

[33] B. Bruegman and J. Pullin, *Nucl. Phys.* **B390**, 399 (1993).

Discussions

Weinstein: At one point you referred to the qualitative predictions of quantum gravity, and I'm wondering how you conceive of the physical significance of the discreteness of, say, the area and the volume operators. And the general problem I'm referring to is that in conventional quantum field theory, we think of measurements as occurring at particular points in space-time, and here that seems to be pulled away from us because we no longer have a fixed background in geometry.

Ashtekar: Right. So for example, what we would like to do, is to look at geometries which correspond to black holes, large black holes, macroscopic black holes, not quantum black holes. In these black holes, the horizon is a diffeomorphism invariant surface. And so one might actually compute its area classically. But there are underlying microscopic spin-network states. So the question is, can you count the states, which will give us this particular black-hole horizon? Can you look at the microscopic description of this macroscopic situation? And then if you can, you count the states, and take the log of that, is that Bekenstein–Hawking Entropy? So we're trying to recover the laws of black-hole thermodynamics from statistical mechanics, i.e. from the fundamental principles. So one can attempt calculations like that.

Secondly, the very last thing I mentioned would have important implications to quantum field theory. If I could actually do quantum field theory on such a quantum space-time background, then hopefully it would be actually free of divergences (since it's really like being on a two-dimensional space-time). I'll give you a geometrical picture just to fix some idea; there is nothing very special about this specific picture. These are some computer simulations of the flat space-time. [Ashtekar shows a transparency.] The idea is to sprinkle points at random, and construct some networks in a particular way, so that the geometry is just excited along the lines of these networks up here. So the idea would be to actually do, say, string theory by just making the strings wrap around these objects and then see if you can get something new. I mean, what is the difference between doing a string theory on a quantum geometry versus doing it on a continuum? And the results of Suskind and other people suggest that perhaps this discretization of the background, not of the target space, should actually not affect the main results. So one would like to test such things.

DeWitt: You seem to be emphasizing the fact that you introduce no background, as if its introduction were a defect of covariant quantization. In my earlier talk this afternoon, I mentioned the effective action as being a functional of background geometry. It doesn't have to be presented that way. In fact a better way is to present it as a functional over averages or in–out matrix elements of geometry. There's no *a priori* background but rather a self-consistency problem to solve.

Ashtekar: You're absolutely right. If this were a relativity talk, I would not emphasize this so much. It was just to bring in one of the lessons we learn about axiomatic quantum field theory from these considerations. That is the reason I was emphasizing it.

DeWitt: It seems to me there should be no more objection to choosing a base point in something like the space of field histories than there is to choosing an origin in a differentiable manifold. I have a couple of questions, one concerning diffeomorphism invariance. You have a background *topology*, other diffeomorphism invariance wouldn't have any meaning.

Ashtekar: Right. A background differential structure, even more than a topology.

DeWitt: You seem to have talked only about three-dimensional diffeomorphism invariance. I have no idea how four-dimensional diffeomorphism invariance would come into this. I do not believe that *area* is an observable unless you tell me intrinsically over what two-dimensional submanifold you're going to calculate the area. You can talk about the area of a physical object, but then you have to bring the physical object in.

Ashtekar: Absolutely.

DeWitt: So what are you talking about here? About observables in a larger than physical Hilbert space?

Ashtekar: Shall I answer the question now?

DeWitt: Okay. [laughter]

Ashtekar: Bryce's question is the following: it's a key question, and the only reason I skipped details on this point is because I was running out of time. As Carlo Rovelli told us just before, even in classical general relativity – forget about quantum theory, even in classical general relativity – if I just take a mathematical surface and ask for its area, that is not a diffeomorphism invariant question. To make it observable, we like to bring in some dynamical quantities, for example, it is meaningful for people to ask, what is the area of this table? Now, why is this question meaningful in general relativity? Because we've got a matter field which is zero here (above and below the table), and non-zero here (on the table). I can define the surface systematically by saying that it is the interface between these two regions, and then I can compute the area of this physically specified surface. However, the way that we actually compute this area in classical general relativity is to use the standard formula from Riemannian geometry which does not depend on the fact that I've specified it by some specific physical description; it doesn't depend on the details of the Hamiltonian of this wood.

DeWitt: Oh, yes, it does.

Ashtekar: Classical relativity? When I compute the area of this, I use the approximately flat metric in this room and integrate the square-root of the determinant of the induced metric on the table. That does not involve the Hamiltonian of the wood!

DeWitt: That table might be oscillating. Even classically.

Ashtekar: Then you give me the detailed classical situation. After that, I'm supposed to go and compute this area. The formula I use in computing area is always the integral of the square-root of the determinant of the two-metric. The two-metric itself is of course dynamically determined by all sorts of other things. But the computation uses just this formula. The same thing is going to happen, I claim, in quantum gravity. We are to solve the constraint equations, we are to extract physical states, but then if you can specify the surfaces dynamically, then the statement is that the formula that I'll use to compute the area of that dynamically specified surface is the operator that I wrote down. The specification had all the conceptual difficulties that you rightly pointed out. But the actual calculation, I would like to say, just as in classical theory, is this calculation. You also said something else about...

DeWitt: Diffeomorphism invariance, three-dimensional versus four-dimensional.

Ashtekar: Right. That's also something I would like to comment on. This, as Bryce knows well, is a canonical approach. As he says, in the context of the Wheeler–DeWitt equation, the weakness of this is that one is taking space-time and dividing it into space and time. I would personally like to do something which is manifestly four-dimensional diffeomorphism covariant. The problem is that we really do not have – we meaning us or anyone else – enough mathematical control on the various functional integrals, or various other things, that you want to do in this four-dimensional approach. However, even in ordinary field theory, say, quantum electrodynamics, the Hamiltonian or the canonical approach is not manifestly Poincaré invariant. But there is Poincaré invariance, both in the phase space and in the Hilbert space of square-integrable functions of connections, because the measure is Poincaré invariant. So here the problem is going to be, when I solve the constraints – there are already some results along these lines – I'd have to construct a measure on the space of the solutions of the constraints. And that, in an appropriate sense, will carry the four-dimensional diffeomorphism invariance. That is a problem and what we have is an awkward way to handle it. But so far this is the only sufficiently mathematically precise way, so that we can actually regularize operators in a rigorous way. So it's really methodology for me.

14. 'Localization' in quantum field theory: how much of QFT is compatible with what we know about space-time?

CARLO ROVELLI

1 How seriously should we take QFT?

One of Newton's most far-reaching intuitions was to break the 'mathematical theory of the world' into two components. One component is given by the various specific force laws,[1] or, in later formulations, specific Lagrangians. The other and more fundamental component is the general theory of motion,[2] which we may denote as 'Mechanics'. Quantum field theory (QFT) is an example of the second. It is a general scheme for treating physical theories, which is not committed to specific systems or to specific Lagrangians.

As a general scheme for treating physical theories, QFT is extraordinarily successful and remarkably flexible. Its impressive effectiveness has been emphasized in this conference by Jackiw, Shankar, and others. QFT had its periods of ill-fortune, for instance in the sixties, at the time of S-matrix theory, recalled in this conference by Kaiser and by Shankar. But then it had its glorious comebacks 'to a glory even greater than before'. Today, our understanding of the world at the fundamental level is based on the Standard Model, which is formulated within the framework of QFT, and on classical general relativity. General relativity cannot be seen as a 'fundamental' theory since it neglects the quantum behavior of the gravitational field, but many of the directions that are explored with the aim of finding a quantum theory of the gravitational field and/or extending the Standard Model – perhaps to a theory of everything – are grounded in QFT. Thus, one may view QFT as *the fundamental theory of motion* at our present stage of knowledge – playing the role that Newtonian mechanics and its Lagrangian and Hamiltonian extensions played in the eighteenth and nineteenth centuries.

Of course, quantum mechanics, of which QFT is just a mature reincarnation, has its notorious interpretative difficulties, and QFT inherits these difficulties. Furthermore, the physical four-dimensional QFTs are plagued by their divergences and lack of sound mathematical foundation; but models of QFTs are well understood in lower dimensions, and, as argued by Jackiw in this conference, the very presence of the divergences is perhaps a key ingredient of QFT's 'unreasonable effectiveness'. Today, in spite of such difficulties, QFT is therefore solidly at the root of our understanding of the physical world.

The problem I am going to address here is the extent to which what I just said is correct – namely, the extent to which we may reasonably take QFT as a good

---

[1] For instance: $F = G\dfrac{m_1 m_2}{r^2}$.

[2] For instance: $F = ma$.

fundamental description of the structure of the physical world at our present level of knowledge. Can we believe that QFT provides or may provide a credible general picture of the physical world, given what we currently know about it? The notion of 'fundamental description of the structure of the physical world' that I am using here does not imply any sort of strong reductionism. Simply, for several centuries Newtonian mechanics – and its later reincarnations as Lagrangian and Hamiltonian mechanics – did provide a consistent, well-founded unifying substratum to the *physical* description of the world. Perhaps this substratum did not exhaust the content of the higher 'cognitive levels', as Rohrlich has argued here. Still, it did provide a kind of accepted common conceptual rock, over which the higher cognitive levels could be built. The problem I want to discuss is whether we can expect QFT to play the same role today. In particular, I will discuss whether or not quantum field theory is compatible with our present knowledge about *space-time*.

Notice that in the past there have been situations in which theories of great empirical success were considered physically 'wrong', at least by some researchers, before the appearance of a substitute theory. More precisely, the boundary of their domain of validity had already been identified *theoretically*. Consider for instance Newton's theory of universal gravitation[3] in 1912. The theory has an immense empirical success. Highly sophisticated perturbation theory has led to triumphs like the discovery of new planets before their actual observation. (This recalls QFT successes, doesn't it?) There are a few open observational puzzles, such as Mercury's perihelion shift, but much hope to resolve them within the theory: the shift may be due to the solar quadrupole moment. The foundations of the theory appear therefore extremely solid. But they are not. Special relativity has already been established as a credible new discovery. And special relativity implies that all interactions propagate with finite speed. Action at distance, which is one of the ingredients of Newton's theory, is compromised as a good assumption about the world. Therefore, in spite of the fact that in 1912 a *field* theory of the gravitational force is still lacking, it is nevertheless clear to the best thinking minds of the period that the foundations of Newton's theory of gravity have already cracked; the theory cannot be taken as a credible description of the physical world at that stage of knowledge.

I believe it is important, in discussing the foundations of QFT, to address the problems of whether or not, among the essential ingredients of QFT, there are any which we may already see are physically wrong, in the light of the ensemble of the present physical knowledge. In the paper presented to this conference, Rohrlich and Hardin use the notion of *established* theory to indicate a theory already 'overthrown' by a more general theory. They argue that only when such a more general theory has taken hold can we see with clarity the domain of validity of a scientific theory; and only at that point can we consider the theory 'established', a term with a positive connotation. Sometimes, in order to detect the boundaries of the domain of validity of a theory one does not have to wait for the new theory to come in – as the example of Newton's gravitation theory in 1912 indicates. The question I am addressing here is: Is QFT 'established', in Rohrlich and Hardin's sense?

It is important to study the foundations of QFT, as this conference aims to do, whatever the general credibility of QFT, since QFT is the best *working* fundamental

---

[3] That is: $F = G \dfrac{m_1 m_2}{r^2}$.

physical theory today. But perhaps something of the tone of the investigation may change, depending on the answer to that question. Imagine that in 1912 (that is, when special relativity was already accepted) there had been a conference of philosophers and physicists, say in Boston, on the foundations of the Newtonian gravity theory. Are we, in some sense, in the same situation?

2 Two meanings of QFT

The first step in order to address the problem is to disentangle an ambiguity of the expression 'QFT'. This expression is used with (at least) two different meanings. Let me illustrate these two different meanings.

Quantum mechanics (QM) is the discovery that Newtonian mechanics and its extensions are not sufficient to describe physical reality at small scale. Novel phenomena appear which cannot be captured by classical mechanics: physical systems can be in probabilistic linear superpositions of distinct states, and some observable quantities exhibit discreteness in certain regimes. Quantum mechanics provides a general, but tightly defined – though perhaps not fully satisfactory – theoretical scheme for describing these phenomena. The scheme admits a variety of formulations (Hilbert space, algebraic, functional integral), which are virtually all equivalent. For the purpose of this discussion, we may take the following as essential features of QM: the space of states of the system is described by a Hilbert space and the observables are described by an algebra of self-adjoint operators, whose physical interpretation is given in the well-known probabilistic way.[4] In order to accommodate finite temperature QFT, or QFT in curved space-time, we may relax the requirement of the Hilbert space, and characterize the theory by a suitable *-algebra of observables, and the states by suitable positive functionals over this algebra.

Fifty years earlier than the discovery of QM, Faraday and Maxwell had shown that a far-reaching extension of traditional Newtonian mechanics was required. In modern terms, their work implied that the world has to be described by systems with an infinite number of degrees of freedom. In other words, constituents of physical reality are *fields*. Twenty years later, special relativity provided an *a priori* justification for the existence of fields.

QM, in the broad terms in which I have characterized it, is a general scheme of physical theories, which can be employed for finite as well as for infinite-dimensional systems. At first, QM was used for finite-dimensional systems. But the need to use it for fields as well became clear very soon, already in the late twenties. Dirac defined the quantum theory of free fields, whose key success is the theoretical understanding of Einstein's photons. After a long struggle, Feynman and his companions constructed non-trivial quantum theories for some interacting field systems. In the hands of scientists such as Weinberg, these theories turned out to describe the world with the fantastic accuracy we admire.

These QFTs are a *peculiar* form of quantum theories for systems with infinite variables. They are characterized by assumptions, such as Poincaré invariance, positivity of the energy, existence of the vacuum, and, foremost, *localizability*. Leaving aside Poincaré invariance, which is already relaxed in QFT on curved space-time, I will

---

[4] If we measure an observable A, represented by an operator A, when the system is in a state ψ, the spectral decomposition of ψ on the eigenbasis of A gives the probability amplitudes for the A's eigenvalues to be obtained as measurement outcomes.

focus on *localizability* as a key ingredient in the foundations of conventional QFT. Localizability will represent a main topic of my presentation. I will soon indicate more precisely what I mean by this expression.

Thus, at the foundation of conventional QFT there are *two* distinct orders of assumptions. The first order of assumptions holds for any quantum theory with an infinite number of variables. These include all the principles of quantum mechanics, and the notion of a system with infinite variables. The second is characteristic of QFT as we know it in the Standard Model, in axiomatic and constructive QFT. Both sets of assumptions are captured by the axiomatic approach.

Now, the two meanings of the expression QFT are the following. In its first meaning, QFT indicates *any quantum theory of a system with an infinite number of variables*. For clarity, I will use the expression *general QFT* to indicate a quantum theory with an infinite number of variables. Being a quantum theory, a general QFT is characterized by the existence of a *-algebra of observables. Being a field theory, that is, having to describe an infinite number of degrees of freedom, this algebra must be infinite-dimensional. The second meaning of QFT indicates the *particular* form of quantum mechanics with infinite degrees of freedom that characterizes the theories used today. This particular form has been developed in three distinct styles. First, there are concrete examples of QFTs that describe the world, such as the Standard Model. Second, the general form of these QFTs has been codified in axioms, following the work of Wightman. Third, mathematically well-defined QFTs have been constructed (so far in lower dimensions only), within the 'constructive' approach. I will denote this particular specific form of QFT as *conventional QFT*.

The reason I am stressing the distinction between general QFT and conventional QFT is that I will argue that the problem of the compatibility of QFT with the present knowledge of space-time is different for the two cases. I will argue that we have learned from general relativity that the notion of localizability on which *conventional* QFT is grounded is physically incorrect. General relativity requires a more subtle and complex notion of localizability than the one on which conventional QFT is grounded. I believe that the lesson about localizability that we learn from GR is so compelling that if it is well understood it is hard to imagine that it will not survive any generalization of the theory. Thus, I will argue there is something at the foundations of *conventional* QFT that we already know is physically wrong – in the sense discussed above. On the other hand, I see so far no reason to suspect that *general* QFT is physically wrong. There are a number of research directions that are currently exploring QFTs that are not conventional QFTs in the sense above, and I will briefly review some of these before closing. I believe that *general* QFT summarizes most of what we have learned so far about the world.

3 Localization in conventional QFT

In conventional QFT the theory is defined over a metric manifold representing space-time. Observables (in the sense of quantum mechanics – namely quantities to which the theory assigns probability distributions once the state of the system is known) are local. A local observable is an observable associated with a finite region of space-time. Its physical interpretation can be given – at least in principle – in terms of physical operations that can be performed inside the finite space-time region. A local observable corresponds to what is measured by a measuring apparatus *located* inside the region. (I should use here the more precise term *quasi-locality* to indicate observables located in *finite* regions of space-time. Locality is indeed often used to indicate observables concentrated in *points*.) Therefore, at the foundations of conventional QFT is the following set of assumptions.

i.   That there is a space-time manifold, which can be described in terms of a metric manifold.
ii. That we can *locate* objects in this space-time manifold in a way that is independent from the quantum field.
iii. That measurements on the quantum field are performed by apparatus *located* in finite regions of the manifold.

For example, in Wightman's formulation, the essential objects of a QFT are the functions

$$W(x_1, \ldots, x_n)$$

where $x_1, \ldots, x_n$ are points in a given, fixed metric manifold.

This implies that the algebra of the observables of the QFT has a structure of a *local algebra*. The importance of the local algebra's structure at the foundation of QFT has been emphasized in this conference by Redhead. A local algebra is a map that associates an algebra $A(S)$ to any finite open set S of the space-time manifold, in such a way that if S is a subset of S', then $A(S)$ is a subalgebra of $A(S')$.

In other words, *observables are indexed by space-time regions*. Following, for instance, Haag, one can pose a number of postulates that regulate the relation between observables and space-time regions. I am not interested in these postulates here, but only in the general form that the theory takes, owing to the assumption of the spatio-temporal localizability of the observables.

4 Localization in GR

I now pick up the second thread of this presentation, which is general relativity (GR), and discuss what we learn about localizability from GR.

In recent years, the pressure for conceptual clarity deriving from the efforts to reconcile GR with quantum mechanics has brought back to the general attention conceptual issues that were much discussed in the sixties – by Peter Bergmann's group for example. In particular, the observability issue is again a subject of debate: what are the observables of the theory? These debates have led to a deeper understanding of GR. Philosophers and historians have contributed to this clarification process. The image of GR that has emerged from these debates is far less simple than the one on which many contemporary textbooks are still based, and than the one held dear by many relativists of the generations that preceded mine – with the notable exception of very few, among whom are certainly Stachel and DeWitt. GR discloses a deep relational[5] core, which perhaps does justice to Einstein's early claims about it, and to its specific philosophical lineage. My interest in this problem started from a remarkable conference that John Stachel gave in Stockholm when I was a student.

---

[5] As in [Earman 1989], I denote as *relational* the position according to which it only makes sense to talk of the motion of objects in relation to one another, as opposed to the motion of objects with respect to space. The term *relational* has the advantage with respect to the term *relativistic* (used in this context) that it does not lead to confusion with Einstein's theories (of which special relativity is certainly very non-relational).

That conference challenged my understanding of GR and is at the root of much of the later work I have done on the subject.

Let me illustrate this recent shift in perspective on GR. According to many accounts, the physical content of GR is captured by the two following tenets:

a. Space-time is (better) described in terms of *four-dimensional curved geometry*.
b. The *geometry is dynamical*, that is, it is affected by the presence of matter, and determined by differential equations (Einstein equations).

I want to claim that the physical content of general relativity reaches much farther than tenets **a** and **b** imply.

Einstein understood the two points **a** and **b** already in 1912, at the time of his paper *Outline of a theory of gravitation*. Still, several years of intense labor on the theory had to elapse before general relativity was born. The 1912 paper lacked the third and major conceptual ingredient of the theory. This third ingredient is a change in the physical meaning of the space-time points of the theory. In Einstein's (pre-coordinate-free differential geometry) language, he had to struggle to understand 'what is the meaning of the coordinates'.

The third and deeper novelty that GR contributes to theoretical physics is a full implementation of the late-Cartesian and Leibnizian program of a completely relational definition of motion,[6] which *discards the very notion of space*:

c. Spatial, as well as temporal, *location* is defined solely in terms of contiguity between the interacting dynamical objects. *Motion* is defined solely as change of contiguity between interacting dynamical objects.

To a philosopher familiar with, say, Aristotle, the Descartes of the *Principles*, Leibniz, Berkeley, or Mach, this assertion may sound somewhat trivial. But modern theoretical physics is borne on the distinction between space and matter, and on the idea that matter moves *in space*. Physics has emerged from the famous Newton's bucket, which was taken as a proof that motion (more precisely: acceleration) is defined intrinsically, and not with respect to other bodies. As Newton himself puts it, in polemic with Cartesian relationism:[7]

...So it is necessary that the definition of *places*, and hence of local motion, be referred to some motionless thing such as extension alone or *space*, in so far as space is seen to be truly distinct from moving bodies.

To a large extent, GR overturns this position. The most radical aspect of this overturn, which, to my knowledge, is *not* anticipated by the philosophical relational tradition, is that this overturn does not refer to space alone, but to time as well. The time 'along which' dynamics develops is discarded from general relativity, as well as the space 'in which' dynamics takes place. I will return to this point later.

To clarify the distinction between the Newtonian absolutist definition of motion, which is essential for the very definition of conventional QFT, and the fully relational one in **c**, allow me to review briefly the pre-relativistic physical conceptualization of space and time.

---

[6] On the relational/absolute controversy, see [Earman 1989] and [Barbour 1992].
[7] This is from the 'De Gravitatione et aequipondio fluidorum', in *Unpublished Scientific Papers of Isaac Newton* [Hall and Hall 1962]. See [Barbour 1992] for a fascinating account of the path that led to the discovery of dynamics.

Newtonian and special-relativistic physics (which I collectively denote here as 'pre-relativistic' physics, meaning pre-*general*-relativistic), describe reality in terms of two classes of entities: space-time and matter. *Space-time* is described as a metric manifold, namely a differential manifold equipped with an additional structure: a metric, i.e. a definition of the distance between any pairs of points. This metric is unique, fixed, and not subject to differential equations. *Matter* is whatever is described according to dynamical equations of motion as moving in space-time.

The physical meaning of this pre-relativistic space-time (that is, how do we individuate space-time points, or how do we know 'where' a space-time point is) is elucidated by the subtle and highly non-trivial notion of *reference-system*, introduced by Neumann and Lange (who coined the expression) at the end of the last century [Lange 1885], precisely as a tool against relational attacks to Newton position on absolute space, and to satisfy the need, which is the core of pre-relativistic mechanics, to combine the relational nature of position and velocity with the absolute nature of acceleration. Pre-relativistic statements about space-time can be detailed as follows.

There exist inertial reference-systems, which move in a uniform motion in relation to one another, and in which mechanics holds. These reference-systems are conceptually thought of, as well as physically realized in any laboratory or in any direct observation of nature, as collections of *physical objects*. The notion of localization is then grounded on the assumptions that:

i. A relation of 'being in the same point, at the same time', or *contiguity*, is physically meaningful and immediately evident (within a relevant approximation scale).
ii. The position of any dynamical material entity is determined *with respect to the physical objects of the reference-system*, in terms of contiguity with these objects.

An essential aspect of this machinery is the clear-cut distinction between physical objects that form the reference-system (laboratory walls) and physical objects whose dynamics we are describing. Thus:

iii. The reference-system objects are *not* part of the dynamical system studied, their motion (better: stasis; by definition) is independent from the dynamics of the system studied.

This clear-cut distinction has the same content as the distinction between space-time and dynamical matter. These first three assumptions are the 'non-metrical' ones. Then we have a metrical assumption:

iv. A notion of congruence exists, such that it is meaningful to say that the distance between a pair of objects *A* and *B* is the same as the distance between a second pair of objects *C* and *D*, and that the time interval between a pair of events **A** and **B** is the same as the time interval between a pair of events **C** and **D**. In other words, we have sensible *rods* and *clocks* in our hands.

This allows the objects of the reference-system to be *labeled* in terms of their physical distance x from an (arbitrary) origin and the events to be labeled in terms of the time t lapsed from an (arbitrary) initial time. Thus, thanks to Neumann and Lange's work, we understand that if we are disturbed by the unobservability of the space-time points as such, we can trade the x and t coordinates which label those points with x and t coordinates which label physical objects of a reference-system, or by physical distances from such reference objects. The other way around, we can read the reference-system construct as the operational device that allows the properties of the background space-time to be manifest.

Whether we wish to ascribe a status of ontological reality to the space-time theoretically and operationally captured in this way, or whether we prefer a strictly empiricist position which relegates space-time to the limbo of the theoretical terms, is irrelevant to the present discussion. The relevant point here is that we may think of reality in terms of objects moving in space and time, and this *works*.

Given these assumptions, pre-relativistic physics asserts that classes of objects that are inertial reference-systems, that is, with respect to which the dynamical equations hold, exist. Once more, the coordinates x and t express the contiguity (i) of given dynamical objects with reference-system objects (ii), which are dynamically independent from the dynamical object (iii), and are labeled in terms of physical distances and time intervals (iv).

This is the theoretical structure that underlies conventional local QFT. The quantum observables are local – that is, they are characterized by their space-time location. Such location is determined with respect to 'space-time', considered as an absolute entity *à la* Newton, or, equivalently but more precisely, with respect to a set of physical objects, which are considered, for the purpose of the measurement, as physically non-interacting with the system, namely with the quantum field (within the relevant measurement's accuracy).

Then came Einstein's struggle with 'the meaning of the coordinates' in gravitational physics. As is well known, in the theory Einstein built, x and t do not indicate physical distances and time intervals (these are given by formulas involving the metric field). An immediate interpretation of x and t is that they are *arbitrary* labels of arbitrary *physical* reference-system objects. That is, assumptions (i), (ii) and (iii) above still hold, but now *any* reference-system, not just inertial ones, is acceptable. This interpretation of x and t is wrong,[8] but unfortunately it still permeates many general relativity textbooks.

The key point is that assumption (iii) fails on physical grounds. This is a consequence of the fact that there is no physical object whose motion is not affected[9] by the gravitational field, but the consequences of this simple fact are far-reaching. The mathematical expression of the failure of (iii) is the invariance of Einstein's equations under active diffeomorphisms. In 1912, Einstein had no difficulty in dropping (iv). Dropping (iii) is a much more serious matter. In fact, if we drop (iii) there is no way we can cleanly separate space-time from dynamical matter. The clear-cut pre-relativistic conceptual distinction between space-time and matter rests on the possibility of separating the reference-system physical objects that determine space-time from the dynamical objects to be studied. If space-time points cannot be determined by using physical objects *external* to the dynamical system we are considering, what are the physical points?

According to a fascinating reconstruction provided by Stachel [Stachel 1989] and Norton [Norton 1989], Einstein struggled with this problem through a very interesting

---

[8] As pointed out to me by John Stachel, such interpretation goes back to Einstein's 'mollusk of reference'. One can make this interpretation correct only at the price of imposing additional (essentially gauge-fixing) conditions on the metric, and relating these conditions to physical motions of the reference system objects.

[9] The motion of every physical object is affected by the gravitational field in a substantial way, not as a perturbation eliminable in principle. Indeed, without knowledge of the gravitational field (that is, of the metric), we have absolutely no clue as to how an object will move.

sequence of twists. In his 1912 *Outline of a theory of gravity*, he claims, to the amazement of every modern reader, that the theory of gravity must be expressed in terms of the metric $g_{\mu\nu}(x,t)$, that the right hand side of the equations of motion must depend on the energy momentum tensor, that the left hand side must depend on $g_{\mu\nu}(x,t)$, and its derivatives up to second order (all this represents three-quarters of general relativity), and that *the equations must not be generally covariant!*

Einstein argues that general covariance implies either that the theory is non-deterministic, which we do not accept, or the failure of (iii), which is completely unreasonable, because if we accept it, we do not know what space-time is. For the details of this argument, which is the famous first version of the 'hole argument', see [Earman and Norton 1987; Stachel 1989; Norton 1989].[10] Later, Einstein changed his mind. He accepted the failure of (iii), and characterized this step as a result

... beyond my wildest expectations.

The significance of the success of this step, I am convinced, has not yet been fully absorbed by part of the scientific community, let alone by the wider intellectual community.

If we are forbidden to define position with respect to *external* objects, what is the physical meaning of the coordinates x and t? The answer is: there isn't one. In fact, it is well known by whoever has applied general relativity to concrete experimental contexts that the theory's quantities that one must compare with experiments are the quantities that are fully *independent* from x and t. Assume that we are studying the dynamics of various objects (particles, planets, stars, galaxies, fields, fluids). We describe this dynamics theoretically as motion with respect to x and t coordinates, but then we restrict our attention solely to the positions of these objects in relation to each other, and not to the position of the objects with respect to the coordinates x and t.

The position of any object with respect to x and t, which in pre-relativistic physics indicates their position with respect to external reference-system objects, is reduced to a computational device, deprived of physical meaning, in general relativity.

But x and t, as well as their later conceptualization in terms of reference-systems, represent *space* and *time*, in pre-relativistic physics: the reference-system construction was nothing but an accurate systematization, operationally motivated, of the general notion of matter moving *on* space and time. In general relativity, such a notion of space and time, or of space-time, has evaporated. Objects do not move with respect to space-time, nor with respect to anything external: they move in relation to one another.

General relativity describes the relative motion of dynamical entities (fields, fluids, particles, planets, stars, galaxies) in relation to one another.

Assumption (i) above (contiguity) still holds in general relativity. Assumption (ii) (need of a reference) is still meaningful, since any position is determined with respect to a reference object. But the reference objects must be chosen *among the dynamical objects themselves that are part of the dynamical system* described by the theory.

---

[10] Einstein was led to reject generally covariant equations by difficulties of recovering the weak field limit, about which he had incorrect assumptions that he corrected later. The general physical argument he provides against generally-covariant equations was probably prompted by those difficulties. But this does not excuse us from taking this argument seriously, particularly from the man who, out of general arguments of this sort, changed our view of the world drastically, and more than once!

A characteristic example of this procedure is the following. Consider a recent experimental work that provided spectacular confirmations of GR predictions: the measurement of the binary pulsar energy decrease via gravitational radiation, which led to the 1994 Nobel prize. The pulsar of the binary system sends a pulsing signal $f(t)$, where t is an arbitrary time coordinate. This signal is further modulated by the pulsar's revolutions around its companion; let $r(t)$ be such a modulation. The two functions $f(t)$ and $r(t)$ have no meaning by themselves, but if we invert $f(t) \to t(f)$ and we insert t in $r(t)$, we obtain $r(f) = r(t(f))$, which expresses the *physically meaningful* evolution of the modulation with respect to the physical time scanned by the pulses of the pulsar.

Disregarding this essential aspect of GR has produced, and continues to produce, serious conceptual mistakes. A very typical example is an argument, rather common in the physical literature, recalled by Weingard in this conference. The argument goes as follows. Let $R(t)$ be the radius of the universe at coordinate time t. In quantum gravity, one can easily see that (at least formally) the expectation value of the operator $R(t)$ is t-independent. Therefore quantum gravity predicts that the radius of the universe is constant, contrary to our observations. This argument is completely wrong. Independence of observables from the coordinate t is totally unrelated to quantum gravity. It is a feature of classical GR. In classical GR, there is no meaning to $R(t)$ (unless we have somehow gauge-fixed t). There is no meaning to the value of the radius of the universe *at some coordinate time t*. What is meaningful is, say, the radius R' of the universe when a given supernova explodes. This quantity R' is well defined, and – in principle – we can ask for its value in quantum gravity.

The *observables* of general relativity are the *relative* (spatial and temporal) positions (contiguity) of the various dynamical entities in the theory, in relation to one another. Localization is only relational within the theory [Rovelli 1991*a, b*].

This is the relational core of general relativity; almost a homage to its Leibnizian lineage.

Fields

The discussion above requires one important clarifying comment. One may wonder what the relation is between the relational character of general relativity that I have emphasized, which implies some kind of disappearance of the very notion of space, and the statement about space contained in point **a**, above.

In order to clarify this point, one must consider the historical evolution of the second horn of the space-time-versus-matter dichotomy. *Matter* was understood in terms of massive particles until the work of Faraday and Maxwell. After Faraday and Maxwell, and particularly after the 1905 paper of Einstein himself, which got definitely rid of the hypothetical material substratum of the Maxwell field, *matter* is no longer described solely in terms of particles, but rather in terms of *fields*.

GR is a product, a main step, and one of the great successes of the process of understanding reality in terms of *fields*. The other steps are Maxwell electrodynamics, the $SU(3) \times SU(2) \times U(1)$ standard model, and Dirac's second quantized field theoretical description of the elementary particles. Einstein's program that led to general relativity, in fact, was to express Newtonian gravity in a *field theoretical* form, which could be compatible with special relativity.

Table 1

Space-time fixed, non-dynamical		Matter dynamical	
Differential manifold	*Metric*	Fields: electromagnetic f. *gravitational* f. other interactions f. electron and quark f. and others	(Massive particles)

Table 2

Fixed, non-dynamical		Dynamical	
Differential manifold	Field: metric f. = gravitational f.	Fields: electromagnetic f. other interactions f. electron and quark f. and others	(Massive particles)

Thus, in contemporary physics *matter* is understood also, and perhaps primarily, in terms of *fields*. The consequences of this fundamental fact on the understanding of general relativity are far-reaching. The pre-relativistic physical ontology is as shown in Table 1.

Let us now see how this ontology is affected by general relativity. The technical core of general relativity is the *identification* of the gravitational field and the metric. Thus, general relativity describes the world in terms of the entities shown in Table 2.

In general relativity, the metric and the gravitational field are the same entity: the metric/gravitational field. In this new situation, illustrated by Table 2 above, what do we mean by *space-time*, and what do we mean by *matter*? Clearly, if we want to maintain the space-time-versus-matter distinction, we have a terminology problem.

In the physical, as well as in the philosophical literature, it is customary to denote the differential manifold *as well as* the metric/gravitational field (see Table 2 above) as 'space-time', and to denote all the other fields (and particles, if any) as matter.[11] But the table above shows that such a terminological distinction loses all its teeth on the picture of the physical world given by general relativity. In general relativity, the metric/gravitational field has acquired most, if not all, the attributes that have characterized matter (as opposed to space-time) from Descartes to Feynman: it satisfies

---

[11] To increase the confusion, the expression *matter* has a different meaning in particle physics and in relativistic physics. In particle physics, one sometimes distinguishes three entities: *space-time, radiation* and *matter*. Radiation includes fields such as the electromagnetic field, the gravitational field, and the gluon fields (bosons); matter includes the leptons and the quarks, which are also described as fields, but have more pronounced particle-like properties than photons or gluons (fermions). In the relativistic physics terminology, which we follow here, one denotes collectively both *radiation* and *matter* (namely both bosons and fermions) as *matter*. What I am arguing here is not that the metric has the same ontological status as the electrons (*matter-matter*); but that the metric/gravitational field has the same ontological status as the electromagnetic field (*radiation-matter*).

differential equations; it carries energy and momentum;[12] in Leibnizian terms: it *can act and also be acted upon*, and so on.

In the quote above, Newton requires motion to be referred to pure extension or *space*

'in so far as *space* is seen to be truly distinct from moving bodies'

that is, insofar as space is pure non-dynamical extension, which is certainly *not* the case for the gravitational/metric field. One cannot dispute terminology, but terminology can be very misleading: we may insist in denoting the metric/gravitational field as *space*, but the gravitational/metric field is much more similar to what Newton would have denoted as matter than to what he would have denoted as space.

A strong burst of gravitational waves could come from the sky and knock down the Rock of Gibraltar, precisely as a strong burst of electromagnetic radiation could. Why should we regard the second burst as ontologically different from the second? Clearly the distinction has become ill-founded.

Einstein's identification between gravitational field and geometry can be read in two alternative ways: i. as the discovery that the gravitational field is nothing but a local distortion of space-time geometry; or: ii as the discovery that *space-time geometry is nothing but a manifestation of a particular physical field*, the gravitational field. The choice between these two points of view is a matter of taste, at least as long as we remain within the realm of non-quantistic and non-thermal general relativity. I believe, however, that the first view, which is perhaps more traditional, tends to obscure, rather than to enlighten, the profound shift in the view of space-time produced by GR.

In the light of the second view, it is perhaps more appropriate to reserve the expression *space-time* for the differential manifold, and to use the expression *matter* for everything which is dynamical, acts and is acted upon, namely all the fields *including the gravitational field* (see Table 2 above). Again, one cannot dispute with terminology. This is not to say that the gravitational field is *exactly* the same object as any other field. The very fact that it *admits* an interpretation in geometrical terms witnesses to its peculiarity. But this peculiarity can be understood as a result of the peculiar way it couples with the other fields.[13]

If, on the other hand, by *space-time* we indicate the differential manifold, then motion with respect to space-time is non-observable. Only motion with respect to dynamical matter is.

Whatever terminology one chooses, the central point is that the pre-relativistic theorization of the structure of physical reality as *space* and *time* on which *dynamical matter* moves has collapsed in general relativistic physics.

*Physical reality is now described as an interacting ensemble of **dynamical** entities (fields) whose localization is only meaningful with respect to one another.*

---

[12] Of course, the expression 'carries energy and momentum' has to be carefully qualified in a general relativistic context, given the technical difficulties with local definitions of gravitational energy-momentum. But the point was made clear by Bondi with his famous pot of water example: a pot of water can be heated up and brought to boil, using gravitational waves.

[13] The gravitational field is characterized by the fact that it is the only field that couples with *everything*. More technically, the fibers of the bundle on which the gravitational connection is defined are *soldered* to the tangent bundle.

The relation among dynamical entities of being *contiguous* (point (i) above) is the foundation of the space-time structure. Among these various entities, there is one, the gravitational field, which interacts with every other one, and thus determines the relative motion of the individual components of every object we want to use as rod or clock. Because of that, it admits a metrical interpretation.

I think it may be appropriate to close this section by quoting Descartes:[14]

> If, however, we consider what should be understood by movement, according to the truth of the matter rather than in accordance with common usage, we can say that movement is the transference of one part of matter or of one body, from the vicinity of those bodies immediately *contiguous* to it, and *considered* at rest, into the vicinity of some others.

The quote by Newton I gave earlier was written directly in polemic with this relational Cartesian doctrine. The amount of energy Newton spent in articulating the details of the technical argument that supported his thesis against Descartes' relationalism, namely the arguments that led to the conclusion quoted above, is a proof of how seriously Newton thought about the issue. The core of Newton's arguments is that *one cannot construct a viable theory of motion out of a purely relational position.* In particular, Newton correctly understood that Descartes' own successful principle of inertia makes no sense if motion is purely relative.[15]

The well-known resistance of continental thinkers to Newtonian physics was grounded in the claimed manifest 'absurdity' of Newton's definition of motion with respect to a non-dynamical ('that cannot be acted upon') and undetectable space. The famous Leibniz–Clarke correspondence and Berkeley's criticisms are the prime examples of these polemics. From the present privileged point of view, everybody was right and wrong. Newton's argument is correct if we add the qualification 'in the seventeenth century' to the *one cannot construct* above. The construction of general relativity disproves the unqualified version of this argument. As often happens in science, progress (Newtonian dynamics) needed to rely on some incorrect assumption, which later further progress (general relativity) was able to rectify.[16]

I often wonder how Leibniz would have reacted by learning that by taking the fully relational nature of motion very seriously, a man living in Switzerland a few centuries down the road would have been able to correct 43 missing seconds-of-arc-per-century in Mercury's motion, which Newton's theory could not account for.

After three centuries of extraordinary success of the 'absurd' Newtonian notion of 'motion with respect to space', general relativity has taken us home, to a respectable but a bit vertiginous understanding of the purely relational nature of space.

It is my conviction that this discovery that came from GR is a lasting contribution. In other words, it may very well be that GR is some sort of low energy approximation of other theories. However, I would be immensely surprised if this discovery of the

---

[14] In the *Principia Philosophiae*, Sec. 11-25, p. 51.

[15] To allow himself to make use of the notion of absolute space, Newton used the extreme resort of calling God to his aid: according to Newton, space has curious ontological properties because it is nothing but the *sensorium of God*.

[16] This entire issue has strong resemblance with the polemics on the notion of *action-at-distance*, which is the other aspect of Newtonian mechanics that scandalized continental natural philosophers: pre-Newtonian physics disliked action-at-distance. Newton's genius introduced it, being aware of its difficulties (he calls it *repugnant*, at one point) but realizing its efficacy. Faraday, Maxwell and Einstein (again) were able to purify the Newtonian prodigiously powerful construct of this 'repugnant' action-at-distance, without impairing the general scheme, but using the costly jewels of a much higher mathematical complexity and of the new idea of field (*tout se tien*).

relational nature of localization turned out to be wrong. I think that there are physical discoveries that are – in a sense – for ever, like Galileo's discovery of the fact that absolute velocity is meaningless, or Einstein's discovery that absolute simultaneity is meaningless. It seems to me that the discovery of the fact that localization is only in reference to dynamical objects will stay in physics for at least quite a while.

From the mathematical point of view, this means that I expect that a credible theory of the world – if formulated in spatio-temporal language at all – must be invariant under active diffeomorphisms. Of course, I may be wrong, but I will take this as a result in the following.

5 Incompatibility

It is time to bring the two threads of this presentation together, namely the discussion of localizability in conventional QFT, and the discussion of localizability in GR. In conventional QFT one assumes that it is meaningful to have observables localized at space-time points in a way that does not depend on the quantum field itself. This notion of localization is incompatible with GR.

Thus:

> *In a QFT compatible with what we know at present about space-time,*
> *we cannot expect observables to be indexed by space-time regions.*

The discussion indicates, I submit, that the very foundations of local, *conventional* QFT are physically unacceptable, on the grounds of what we know about space-time.

Let me give a few examples of this incompatibility, in order to emphasize this point. We have seen for instance that the content of a conventional QFT can be expressed in terms of Wightman's *n*-point functions

$$W(x_1, \ldots, x_n).$$

Take as the simplest example of these, the propagator

$$W(x, y).$$

Now, $W(x, y)$ is a function of the *distance* between x and y. But this is meaningless in a general relativistic context, where the theory is defined over a differential manifold. There is no distance between two points on a differential manifold. One may require x and y to be coordinates in the sense of differential geometry, and thus require $W(x, y)$ to be diffeomorphism invariant. If so, $W(x, y)$ can take only two values: one value when x and y overlap and the other value otherwise! Essentially the same is true for the other *n*-point functions. The point is that the *n*-point functions are not an appropriate way of capturing the physical content of a QFT compatible with GR.

Indeed, it has been remarked that out of the five axioms of axiomatic QFT, four axioms cannot even be formulated in a general relativistic framework!

Conventional QFT is constructed over a metric manifold. The metric structure of the manifold is an essential ingredient for the QFT machinery. We learn from GR that the metric structure of the space-time manifold cannot be taken as independent from the quantum field itself. Indeed, it is a quantum field itself. Therefore, the sole spatio-temporal background structure that we are allowed to consider is the differential manifold. Even the differential manifold, then, disappears from the actual physical predictions that the theory provides, because observables must be diffeomorphism invariant, namely non-localized on the manifold.

The requirement that the theory be defined over a manifold rather than on a metric space is highly non-trivial. The jump from a QFT over a metric space to a QFT on a manifold is a big jump, requiring a renewal of most of the technical machinery that forms the theory. Indeed, the traditional technical and conceptual tools in terms of which we understand conventional QFT do not make sense any more in the absence of a metric structure. Analogous considerations apply for the notions of vacuum, *n*-particle states, time evolution, Hamiltonian and unitarity. All these notions cease to make sense at all on a differentiable manifold.

Thus, if we take the physical content of general relativity seriously, it follows that *conventional* QFT is physically incorrect as a fundamental theory.

6 Diffeomorphism invariant QFT

Does the above discussion imply that there is some kind of intrinsic incompatibility between QFT and GR? I do not believe so. As I observed at the beginning of this talk, localizability is an aspect of *conventional* QFT, but not necessarily of *general* QFT, namely of a quantum theory of a system with infinite variables.

I am convinced – I will not enter into details here – that nothing of GR challenges the essential physical tenets of QM itself, or of *general* QFT. Of course, for the moment this is only a statement of belief, since a well-defined general QFT explicitly compatible with GR does not exist yet. Still, there is a variety of research programs whose aim is precisely to construct *a general* QFT *which is not a conventional* QFT, and in which conventional localizability is replaced by a general relativistic notion of localizability.

A QFT of this kind is defined over a differential manifold and not over a metric space. A metric structure, if any, will appear only in the solutions of the theory as a special feature of certain quantum states. Notice that the need to accommodate the superposition principle, and thus the physical possibility of quantum superposition of distinct geometries, prevents the possibility that a well-defined metric structure be well defined for all quantum states of the theory. Therefore, a general QFT of this sort is also *a theory of quantum geometry.*

From the technical point of view, a theory defined over a manifold (respecting the invariances of the manifold) is invariant (gauge invariant) under the action of the active diffeomorphism group. In other words, the general relativistic notion of localizability is technically captured by the invariance of GR under active diffeomorphisms. Thus, a QFT compatible with what we currently know about space-time has to be a *diffeomorphism invariant quantum field theory.*

For the reasons I discussed above, a 'diff-invariant' (diffeomorphism invariant) QFT, or more loosely speaking, a general covariant QFT, is necessarily an object profoundly different from conventional QFT. It cannot be formulated in terms of propagators and *n*-point functions. There is no evolution in an external geometrical time. Its observables are not indexed by space-time regions, and localization cannot be defined extrinsically from the theory itself.

A large number of research directions which are currently active can be seen as attempts to construct non-trivial examples of such a theory. These research directions are characterized by a great variety of styles, motivations, and levels of mathematical rigor. However, they can all be seen as attempts to define a diffeomorphism invariant QFT.

I believe that constructing and understanding a diffeomorphism invariant QFT is one of the major challenges of contemporary theoretical physics, and it is perhaps the essential step we need to make in order to bring to a conclusion the great revolution started in physics with QM and relativity, and to recover a new synthesis of what we know of the world at the fundamental level.

7 Directions of investigation and examples of diff-invariant QFTs

I will now give a very sketchy overview of a number of research directions currently pursued, attempting the construction of variants of QFT compatible with GR.[17]

1 Topological quantum field theories

A particular kind of diffeomorphism invariant QFT has been effectively and completely constructed, and investigated in detail: the topological QFT or TQFT. Thus, TQFTs are examples of *general* QFT that are not conventional QFT and do not require an underlying metric manifold to be defined. These are very simple theories characterized by the fact that there are as many gauges as variables, so that in spite of the field theoretical formulation the number of physical degrees of freedom is finite. They are interesting as theoretical models of phenomena that may happen in the presence of diffeomorphism invariance (and have great mathematical interest), but have no direct physical use *per se*. They are interesting for the present discussion because, in spite of their simplicity, they are concrete examples of QFT which are not conventional QFT. For instance, in one of the prototypes of TQFT, the CSW theory, observables are indexed by loops in the space manifold, where the actual *localization* of the loop is irrelevant – the only relevant aspect of the loop is the way it is knotted, namely its features up to arbitrary smooth transformations. This example shows concretely how a quantum field theory may have a profoundly non-local character, still retaining its quantum mechanical and field theoretical nature. Axioms for TQFT have been formulated by Atiyah. Remarkably, these are compatible with diffeomorphism invariance and do not include any notion of localizability. It has been pointed out by many, for instance by Atiyah himself, that theories of this kind may have the right structure as fundamental theories of reality. A certain confusion in this regard has been generated by a common prejudice, according to which any field theory invariant under diffeomorphisms must have a finite number of degrees of freedom. This is wrong. On the contrary, it is reasonable to suspect that a fundamental theory consistent with everything we currently know about the world could have the form of a TQFT with an infinite number of degrees of freedom.

2 Euclidean quantum gravity

An inspiring, but so far only purely formal attempt to define a diff-invariant QFT is Hawking's Euclidean quantum gravity program.

---

[17] I will not attempt to provide comprehensive references, but only a few indications. For an overview of present ideas on quantum geometry and diffeomorphism invariant QFT, see [JMP 1995]. On quantum gravity, see [Isham 1995].

3 Simplicial and discrete quantum gravity

These are ongoing attempts to formulate non-perturbative quantum gravity on the model of lattice QCD.

4 Non-commutative geometry

This is an algebraic approach to QFT that attempts to incorporate QMs non-commutativity within the very definition of space-time. The hope is to use spectra of spatio-temporal operators (as the Dirac operator) to capture the geometry in a diffeomorphism invariant manner.

5 Non-perturbative string theory

Perturbative string theory is a successful attempt to construct (an extension of) a QFT that includes the gravitational field, as well as any other known field, in terms of perturbations around a fixed background geometry. The open problem of string theory is to understand the *non-perturbative* formulation of the theory, which, if found, would allow us to use the theory for describing physics in the genuine quantum-gravitational regime – where the perturbative series diverges – and to get information about the dynamically selected vacuum of the theory, without which little information on Planck scale physics can be extracted from the theory. There have been several attempts to construct non-perturbative string theory. Some of these have been mentioned by Weingard in this conference. As far as I know, no solution is in view for the moment. It is my opinion that the difficulties of defining non-perturbative string theory are deep conceptual and foundational difficulties: non-perturbative string theories drive us towards a regime in which no space and time are defined at all. Others in this conference are more competent that I am on this topic, and I will not pursue it further.

6 Loop quantum gravity

This is a concrete, although incomplete, example of diff-invariant QFT with infinite degrees of freedom. Loop quantization is the program of canonical quantization of GR based on the Ashtekar formulation of GR and on the hypothesis that the fundamental excitations of the quantum theory are one-dimensional loop-like quantum states [Rovelli and Smolin 1988]. (More precisely, spin-network-like states.) This approach is a concrete example of an attempt to retain most of the machinery and the physical assumptions of QFT, while discarding the underlying metric manifold and the locality assumptions of conventional QFT. A concrete diff-invariant QFT is constructed. The program is not completed, and there are still missing elements in the theory. Still, a number of results have been obtained, and applications to concrete physical problems have been already considered [Rovelli 1996].

I briefly summarize here the physical assumptions of this research program, and its results. The aim is to construct a consistent quantum theory whose classical limit is GR. A unitary representation of a suitable algebra of GR observables (the loop algebra) has been constructed. This is defined on a Hilbert space. Operators corresponding to area and volume of space-time regions have been defined by means of regularization techniques, and shown to be finite, self-adjoint, and generally covariant. These operators have been shown to have a discrete spectrum, and the spectrum has been recently computed explicitly. The discreteness of the spectrum of area and volume implies a discrete structure of physical space at the Plank scale [Ashtekar, Rovelli and Smolin 1992]. Notice that this discreteness is not imposed on the theory, but simply derived by a conceptually straightforward application of quantum theory to GR. Area and volume operators are not physical observables as such, but there are arguments suggesting that the result extends immediately to the spectra of physical area and volumes of spatial regions determined by dynamical matter, in theories in which GR is coupled with matter. To the extent that these arguments are correct, the spectra of area and volume computed from the theory can be taken as detailed quantitative predictions from a diff-invariant QFT.

The theory allows a rather good control of spatial diff-invariance. Physical states of the quantum gravitational field turn out to be labeled by abstract colored graphs, with no relation to their spatial localization. Indeed, there is no spatial localization of a physical state of quantum gravity. It is localization that has to be defined with respect to such a state. Something is known with regard to four-dimensional diff-invariance, or, equivalently, time evolution, but only at the level of a tentative strong coupling perturbation scheme. An infinite class of exact solutions of the quantum dynamics is known, but its physical relevance is still unclear. Physical observables in the theory are required to commute with the canonical constraints and are therefore all nonlocal in space as well as in time.

From the mathematical point of view, the theory exists now in two unitarily isomorphic forms, the loop representation [DePietri and Rovelli 1996] and the connection representation [Ashtekar and Lewandowski 1996]. Ashtekar has described in detail the connection representation approach in this conference, and thus I will not comment further on this approach.

In virtually all these attempts to extend QFT to a domain in which conventional locality does not hold, something profoundly new happens to space-time. Indeed, virtually all the above attempts lead to some form of *discretized quantum geometry at small scales*. This is the perhaps inevitable result of subjecting the dynamical space-time geometry to quantum mechanics. Understanding what is a *quantum geometry* is the physical version of the mathematical problem of constructing a general covariant quantum field theory.

8 Common issues and foundations of diff-invariant QFT

Let me briefly illustrate some of the problems that emerge in dropping the assumption of localizability from QFT.

1 Observables

Observable quantities are obtained in classical GR by means of a complicated process of elimination of the space-time coordinates x and t from solutions of the equations of motion. The same process does not make sense in quantum theory, since the quantum theory – in a sense – needs to know what the observables are that are to be formulated. It is to observable quantities that we must associate probability distributions, and thus self-adjoint operators, or elements of the physical *-algebra. Thus, if the theory is formulated in terms of elementary fields, the observables must be expressed as functions (more precisely functionals) of fields. This is in general extremely difficult in GR. In the canonical formulation of the theory, a diffeomorphism invariant quantity is a function on the phase space that commutes with the constraints. Virtually no such function is explicitly known in GR. The issue is subtle: one can define diffeomorphism invariant quantities, for instance by gauge fixing, but the dynamics become intractable in the gauge-fixed scheme. For instance, in the sixties the idea of using curvature invariants as physically defined coordinates was investigated (Kretschmann–Komar coordinates). This is a conceptually correct way of approaching the observability problem, but has two drawbacks: it is very removed from the actual experimental practice, and it leads to a great complication in the dynamics. These complications prevented further developments. Similarly, attempts were made to discuss the quantum theory in gauge-fixed formulations, but this seems to work, at least so far, only in simplified models of the theory.

The technical task of writing observables in a gravitational theory becomes a bit more readily treated in the presence of matter, where one can use the matter as a natural 'locator'. This is the idea of incorporating the *Lange–Neumann* material systems in the quantum theory.

GR is constructed in terms of an ensemble of variables redundant with respect to the quantities connected with observations. The non-physical redundant (gauge) part of the field is given by the dependence on the (x, t) coordinates themselves. The physical non-redundant component of the field is given by the coordinate independent quantities, namely the quantities that express the value of physical quantities *at the space-time location* determined by other quantities. These quantities, which become quantum entities, are the ones subjected to quantum fluctuations, and are observable. If we believe that the core discoveries of QM and GR are correct (I do) we must accept something like the following about the theory that describes the world:

> *The content of the theory is in the probability distributions of the values of physical quantities at space-time locations determined by other dynamical entities.*

I suspect that this rather ethereal structure is what we have to learn to master, if we want to reach a unified picture of what we have learned so far of the physical world.

2 Time

In the evolution from Cartesian relationalism to the modern one, we have gained a strange extra bonus: the resurrected relationalism, combined with 'relativity', i.e., with the discovery of the intertwining of space with time (discovered, of course, by Einstein himself), leads us to the relational nature of time. It is easy to talk about space-time, and think of it as a sophisticated version of our familiar space. But of course space-time includes, for better or for worse, *time*. The relational nature of time, I believe, is essentially unexplored land. The notion of time is certainly among the most problematic ones. The difficulty of physics in dealing with the various layers of the notion of time has been repeatedly emphasized. For instance, the lack of any concept of 'becoming' in physics [Reichenbach 1956, Grunbaum 1963], the loss of the notion of *present*, and the strident friction between physical time and phenomenological time [see for instance the interesting and comprehensive discussion in Pauri

1994] are embarrassing facts that have made rivers of ink flow. [The following book is a collection of thousands of references on this subject: Macey 1991.] The physics of this century, general relativity, quantum mechanics and their hypothetical combination make the matter much more serious [see for instance Ashtekar and Stachel 1991].

If the space-time continuum geometry breaks down at short scales (in time units, around 10^{-40} seconds) as indicated by virtually any attempt to combine GR with QM, there is no way to think of the world as 'flowing through time'. This is not going to affect our description or understanding of temporal perception, or mental processes, which are insensitive to such small time scales. But it is going to affect our basic description of the world. In a sense, 'there is no time' at the fundamental level [see Rovelli 1993].

The disappearance of time is dramatically evident in the mathematical formalism of the theory, both in non-perturbative canonical gravity and in non-perturbative string theory. In both theories, indeed, the fundamental equations essentially *do not contain a time variable*. In canonical quantum gravity, this is due to the fact that since the dependence on the coordinates x and t is a non-observable gauge, the physical quantities in terms of which the quantum theory is constructed are independent from t. We still have clock-times, which are phenomenologically connected to what we usually call time, but the different clock-times are only probabilistically related at very short scale, and cannot be thought of as approximating a universal absolute time variable. In non-perturbative string theory, the very space-time manifold is born only as a particular vacuum solution of the full theory. In general, there isn't even a space-time. In both cases, a reasonable evolution in time is hypothetically recovered only within particular regimes.

In other words, as noticed by DeWitt in this conference, in a quantum general relativistic theory *time* is only a phenomenological concept, not an absolute one.

But can we deal with a fundamental physical theory radically without time? I believe we can. From the mathematical point of view, there are some technical issues to be solved, on which the debate is still open [see Rovelli 1991c, d]. But from the physical point of view, a profound re-thinking of reality has to accompany such a step. This is not easy. Even the comparatively simple and well-established alteration of the notion of time brought by special relativity is difficult to digest. Many physicists, even many relativists, agree that the flow of time depends on the state of motion of the observer, but still cannot free themselves from thinking of the great big clock that clicks the universal time in which everything happens. Much more difficult is to fully familiarize oneself with the general relativistic relational notion of time: while it is easy to have a good intuition of physics on a given curved manifold, it is, on the other hand, rather difficult to have good intuition of the dynamics of the gravitational/metric field itself. The hypostatization of unphysical structures such as the ADM surface is a good example of the traps along the way, into which too many good physicists keep falling. But the radical step that quantum gravity seems to ask us to take, i.e., learning how to think of the world in completely non-temporal terms, is definitely the hardest.

3 Relational aspects

Notice that there are *two levels* at which a diffeomorphism invariant general QFT is relational. Because of quantum mechanics, the observations described or predicted by the theory refer to the relation between an observing system and an observed system; because of general relativity, observations refer to the contiguity between an entity considered as dynamical and an entity considered as a spatio-temporal reference.

Is there any connection between these two levels? If *localization* is to be included in the theory in some fundamental form, *contiguity* should also be an aspect of the observer/observed relation. The contiguity relation is indeed the only remnant of the space-time structuring of the events of pre-relativistic physics. I suspect a connection between the general relativistic relationalism and the quantum mechanical one should exist, but no theoretical effort, to my knowledge, has yet succeeded in grasping such a connection. The works of Baez, Crane, Smolin and others in TQFT [JMP 1995] are attempts to incorporate these ideas into the theory.

9 Conclusion

General relativity on the one hand and quantum mechanics on the other have dismantled the concepts on which the grand synthesis of the Cartesian–Newtonian physics rested for three centuries. *We do not have a new synthesis.* We have collected an impressive amount of fragmentary knowledge about the physical world but we lack the general picture. We do not know how to think of space, time, matter and causality in a way which is coherent and consistent with everything we have learned so far.

To find a period that questioned accepted assumptions as deeply as now, one must go back four centuries. Between the publication of Copernicus' *De Revolutionibus*, which opened *the* scientific revolution *by autonomasia*, and the publication of Newton's *Principia*, which brought it to a spectacularly successful conclusion, 150 years elapsed. Along those 150 years, the western conceptions about time, space, cause, matter, and rationality were profoundly shaken, bitterly discussed, and finally reshaped in a new grand synthesis. The personae of this intellectual adventure – Bacon, Galileo, Descartes, Kepler, Huygens,· Leibniz – had acute awareness of being in the middle of a major conceptual change, but little clue as to where the change was leading. Each of them saw fragments of the final fresco. The fragments looked strange and sometimes inconsistent. And each of them struggled to combine those fragments, sometimes falling back to medieval views which in retrospect make us smile, sometimes with visionary intuitions. Newton saw the final picture, which then remained one of the wonders of Europe for over three centuries.

I have little doubt that we are in the middle of a similar process. At the beginning of this century, GR altered the classical understanding of the concepts of space and time in a way which, I have argued, is far from being fully understood yet. A few years later, QM challenged the classical account of matter and causality, to a degree which is still the subject of controversies. After the discovery of GR we are no longer sure of what is space-time and after the discovery of QM we are no longer sure of what is matter. The very distinction between space-time and matter is likely to be ill-founded, as I have argued. The picture of the physical world that had been available for three centuries had problematic aspects and was repeatedly updated, but it was consistent and the core of its conceptual structure remained clear and stable from the formulation of the Cartesian program to the end of the last century. I think it is fair to say that today we *do not have* a consistent picture of the physical world at all.

This incongruity between quantum mechanics and general relativity has taken us back to a pre-Newtonian situation in which the physics of the sky and physics of the Earth were distinct, and the laws of motion of terrestrial objects were assumed to be different from the laws of motion of celestial objects. Physics has been in a schizophrenic state of mind for more than half a century, which, after all, is roughly the same time for which Kepler's ellipses and Galileo's parabolas (quite incompatible, if taken as fundamental laws of motion) were forced to cohabit within physics. The pragmatism of the middle of this century, a pessimistic diffidence toward major synthetical efforts and the enormous (terrifying, sometimes) effectiveness of the modern *Scientia Nova* has restrained the majority of the physicists from confronting such an untenable situation directly. But I do not think that this utterly pragmatic attitude may be satisfactory or productive in the long run.

The conceptual notions on which the Newtonian synthesis was grounded emerged from a century of eminently *philosophical* debate on the extraordinary Copernican discovery that the solar system looks very much simpler if seen from the point of view of the sun. Galileo asked, 'If the Earth really moved, could we perceive such a motion?'; Descartes asked, 'What is motion?'; Kepler asked, 'If planets revolve around the sun, isn't there maybe an influence, a *force*, due to the sun, that drives the planets?', and so on. It is difficult to underestimate the role that a purely speculative and *philosophical* quest has had on the shaping of the classical scientific image of the world. Consider just a few examples: the idea that the world admits a description in terms of bits of solid matter moving through a background space and mechanically affecting each other; the idea that simple mathematical equations can describe any such motion and any such influence; determinism; the relativity of velocity; the distinction between primary and secondary qualities... Classical science would not have developed without these philosophical doctrines, which were elaborated, or revived, and integrated, in the debate that followed Copernicus's discovery.

General relativity and quantum mechanics are discoveries as extraordinary as the Copernican discovery. I believe they are, like Kepler's ellipses and Descartes' principle of inertia, fragments of a future science. It is my conviction that (general) QFT, and GR are – if properly understood – correct. Perhaps it is time to take them seriously, to try to understand what we have actually learned about the world by discovering relativity and quantum theory, and to find the fruitful questions. The task is simply to be smart enough to see how they fit together; that is, I suspect, to understand how to construct a diffeomorphism invariant QFT.

Maybe the Newtonian age has been an accident and we will never again reach a synthesis. If so, a major project of natural philosophy has failed. This possibility has been suggested, but before committing ourselves to an excessively radical form of scientific pluralism, I think we should ask ourselves whether we are not mistaking a vital moment of confusion with a permanent loss of clarity.

If a new synthesis is to be reached, I believe that philosophical thinking will be, once more, one of its ingredients. Owing to the conceptual vastness of the problematic involved, the generality and accuracy of philosophical thinking, and its capacity to clarify conceptual premises, are probably necessary to help physics out of a situation in which we have learned so much about the world, but we no longer know what matter, time, space and causality are. As a physicist involved in this effort, I wish the philosophers who are interested in the scientific description of the world would not confine themselves to commenting and polishing the present fragmentary physical theories, but would take the risk of trying to look *ahead*.

My belief is that the extraordinary physical discoveries of the twentieth century, GR and QM in particular, are pointing towards a profoundly new and deeper way of understanding physical reality. It is almost as if they are screaming something to us, and we do not understand what. It has been repeatedly claimed that we may be close to the end of physics. We are not close to the end of physics, not to the final theory of everything, I believe. We are very much in the dark. We left the sunny grasses of Cartesian–Newtonian physics, and we are only halfway through the woods.

References

Ashtekar A. and Stachel J. eds. 1991, *Conceptual Problems of Quantum Gravity*, Birkhauser, New York.

Ashtekar A., Rovelli C. and Smolin L. 1992, 'Weaving a classical metric with quantum threads', *Physical Review Letters* **69**, 237.

Ashtekar A. and Lewandowski J. 1996. 'Quantum theory of geometry: area operator', gr-qc/9602046.

Barbour J. 1992, *Relative and Absolute Space*, Oxford University Press.

Connes A. and Rovelli C. 1994, 'Von Neumann Algebra and Automorphisms and Time – Thermodynamics Relation in General Covariant Quantum Theories' *Classical and Quantum Gravity*, **11**, 2899.

Descartes R. 1644, *Principia Philosophiae*.

DePietri R. and Rovelli C. 1996, 'Geometry eigenvalues and scalar product in loop quantum gravity from recoupling theory', gr-qc/9602023, to appear in *Phys. Rev. D*.

Earman J. 1989, *World Enough and Spacetime: Absolute vs. Relational Theories of Space and Time*, MIT Press, 1989.

Earman J. and Norton J. 1987, *British Journal of the Philosophy of Science*, **38**, 515.

Grunbaum A. 1963, *Philosophical Problems of Space and Time*, Knopf, New York.

Hall A.R. and Hall M.B. eds. 1962, *Unpublished Scientific Papers of Isaac Newton*, Cambridge University Press, Cambridge.

Isham C. 1995, 'Structural issues in quantum gravity', to appear in Proceedings of the GR14 (Florence 1995).

JMP 1995, 'Diffeomorphism invariant quantum field theory and quantum geometry', Special Issue of *Journal of Mathematical Physics, J. Math. Phys.* **36**.

Lange L. 1885, 'Über das Beharrungsgesetz', *Berichte der math.-phys.*

Macey S.L. 1991, *Time: a Bibliographic Guide*, Garland, New York.

Norton J. 1989, 'How Einstein found his field equations, 1912-1915' in *Einstein Studies, Vol. 1 Einstein and the History of General Relativity*, Horward D. and Stachel J. eds., Birkäuser, Boston.

Pauri 1994, 'Physical time, experiential time, real time', in *International Colloquium on Time in Science and Philosophy*, Istituto Suor Orsola Benincasa, Napoli 1992. See also: *Dizionario delle Scienze Fisiche; Voce quadro: Spazio e Tempo*. Istituto Treccani, Italy, in print.

Reichenbach H. 1956, *The Direction of Time*, University of California Press, Berkeley.

Rovelli C. 1991*a*, 'What is observable in classical and quantum gravity?' *Classical and Quantum Gravity* **8**, 297; 1991*b* 'Quantum reference-systems', *Classical and Quantum Gravity* **8**, 317.

Rovelli C. 1991*c*, 'Quantum mechanics without time: a model', *Physical Review* **D42**, 2638; 1991*d* 'Time in quantum gravity: a hypothesis', *Physical Review* **D43**, 442.

Rovelli C. 1993, 'Statistical mechanics of gravity and thermodynamical origin of time', *Classical and Quantum Gravity* **10**, 1549; 'The statistical state of the universe', *Classical and Quantum Gravity* **10**, 1567; Connes A. and Rovelli C. 1994, *Classical and Quantum Gravity* **11**, 2899.

230   *Carlo Rovelli*

Rovelli C. 1996, 'Black hole entropy from loop quantum gravity', gr-qc/9603063.

Rovelli C. and Smolin L. 1988, 'Knot theory and quantum gravity', *Physical Review Letters* **61**, 1155; 'Loop space representation for quantum general relativity', *Nuclear Physics* **B331**, 80. For various perspectives on loop quantum gravity, see: C. Rovelli, *Classical and Quantum Gravity* **8**, 1613 (1991); A. Ashtekar in *Gravitation and Quantization, Les Houches 1992*, eds. Julia B. and Zinn-Justin J. (Elsevier Science, Paris, 1995); L. Smolin in *Quantum Gravity and Cosmology*, Perez-Mercader J., Sola J. and Verdaguer E. eds. (World Scientific: Singapore, 1992).

Stachel J. 1989, 'Einstein's search for general covariance 1912–1915' in *Einstein Studies, Vol. 1 Einstein and the History of General Relativity*, Horward D. and Stachel J. eds., Birkhäuser, Boston.

Discussions

Audience: You based your difference between the pre-GTR and post-GTR ontologies on the invariance under diffeomorphisms, right? But is this invariance not a trivial thing, true for every physical theory, for pre-relativistic, special relativistic, and post-relativistic? Isn't it just model theoretical fact that a theory can fix a model only up to isomorphism?

Rovelli: No, it is not. This issue has been much debated (for instance by Earman and Norton, Maudlin, Liu, Rynasiewicz, Leeds, and recently in a most extended and convincing manner by Belot), and it is subtle. It is certainly true, as you say, that a theory can fix a model only up to isomorphism. In this context, in particular, it has been repeatedly noticed that any theory (including Newton's theory) can be reformulated in a diffeomorphism invariant form (just by using arbitrary coordinates). In this respect, general relativity is not a novelty. What is the peculiar novelty of general relativity, is that the theory includes active diffeomorphism invariance. This is a physical fact, not a matter of metaphysics or of philosophy of language. The physical fact is the following. Before GTR, we thought we could choose suitable reference bodies external to the gravitational system. Doing so, and expressing location as 'position with respect to these bodies', we thought we would have a deterministic theory. Now we know that this procedure is not viable. It follows that before GTR we could interpret coordinates as absolute location (in some suitable sense). After GTR, we are forced to retreat to a purely relational interpretation of location. Therefore, there are two distinct issues here: one is the definitive physical novelty of GTR, which I just described. The other is the ambiguity connected to the inscrutability of reference, common to all theories, to which you referred. Unfortunately, these two facts get very subtly intertwined in the analysis of the mathematical formalism of GTR, and one can easily mistake one for the other.

Stachel: In some general covariant formulations of the Newtonian theory there are nondynamical fields. This shows that general covariance is 'artificial' in these cases. The big difference is whether or not there are any nondynamical fields in the theory, that is, fields given *a priori* independent of (although they enter into) any field equation for dynamic fields.

Rovelli: There are. But this does not change the main point. A theory in which there is a given (fixed, nondynamical) metric tensor does not have active diffeomorphism invariance, even if expressed in general coordinates. On the other hand, you are right that there are also formulations of Newtonian theory in which all fields 'look' dynamical, in the sense that they all obey equations of motion. For instance, one may have a metric field satisfying the equation: Riemann curvature equal to zero. But if you go through the complete dynamical analysis of the theory (for instance following Dirac's procedure), and you do the correct counting of the physical degrees of freedom of the theory, then you see that such a field is not dynamical.

Dickson: I do think it's important to try to suggest how physical theory might look in the future, but an essential part of doing so is trying to understand those physical theories that we have now, and even those that we now believe to be false. The reason was mentioned by Redhead. One of the things philosophers try to do is to understand the success of a theory, even if we believe the theory is false, and in fact *especially* if we believe the theory is false, because then it's even more difficult to understand how the theory in question can be successful in light of the fact that, for example, it seems to be incoherent, or its axioms seem to be inconsistent or intractable. And as Redhead pointed out, what's essential to understanding how a theory we believe to be intractable or incoherent is nonetheless successful, is trying to find those structures or principles of the theory that are leading to its success in spite of its incoherence. Indeed, one might reasonably suppose that those structures or general principles will carry over to a successor theory. Hence understanding them in the context of present theories might be essential to pointing the way to future development.

Rovelli: Yes! I don't disagree with anything you said. The attitude and the program you describe would be of great interest for physics.

Fisher: I had a general question: What is the degree to which the problems you raise are not just a matter of shying away from what might be called the measurement problem? In a rather trivial way this morning, I exhibited the two-point correlation function and said: Here it is, and it varies as $1/r^{d-2}$ (although actually that is not quite correct). However, I did take the trouble to mention that there were experiments which are done – scattering experiments – and I was prepared to describe them. My impression is that if you want to understand what you really mean by the Wightman functions or the correlation functions, etc., you must go back and ask: What do we do in the laboratory? Usually we set up some situation in which the objects of interest are 'free' or 'weakly interacting'. We know this is only an approximation but, within that sense, one hopes to limit the effects of faraway influences. Such considerations seemed to me to be missing from your discussion; or maybe I just missed the point entirely? However, at least for experimental physicists and those theorists among us concerned with turning our theories into something verifiable (or falsifiable), that is what has to be looked at carefully.

Rovelli: Great! Very very good! You put your finger on the key point, and I am very pleased. We have to look at what we do in experiments. If we want to understand the foundations of a subtle theory like general relativity, we have to look at the kind of experiments that are performed in reality in relativistic experimental gravitation. Indeed, this is precisely the path that leads me to my comments on the meaning of space-time and location in general relativity. You see, if you analyze realistic gravitational experiments, you immediately discover that you can never express the quantity measured as a quantity that depends on given space-time coordinates. In particle physics, you put a detector in the point x, measure something at time t, and you get a quantity 'at the space-time point (x, t)'. Your detector sits there, whatever the particles under consideration are doing.

It is not so in GR (or GTR, as philosophers seem to like better). Let me briefly mention some gravitational experiments. The prototype gravitational measurement is the following: how many degrees over the horizon is Venus, seen from Alexandria, on the first day of spring of the year Ptolemy is 40? The point to be noticed is that the spatio-temporal determination is given by a certain position of the Earth, whose motion is precisely one of the dynamical variables considered.

Another example: recently Taylor got the Nobel prize in experimental gravitation for an indirect measurement of gravitational wave emission which confirms GTR predictions to a spectacularly good precision. What did Taylor measure? He observed a binary pulsar, which is a system in which a star emitting a pulsating signal (beep, beep, beep, ...) revolves (rotates) around another star. Because of these revolutions, we get the beeps sometimes faster (when the star moves towards us) sometimes slower (when the star moves away from us). Thus the beeps are modulated by a period which is the revolution period of the star. Now, the star emits gravitational waves, it slowly loses energy, and therefore its period changes. Over many years, Taylor sees the change in the revolution period of the star. More precisely he sees that the number of bits in each period is (very slowly) changing. What is meaningful here is the number of bits per period. That is, the bits play the role of a clock, and the period is the quantity measured: temporal location is only relational in GR. This is what we learn from all realistic gravitational experiments.

Stachel: I like the Feynman approach to quantum mechanics and quantum field theory, in which you talk about a probability amplitude. In general we have an amplitude for some process. The question is: if you measure *this* here and now, what result will you get if you measure *that* there and then? And the whole problem of quantum mechanics is to compute the amplitude for the process, and then you square it to get the probability of this process. The point is that in general relativity, even at the classical level, the 'here' and 'there' and the 'now' and 'then' are part of the answer, not part of the question, as they are traditionally in every other part of physics. So all you can say is that somehow a quantum gravity theory must give you an amplitude for connecting two 'somethings'. And only after you answer the question will you know what the two 'somethings' are.

15. Comments

JOHN STACHEL

During the almost 40 years that I have been following it, there has been an amazing (to me, at any rate) change in the tenor of the discussion about the relation between quantum field theory and general relativity. In 1957, I started graduate studies at Stevens Institute of Technology, then a world center of relativity research: there were actually three people there who worked on such problems![1] I soon started attending the famous informal Stevens relativity meetings, getting to know many of the leading figures in the field, and meeting most of the others at the 1959 Royaumont GRG meeting.

This was a time of high tension, of struggle between two rival imperialisms, one clearly much stronger than the other. I am referring, of course, to the dominant quantum field theory paradigm, which was stubbornly resisted by the much weaker unified field theory program. I call these two programs imperialisms because each had a universalist ideology used to justify an annexationist policy. Einstein's unified field program aimed to annex quantum phenomena by means of some generally covariant extension of general relativity that would include electromagnetism, somehow miraculously bypassing quantum mechanics. The structure of matter and radiation, including all quantum effects, would result from finding non-singular solutions to the right set of non-linear field equations. In spite of repeated failures over 30–40 years, Einstein persisted in working toward this goal, though not without increasing doubts, particularly towards the end of his life.[2]

Advocates of the quantum field theory approach, insofar as they were interested in gravitation at all, were convinced that, in order to annex gravitational phenomena, one just had to apply the techniques being developed for special-relativistic quantum field theories – in particular, for the then-triumphant quantum electrodynamics – to the Einstein equations. Since gravitation is even weaker than electromagnetism, one could proceed by treating general relativity as if it were a Poincaré-invariant theory that happens to be non-linear (self-interacting) and to have a nasty gauge group. Starting from Minkowski space-time and quantizing only deviations from its metric, they literally aimed at flattening out general relativity with the steamroller

---

[1] They were Jim Anderson, Dave Finkelstein and Ralph Schiller. I took my only formal course in general relativity with Huseyn Yilmaz, who soon left. I started research on relativistic plasma theory with Dave, but soon found that the relativity interested me much more than the plasmas.

[2] See John Stachel, 'Einstein and the quantum: fifty years of struggle,' in *From Quarks to Quasars/Philosophical Problems of Modern Physics* (University of Pittsburgh Press, 1986), pp. 349–85; and 'The other Einstein: Einstein contra field theory,' in Mara Beller, Robert S. Cohen, Jürgen Renn, eds., *Einstein in Context* (Cambridge University Press, 1993), pp. 275–90.

of perturbation theory in order to calculate the incredibly small cross sections for things like graviton–graviton and graviton–photon scattering.

I soon concluded that neither imperialism offered a particularly attractive alternative and, to continue my Cold War metaphor, what was needed was a period of peaceful coexistence and competition between the two contending programs, with the prospect that in the long run neither ideology would survive unchanged if a really successful program were to combine the two into one coherent world-picture. My sympathies were with the 'third-worlders', the small band of hardy souls trying to survive in the interstices between the two imperialisms, looked at askance by the advocates of both.

I recently reread my first comments on the subject, written just about 30 years ago.[3] Pointing out 'that we have no direct knowledge of the range of validity of our concepts of the space-time continuum and its metric structure', I concluded '... one must recognize that the breakdown of the continuum-metric approach to space-time is not at all impossible when we come to extend our research into the microworld.' After giving some of the standard 'third-worlder' arguments (such as the infinities of quantum field theory and the singularity theorems of general relativity) suggesting that we might be approaching the limits of validity of our current macroscopic space-time concepts, I went on to list five types of theoretical program that might take us beyond those limits.

These various possible speculations have been listed in more or less descending order of distance from our present-day ideas.... If we base our guess on the past development of physics it would seem likely that the latter possibilities will be explored first; and only repeated failure of such attempts ... will lead to serious consideration of the former possibilities, representing a much sharper break with current ideas.

(p. 193)

At that time, as I mentioned, field-theoretical imperialism reigned supreme, and I didn't expect to see the day when the prevalent opinion among physicists – even quantum field theorists – would be that drastically new concepts are needed to make real progress in quantum gravity, or indeed in elementary particle theory itself. But today the ideology of effective field theories reigns (almost) supreme, and research programs that would have been dismissed as crankish 30 years ago are now received with all due respect. In his recent book, rapidly becoming the canonical text on quantum field theory, Steve Weinberg writes:[4]

The point of view of this book is that quantum field theory is the way it is because (aside from theories like string theory that have an infinite number of particle types) it is the only way to reconcile the principles of quantum mechanics (including the cluster decomposition property) with those of special relativity. This is a point of view I have held for many years, but it is also one that has become newly appropriate. We have learned in recent years to think of our successful quantum field theories, including quantum electrodynamics, as 'effective field theories,' low-energy approximations to a deeper theory that may not even be a field theory, but something different like a string theory. On this basis, the reason that quantum field theories describe physics at accessible energies is that *any* relativistic quantum theory will look at sufficiently low energy like a quantum field theory.

(p. xxi)

[3] John Stachel, 'Comments on "Causality requirements and the theory of relativity",' in *Boston Studies in the Philosophy of Science*, vol. 5 *Proceedings of the Boston Colloquium for the Philosophy of Science 1966/1968*, Robert S. Cohen and Marx Wartofsky, eds. (Reidel, Dordrecht, 1968), pp. 179–98.
[4] Steven Weinberg, *The Quantum Theory of Fields*, vol. 1, *Foundations* (Cambridge University Press, 1995).

I recently heard talks by two leading string theorists (Strominger and Witten), suggesting that string theory itself may be just a low-energy approximation to something entirely different. In this open-minded atmosphere, non-standard approaches to quantum gravity have finally achieved respectability – which is a good thing, I suppose, as long as respectability does not become an aim in itself.

Now for some comments on the papers. Bryce DeWitt expressed himself rather categorically on two points, about which I feel the need for more caution. He argues: 'Since it interacts with all other dynamical systems the gravitational field simply *must* be quantized', although he immediately softens the impact of these words by adding 'whether by string theory or through some as yet unsuspected approach'. The first part of his remark harks back to an earlier discussion with Leon Rosenfeld about the significance of the Bohr–Rosenfeld analysis of the measurability of the components of the electromagnetic field strengths, and its extension to an analysis of the measurability of the components of the gravitational field.[5] DeWitt argued that, since the uncertainty relations apply to the material sources of the fields, the fields themselves cannot escape quantization. Rosenfeld replied (quite emphatically) that the aim of the Bohr–Rosenfeld paper was a proof of the consistency, not of the necessity, of electromagnetic field quantization. Concerning the extension of their argument to the gravitational case, the judicious conclusions of Kalckar are worth recalling:[6]

Extensive efforts have been displayed, especially by DeWitt, with the purpose of deducing the structure of the commutation relations among gravitational field quantities from an analysis of the measurability of such quantities within the framework of general relativity. For a proper judgement of the results obtained, it should be realized that the mentioned attempts rest crucially on the assumption that the coupling between field and test body is so weak, that it can be treated as a small perturbation. Whatever intuitive reasons may be given to defend such a procedure, the fact that gravitation theory by its very nature is a nonlinear theory makes it somewhat dangerous to base rather far-reaching conclusions on the shaky assumption of the free field commutators being a reasonable starting point for a discussion of the measuring problem... While Rosenfeld's critique is certainly justified insofar as the attempts by DeWitt and others to analyse the measurability of gravitational fields are based on the *ad hoc* assumption of the possibility of a correspondence treatment, which is not at all so innocent or plausible as it looks, nevertheless Rosenfeld appears to push his case too far, in as much as there can be no doubt that, once the actual occurrence of gravitational radiation was experimentally demonstrated, the existence of all relevant quantal features in the gravitational field indeed [may] be immediately inferred.

(pp. 157, 159)

The second categorical statement that I question is: 'What is certain is that gravity leads unavoidably to a consideration of the whole universe as a quantum object.... The universe...is an *isolated* system. We are not going to understand quantum gravity until we are able to understand the quantum mechanics of isolated systems'.

[5] For this discussion, see Leopold Infeld, ed., *Proceedings on Theory of Gravitation/Conference in Warszawa and Jablonna 25–31 July 1962* (Gauthier-Villars, Paris and PWN, Warszawa 1964): Bryce S. DeWitt, 'The quantization of geometry,' pp. 131–43, and the discussion comments by Leon Rosenfeld, pp. 144–5, 219–22; and Leon Rosenfeld, *Selected Papers*, Robert S. Cohen and John Stachel, eds. (Reidel, Dordrecht/Boston/London 1979): 'On quantization of fields,' pp. 442–4 and 'Quantum theory and gravitation,' pp. 599–608.
[6] J. Kalckar, 'Measurability problems in the quantum theory of fields,' in *Foundations of Quantum Mechanics/Rendiconti della Scuola Internazionale 'Enrico Fermi' IL Corso*, Bernard d'Espagnat, ed. (Academic Press, New York/London, 1971), pp. 127–69.

I think it would be a very grave situation indeed if, at some physically meaningful level, one could not separate the problems of quantum gravity from the quite different ones of cosmology. If the only meaningful quantum general relativity were a quantum cosmology, then we could not achieve any progress in quantum gravity based on the usual quantum mechanics of *non*-isolated systems. Fortunately, I believe that such progress has already been demonstrated in (at least) one case: the development of a quantization technique for asymptotically flat gravitational fields such that Penrose's Scri can be defined for them.[7] This technique differs from the usual Minkowski-space quantization in a number of significant ways: the gravitational field in the interior region can be arbitrarily strong, the asymptotic symmetry group is the Bondi–Metzer–Sachs group, gravitons are only defined on Scri, to name three major ones. In principle, the use of physical apparatuses external to the system and placed very far from the strong-field (interior) region would enable one to physically interpret this formalism in much the usual quantum-mechanical way; e.g., in terms of the preparation of incoming graviton states and measurement of cross sections for various outgoing graviton states. Much more could be said about other circumstances in which a non-cosmological gravitational quantization could make sense, but one counter-example suffices to refute a universal claim.

So on to the problem of time in quantum gravity. DeWitt states:

> In a general relativistic theory time is a phenomenological concept, not an absolute one. It is what a clock measures and can never be infinitely precise. Probability too, in the end, is a phenomenological concept. It can never be infinitely precise.

I think what he is saying is that the concept of time is inherent in a system consisting of gravitational field-plus-matter, but only under conditions such that some part of the matter has the appropriate structure to function as a clock.[8] Not being an instrumentalist, I would say that a clock measures (as opposed to defining) what I call a 'local time', which brings me to my main point.

In discussions of 'the problem of time in general relativity', an important distinction often is not made between two quite distinct concepts: *global* time and *local* time. Classically (i.e., non-quantum-mechanically), a local time is associated with any timelike world-line, which in relativity is the proper time along such a world-line, measured by a 'good clock' traveling along it.[9] Since any clock has to be of finite size (and so cannot be confined to a single world-line) and since one may have to compensate for the effects on its mechanism of any acceleration along its world-line, such a measurement of the proper time can never be infinitely precise.

---

[7] For a review, see Abhay Ashtekar, *Asymptotic Quantization/Based on 1984 Naples Lectures* (Bibliopolis, Naples 1987), Part II. *Asymptotic Quantization*, pp. 35–107.

[8] I am not sure what meaning Bryce attaches to the word 'phenomenological'. He contrasts it with 'absolute', but here it does not seem to mean 'relative'. In physics, it is usually contrasted with 'fundamental' (see, for example, Carl M. Bender, 'Theoretical physics,' *Macmillan Encyclopedia of Physics*, John S. Rigden, ed. (Simon and Schuster Macmillan, New York 1996), vol. 4, pp. 1585–8; the section on 'Phenomenological versus theoretical theory,' p. 1587). But I presume Bryce calls something phenomenological if it is intrinsic to a system, but can only be meaningfully associated with the system under some appropriate set of conditions.

[9] I emphasize that we are discussing model clocks: ideal entities constructed within some theoretical framework. The question of the relation between these models and actual clocks used in laboratory measurements falls under the general question of the relation between such models and the real objects they are supposed to model – an important philosophical question, but one not peculiar to quantum gravity!

But, within the framework of classical general relativity, this is a consistency problem, of the kind that arises within any physical theory.

In general relativity they take the form: can we construct model instruments consistent with relativistic dynamics that give us access to a kinematic entity (or relation) whose existence is demanded by the theory?[10] In other words, is there any obstacle in principle to constructing a device that can measure proper time with any preassigned level of precision? At the classical level of general relativity, there is not: there is consonance between relativistic kinematics and dynamics as far as proper time measurement goes.

In the absence of a complete theory of quantum gravity, attempts to answer the corresponding question can help us to probe the possible structure of the future theory.[11] It is certainly meaningful to talk about quantization of the proper time: or, at any rate, about quantum limits on the concept of proper time. Salecker and Wigner carried out such an investigation of limits on the measurability of proper time resulting from the quantum structure of a material clock,[12] which suggested that there could be such quantum limits.[13]

Global time is a quite different concept, based on some simultaneity relation between distant events.[14] This is an equivalence relation among events, dividing them into a one-parameter family of equivalence classes: by definition (i.e., by stipulation or convention), all events in the same equivalence class occur at the same global time.[15] When it is said that in non-relativistic quantum mechanics time is an ordering parameter that can have no operator associated with it, it is the global time which is meant. And indeed, why should the global time have an operator associated with it? As Einstein emphasized in his first (1905) paper on special relativity, any distant simultaneity relation *always* involves a conventional element, thus has no direct

---

[10] For example, one may postulate rigid bodies in Newtonian kinematics, because there is nothing in Newtonian dynamics to prevent us from constructing models of such bodies. To postulate rigid bodies in relativistic kinematics is not consistent with relativistic dynamics (special or general). Of course, in special relativity, certain classes of rigid *motions* (of non-rigid bodies) are compatible with relativistic dynamics – and that is all that is needed to establish the consistency of special relativistic kinematics.

[11] Indeed, that is just what Wigner and Salecker did when they discussed measurement of space-time intervals using a clock that did not violate the uncertainty principle. See Eugene Wigner and Edward Salecker, 'Quantum limitations of the measurement of space-time distances,' *Physical Review* 109 (1958): 571–7, reprinted in Eugene Wigner, *The Collected Works of Eugene Wigner Part A: The Scientific Papers*, vol. 3, Part I: *Particles and Fields* (Springer, Berlin/Heidelberg/New York, 1997), pp. 148–54.

[12] See footnote 11; ibid.

[13] Wightman recently commented that 'this paper's full physical implications have probably not yet been realized' (Arthur S. Wightman, 'Wigner on Particles and Fields,' in *The Collected Works of Eugene Wigner, Part A: The Scientific Papers*, vol. 3, Part I: *Particles and Fields* (Springer-Verlag, Berlin/Heidelberg/New York, 1997), pp. 3–23, citation from p. 22).

[14] The situation is made somewhat confusing by the numerical identity of local and global time in Newtonian space-time, where the convention of identifying the global time with the absolute time is so 'obviously' natural that it is usually not remarked. Similarly, in special relativity, there is a numerical identity between the global time in an inertial frame of reference, defined using the standard Einstein simultaneity convention, and the proper time, measured by a clock at rest in such an inertial frame of reference. Yet it is only the local clock times that have a non-conventional physical significance. In general relativity, of course, there is no natural convention for the global time, except in certain special cases, such as static space-times; and even in that case, the global time function does not coincide with the proper time of a clock at rest (i.e., one whose world-line coincides with one of the static Killing vector trajectories).

[15] Mathematically, this amounts to introducing a global foliation of space-time. As is well known, not all space-times permit a global foliation. The issue of definitions of simultaneity in various space-time theories is discussed in Sahotra Sarkar and John Stachel, 'Simultaneity revisited' (forthcoming).

physical significance; and hence cannot be associated with an observable.[16] That means that the answer to *no* physically meaningful question can depend on the convention chosen: the question to be asked about a convention is not 'Is it correct?', but 'Is it convenient?'.

So far what I have said refers primarily to classical relativity. But it seems clear that no meaningful quantum-mechanical question – nor the answer to it – can depend upon the choice of a global time function. It would be meaningless to talk about the quantization of anything corresponding to a global time, nor is it clear what the quantum analog of the global time (if any) should be.

This is one big reason why I share Bryce's unease with canonical quantization, which is based on singling out a global time function. Of course, in the end one hopes to show that the resulting quantum theory is independent of this choice; but I have the feeling it would be better never to paint oneself into this corner, out of which one must later so painfully try to paint oneself. But (let me emphasize once again) the local (i.e., the proper) time is a physically meaningful concept classically, and it must have *some* quantum-mechanical aspect or analog insofar as any sort of more-or-less conventional quantization of general relativity is contemplated.

I think some of Carlo Rovelli's comments on 'time' could be made more precise if he were to distinguish between places where he means 'local time' and places where he means 'global time'. For example, when he speaks of labeling 'events ... in terms of the time t elapsed from an (arbitrary) initial time', it is not clear which concept he has in mind. When he speaks of 'rods and clocks', this might suggest a measurement of the local (proper) time between the two events along some timelike world-line connecting them. However, when he speaks of constructing a reference frame using such rods and clocks, this suggests the adoption of a convention for defining a global time that connects it with physical measurements of certain proper times (the only kind of time intervals that can be measured) as well as (proper) spatial intervals.

The main point that Carlo makes about the lesson of the hole argument can be put in the following philosophical context. Traditional ontology, still widely accepted today, makes a sharp distinction between *things* and *relations* between things. It is assumed that it makes no sense to speak of relations without introduction of a previous category of things, whose existence is presumed independent of any relations that may exist between them. The philosophical lesson of the way of thinking about general relativity that Carlo has been advocating is that general relativity is best understood as fundamentally a relational theory. That is, the entities treated by the theory, such as points of space-time ('events'), do not exist apart from the relational structure (the metric tensor field) connecting them. Philosophically, this is an instance of what is called the theory of internal relations. This theory has a long philosophical pedigree including names such as Leibniz, Hegel and Marx – at least in some interpretations, such as that of Ollman, who has discussed this theory at some length.[17] In this approach, there is no

---

[16] The non-conventional relation among events, in both special and general relativity, is the partial ordering induced by the causal relation between events; i.e., the light-conoid structure associated with each event. One could adopt the convention that the light-conoid structures associated with the events along some timelike world-line define a global time (Einstein actually mentioned this possibility in 1905). But again, the physically significant thing is the structure of the light conoids, and not the global time convention adopted, making the conoids hypersurfaces of simultaneity.

[17] See Bertell Ollman, *Alienation/Marx's Conception of Man in Capitalist Society* (Cambridge University Press, 1971), Part I 'Philosophical introduction,' Section 3, 'The philosophy of internal relations,' pp. 27–42; and the Appendix, 'The philosophy of internal relations – II,' pp. 249–54.

rigid separation between things and relations. What looks like a thing from one point of view may look like a relation from another. I don't want to enter here into a long discussion of this point, except to say that we should be trying to develop ways of presenting general relativity that avoid the pitfalls of the things first, relations afterwards approach that the usual manifold-metric tensor presentations encourage.[18]

Carlo himself sometimes falls into this trap. For example, he says: '...it is perhaps more appropriate to reserve the expression space-time for the differential manifold...'. But without the introduction of the metric tensor field on a differentiable manifold, the manifold does not represent a space-time, and its points do not represent events – they have no spatio-temporal, indeed no physical significance at all. To call the bare manifold 'space-time' is therefore quite misleading. I am sure Carlo does not disagree; it is just a matter of finding a way of developing the theory that does not allow us to speak of space-times without speaking of metric fields.[19]

His inclusion of the differentiable manifold among the 'Fixed, non-dynamical' elements of general relativity, while traditional, is also in a sense misleading. How do we actually establish the differentiable manifold that is to be associated with a solution to the gravitational field equations? We first solve these equations locally, and then look for the maximal extension of this local patch to a global manifold that is compatible with certain criteria that we impose on global solutions to the field equations (e.g., that it be geodesically complete). Of course, different criteria may lead to different global manifolds (e.g., we might require that it be only timelike, or timelike and null, geodesically complete). This observation only reinforces my point, which is that the global structure of the manifold is not fixed, and depends on the dynamics of the theory in the sense that it cannot be specified before specifying a particular solution to the field equations.[20]

Another point where I feel Carlo's formulation might be misleading is his assertion that 'In general relativity, the metric and the gravitational field are the same entity'. He fails to distinguish between two conceptually quite distinct space-time structures: the metric, which represents the chronogeometric structure of space-time, and the affine connection, which represents its inertio-gravitational structure. Any gravitational theory that incorporates the equivalence principle must represent the gravitational field by a connection. The peculiarity of general relativity is that this connection is the unique symmetric one that is compatible with the metric: in more physical terms, this means that the inertio-gravitational and chronogeometrical structures of space-time are intimately related.[21] But when we speak of an (inertio-)gravitational

[18] I find the approach to general relativity sketched in Robert Geroch, 'Einstein algebras,' *Communications in Mathematical Physics* 26, 272–3 (1972), especially promising. One wants an abstract algebra of tensor fields, and a representation theory for that algebra, such that differentiable manifolds constitute one possible type of representation; but hopefully not the only type.

[19] I suggested a possible way in John Stachel, 'What a physicist can learn from the history of Einstein's discovery of general relativity.' In *Proceedings of the Fourth Marcel Grossmann Meeting on General Relativity*. Remo Ruffini, ed. Amsterdam: Elsevier, 1986, pp. 1857–1862.

[20] See John Stachel, 'How Einstein discovered general relativity: a historical tale with some contemporary morals.' In *General Relativity and Gravitation – Proceedings of the 11th International Conference on General Relativity and Gravitation*. M.A.H. MacCallum, ed. Cambridge: Cambridge University Press, 1987, pp. 200–208.

[21] For a discussion of these space-time structures and their relations in Newtonian theory, special relativity, and general relativity, see John Stachel, 'Changes in the concepts of space and time brought about by relativity,' in Carol C. Gould and Robert S. Cohen, eds., *Artefacts, Representations, and Social Practice* (Kluwer, Dordrecht, 1994), pp. 141–62; and 'The story of Newstein or: is gravity just another pretty force?' (forthcoming).

field, we should mean primarily an affine connection. The fact is that in general relativity this connection can be uniquely derived from a (dynamical) metric with Lorentzian signature that carries all the chronogeometric information, while in Newtonian gravitation theory it is derived from a connection potential that is independent of the (non-dynamical) chronogeometric structures; but these facts are secondary to the fact that in both cases the inertio-gravitational structure is represented by a connection. Again, I think our presentations of general relativity should emphasize the crucial role of the connection from the start.

Which brings me to Ashtekar and Lewandowski's paper. I think the emphasis Ashtekar's work places on the role of the connection in general relativity has been a major factor in bringing general relativity back into the mainstream of field theory in a way that does not sacrifice any of its unique features. Once we view general relativity as the theory of a connection on a principal fibre bundle, it takes its place alongside the other gauge theories of fundamental interactions, without losing sight of the principal difference: as Trautman has emphasized for the last 20 years, in general relativity, the principal bundle is a sub-bundle of the bundle of linear bases, with a solder form that ties it more tightly to the base manifold than is the case in a general bundle.[22]

If some more-or-less direct generalization of conventional quantum-field-theoretical quantization schemes is going to work, I believe it will be based on the connection approach. On the other hand, I share Bryce's unease with the heavy emphasis on the canonical approach, with its inherent three-plus-one breakup of space-time, which has so far also characterized the Ashtekar approach. One would hope that a more intrinsically four-dimensional variant of that approach could be found, but still based on the connection. I have been more-and-more impressed by the profundity and power of the path integral approach to non-relativistic quantum mechanics, with its inherently space-time viewpoint.[23] I would hope that ultimately some way of extending Ashtekar's emphasis on the three-connection and triad variables to the four-connection and tetrads could be found by combining it with the path integral approach to quantization. Meanwhile, one can only applaud and admire the new insights that those working on the Ashtekar program are bringing to the many problems of quantum gravity.

---

[22] See, for example, Andrzej Trautman, 'Fiber bundles, gauge fields, and gravitation,' in Alan Held, ed., *General Relativity and Gravitation One Hundred Years After the Birth of Albert Einstein* (Plenum Press, New York/London, 1980), vol. 1, pp. 287–308.

[23] See John Stachel, 'Feynman paths and quantum entanglement: is there any more to the mystery?' in Robert S. Cohen, Michael Horne and John Stachel, eds., *Potentiality, Entanglement and Passion-at-a-Distance/Quantum Mechanical Studies for Abner Shimony*, Vol. Two (Kluwer, Dordrecht/Boston/London, 1997), pp. 245–56.

Part Six

16. What is quantum field theory, and what did we think it was?

STEVEN WEINBERG*

Quantum field theory was originally thought to be simply the quantum theory of fields. That is, when quantum mechanics was developed physicists already knew about various classical fields, notably the electromagnetic field, so what else would they do but quantize the electromagnetic field in the same way that they quantized the theory of single particles? In 1926, in one of the very first papers on quantum mechanics,[1] Born, Heisenberg and Jordan presented the quantum theory of the electromagnetic field. For simplicity they left out the polarization of the photon, and took space-time to have one space and one time dimension, but that didn't affect the main results. (Comment from audience.) Yes, they were really doing string theory, so in this sense string theory is earlier than quantum field theory. Born *et al.* gave a formula for the electromagnetic field as a Fourier transform and used the canonical commutation relations to identify the coefficients in this Fourier transform as operators that destroy and create photons, so that when quantized this field theory became a theory of photons. Photons, of course, had been around (though not under that name) since Einstein's work on the photoelectric effect two decades earlier, but this paper showed that photons are an inevitable consequence of quantum mechanics as applied to electromagnetism.

The quantum theory of particles like electrons was being developed at the same time, and made relativistic by Dirac[2] in 1928–1930. For quite a long time many physicists thought that the world consisted of both fields and particles: the electron is a particle, described by a relativistically invariant version of the Schrödinger wave equation, and the electromagnetic field is a field, even though it also behaves like particles. Dirac I think never really changed his mind about this, and I believe that this was Feynman's understanding when he first developed the path integral and worked out his rules for calculating in quantum electrodynamics. When I first learned about the path-integral formalism, it was in terms of electron trajectories (as it is also presented in the book by Feynman and Hibbs[3]). I already thought that wasn't the best way to look at electrons, so this gave me a distaste for the path-integral formalism, which although unreasonable lasted until I learned of 't Hooft's work[4] in 1971. I feel it's all right to mention autobiographical details like that as long as the story shows how the speaker was wrong.

In fact, it was quite soon after the Born–Heisenberg–Jordan paper of 1926 that the idea came along that in fact one could use quantum field theory for everything, not just for electromagnetism. This was the work of many theorists during the period

* Research supported in part by the Robert A. Welch Foundation and NSF grant PHY 9009850. E-mail address: weinberg@utaphy.ph.utexas.edu

1928–1934, including Jordan, Wigner, Heisenberg, Pauli, Weisskopf, Furry, and Oppenheimer. Although this is often talked about as second quantization, I would like to urge that this description should be banned from physics, because a quantum field is not a quantized wave function. Certainly the Maxwell field is not the wave function of the photon, and for reasons that Dirac himself pointed out, the Klein–Gordon fields that we use for pions and Higgs bosons could not be the wave functions of the bosons. In its mature form, the idea of quantum field theory is that quantum fields are the basic ingredients of the universe, and particles are just bundles of energy and momentum of the fields. In a relativistic theory the wave function is a functional of these fields, not a function of particle coordinates. Quantum field theory hence led to a more unified view of nature than the old dualistic interpretation in terms of both fields and particles.

There is an irony in this. (I'll point out several ironies as I go along – this whole subject is filled with delicious ironies.) It is that although the battle is over, and the old dualism that treated photons in an entirely different way from electrons is I think safely dead and will never return, some calculations are actually easier in the old particle framework. When Euler, Heisenberg and Kockel[5] in the mid-1930s calculated the effective action (often called the Euler–Heisenberg action) of a constant external electromagnetic field, they calculated to all orders in the field, although their result is usually presented only to fourth order. This calculation would probably have been impossible with the old fashioned perturbation theory techniques of the time, if they had not done it by first solving the Dirac equation in a constant external electromagnetic field and using those Dirac wave functions to figure out the effective action. These techniques of using particle trajectories rather than field histories in calculations have been revived in recent years. Under the stimulus of string theory, Bern and Kosower,[6] in particular, have developed a useful formalism for doing calculations by following particle world lines rather than by thinking of fields evolving in time. Although this approach was stimulated by string theory, it has been reformulated entirely within the scope of ordinary quantum field theory, and simply represents a more efficient way of doing certain calculations.

One of the key elements in the triumph of quantum field theory was the development of renormalization theory. I'm sure this has been discussed often here, and so I won't dwell on it. The version of renormalization theory that had been developed in the late 1940s remained somewhat in the shade for a long time for two reasons: (1) for the weak interactions it did not seem possible to develop a renormalizable theory, and (2) for the strong interactions it was easy to write down renormalizable theories, but since perturbation theory was inapplicable it did not seem that there was anything that could be done with these theories. Finally all these problems were resolved through the development of the standard model, which was triumphantly verified by experiments during the mid-1970s, and today the weak, electromagnetic and strong interactions are happily all described by a renormalizable quantum field theory. If you had asked me in the mid-1970s about the shape of future fundamental physical theories, I would have guessed that they would take the form of better, more all-embracing, less arbitrary, renormalizable quantum field theories. I gave a talk at the Harvard Science Center at around this time, called 'The renaissance of quantum field theory,' which shows you the mood I was in.

There were two things that especially attracted me to the ideas of renormalization and quantum field theory. One of them was that the requirement that a physical theory be renormalizable is a precise and rational criterion of simplicity. In a sense, this requirement had been used long before the advent of renormalization theory. When Dirac wrote down the Dirac equation in 1928 he could have added an extra 'Pauli' term[7] which would have given the electron an arbitrary anomalous magnetic moment. Dirac could (and perhaps did) say 'I won't add this term because it's ugly and complicated and there's no need for it.' I think that in physics this approach generally makes good strategies but bad rationales. It's often a good strategy to study simple theories before you study complicated theories because it's easier to see how they work, but the purpose of physics is to find out why nature is the way it is, and simplicity by itself is I think never the answer. But renormalizability was a condition of simplicity which was being imposed for what seemed after Dyson's 1949 papers[8] like a rational reason, and it explained not only why the electron has the magnetic moment it has, but also (together with gauge symmetries) all the detailed features of the standard model of weak, electromagnetic, and strong interactions, aside from some numerical parameters.

The other thing I liked about quantum field theory during this period of tremendous optimism was that it offered a clear answer to the ancient question of what we mean by an elementary particle: it is simply a particle whose field appears in the Lagrangian. It doesn't matter if it's stable, unstable, heavy, light – if it's field appears in the Lagrangian then it's elementary, otherwise it's composite.**

Now my point of view has changed. It has changed partly because of my experience in teaching quantum field theory. When you teach any branch of physics you must motivate the formalism – it isn't any good just to present the formalism and say that it agrees with experiment – you have to explain to the students why this is the way the world is. After all, this is our aim in physics, not just to describe nature, but to explain nature. In the course of teaching quantum field theory, I developed a rationale for it, which very briefly is that it is the only way of satisfying the principles of Lorentz invariance plus quantum mechanics plus one other principle.

Let me run through this argument very rapidly. The first point is to start with Wigner's definition of physical multi-particle states as representations of the inhomogeneous Lorentz group.[9] You then define annihilation and creation operators $a(\vec{p}, \sigma, n)$ and $a^{\dagger}(\vec{p}, \sigma, n)$ that act on these states (where $\vec{p}$ is the three-momentum, σ is the spin z-component, and n is a species label). There's no physics in introducing such operators, for it is easy to see that any operator whatever can be expressed as a functional of them. The existence of a Hamiltonian follows from time-translation invariance, and much of physics is described by the S-matrix, which is given by the well known Feynman–Dyson series of integrals over time of time-ordered products of the interaction Hamiltonian $H_I(t)$ in the interaction picture:

$$S = \sum_{n=0}^{\infty} \frac{(-i)^n}{n!} \int_{-\infty}^{\infty} \mathrm{d}t_1 \int_{-\infty}^{\infty} \mathrm{d}t_2 \cdots \int_{-\infty}^{\infty} \mathrm{d}t_n \times T\{H_I(t_1)H_I(t_2)\cdots H_I(t_n)\}. \quad (1)$$

This should all be familiar. The other principle that has to be added is the cluster decomposition principle, which requires that distant experiments give uncorrelated results.[10] In order to have cluster decomposition, the Hamiltonian is written not

---

** We should not really give quantum field theory too much credit for clarifying the distinction between elementary and composite particles, because some quantum field theories exhibit the phenomenon of bosonization: at least in two dimensions there are theories of elementary scalars that are equivalent to theories with elementary fermions.

just as any functional of creation and annihilation operators, but as a power series in these operators with coefficients that (aside from a *single* momentum-conservation delta function) are sufficiently smooth functions of the momenta carried by the operators. This condition is satisfied for an interaction Hamiltonian of the form

$$H_I(t) = \int d^3x \, \mathcal{H}(\vec{x}, t), \tag{2}$$

where $\mathcal{H}(x)$ is a power series (usually a polynomial) with terms that are local in annihilation fields, which are Fourier transforms of the annihilation operators:

$$\psi_\ell^{(+)}(x) = \int d^3p \sum_{\sigma,n} e^{ip \cdot x} u_\ell(\vec{p}, \sigma, n) a(\vec{p}, \sigma, n), \tag{3}$$

together of course with their adjoints, the creation fields.

So far this all applies to non-relativistic as well as relativistic theories.[†] Now if you also want Lorentz invariance, then you have to face the fact that the time-ordering in the Feynman–Dyson series (1) for the S-matrix doesn't look very Lorentz invariant. The obvious way to make the S-matrix Lorentz invariant is to take the interaction Hamiltonian density $\mathcal{H}(x)$ to be a scalar, and also to require that these Hamiltonian densities commute at spacelike separations

$$[\mathcal{H}(x), \mathcal{H}(y)] = 0 \quad \text{for spacelike } x - y, \tag{4}$$

in order to exploit the fact that time ordering *is* Lorentz invariant when the separation between space-time points is timelike. In order to satisfy the requirement that the Hamiltonian density commute with itself at spacelike separations, it is constructed out of fields which satisfy the same requirement. These are given by sums of fields that annihilate particles plus fields that create the corresponding antiparticles

$$\psi_\ell(x) = \sum_{\sigma,n} \int d^3p [e^{ip \cdot x} u_\ell(\vec{p}, \sigma, n) a(\vec{p}, \sigma, n) + e^{-ip \cdot x} v_\ell(\vec{p}, \sigma, n) a^\dagger(\vec{p}, \sigma, \bar{n})], \tag{5}$$

where $\bar{n}$ denotes the antiparticle of the particle of species n. For a field ψ_ℓ that transforms according to an irreducible representation of the homogeneous Lorentz group, the form of the coefficients u_ℓ and v_ℓ is completely determined (up to a single over-all constant factor) by the Lorentz transformation properties of the fields and one-particle states, and by the condition that the fields commute at spacelike separations. Thus the whole formalism of fields, particles, and antiparticles seems to be an inevitable consequence of Lorentz invariance, quantum mechanics, and cluster decomposition, without any ancillary assumptions about locality or causality.

This discussion has been extremely sketchy, and is subject to all sorts of qualifications. One of them is that for massless particles, the range of possible theories is slightly larger than I have indicated here. For example, in quantum electrodynamics, in a physical gauge like Coulomb gauge, the Hamiltonian is not of the form (2) – there is an additional term, the Coulomb potential, which is bilocal and serves to cancel a non-covariant term in the propagator. But relativistically invariant quantum theories

---

[†] By the way, the reason that quantum field theory is useful even in non-relativistic statistical mechanics, where there is often a selection rule that makes the actual creation or annihilation of particles impossible, is that in statistical mechanics you have to impose a cluster decomposition principle, and quantum field theory is the natural way to do so.

always (with some qualifications I'll come to later) do turn out to be quantum field theories, more or less as I have described them here.

One can go further, and ask why we should formulate our quantum field theories in terms of Lagrangians. Well, of course creation and annihilation operators by themselves yield pairs of canonically conjugate variables: from the as and $a^\dagger$s, it is easy to construct qs and ps. The time-dependence of these operators is dictated in terms of the Hamiltonian, the generator of time translations, so the Hamiltonian formalism is trivially always with us. But why the Lagrangian formalism? Why do we enumerate possible theories by giving their Lagrangians rather than by writing down Hamiltonians? I think the reason for this is that it is only in the Lagrangian formalism (or more generally the action formalism) that symmetries imply the existence of Lie algebras of suitable quantum operators, and you need these Lie algebras to make sensible quantum theories. In particular, the S-matrix will be Lorentz invariant if there is a set of 10 sufficiently smooth operators satisfying the commutation relations of the inhomogeneous Lorentz group. It's not trivial to write down a Hamiltonian that will give you a Lorentz invariant S-matrix – it's not so easy to think of the Coulomb potential just on the basis of Lorentz invariance – but if you start with a Lorentz invariant Lagrangian density then because of Noether's theorem the Lorentz invariance of the S-matrix is automatic.

Finally, what is the motivation for the special gauge invariant Lagrangians that we use in the standard model and general relativity? One possible answer is that quantum theories of mass zero, spin one particles violate Lorentz invariance unless the fields are coupled in a gauge invariant way, while quantum theories of mass zero, spin two particles violate Lorentz invariance unless the fields are coupled in a way that satisfies the equivalence principle.

This has been an outline of the way I've been teaching quantum field theory these many years. Recently I've put this all together into a book,[11] now being sold for a negligible price. The bottom line is that quantum mechanics plus Lorentz invariance plus cluster decomposition implies quantum field theory. But there are caveats that have to be attached to this, and I can see David Gross in the front row anxious to take me by the throat over various gaps in what I have said, so I had better list these caveats quickly to save myself.

First of all, the argument I have presented is obviously based on perturbation theory. Second, even in perturbation theory, I haven't stated a clear theorem, much less proved one. As I mentioned there are complications when you have things like mass zero, spin one particles for example; in this case you don't really have a fully Lorentz invariant Hamiltonian density, or even one that is completely local. Because of these complications, I don't know how even to state a general theorem, let alone prove it, even in perturbation theory. But I don't think that these are insuperable obstacles.

A much more serious objection to this not-yet-formulated theorem is that there's already a counter example to it: string theory. When you first learn string theory it seems in an almost miraculous way to give Lorentz invariant, unitary S-matrix elements without being a field theory in the sense that I've been using it. (Of course it is a field theory in a different sense – it's a two-dimensional conformally invariant field theory, but not a quantum field theory in four space-time dimensions.) So before even being formulated precisely, this theorem suffers from at least one counter example.

Another fundamental problem is that the *S*-matrix isn't everything. Space-time could be radically curved, not just have little ripples on it. Also, at finite temperature there's no *S*-matrix because particles cannot get out to infinite distances from a collision without bumping into things. Also, it seems quite possible that at very short distances the description of events in four-dimensional flat space-time becomes inappropriate.

Now, all of these caveats really work only against the idea that the final theory of nature is a quantum field theory. They leave open the view, which is in fact the point of view of my book, that although you cannot argue that relativity plus quantum mechanics plus cluster decomposition necessarily leads only to quantum field theory, it is very likely that any quantum theory that at sufficiently low energy and large distances looks Lorentz invariant and satisfies the cluster decomposition principle will also at sufficiently low energy *look* like a quantum field theory. Picking up a phrase from Arthur Wightman, I'll call this a folk theorem. At any rate, this folk theorem is satisfied by string theory and we don't know of any counter examples.

This leads us to the idea of effective field theories. When you use quantum field theory to study low-energy phenomena, then according to the folk theorem you're not really making any assumption that could be wrong, unless of course Lorentz invariance or quantum mechanics or cluster decomposition is wrong, provided you don't say specifically what the Lagrangian is. As long as you let it be the most general possible Lagrangian consistent with the symmetries of the theory, you're simply writing down the most general theory you could possibly write down. This point of view has been used in the last 15 years or so to justify the use of effective field theories, not just in the tree approximation where they had been used for some time earlier, but also including loop diagrams. Effective field theory was first used in this way to calculate processes involving soft π mesons,[12] that is, π mesons with energy less than about $2\pi F_\pi \approx 1200 \, \text{MeV}$. The use of effective quantum field theories has been extended more recently to nuclear physics,[13] where although nucleons are not soft they never get far from their mass shell, and for that reason can be also treated by similar methods as the soft pions. Nuclear physicists have adopted this point of view, and I gather that they are happy about using this new language because it allows one to show in a fairly convincing way that what they've been doing all along (using two-body potentials only, including one-pion exchange and a hard core) is the correct first step in a consistent approximation scheme. The effective field theory approach has been applied more recently to superconductivity. Shankar, I believe, in a contribution to this conference is talking about this. The present educated view of the standard model, and of general relativity,[14] is again that these are the leading terms in effective field theories.

The essential point in using an effective field theory is you're not allowed to make any assumption of simplicity about the Lagrangian. Certainly you're not allowed to assume renormalizability. Such assumptions might be appropriate if you were dealing with a fundamental theory, but not for an effective field theory, where you must include all possible terms that are consistent with the symmetry. The thing that makes this procedure useful is that although the more complicated terms are not excluded because they're non-renormalizable, their effect is suppressed by factors of the ratio of the energy to some fundamental energy scale of the theory. Of course, as you go to higher and higher energies, you have more and more of these suppressed terms that you have to worry about.

On this basis, I don't see any reason why anyone today would take Einstein's general theory of relativity seriously as the foundation of a quantum theory of gravitation, if by Einstein's theory is meant the theory with a Lagrangian density given by just the term $\sqrt{g}R/16\pi G$. It seems to me there's no reason in the world to suppose that the Lagrangian does not contain all the higher terms with more factors of the curvature and/or more derivatives, all of which are suppressed by inverse powers of the Planck mass, and of course don't show up at any energy far below the Planck mass, much less in astronomy or particle physics. Why would anyone suppose that these higher terms are absent?

Likewise, since now we know that without new fields there's no way that the renormalizable terms in the standard model could violate baryon conservation or lepton conservation, we now understand in a rational way why baryon number and lepton number are as well conserved as they are, without having to assume that they are exactly conserved.[††] Unless someone has some *a priori* reason for exact baryon and lepton conservation of which I haven't heard, I would bet very strong odds that baryon number and lepton number conservation are in fact violated by suppressed non-renormalizable corrections to the standard model.

These effective field theories are non-renormalizable in the old Dyson power-counting sense. That is, although to achieve a given accuracy at any given energy, you need only take account of a finite number of terms in the action, as you increase the accuracy or the energy you need to include more and more terms, and so you have to know more and more. On the other hand, effective field theories still must be renormalizable theories in what I call the modern sense: the symmetries that govern the action also have to govern the infinities, for otherwise there will be infinities that can't be eliminated by absorbing them into counter terms to the parameters in the action. This requirement is automatically satisfied for unbroken global symmetries, such as Lorentz invariance and isotopic spin invariance and so on. Where it's not trivial is for gauge symmetries. We generally deal with gauge theories by choosing a gauge before quantizing the theory, which of course breaks the gauge invariance, so it's not obvious how gauge invariance constrains the infinities. (There is a symmetry called BRST invariance[15] that survives gauge fixing, but that's non-linearly realized, and non-linearly realized symmetries of the action are not symmetries of the Feynman amplitudes.) This raises a question, whether gauge theories that are not renormalizable in the power-counting sense are renormalizable in the modern sense. The theorem that says that infinities are governed by the same gauge symmetries as the terms in the Lagrangian was originally proved back in the old days by 't Hooft and Veltman[16] and Lee and Zinn-Justin[17] only for theories that are renormalizable in the old power-counting sense, but this theorem has only recently been extended to theories of the Yang–Mills[18] or Einstein type with arbitrary numbers of complicated interactions that are not renormalizable in the power-counting sense.[‡] You'll be reassured to know that these theories are renormalizable in the modern sense, but there's no proof that this will be true of all quantum field theories with local symmetries.

[††] The extra fields required by low-energy supersymmetry may invalidate this argument.
[‡] I refer here to work of myself and Joaquim Gomis,[19] relying on recent theorems about the cohomology of the Batalin–Vilkovisky operator by Barnich, Brandt and Henneaux.[20] Earlier work along these lines but with different motivation was done by Voronov, Tyutin and Lavrov;[21] Anselmi;[22] and Harada, Kugo and Yamawaki.[23]

I promised you a few ironies today. The second one takes me back to the early 1960s when S-matrix theory was very popular at Berkeley and elsewhere. The hope of S-matrix theory was that, by using the principles of unitarity, analyticity, Lorentz invariance and other symmetries, it would be possible to calculate the S-matrix, and you would never have to think about a quantum field. In a way, this hope reflected a kind of positivistic puritanism: we can't measure the field of a pion or a nucleon, so we shouldn't talk about it, while we do measure S-matrix elements, so this is what we should stick to as ingredients of our theories. But more important than any philosophical hang-ups was the fact that quantum field theory didn't seem to be going anywhere in accounting for the strong and weak interactions.

One problem with the S-matrix program was in formulating what is meant by the analyticity of the S-matrix. What precisely are the analytic properties of a multiparticle S-matrix element? I don't think anyone ever knew. I certainly didn't know, so even though I was at Berkeley I never got too enthusiastic about the details of this program, although I thought it was a lovely idea in principle. Eventually the S-matrix program had to retreat, as described by Kaiser in a contribution to this conference, to a sort of mix of field theory and S-matrix theory. Feynman rules were used to find the singularities in the S-matrix, and then they were thrown away, and the analytic structure of the S-matrix with these singularities, together with unitarity and Lorentz invariance, was used to do calculations.

Unfortunately to use these assumptions it was necessary to make uncontrolled approximations, such as the strip approximation, whose mention will bring tears to the eyes of those of us who are old enough to remember it. By the mid-1960s it was clear that S-matrix theory had failed in dealing with the one problem it had tried hardest to solve, that of pion–pion scattering. The strip approximation rested on the assumption that double dispersion relations are dominated by regions of the Mandelstam diagram near the fringes of the physical region, which would only make sense if π–π scattering is strong at low energy, and these calculations predicted that π–π scattering is indeed strong at low energy, which was at least consistent, but it was then discovered that π–π scattering is *not* strong at low energy. Current algebra came along at just that time, and not only was used to predict that low-energy π–π scattering is not strong, but also successfully predicted the values of the π–π scattering lengths.[24] From a practical point of view, this was the greatest defeat of S-matrix theory. The irony here is that the S-matrix philosophy is not that far from the modern philosophy of effective field theories, that what you should do is just write down the most general S-matrix that satisfies basic principles. But the practical way to implement S-matrix theory is to use an effective quantum field theory – instead of deriving analyticity properties from Feynman diagrams, we use the Feynman diagrams themselves. So here's another answer to the question of what quantum field theory is: it is S-matrix theory, made practical.

By the way, I think that the emphasis in S-matrix theory on analyticity as a fundamental principle was misguided, not only because no one could ever state the detailed analyticity properties of general S-matrix elements, but also because Lorentz invariance requires causality (because as I argued earlier otherwise you're not going to get a Lorentz invariant S-matrix), and in quantum field theory causality allows you to derive analyticity properties. So I would include Lorentz invariance, quantum mechanics and cluster decomposition as fundamental principles, but not analyticity.

As I have said, quantum field theories provide an expansion in powers of the energy of a process divided by some characteristic energy; for soft pions this characteristic energy is about a GeV: for superconductivity it's the Debye frequency or temperature; for the standard model it's 10^{15} to 10^{16} GeV; and for gravitation it's about 10^{18} GeV. Any effective field theory loses its predictive power when the energy of the processes in question approaches the characteristic energy. So what happens to the effective field theories of electroweak, strong, and gravitational interactions at energies of order 10^{15}–10^{18} GeV? I know of only two plausible alternatives.

One possibility is that the theory remains a quantum field theory, but one in which the finite or infinite number of renormalized couplings do not run off to infinity with increasing energy, but hit a fixed point of the renormalization group equations. One way that can happen is provided by asymptotic freedom in a renormalizable theory,[25] where the fixed point is at zero coupling, but it's possible to have more general fixed points with infinite numbers of non-zero non-renormalizable couplings. Now, we don't know how to calculate these non-zero fixed points very well, but one thing we know with fair certainty is that the trajectories that run into a fixed point in the ultraviolet limit form a finite-dimensional subspace of the infinite-dimensional space of all coupling constants. (If anyone wants to know how we know that, I'll explain this later.) That means that the condition, that the trajectories hit a fixed point, is just as restrictive in a nice way as renormalizability used to be: it reduces the number of free coupling parameters to a finite number. We don't yet know how to do calculations for fixed points that are not near zero coupling. Some time ago I proposed[26] that these calculations could be done in the theory of gravitation by working in $2 + \epsilon$ dimensions and expanding in powers of $\epsilon = 2$, in analogy with the way that Wilson and Fisher[27] had calculated critical exponents by working in $4 - \epsilon$ dimensions and expanding in powers of $\epsilon = 1$, but this program doesn't seem to be working very well.

The other possibility, which I have to admit is *a priori* more likely, is that at very high energy we will run into really new physics, not describable in terms of a quantum field theory. I think that by far the most likely possibility is that this will be something like a string theory.

Before I leave the renormalization group, I did want to say another word about it because there's going to be an interesting discussion on this subject here tomorrow morning, and for reasons I've already explained I can't be here. I've read a lot of argument about the Wilson approach[28] vs. the Gell-Mann–Low approach,[29] which seems to me to call for reconciliation. There have been two fundamental insights in the development of the renormalization group. One, due to Gell-Mann and Low, is that logarithms of energy that violate naive scaling and invalidate perturbation theory arise because of the way that renormalized coupling constants are defined, and that these logarithms can be avoided by renormalizing at a sliding energy scale. The second fundamental insight, due to Wilson, is that it's very important in dealing with phenomena at a certain energy scale to integrate out the physics at much higher energy scales. It seems to me these are the same insight, because when you adopt the Gell-Mann–Low approach and define a renormalized coupling at a sliding scale and use renormalization theory to eliminate the infinities rather than an explicit cutoff, you are in effect integrating out the higher-energy degrees of free-dom – the integrals converge because after renormalization the integrand begins to fall off rapidly at the energy scale at which the coupling constant is defined. (This

is true whether or not the theory is renormalizable in the power-counting sense.) So in other words instead of a sharp cutoff *à la* Wilson, you have a soft cutoff, but it's a cutoff nonetheless and it serves the same purpose of integrating out the short distance degrees of freedom. There are practical differences between the Gell-Mann–Low and Wilson approaches, and there are some problems for which one is better and other problems for which the other is better. In statistical mechanics it isn't important to maintain Lorentz invariance, so you might as well have a cutoff. In quantum field theories, Lorentz invariance is necessary, so it's nice to renormalize *à la* Gell-Mann–Low. On the other hand, in supersymmetry theories there are some non-renormalization theorems that are simpler if you use a Wilsonian cutoff than a Gell-Mann–Low cutoff.[30] These are all practical differences, which we have to take into account, but I don't find any fundamental philosophical difference between these two approaches.
Fisher: Stay tomorrow and you'll hear the error of your ways.
Weinberg: Okay, well maybe, maybe not, you never know. But I can't stay.

On the plane coming here I read a comment by Michael Redhead, in a paper submitted to this conference: 'To subscribe to the new effective field theory programme is to give up on this endeavour' [the endeavor of finding really fundamental laws of nature], 'and retreat to a position that is somehow less intellectually exciting.' It seems to me that this is analogous to saying that to balance your checkbook is to give up dreams of wealth and have a life that is intrinsically less exciting. In a sense that's true, but nevertheless it's still something that you had better do every once in a while. I think that in regarding the standard model and general relativity as effective field theories we're simply balancing our checkbook and realizing that we perhaps didn't know as much as we thought we did, but this is the way the world is and now we're going to go on to the next step and try to find an ultraviolet fixed point, or (much more likely) find entirely new physics. I have said that I thought that this new physics takes the form of string theory, but of course, we don't know if that's the final answer. Nielsen and Oleson[31] showed long ago that relativistic quantum field theories can have string-like solutions. It's conceivable, although I admit not entirely likely, that something like modern string theory arises from a quantum field theory. And that would be the final irony.

References

1. M. Born. W. Heisenberg, and P. Jordan, *Z. Phys.* **35**, 557 (1926).
2. P.A.M. Dirac, *Proc. Roy. Soc.* **A117,** 610 (1928); *ibid.* **A118,** 351 (1928); *ibid.* **A126,** 360 (1930).
3. R.P. Feynman and A.R. Hibbs, *Quantum Mechanics and Path Integrals* (McGraw-Hill, New York, 1965).
4. G. 't Hooft, *Nucl. Phys.* **B35**, 167 (1971).
5. H. Euler and B. Kockel. *Naturwiss.* **23**, 246 (1935); W. Heisenberg and H. Euler, *Z. Phys.* **98,** 714 (1936).
6. Z. Bern and D.A. Kosower, in *International Symposium on Particles, Strings, and Cosmology,* eds. P. Nath and S. Reucroft (World Scientific, Singapore, 1992): 794; *Phys. Rev. Lett.* **66**, 669 (1991).
7. W. Pauli, *Z. Phys.* **37**, 263 (1926); **43**, 601 (1927).
8. F.J. Dyson, *Phys. Rev.* **75**, 486, 1736 (1949).
9. E.P. Wigner, *Ann. Math.* **40**, 149 (1939).
10. The cluster decomposition principle seems to have been first stated explicitly in quantum field theory by E.H. Wichmann and J.H. Crichton, *Phys. Rev.* **132**, 2788 (1963).

11. S. Weinberg, *The Quantum Theory of Fields – Volume I: Foundations* (Cambridge University Press, Cambridge, 1995).

12. S. Weinberg, *Phys. Rev. Lett.* **18**, 188 (1967); *Phys. Rev.* **166**, 1568 (1968); *Physica* **96A**, 327 (1979).

13. S. Weinberg, *Phys. Lett.* **B251**, 288 (1990); *Nucl. Phys.* **B363**, 3 (1991); *Phys. Lett.* **B295**, 114 (1992). C. Ordóñez and U. van Kolck, *Phys. Lett.* **B291**, 459 (1992); C. Ordóñez, L. Ray and U. van Kolck, *Phys. Rev. Lett.* **72**, 1982 (1994); U. van Kolck, *Phys. Rev.* **C49**, 2932 (1994); U. van Kolck, J. Friar and T. Goldman, to appear in *Phys. Lett. B*. This approach to nuclear forces is summarized in C. Ordóñez, L. Ray and U. van Kolck, Texas preprint UTTG-15-95, nucl-th/9511380, submitted to *Phys. Rev. C*; J. Friar, *Few-Body Systems Suppl.* **99**, 1 (1996). For application of these techniques to related nuclear processes, see T.-S. Park, D.-P. Min and M. Rho, *Phys. Rep.* **233**, 341 (1993); Seoul preprint SNUTP 95-043, nucl-th/9505017; S.R. Beane, C.Y. Lee and U. van Kolck, *Phys. Rev.* **C52**, 2915 (1995); T. Cohen, J. Friar, G. Miller and U. van Kolck, Washington preprint DOE/ER/40427-26-N95, nucl-th/9512036.

14. J.F. Donoghue, *Phys. Rev.* **D50**, 3874 (1994).

15. C. Becchi, A. Rouet and R. Stora, *Comm. Math. Phys.* **42**, 127 (1975); in *Renormalization Theory*, eds. G. Velo and A.S. Wightman (Reidel, Dordrecht, 1976); *Ann. Phys.* **98**, 287 (1976); I.V. Tyutin, Lebedev Institute preprint N39 (1975).

16. G. 't Hooft and M. Veltman, *Nucl. Phys.* **B50**, 318 (1972).

17. B.W. Lee and J. Zinn-Justin, *Phys. Rev.* **D5**, 3121, 3137 (1972); *Phys. Rev.* **D7**, 1049 (1972).

18. C.N. Yang and R.L. Mills, *Phys. Rev.* **96**, 191 (1954).

19. J. Gomis and S. Weinberg, *Nucl. Phys.* B **469**, 475–487 (1996).

20. G. Barnich and M. Henneaux, *Phys. Rev. Lett.* **72**, 1588 (1994); G. Barnich, F. Brandt, and M. Henneaux, *Phys. Rev.* **51**, R143 (1995); *Comm. Math. Phys.* **174**, 57, 93 (1995); *Nucl. Phys.* **B455**, 357 (1995).

21. B.L. Voronov and I.V. Tyutin, *Theor. Math. Phys.* **50**, 218 (1982); **52**, 628 (1982); B.L. Voronov, P.M. Lavrov and I.V. Tyutin, *Sov. J. Nucl. Phys.* **36**, 292 (1982); P.M. Lavrov and I.V. Tyutin, *Sov. J. Nucl. Phys.* **41**, 1049 (1985).

22. D. Anselmi, *Class. and Quant. Grav.* **11**, 2181 (1994); **12**, 319 (1995).

23. M. Harada, T. Kugo, and K. Yamawaki, *Prog. Theor. Phys.* **91**, 801 (1994).

24. S. Weinberg, *Phys. Rev. Lett.* **16**, 879 (1966).

25. D.J. Gross and F. Wilczek, *Phys. Rev. Lett.* **30**, 1343 (1973); H.D. Politzer, *Phys. Rev. Lett.* **30**, 1346 (1973).

26. S. Weinberg, in *General Relativity*, eds. S.W. Hawking and W. Israel (Cambridge University Press, Cambridge, 1979): p. 790.

27. K.G. Wilson and M.E. Fisher, *Phys. Rev. Lett.* **28**, 240 (1972); K.G. Wilson, *Phys. Rev. Lett.* **28**, 548 (1972).

28. K.G. Wilson, *Phys. Rev.* **B4**, 3174, 3184 (1971); *Rev. Mod. Phys.* **47**, 773 (1975).

29. M. Gell-Mann and F.E. Low, *Phys. Rev.* **95**, 1300 (1954).

30. V. Novikov, M.A. Shifman, A.I. Vainshtein and V.I. Zakharov, *Nucl. Phys.* **B229**, 381 (1983); M.A. Shifman and A.I. Vainshtein, *Nucl. Phys.* **B277**, 456 (1986); and references quoted therein. See also M.A. Shifman and A.I. Vainshtein, *Nucl. Phys.* **B359**, 571 (1991).

31. H. Nielsen and P. Oleson, *Nucl. Phys.* **B61**, 45 (1973).

17. Comments

LAURIE M. BROWN AND FRITZ ROHRLICH

1 Comments by Laurie M. Brown

I shall assume that I was invited to contribute to this symposium-workshop as an *historian* of recent physics; any 'philosophical' inferences will be unintended. In order to comment on Steve Weinberg's talk, I expected to have the text a few weeks in advance, which did not happen; but Steve kindly supplied a list of some writings on which he would base his talk:

1. Chapter 1 of the recently published (1995) first volume of '*The Quantum Theory of Fields*' (QFT).
2. The well-known article in *Daedalus* (1977) called 'The search for unity: Notes for a history of quantum field theory.'
3. 'The ultimate structure of matter,' in *A Passion for Physics – Essays in Honor of Geoffrey Chew* (1985).
4. The (1993) book '*Dreams of a Final Theory*,' especially Chapters 5 and 8.
5. 'Night thoughts of a quantum physicist,' *Bull Amer. Acad. Arts and Sciences* **69**, No. 3, p. 51 (1995).
6. 'Nature itself,' an essay in Chapter 27 of *Twentieth Century Physics*, a recently (1995) published compendium. The same chapter has parallel essays by Philip Anderson and John Ziman.
7. 'Conference summary,' in *Proc. XXVI Intern. Conf. on High Energy Physics* (AIP 1993).

In general, I did not detect appreciable differences in the writings I consulted, which extend from 1977 to 1995, over nearly two decades. This may say something about the progress (or lack thereof) in fundamental physics during that time interval. I read Steve's '*Dreams*' book relatively recently, being somewhat put off by its title – for me one of the attractive features of science is its open-ended character – but it is hard to dislike a book with chapter titles like 'Against philosophy' and 'Two cheers for reductionism'.

Commenting on Steve Weinberg has the drawback that I generally agree with what Steve has to say and how he says it – after all, we come from similar physics backgrounds – but I will try nevertheless to put a little different spin on the historical roots of quantum field theory.

In the introductory chapter of Steve's '*QFT*' he points out that 'QFT grew in part out of the study of *relativistic* wave equations, including the Maxwell, Klein–Gordon, and Dirac equations.' The *Daedalus* article also emphasized relativity, for it began:

'Quantum field theory is the theory of matter and its interactions, which grew out of the fusion of quantum mechanics and special relativity in the late 1920s.' The first phrase, 'matter and its interactions' sounds like a good description of physics itself (and perhaps of all the natural sciences), but I will pass over that point and focus on the second phrase 'the fusion of *quantum mechanics and relativity.*' In a recently translated book of Sin-itiro Tomonaga (*The Story of Spin*, 1997), he gives the impact of field theory differently, without bringing in relativity. Tomonaga and his future co-Nobelist classmate Hideki Yukawa (a pair of theorists reminiscent of Glashow and Weinberg) learned quantum mechanics at Kyoto University not from their professors but from the journals as they arrived in Kyoto during the formative period 1927 to 1930. Tomonaga recalled that the Schrödinger wave function ψ was at first considered to be a real 'wave', then a complex wave, in ordinary three-dimensional space. On the other hand, in Dirac's transformation theory ψ became the state vector, a function of the set of generalized coordinates describing the mechanical system – and he said 'therefore the wave expressed by ψ exists in the abstract coordinate space and not in our three-dimensional space.' (For two particles, for example, it is a wave in six dimensions.) 'The idea that the radiation wave does indeed exist in our three-dimensional space, but the matter wave does not, was about to become orthodox. However, this situation suddenly changed when the idea of quantization of the field was introduced.'

The first articles on *Einstein–Bose* quantization (Dirac 1927, Jordan and Klein 1927) said that one must *eventually* take into account the finite light velocity to treat the quantum interactions of particles consistently, but they actually set up the quantized matter field in instantaneous interaction with itself. In the first paper on *Fermi–Dirac* quantization (Jordan and Wigner 1927), there is *no mention at all* of relativity. Its abstract reads as follows:

[This article] is concerned with describing a gas, ideal or non-ideal, obeying the Pauli Exclusion Principle with concepts which make no use of the abstract coordinate space of all the atoms (*Atomgesamtheit*) of the gas, but uses only the usual three-dimensional space.

A second example of a small divergence in my viewpoint from Steve's is suggested by his 'Particle physics: Past and future' (1986), in which Steve said that:[1]

...a good deal of the history of particle physics in the last thirty years or so has been a story of oscillation between two broad lines of approach to the underlying laws, two approaches that might briefly be called quantum field theory and S-matrix theory. Speaking more loosely, we oscillate between a view of nature as a world of fields, in which the particles are just quanta of the fields, and as a world of particles, of which the fields are just macroscopic coherent states.

I am reasonably content with Steve's view of this issue as expressed above – restricted to the three decades that began in the mid-1950s. However, I note that in the historical introduction to *Quantum Theory of Fields*, Steve has treated hole theory and quantum field theory as distinct fundamental viewpoints. I also notice that philosophers are taking the particle–field dichotomy or 'oscillation' very seriously at present. At tomorrow's workshop a whole session is scheduled on this issue, and I may learn there the error of my ways. Perhaps misguidedly, I thought this whole matter was settled by about 1930, when many quantum physicists realized that the wave–particle duality was an example of Bohr's Principle of Complementarity.

[1] *International Journal of Modern Physics* **A1**, 134–145 (1986).

Now I am aware that complementarity is regarded by most philosophers as vague, sentimental, feel-good popularization – something that says 'It's really OK, don't worry – just accept it.' I do not share that philosophical judgement, and I feel that complementarity is one of this century's most important (and overlooked) insights.

As an authority on waves and particles, I shall quote from the first, and justly so, item in Julian Schwinger's 1958 historical collection of selected papers on QED, namely Dirac's 1927 article in *Proc. Roy. Soc.* entitled 'The quantum theory of the emission and absorption of radiation.' Near the beginning Dirac wrote:

It will...be shown that the Hamiltonian which describes the interaction of the atom and the electromagnetic waves can be made identical with the Hamiltonian for the problem of the interaction of the atom with an assembly of particles moving with the velocity of light and satisfying the Einstein–Bose statistics, by a suitable choice of the interaction energy for the particles. The number of particles having any specified direction of motion and energy, which can be used as a dynamical variable in the Hamiltonian for the particles, is equal to the number of quanta of energy in the corresponding wave in the Hamiltonian for the waves. *There is thus a complete harmony between the wave and light quantum descriptions of the interaction.* [My emphasis]

Well, if we grant that this is not really an issue for bosons, what about the case of fermions? Dirac himself always regarded the electron as an especially particle-like particle (in which he seemed to take a proprietary view) and the hole-theory interpretation of his famous relativistic electron equation is often regarded as a particle-like alternative to a wave-like quantum field theory. For example, in Fermi's first paper on β-decay he tried to distance himself from Dirac's at the time *not* highly regarded theory – Fermi insisting that *his* β-decay electron was truly created and not just knocked out of a hole in the vacuum. However, all features of the hole theory and the QFT – the infinite numbers of degrees of freedom, the necessity for infinite renormalizations, etc., and the marvellously accurate radiative corrections – are identical. This similarity of predictions and problems suggested very strongly, and in fact it was mathematically proven by Furry and Oppenheimer in 1934 that the *two theories* are in fact identical, and not dichotomous.

The situation with regard to the S-matrix is rather different. I will say very little about this, as Fritz Rohrlich was very much involved in some of the pioneering work and Jim Cushing has written a fine work on the subject.[2] As stated, e.g., in Helmut Rechenberg's article on early S-matrix theory (in *Pions to Quarks* 1989), Heisenberg began in 1942 (independently of an earlier similar approach taken to nuclear scattering by John Wheeler) to consider a more positivistic approach to elementary particle physics than QFT, having despaired of incorporating into the latter two features that he considered indispensable and connected – namely a new fundamental length and nonlinearity. He felt this was necessary to overcome the divergence difficulties – that is the infinities – of QFT. He was not alone in thinking of QFT as badly flawed. According to Weinberg's *The Quantum Theory of Fields* (on p. 38):

...it had become accepted wisdom in the 1930s, and a point of view especially urged by Oppenheimer, that QED could not be taken seriously at energies of more than about 100 MeV, and that new solutions to its problems could be found only in adventurous new ideas.

---

[2] J.T. Cushing, *Theory Construction and Selection in Modern Physics*, Cambridge Univ. Press, 1990.

Steve cited Oppenheimer's associate Robert Serber for reporting this mistaken 'accepted wisdom', but what Serber said (in *Birth of Particle Physics* 1983, p. 206) was even stronger:

Another thing that struck me [beside the isolation of the California group] was Oppenheimer's conviction that QED was wrong, not only in its divergences but also in its finite predictions... Oppie at first disbelieved at mc^2 (the rest energy of the electron), then retreated to $137\,mc^2$, but could hardly write a paper without a lament.

This was all before the successful application of renormalization theory to QED, of course. Also before renormalization, in 1942 Heisenberg set out to abstract from QED what he considered to work successfully and to limit the formulation (and the questions to be asked of the theory) to what could be actually observed. In this way he hoped to incorporate some of the features that would still be present in a future finite QFT. In an essay in Heisenberg's *Collected Works*, Rheinhard Oehme wrote:

At the time when Heisenberg wrote the [S-matrix] papers, he viewed this matrix as a primary quantity to be calculated directly. It was to replace the Hamiltonian, which had led to the divergence difficulties. Later after it had been learned how to handle covariant perturbation theory, the S-matrix became the quantity to be *calculated* from quantum theory. More generally, it became the physical quantity to be extracted from a field theory formalism, which itself may contain many unphysical elements as a price to pay for more mathematical simplicity.

Thus in the *later* 1940s, after renormalized QED, the S-matrix became something to be obtained *from* the field theory, not an alternative to it. Think of the title of Freeman Dyson's 1949 article: 'The S-matrix in quantum electrodynamics'. A decade later, however, there grew up a radical school around Geoffrey Chew that tried to supplant field theory altogether. But it must be pointed out that some of the main contributors to dispersion theory and S-matrix theory (e.g., Richard Eden and Yoichiro Nambu), tried to derive the analytic S-matrix from QFT and for many others the new methods were merely a stopgap to use while waiting for a better QFT. And let us not forget the great pressure at that time to do curve-fitting for the large experimental groups working at expensive machines, and to turn out plenty of new Ph.D.s (i.e., the forces of 'social construction' at work).

I first mentioned the attractiveness of QFT for describing nature in its actual three-dimensional space as a complement to the need to include relativity. Then I discussed the supposed 'oscillation' between particle and field as a different type of complementarity. I would like to comment on two other pairs of concepts – dichotomous or complementary – I am not sure which.

In 'Nature itself,' his essay in *Twentieth Century Physics*, Steve wrote that 'we have come in this century to understand the very pattern of scientific knowledge.' And after some words about nuclear physics, chemistry, and biology, he added: 'This has truly been the century of the triumph of reductionism.' By contrast with this, the same chapter contains an essay by Philip Anderson that also speaks of profound changes in the scientific world-view of this century. The first main theme Anderson mentions is 'the flight from common sense'; i.e., 'the structure of physical law can no longer be assumed to correspond in any way with our direct perception of the world.' However, it is Anderson's second major theme that I want to contrast with Steve's reductionism. 'The second theme is that the process of "emergence" is in fact the key to the structure of twentieth century science on all scales.' And again:

This then is the fundamental philosophical insight of twentieth century science: everything we observe emerges from a more primitive substrate, in the precise meaning of the term 'emergent', which is to say obedient to the laws of the more primitive level, but not conceptually consequent from that level. Molecular biology does not violate the laws of chemistry, yet it contains ideas which were not (and probably could not have been) deduced directly from those laws; nuclear physics is assumed to be not inconsistent with QCD, yet it has not even yet been reduced to QCD, etc.

I have not time to do more than mention these dichotomous (or are they complementary?) views, but I would like to suggest that there really is an oscillation and complementarity in the progress of science between what I refer to as unification and diversification. Science is *exploration* and we keep discovering more and more diverse things and processes. Eventually we need to simplify and order those things and processes in terms of simpler objects, concepts and laws. Thus the more we diversify, the more we need to unify. The miracle, of course, as Einstein liked to say, is that to a great extent we can do so.

And now a few words on renormalizability as a criterion (or not) for a successful field theory and on low-energy effective field theories, on which Steve has been most eloquent. I would simply like to point out that in spite of some authorities (like Freeman Dyson) who wanted to make renormalizability a restrictive principle, like relativistic invariance, the inventors of renormalization had no such intention. Well aware that physical QED would have to include in its radiative corrections, as one approached infinite energy, not only electron pairs but all sorts of hadrons, known and unknown, uranium atoms and their anti-atoms, large buildings and anti-buildings, etc., how could one possibly know how such an open-ended system would behave? Thus they felt that they were really (as Feynman, for example, often said) sweeping the problem under the rug. And as for present theory, even the glorious Standard Model, being justified as a low-energy approximation, with corrections coming only from terms with coefficients containing powers of the inverse Planck mass (and hence negligible practically forever), it is a really long way to the Planck mass – and how do we know what new continents we may encounter on that journey?

2 Comments by Fritz Rohrlich

In the absence of an advanced copy of this talk by Steven Weinberg, I decided to comment briefly on some philosophical issues which he has been interested in, and on which he has written repeatedly. Since his actual talk was primarily historical, this will provide a philosophical complement to it.

Philosophy of science examines what scientists do (epistemological issues), how they do it (methodology), and what their results tell us about what there is in the world (ontology). Philosophers of science ask many questions that scientists don't usually ask; they are often more precise in their use of words than scientists are; and they are concerned with issues that most scientists pay little attention to because of the pragmatic orientation of most practitioners. To be specific, let me take quantum electrodynamics (QED) as a specific theory in point.

How good is QED as a theory? Its agreement with experiments (in philosophical jargon its 'empirical adequacy') is unsurpassed. No other physical theory can pride itself of ten and more significant figure accuracy in agreement with measurements.

And how good is its logical consistency? If mathematics is a guide for that, then we must admit that the theory as actually applied in practice involves divergent integrals which must be removed by suitable methods. QED is not a mathematically clean theory: it takes a physicist's intuition to devise the appropriate recipes; a pure mathematician could never have produced them. With apologies to Eugene Wigner, one must marvel at the 'unreasonable effectiveness – not of mathematics – but of the physicist's intuition.' From the point of view of the philosopher of science, this is a methodologist's nightmare.

But there is more to the demands by philosophers than internal consistency. There is also the question of the coherence between different theories. Thus, QED is supposed to be a generalization of classical electrodynamics. Does it reduce to it in a suitable limit? Dirac has pointed out several times that he finds the QED perturbation expansion unsatisfactory: in the classical limit (crudely speaking in the limit as Planck's constant vanishes), QED should reduce to the classical electrodynamics of charged particles; but instead, the perturbation expansion diverges term by term! This expansion is an asymptotic expansion in a double sense; it does not have a limit to the classical theory, and the perturbation expansion also diverges, converging only for the first few terms.

Let us therefore consider a more modest and more specific demand: require that QED should yield (in the limit but not necessarily from a perturbation expansion) a relativistic equation of motion for a classical charged particle such as the Lorentz–Dirac equation (Dirac 1938). We can find no such derivation in the literature.[3] This shows that even our empirically best theory, QED, does not provide us with the properties demanded by a *reductionist view*: the deeper, more fundamental theory does not permit us to deduce the less fundamental theory.

But we should not be surprised at that: while there are numerous examples of physical theories that are approximations to more fundamental theories in the sense of regular perturbation expansions (special relativistic dynamics reduces to Newtonian dynamics in the limit of vanishing $(v/c)^2$), there are many other examples in which the less fundamental theory is obtained only as a *singular* expansion. Thus, everyone knows that geometrical optics is a special case of Maxwellian electrodynamics; yet, its fundamental equation, the eikonal equation, is derivable only in singular perturbation expansion for small wave lengths. One cannot go from wave optics to ray optics in the smooth limit of vanishing wave lengths. This is an example of a situation often encountered in physics: the more fundamental theory is *not* analytic at $p = 0$ where p is the small parameter that characterizes the approximation to a less fundamental theory.

This is a mathematical blow to theory reduction. But the fatal blow is of a different nature. It has to do with the *incommensurability of concepts* between a more fundamental and a less fundamental theory: even if the equations of the latter are obtained from the former in some limit, the *interpretation* of the newly derived equations is not implied by this mathematical derivation: some of the symbols of the deduced equations have a *different meaning* that is *not derivable* from their meaning in the more fundamental theory. Entirely new concepts are necessary that have nothing

---

[3] There is actually a good reason for that: QED deals with point charges while a classical theory would be beyond its domain of validity to deal with point charges: no lengths below a Compton wave length is necessarily permitted. The Lorentz–Dirac equation is for classical point charges and not surprisingly has therefore various pathologies.

to do (are incommensurable) with the concepts of the more fundamental theory. For example, the fundamental concept of Newtonian gravitation theory (the concept of force) is not implied by the equations of the (geometric) Einstein theory of gravity which deal with space-time curvature.

The demise of reductionism is now widely accepted in the philosophy of science literature, and there are also physicists and biologists who have realized this situation for some time.[4] The less fundamental theories have a great deal of autonomy and are not made superfluous by the more fundamental ones; research on those scales of length, time, and energy are just as important as those on deeper levels. The fact that some (but by far not all!) specific problems on a given level may require a deeper level for their explanation (reductive explanation) does not imply the need or even the possibility of reduction of the whole theory.

Finally, a word about the question of ontology which after all is in the title of our conference, i.e. the question: which objects, properties, and processes described by a physical theory such as QFT should be given credence as 'really there'? Many physicists are realists (or claim to be). They attach real existence not only to chairs and billiard balls but also to atoms, electrons, and nucleons. But what about quarks (which are never free particles) or virtual particles (which are not on the mass shell), or ghosts in QFT? Here it is important to be sure that the putative object is not tied to a particular way in which the theory is formulated.

But it may also happen that two theories are empirically equivalent while they differ in their ontology. For example, the Bohm–Vigier formulation of QM cannot be distinguished from the former by experiments. Yet, that formulation suggests a very different ontology. The often heard assertion 'I don't like it' is hardly sufficient justification for rejecting it.[5]

It follows that great care must be taken in the decision of the ontic status of the entities and properties suggested by a theory. Just as a deeper understanding of quantum mechanics was obtained from the discovery that Heisenberg's matrix mechanics and Schrödinger's wave mechanics (each suggesting different ontologies) are representations of a more general theory, so has the work of Dyson yielded much greater insight into the theory when he showed the equivalence of the Schwinger–Tomonaga formulation and the Feynman formulation of QED. We now know that the operator formulation and the path integral formulation of QFT are alternatives, and we must take care that our assignments of reality to objects or properties do not depend on a particular formulation.

Mathematically, the insight gained through the rigorous work by Osterwalder and Schrader is most valuable; but there still remain important mathematical problems in the theory (associated with the presence of massless particles). Their solution may correct our ontological picture.

I hope that these remarks will help the physicist understand the kinds of problems that are of interest to the philosopher of science. Steven Weinberg has been interested in some of these for a long time.

[4] Anderson, P.W., 'More is different', *Science*, **177**, 393 (1972); Mayr, E. 'The limits of reductionism', *Nature*, **331**, 475 (1988). This latter paper is written in response to a paper by Weinberg, *Nature*, **330**, 433 (1987), in which theory reduction and reductive explanation are conflated.

[5] J.T. Cushing, *Quantum Mechanics: historical contingency and the Copenhagen hegemony*, Chicago: Chicago University Press, 1994.

References

P.A.M. Dirac, 'The quantum theory of the emission and absorption of radiation,' *Proc. Roy. Soc. (London)* **A114** (1927), 243–65.

P.A.M. Dirac, 'Classical theory of radiating electrons,' *Proc. Roy. Soc. (London)* **A167** (1938), 148–69.

F.J. Dyson, 'The S-matrix in quantum electrodynamics,' *Phys. Rev.* **75** (1949), 292–311.

P. Jordan and O. Klein, 'Zur Mehrkoerperproblem der Quantentheorie,' *Z. Phys.* **47** (1927), 751–63.

P. Jordan and E. Wigner, 'Ueber das Paulische Aequivalentverbot,' *Z. Phys.* **47** (1927), 631–51.

R. Oehme, 'Theory of the scattering matrix,' Annotation in *W. Heisenberg, Collected Works*, Vol. AII (1989), 605–10.

H. Rechenberg, 'The early S-matrix and its propagation,' in *Pions to Quarks*, eds. L.M. Brown *et al.* (Cambridge University Press 1989), 551–78.

J. Schwinger, ed., *Quantum Electrodynamics* (Dover, New York 1958).

R. Serber, 'Particle physics in the 1930s,' in *The Birth of Particle Physics*, eds. L.M. Brown and L. Hoddeson (Cambridge University Press 1983), 206–21.

Sin-itiro Tomonaga, *The Story of Spin*, transl. Takeshi Oka (University of Chicago Press 1997).

Session discussions

Weinberg: I'll first touch on some minor points, and then I want to come to the big issue of reductionism.

With regard to effective field theory, what's new about it? It's certainly true that, as Laurie Brown has said, many physicists from time to time must have thought that maybe there are nonrenormalizable terms in the Lagrangian. (By the way, Laurie, it's not just that there are new terms appearing in calculations from radiative corrections involving things like uranium and anti-uranium, it's that there are new nonrenormalizable terms in the Lagrangian itself. In a renormalizable theory of strong interactions, like quantum chromodynamics, the existence of uranium in intermediate states does not lead to new terms in the Lagrangian.) Others, including me, and I gather also Dyson, thought that it was a useful hypothesis to assume there weren't. The new thing it seems to me about effective field theories as we now use them is the realization that you could take a theory with an infinite number of infinitely complicated terms, and use it effectively to do useful calculations beyond the tree approximation. As I said, in effect you're just doing S-matrix theory in a modern and efficient way. And that is what I think is new about effective field theory.

With regards to quantum field theory vs. hole theory, first of all I want to say that although, as Laurie Brown pointed out, they are equivalent in the theory of electrons and photons, there were other early theories that couldn't be formulated in terms of hole theory, such as the 1934 Fermi theory of nuclear beta decay.[32] (Don't think that this has anything to do with wave–particle duality, because a quantum field has nothing to do with Schrödinger wave functions.) I would like to add that in 1934 a sensible person might have read the papers by Fermi, Pauli and Weisskopf,[33] and Furry and Oppenheimer,[34] and said: 'Well, it's over. Quantum field theory is clearly the answer; hole theory gives the same results in electrodynamics, but the real truth is quantum field theory,' but in fact the issue was not settled then; lots of intelligent people like Dirac and Feynman went on thinking that the fundamental ingredients of quantum electrodynamics are electron and positron particles and the electromagnetic field. I believe that Dirac thought that this difference between electrons and photons is inevitable, because photons are how we learn out about electrons.

Now, let me finally come to an issue where we might have an interesting argument: the issue of reductionism. As Laurie Brown hinted, there really isn't that much difference between Phil Anderson and me. After all if Anderson and I looked at any physical problem, and if both of us knew enough about it, we would probably think about it in similar ways. So what are we arguing about? Is the argument about reductionism just an argument about words? Before we talk about reductionism, we should find some way of making it matter. I think there is one sense in which this argument does matter: it's funding [laughter]. That is, particle physics has no claim to be solving the world's technological or medical problems. It is very esoteric. Why should so much money be spent on particle physics? Notice I'm not saying why should *all* research money be spent on particle physics. The particle physicist doesn't want to be the only recipient of government funds. But particle –

Fisher: They don't any longer!

Weinberg: No, we never did. We were always interested in what you were doing. What we felt was that, just as the condensed matter physicists can point to mathematical ingenuity and sophistication of their field and also its practical value, we thought we ought to have something to offer the public. What we offer is the fundamental character of what we do. In what sense is it fundamental? That takes some explanation, and I think here I disagree very strongly with most of the things Fritz Rohrlich said. We know very well that in solving problems of hydrodynamics, it isn't helpful to keep going back to the atomic level.

Treiman: What about viscosity?

Weinberg: Yes, that involves atomic physics, but I want to make a stronger point. Where do we think the Navier–Stokes equation comes from? Do we think it is just a fundamental law of nature, coming down from Mount Sinai? No, the Navier–Stokes equation is what it is because of the laws of energy and momentum conservation applied to fluids, which are defined by the condition that they are almost isotropic in co-moving frames. And that condition is explained by what a fluid is, and ultimately in terms of the elementary particles of which it is composed. It's in this sense that reductionism works, not as a practical research program, because in many cases the reductive perspective has no practical advantages, but as a guide to why our higher-level theories like hydrodynamics are what they are. To take an example far removed from physics, why do the kind of markets that economists talk about work the way they do? Well, I don't really know if anyone understands why they work the way they do, but I'm absolutely confident that whenever it's explained, it will be explained in terms of the psychology of individual human beings acting in large numbers. Does anyone really think that there are laws of market behavior which stand apart as fundamental free-standing autonomous laws? Or does anyone think that there are laws of superconductivity that are autonomous, not needing an explanation in terms of deeper theories?

On this issue, I think Fritz makes a fundamental logical error. He says, correctly, that there are concepts at a higher emergent level which do not exist at the lower level. For example, the concept of viscosity does not exist in elementary particle physics. The concept of supply and demand doesn't exist for an individual person. But the reason why these concepts work the way they do, the reason why the principles that govern them are what they are, has to be found at a deeper level. And the deepest level to which we have plumbed is what we call, roughly speaking, the theory of elementary particles. I think that it is a non-sequitur to say that because certain concepts at one level don't exist at a more fundamental level, that the laws governing these concepts cannot be *deduced* from the more fundamental level. I don't think that this needs a long argument, because it is something we have seen clearly many times. The concept of temperature does not exist at the level of individual particles, and neither does the concept of entropy. In the late nineteenth century many physicists thought that the laws of thermodynamics were independent, free-standing laws of nature. In fact, as I mentioned in *Dreams of a Final Theory*,[35] Zermelo actually thought that thermodynamics was an argument against the existence of atoms, because if atoms existed, then temperature and entropy and so forth would have no meaning at that level, therefore they couldn't exist, because he thought that the laws of thermodynamics were fundamental physical laws. Well, we know better now. Even though the concept of temperature does not exist at the particle level, we understand why there is such a concept and we understand how it works, in terms of a deeper level. Namely, the level that says that energy is conserved and that energy can be distributed among a large number of subsystems which we now understand to be atoms or molecules or whatever. I've said what I wanted to say. Thank you [applause].

Yes, I'll be happy to take questions until it's dinner time. David Gross had his hand up, and he told me when I was sitting next to him that I almost covered myself with the caveats, but not quite, so now I'm going to find out what I left out.

David Gross: Well, this has to do with the issue as to whether quantum field theory is the inevitable consequence of Lorentz invariance, causality, unitarity, etc. You almost covered yourself. But it seems to me there are two issues that I think I would disagree with. First, the argument for Lagrangians has to be a little stronger than you made. It's not just a way of getting symmetries, you don't need Lagrangians to get symmetries, you could postulate them.

Weinberg: No, I didn't say that you needed Lagrangians to get symmetries, but to get Lie algebras of quantum operators from the symmetries you postulate.

Gross: Right. But there are other things we understand. I mean the modern approach to field theory, based on path integrals, tells us that the Lagrangian is in some sense measurable, and is more than just a way of deriving the equations of motion, or the Hamiltonian. For example, topological terms in the action, which would never appear in the perturbative expansion of the S-matrix that you discussed, you could never really get from the perturbatively expanded S-matrix. You covered yourself by saying that your argument was based on perturbation theory, but I think that's a very serious limitation. Another technical assumption that I think you make, that you didn't spell out, was that you were assuming implicitly that the number of fields, the number of creation–annihilation operators, was if not finite, at least bounded.

Weinberg: Well, that's what makes the difference with the string theory.

Gross: That's what most clearly makes the difference with the string theory. I think if you set out in your way of trying to derive the most perfect of all quantum field theories, which I argued yesterday was QCD, in this fashion, you would have a lot of problems. You would start with asymptotic states, like the proton [Weinberg laughs] and the neutron. And by this perturbative technique try to rederive a quantum field theory. Of course you would never succeed, since the states themselves are a nonperturbative object.

Actually, you might succeed in finding a string theory. If you were to follow that procedure, you would find it useful to introduce a highly infinite number of states, say all the hadrons, in the approximation where they were stable, which in fact is a mathematically good approximation to QCD – for a large number of colors. And then what you probably would derive in that process, given now a theory with an infinite number, a highly infinite number of particles, I believe would be a string theory. You would derive a theory of hadronic flux tubes, which I believe has a very good chance of existing although it hasn't yet been constructed or proved to be equivalent to QCD. This would be a counter-example to your modified statement that every Lorentz invariant, unitary, causal theory at low energies looks like a field theory. I believe, in fact, that there could very well exist, and I think it's a reasonable hypothesis that people are working on it, an equally fundamental description of QCD as a theory of noncritical strings. This would be not the high energy, short distance theory, but would be on an equal footing with QCD. There would be no low energy effective field theory for QCD, perhaps except for chiral Lagrangians. So I disagree even with your modified statement.

Weinberg: Well, we don't know. One good thing about folk theorems is you get to tinker around with them as things change. We don't know whether or not there's a formulation of quantum chromodynamics like that. But suppose that it were possible to formulate quantum chromodynamics as a string theory. It would still be true that at sufficiently low energies it would look like a field theory. As you say, it would not be a field theory of quarks and gluons, because as you said you don't have quarks and gluons in the initial and final states of collision processes. That was probably another thing I should have added to the list of caveats; I think that point is very well taken. Rather, it is a quantum field theory of pions. When we write down the action for soft pions, we don't really doubt it for any of the reasons you're describing.

Gross: Because the symmetry that you postulate, the spontaneously broken symmetry predicts, even before you get started, that there are only a finite number of particles.

Weinberg: I agree with that. I mentioned string theory as a counter-example to the idea that any relativistic quantum theory must be a quantum field theory at all energies. You can ask what is it about string theory that allows it to escape from the argument for this idea that I gave previously. It's not some technical gap in the rigor of my previous argument. It is that, precisely as David said, string theory has an infinite number of particle types. And in particular, whereas in quantum field theory the singularities in any variable always come from intermediate states in which that variable takes a physical value, in string theory the singularities can come from the exchange of infinite numbers of types of other particles. That's amazing, and as you know we're all still learning the implications of that discovery.

Rovelli: Steve, your synthesis of the foundations of quantum field theory is remarkable. You clearly like it very much; you even conclude, with a possible irony, that we might go back to such a quantum field theory in the 'final theory.' But there is a problem. One of the four or five founding stones of your foundation of quantum field theory is Poincaré invariance. Now, if I may advertise another book of yours, I learned general relativity from your '*Gravitation and Cosmology*' – great book. I learned from your book that the world in which we happen to live is not Poincaré invariant, and is not described by Poincaré invariant theory. There is no sense in which general relativity is Poincaré invariant. (If it were, what would be the Poincaré transform of the closed Friedman–Robertson–Walker solution of the Einstein equation?) Thus, Poincaré invariance is neither a symmetry of our universe, nor a symmetry of the laws governing our universe. Don't you find it a bit disturbing basing the foundation of our understanding of the world on a symmetry which is not a symmetry of the world nor of its laws?

Weinberg: Well, I think there's always been a distinction that we have to make between the symmetries of laws and the symmetries of things. You look at a chair; it's not rotationally invariant. Do you conclude that there's something wrong with rotation invariance? Actually, it's fairly subtle why the chair breaks rotational invariance: it's because the chair is big. In fact an isolated chair in its ground state in empty space, without any external perturbations, will not violate rotational invariance. It will be spinning in a state with zero rotational quantum numbers, and be rotationally invariant. But because it's big, the states of different angular momentum of the chair are incredibly close together (since the rotational energy differences go inversely with the moment of inertia), so that any tiny perturbation will make the chair line up in a certain direction. That's why chairs break rotational invariance. That's why the universe breaks symmetries like chiral invariance; it is very big, even bigger than a chair. This doesn't seem to me to be relevant to what we take as our fundamental principles. You can still talk about Lorentz invariance as a fundamental law of nature and live in a Lorentz non-invariant universe, and in fact sit on a Lorentz non-invariant chair, as you are doing. [Added note: Lorentz invariance *is* incorporated in general relativity, as the holonomy group, or in other words, the symmetry group in locally inertial frames.]

Holton: Let me thank both the speaker and the commentators for their attention [applause]. And let me just add one more point. Some of the commentators have deplored that they did not see a paper. I was very glad that they didn't, because I think that per second per square centimeter we got more enjoyment of this session than we would have otherwise. Thank you.

Weinberg: I should apologize to Laurie and Fritz for not having supplied a written version of my talk. I have only once in my life prepared a written version of a talk in advance, and although the written version was pretty good, the actual talk was terrible.

References

32. E. Fermi, *Z. Phys.* **88**, 161 (1934).
33. W. Pauli and V. Weisskopf, *Helv. Phys. Acta* **7**, 709 (1934), reprinted in English translation in A.I. Miller, *Early Quantum Electrodynamics* (Cambridge University Press, Cambridge, 1994).
34. W.H. Furry and J.R. Oppenheimer, *Phys. Rev.* **45**, 245 (1934).
35. S. Weinberg, *Dreams of a Final Theory* (Pantheon, New York, 1993: Hutchinson, London, 1993).

Part Seven. Renormalization group

18. What is fundamental physics?
A renormalization group perspective

DAVID NELSON

The renormalization group, as embodied in the Callan–Symanzik equation, has had a profound impact on field theory and particle physics. You've heard from Ramamurti Shankar and Michael about some of its impact on condensed matter physics. I'd like to tell you my view of what the renormalization group has meant to practising condensed matter physicists.

What have we learned from renormalization theory? We learned that the detailed physics of matter at microscopic length scales and high energies is *irrelevant* for critical phenomena. Many different microscopic theories lead to exactly the same physical laws at a critical point. As Michael Fisher explained, one can even make precise quantitative predictions about certain 'universal' critical exponents without getting the microscopic physics right in detail. What is important is symmetry, conservation laws, the range of interactions, and the dimensionality of space. The physics of the diverging fluctuations at a critical point, which take place on scales of a micron or more, that's 10^4 angstrom, is largely 'decoupled' from the physics at angstrom length scales.

This story about scaling laws at a critical point tells us something about the meaning of a 'fundamental physics'. Fundamental physics is *not* necessarily the physics of smaller and smaller length scales, to the extent that these length scales decouple from the physics that we're interested in at the moment. To elaborate on this point, I'd like to refer to a short paper, which influenced me a lot as a graduate student, that Ken Wilson wrote in 1972. He argued that Landau's hydrodynamic treatment of magnets in the 1930s was not about critical phenomena, but just about magnets in general at any temperature, possibly quite far from the Curie temperature, the critical point. He argued that hydrodynamic theory was itself representative of a renormalization group fixed point. A rather simple one that prompted me to think a little more in detail about generic classes of theories away from critical points, and how they might be viewed in the context of renormalization theory. You can make similar statements about the hydrodynamic laws derived for fluids in the nineteenth century.

The idea of nineteenth century hydrodynamic physics was that, upon systematically integrating out the high frequency, short wavelength modes associated with atoms and molecules, one ought to be able to arrive at a universal long wavelength theory of fluids, say, the Navier–Stokes equations, regardless of whether the fluid was composed of argon, water, toluene, benzene.

One does *not* have to be at a critical point in order to have universal physical laws, which are insensitive to the microscopic details. We now have many concrete calculations, well away from critical points, which support this point of view. Ignorance about microscopic details is typically packaged into a few phenomenological parameters characterizing the 'fixed point', such as the density and viscosity of an incompressible fluid like water. The extreme insensitivity of the hydrodynamics of fluids to the precise physics at higher energies and shorter distances is highlighted when we remember that the Navier–Stokes equations were actually derived in the early nineteenth century. They were completed in about 1845, at a time when the discrete atomistic nature of matter was still in doubt. The same equations would have resulted had matter been continuous at all length scales. The existence of atoms and molecules is irrelevant to the profound and, some might say, even fundamental problem of understanding the Navier–Stokes equations at high Reynold's numbers. We would face almost identical problems in constructing a theory of turbulence if quantum mechanics did not exist, or if matter first became discrete at length scales of Fermis instead of Angstroms.

Many problems in condensed matter physics, by which I mean the study of matter at everyday length or energy scales, do, of course, depend crucially on quantum mechanics and particulate nature of matter: we can't begin to understand phonons in solids, the specific heat of metals, localization in semiconductors, the quantum Hall effect, and high temperature superconductors without knowing about quantum mechanics of protons, neutrons, electrons, and occasionally even muons and positrons. There comes a point, however, when conventional 'fundamental' particle physics burrows down to such short length scales and high energies that its conclusions, however beautiful, become largely irrelevant to the physics of the world around us. That is why many of my colleagues in condensed matter are not aiming to discover 'fundamental' laws at the smallest accessible length scales. High energy physics, while remaining a noble intellectual enterprise, is now virtually 'decoupled' from physics at Angstrom scales just as atomic physics is decoupled from the Navier–Stokes equations. New particles discovered in high energy physics will *not* help us understand turbulence, or how itinerant magnetism arises from the Hubbard model, nor will they unravel the complexities of repetition dynamics in entangled polymer melts. Completing the 'periodic table of quarks' by discovering a sixth top quark in a particle accelerator was exciting, but so were the discoveries of the exotic and short-lived atoms berkelium, californium, einsteinium, fermium, mendelevium, nobelium and lawrencium in their day. Nowadays, it's hard to get excited about the discovery of element number 119. David Mermin has compared the experiments at the new generation of particle accelerators to an archeological dig into the remote early history of the universe. The connection with cosmology is thrilling, but the results are about as relevant to the way matter behaves today as newly discovered shards of ancient Sumerian pottery would be to the next presidential election.

I don't want to overstate the case I am trying to make here. Research into particle physics has great intellectual value, and many condensed matter physicists, including myself, shamelessly exported its beautiful mathematical ideas into our own research. There's a wonderful cross-fertilization between the two disciplines which dates back at least to Ken Wilson and continues to this day.

I have only tried to argue that the physics at the length scales of the more exotic quarks and leptons, despite its intrinsic interest, is decoupled from physics at Angstrom length scales, just as atomic and molecular physics are decoupled from hard physics problems, like turbulence, which occur at even larger distances. The precise nature of physics at even shorter length scales is unlikely to have a profound impact on deep unresolved problems at much larger scales.

Of course I do not wish to imply that the correct short distance theory is useless. A first principle calculation of the viscosity and density of water, for example, would require an atomic or molecular starting point. Deriving hydrodynamics from an atomistic framework is the task of kinetic theory, which has made significant progress in the past century, at least for weakly interacting gases. We are all delighted that lattice gauge theories of QCD may now be fulfilling their promise, and correctly predicting the mass ratios of pions, protons, and neutrons, just as we are thrilled, at least condensed matter people are, when *ab initio* band structure experts are able to predict the lattice constant and correct crystal structure of silicon simply by solving Schrödinger's equation. The point is that there are always important problems, such as turbulence in fluids or the stability of quasi-crystals, which a 'constructionist' approach based on a more fundamental microscopic theory is unlikely to resolve. It may even be that we are even now far from truly 'fundamental' theory, with new particles and fields, such as preons inside quarks, appearing as we probe ever more deeply into nature. *All* current physics would then be phenomenological, with our ignorance about the detailed physics at smaller length scales packaged into a small number of phenomenological coupling constants.

Discussions

Fisher: Thank you, David, for a beautiful demonstration of what I was preaching yesterday, namely, that the renormalization group within condensed matter physics is an approach of enormous richness. Part of that richness arises because there are many, many different systems we can look at, and many, many puzzles to sort out concerning them!

It's also important for the philosophers to realize that these field theories, like Navier–Stokes equation, many of them are very old and were found in easy ways. Generally these are the ones we characterize as trivial these days, and this latest example I think is particularly striking because of the instability of the von Carmen fixed point.

I'm going to stress one other thing, which I know David will agree with. But if you had doubts that he had done everything he should have done here, you could say, well, what about the such and such term that you would have wanted to put in that he left out. He's probably put it in and checked that it's irrelevant. But if not, it's a calculation that you can do. So although he has largely put down what are sometimes called the minimal theories, you can check that the things that are really there need to be there.

Audience: I would like to hear you elaborate on what you said last night after Steven Weinberg made his remarks about renormalization. In effect, you challenged him to come and listen to a refutation.

Fisher: Yes, that is correct. The point is that Steven Weinberg was addressing the topic of the symposium and, as I pointed out in my own introductory remarks, one can hold that the renormalization group does not matter much for quantum field theory in a way that is really interesting. Thus, in the context of quantum field theory, Weinberg said: 'Well, look, we had the Gell-Mann–Low equation and that basically enabled us to study logarithmic factors and what happens when we change the scale on which we study the physics.' What was it, from that viewpoint, that Wilson added that was really new? The answer is: the integration out of the less important degrees of freedom. That was the point that I felt had to be made. I had a brief word with Weinberg after his talk and I believe he does not disagree.

On the other hand, David Gross took pains to actually say, admittedly in one fast sentence, that the essence of the revolution which Wilson inaugurated that has changed the whole way we look at condensed matter theory was to introduce the *space*, if you will, *of all possible Hamiltonians* and *flows* induced within that space by the action of the renormalization group.

I also wanted to emphasize that if, say, Professor Treiman did not happen to like what David Nelson wrote down for his basic Hamiltonian, he could add his own chosen terms. Then the renormalization group approach provides a philosophy, a conceptual way of calculating and testing to see if the new terms are relevant or irrelevant for the behavior under study.

So when David Nelson first started thinking about membranes and tethered surfaces, he told me: 'This theory of elastic shells is a wonderful thing and it is already in the books.' But now he has told you, in effect, that even if he had not got the von Carmen theory and did not know of this particular, corresponding fixed point, it would not matter anyway since, on the long length scales of interest, it is unstable. And so, to put it another way, it is not relevant for real fluctuating surfaces or networks. Now, unless you have this perspective of stable and unstable fixed points, relevance and irrelevance, marginality, and so on, you are really lost. Hence, from the viewpoint of condensed matter physics, the crucial point made by Wilson was that universality and other fundamental, novel concepts arise naturally once you allow yourself a big enough space of Hamiltonians in which to follow the renormalization group flows.

Now from a practical angle, as David Nelson brought out, one can often expect, optimistically, that there are only three or four parameters that you really need to know about in the appropriate effective field theory. If you are Landau or someone with comparable talents, you may succeed in guessing the correct effective theory. It might even be a more or less trivial theory that turns out, mainly for reasons of 'luck', to be the correct one. Then, if only a few parameters are relevant, a practical theoretical physicist (who wants to explain some experiments) would like an efficient method of calculating with those essential parameters. If there is only one variable, she might use the Gell-Mann–Low type of formulation. If there are only a couple of parameters, then the Callan–Symanzik equation or similar approaches provide efficient ways of computing explicit results. But if, as often happens in condensed matter physics, one does not really know the answer (and if proof by Shankar's method of using as small parameter the inverse ego of the investigator fails to go through perhaps because of inadequate hero worship in the field) then you need some calculational approach and some basic philosophy in which you can have a reasoned argument about the issues, about certain terms being left out, about the difference between tethered and nontethered membranes, and so on. Thanks to Wilson, such discussions have waxed fast and furious and for very good reasons.

19. Renormalization group: an interesting yet puzzling idea

TIAN YU CAO

Very few physicists nowadays would challenge the statement that the idea of renormalization group occupies a central place in our theoretical understanding of the physical world. Not only are many crucial components of the standard model, such as asymptotic freedom and quark confinement, somewhat justified by the idea of renormalization group, but our understanding of quantum field theory itself, from the nature of parameters that characterize the system it describes and their renormalization, to the justification of effective field theories and nonrenormalizable interactions and their theoretical structures, have been substantially dependent on the idea of renormalization group. More profoundly, this idea has also shaped our understanding of the relationship between fundamental and effective theories, which in turn has suggested a hierarchical structure of the physical world. Since all these issues have been discussed elsewhere in the past few years, there is no point in repeating them again.[1]

What I am going to do now is to lay out the difficulties that I have felt in my desire to justify the idea of renormalization group. As is well known, a big difference between the old idea of renormalization in quantum field theory and the idea of renormalization group, which was first developed in the context of quantum field theory in the early 1950s and then transferred to condensed matter physics by Ken Wilson during the period of 1965–1972,[2] is that while the former exclusively focuses on the high energy behavior of local fields and the ways of circumventing ultraviolet divergences, the latter's concern is with the finite variations of physical parameters with finite changes of energy scales, or more generally with the elimination of irrelevant degrees of freedom. It is possible to argue that the former is a special case of the latter if we restrict our attention only to one or few parameters and take the cutoff

---

[1] T. Y. Cao and S. S. Schweber (1993): 'The conceptual foundations and philosophical aspects of renormalization theory,' *Synthese*, **97**: 33–108; T. Y. Cao (1997): *Conceptual developments of 20th century field theories*, Cambridge University Press.

[2] E. C. G. Stueckelberg and A. Peterman (1953): 'La normalisation des constances dans la théorie des quanta', *Helv. Phys. Acta* **26**: 499–520.
M. Gell-Mann and F. E. Low (1954): 'Quantum electrodynamics at small distances', *Phys. Rev.* **95**: 1300–12.
K. G. Wilson (1965): 'Model Hamiltonians for local quantum field theory', *Phys. Rev.* **B140**: 445–57; (1969): 'Non-Lagrangian models of current algebra', *Phys. Rev.* **179**: 1499–1512; (1970a): 'Model of coupling-constant renormalization', *Phys. Rev.* **D2**: 1438–72; (1970b): 'Operator-product expansions and anomalous dimensions in the Thirring model', *Phys. Rev.* **D2**: 1473–7; (1970c): 'Anomalous dimensions and the breakdown of scale invariance in perturbation theory', *Phys. Rev.* **D2**: 1478–93; (1971a): 'The renormalization group and strong interactions', *Phys. Rev.* **D3**: 1818–46; (1971b): 'Renormalization group and critical phenomena. I. Renormalization group and the Kadanoff scaling picture', *Phys. Rev.* **B4**: 3174–83; (1971c): 'Renormalization group and critical phenomena. II. Phase-space cell analysis of critical behavior', *Phys. Rev.* **B4**: 3184–205; (1972): 'Renormalization of a scalar field in strong coupling', *Phys. Rev.* **D6**: 419–26.

to infinity and the reference scales of energy to those where the renormalized parameters are defined. However, a significant difference in theoretical structure between quantum field theory and condensed matter physics has obscured our understanding of the relationship between renormalization and renormalization group, and any clarification in this regard requires, it seems to me, a clarification of the difference between two versions of renormalization group: one in quantum field theory and the other in condensed matter physics.

The difference in guiding ideas can be briefly summarized as follows. In quantum field theory, the idea of the scale dependence of physical parameters on energy scales being probed justifies the idea of renormalization, and provides a ground for the idea of renormalization group. The idea was first elaborated by Freeman Dyson in 1951[3] and then by Lev Landau and his collaborators in the early 1950s.[4] Later, in 1971, when Wilson was developing his renormalization approach to the strong interactions, he particularly acknowledged that 'a rather similar approach was suggested many years ago by F. J. Dyson'.[5] This idea, together with the idea of the scale independence of physics, which was first suggested by Dyson in his (1951) paper, logically led to the idea of invariance of physical quantities under renormalization transformations as ways of imposing boundary conditions, which was mathematically embodied in the idea of renormalization group and expressed as renormalization group equations, first formulated in the 1950s by Stueckelberg and Petermann and by Gell-Mann and Low.[6]

In condensed matter physics, however, the change of parameters with the change of scale is accompanied by the change of physics. In contrast with the situation in quantum field theory, the guiding idea here is not the scale independence of physics, but rather the scale dependence of physics. As a consequence of this, as I will argue shortly, the scale dependence of parameters is only the exception rather than the rule. It is interesting to note that the very idea of renormalization group, when it was transplanted by Wilson from quantum field theory to condensed matter physics, has experienced a genetic mutation: in the new theoretical context, what the condensed matter physicists usually call the renormalization group is in fact only a semi-group. Without the underlying idea of scale dependence of parameters, this is quite understandable. But even in quantum field theory that is the main context of my discussion today, the physical justification for the assumed idea of scale dependence is quite obscure—

Fisher: I think you should say that again, because I think I disagree profoundly with it. So do you want to read those couple of sentences again, this first difficulty?

[3] F. J. Dyson (1951): 'Renormalization method in quantum electrodynamics', *Proc. Roy. Soc.* **A207**: 395–401.

[4] L. D. Landau, A. A. Abrikosov, and I. M. Khalatnikov (1954a): 'The removal of infinities in quantum electrodynamics', *Dokl. Acad. Nauk* **95**: 497–9; (1954b): 'An asymptotic expression for the electro Green function in quantum electrodynamics', *Dokl. Acad. Nauk* **95**: 773–6; (1954c): 'An asymptotic expression for the photon Green function in quantum electrodynamics', *Dokl. Acad. Nauk* **95**: 1117–20; (1954d): 'The electron mass in quantum electrodynamics', *Dokl. Acad. Nauk* **96**: 261–3; (1956): 'On the quantum theory of fields', *Nuovo Cimento (Suppl.)* **3**: 80–104.
L. D. Landau (1955): 'On the quantum theory of fields', in *Niels Bohr and the Development of Physics*, ed. Pauli, W. (Pergamon, London), 52–69.
L. D. Landau and I. Pomeranchuck (1955): 'On point interactions in quantum electrodynamics', *Dokl. Acad. Nauk* **102**: 489–91.

[5] See Wilson (1971a) in note 2 and Dyson (1951) in note 3.

[6] See note 2.

Cao: First difficulty? That's not the first difficulty. It's just background. I say that if there is no scale dependence, then the whole idea of renormalization group will collapse.

Fisher: Okay, now you—

Cao: You disagree with me?

Fisher: . . . the organizer, you want to argue this one?

Cao: Oh, no, I'm not arguing that. It's just a fact. [laughter]

Fisher: I think that's just a misunderstanding.

Cao: Oh, can I finish it? We can discuss.

Fisher: That is what I'm asking you.

Cao: Oh, yes?

Fisher: But I mean your thesis is that if there were no scale dependence of the physical parameters, the idea of the renormalization group would collapse. Is that your thesis?

Cao: Yes.

Fisher: Okay. It's incorrect, but we'll discuss it when you choose.

Cao: I would say that only in quantum field theory the idea of scale dependence is explicitly assumed, and in statistical mechanics, it is impossible because of the way the so-called renormalization transformations are operating there. As a result, in statistical mechanics there can be no renormalization group, the most you can get is a semi-group. Because once you lost your memory, you cannot recover your memory. You can go from the high energy, through rescaling lattices, to the critical point. But then you lose all the memory. It's very effective, you can get universality. But once you get universality at or near the critical point, can you go back to know all the initial (molecular or atomic) details? There is no reversible operation. Thus, not only in the mathematical sense there is no group, only semi-group, but also there is no physical mechanism to carry the reverse operations in the context of condensed matter physics. I will elaborate this point later.

Fisher: Okay, well, I don't know if it's useful to have it in dialogue. I suspect it may be more useful. One of the points which Wilson originally made, and which David Nelson referred to again, which I didn't stress so much, was that if you're interested near a critical point, you can integrate outwards and compute in another region. Now that actually tells you what happens initially. So although it's a semi-group—

Cao: So you agree with me that it's a semi-group.

Fisher: It's a semi-group.

Cao: One-way street.

Fisher: That was on my first transparency. But the consequence you seem to be drawing from that is the wrong consequence. And because the point is, we can, as it were, run backwards in the semi-group, provided there is no crossing of trajectories. And the only crossing of trajectories occurs at a fixed point, not in the vicinity. So in fact we can go backwards. In fact we do, and the original Wilson hypothesis always was that you have a difficult region, you integrate out, but there's a conservation which was illustrated in David's case of ρ_s, on my transparencies I didn't emphasize it, but the free energy is a conserved quantity. So if you're interested in the original free energy, you can map it into a region and you don't lose information. So I mean I think there's a misunderstanding. I don't want to carry it on, but I think it may help people to pick up your points as you say them.

Cao: But I will say that the so-called scale dependence in statistical mechanics was based on the idea of thinning out, or averaging over the degrees of freedom. But once you get the average, you cannot get the original one.

Fisher: No. Because the error there is when we write down a free energy, we put in as many parameters as we're interested in, field parameters, parameters that control the other ones. It's that information that we want, and that is what is retained. So we don't really lose in that sense. It looks as though we're losing. It looks as though we're integrating out. But for the information we want for experiment, that's not the situation.

Cao: Even if you are right that some information, such as free energy, which is conserved anyway, will not be lost, the question remains: if you have no idea at all about the initial conditions, can you get something initially unknown simply by running backwards from information at the critical point? Is it possible to know asymptotic freedom of the ingredients of hadrons by the back-running of averaging out, which is something I find difficult to understand?

But this is not the puzzlement I wish to discuss. This is only a background. One of the things that I want to discuss in this talk is that there are two different interpretations of renormalization group. People mention the unreasonable effectiveness of mathematical formulation, which in this case implies that the same renormalization group equations can be applied to both statistical mechanics and quantum field theory. I would challenge this idea. In fact, equation is not understanding. In different physical systems with different basic entities, if the mechanisms for carrying out the mathematical operations are different, then interpretations are different too. I will take this case, the difference between statistical mechanics and quantum field theory, to make a philosophical point. May I continue?

Fisher: I'm happy to let him finish. But I wanted to try and isolate some of the points. Because the physicist will eventually say, we have a certain arrogance, well, if he's misunderstood line one, why should I bother to listen to his line three, four and five. And, so, at least I would like to say that there's a rather profound disagreement—

Nelson: Just a brief interruption. There's certainly something you could say, and maybe you did say it, and we didn't understand it properly, which I would agree with. And that is, if I take a condensed matter renormalization group, like the one I talked about, the Navier–Stokes equation, and ran it backwards, I certainly couldn't deduce the laws of atomic physics, or the existence of quarks by going backwards—

Cao: Yes, that's the point.

Nelson: —and in that sense I certainly—

Cao: You agree with me, mapping in condensed matter physics is a one-way street, only one direction?

Nelson: In a certain sense, but I also agree—

Cao: That's enough for me because—

Fisher: Well what I will say [laughter], and Gene Stanley at an early stage, and Wegner at an early stage, pointed out, that if I take a big enough space, and that will therefore include the microscopics and the quarks, then I can in fact run forwards and backwards because the flows in fact don't cross. We're normally happy to get rid of these details and they become smaller and smaller. But if I develop the full spectrum about the fixed point, with all the corrections, I can run them backwards. Normally it's not feasible and normally we don't want to.

But in that case, and one can demonstrate it in one or two cases, one can run all the way backwards and explore the full space.

Audience: The fixed point, no?

Fisher: Yes, but we never go to the fixed point. We always analyze in the vicinity of the fixed point. The fixed point is that organizing center, an organizing structure around which a linear analysis is described. What Wegner stressed was that if you want to go further, you need a nonlinear analysis, and if you extend this non-linear analysis back by analyticity and other, you know there are always special problems, as we heard from Weinberg yesterday. Then in principle you can run back to large areas, and you would eventually discover some of the details you came from, from all these spectra of the irrelevant operators. Normally we—

Cao: Can you get some details initially you didn't have?

Stachel: Well, now that we've had the discussion maybe we could have the talk.

Fisher: I think we should go on, yes.

Cao: Thank you.

Of course this idea of scale dependence has a long history. As early as 1933, when Dirac published the first paper on charge renormalization, he expressed the observable electron charge as the function of both the cutoff momentum and the momentum being probed, an expression of the scale dependence of the electron charge.[7] Eighteen years later, in 1951, when Dyson worked on his smoothed interaction representation, he introduced the concept of a smoothly varying charge of the electron and a smoothly varying interaction with the help of a smoothly varying parameter g, and argues that when g is varied, some modification had to be made in the definition of the g-dependent interaction, so that the effect caused by the change of the g-dependent charge can be compensated. Perhaps stimulated by Dyson's idea, Landau and his collaborators developed a similar concept of smeared out interaction, according to which the electron's charge must be regarded not as a constant, but as an unknown function of the radius of interaction. With the help of this concept, Landau studied the short distance behavior of quantum electrodynamics, and developed his famous zero charge argument against the conceptual consistency of quantum field theory, although many physicists have argued afterwards that Landau's argument is not conclusive.[8] Also, in all these early speculations of scale dependence, the idea was simply assumed, and no attempt was made to give it any convincing justification.

In the classic work of renormalization group since the 1950s (the Gell-Mann–Low equations and other formulations, the Russian school and the American school), the scale-dependent character of physical parameters and the connections between parameters at different scales have been elaborated in terms of renormalization group transformations, and the scale independence of physics is embodied in renormalization group equations. This mathematical formulation seemed to many people to have unreasonable effectiveness because it is assumed to be applicable to both quantum field system and to statistical mechanical system. At least it was Wilson's initial hunch when he tried to apply Gell-Mann and Low's idea to critical phenomena.

However, mathematical formulations can have different physical interpretations if in different conceptual structures theoretical entities and underlying mechanisms for

[7] P. A. M. Dirac (1933): 'Théorie du positron', in *Rapport du Septieme Conseil de Solvay Physique, Structure et Propriétés des noyaux atomiques (22–29 Oct. 1933)* (Gauthier-Villars, Paris), 203–12.
[8] See, for example, S. Weinberg (1973): *Phys. Rev.* **D8**: 3497.

mathematical operations are different. For example, in the study of critical phenomena in spin lattice systems, physicists' interest is in the continuous limit rather than their lattice structure. Thus the renormalization group transformations developed by Kadanoff and Wilson in this context,[9] which are usually presented as trajectories in the Wilson space of parameters, amount to nothing but an averaging over some set of degrees of freedom at the smaller distances. With this manipulation, the degrees of freedom are diminished, the system is simplified, the parameters obtain finite renormalization, the long range properties are preserved, but there is no real scale dependence. Why? Because for a transformed system to be similar to the initial one, some terms that are not important to the long range properties have to be omitted. Thus, for example, a transformed correlation function and the initial one do not refer to the same system as being probed at different scales, although they sit at different positions of the same Wilson trajectory. Rather, they refer to different physical systems characterized by different sets of parameters, or more exactly, to different models for a given problem, whose Hamiltonians in general have different forms of structure. As we shall see, this feature of renormalization group in condensed matter physics is quite different from that in quantum field theory.

A closely related difference is even more striking. Although it is easy to do the averaging, which in general results in a different model in each step of averaging, no inverse operation would be physically possible. That is, the procedure of Kadanoff transformations is not invertible. Of course, in the study of critical phenomena, it is not even desirable. Thus the label renormalization group in this context is somewhat misleading. What we have here is only a semi-group rather than a group in the rigorous mathematical sense.

Then what is the situation in quantum field theory? That's the major topic of this talk. The basic entities here are local fields. A local field is defined as a local mean value of some fluctuating quantities over a volume of a certain size. The locality assumption implies that the size is approaching to zero, which, according to the uncertainty relations, can be translated into a statement involving infinite momentum scale. But a field can only be measured at some finite distance, which corresponds to a much lower momentum scale. The system described by quantum field theory shares one feature with the system described by statistical mechanics, that is that they are systems involving many scales, in fact there are an infinite number of scales in quantum field theory, and the fluctuations at these scales are coupled to each other, having equal contributions to physical quantities. Thus, in quantum field theory calculations, all quantities of interest are expressed as a divergent series in powers of the coupling constant, which reflects the infinite number of degrees of freedom of a field theoretical system. The task is to find a feasible way to carry out the summation of the series, which, by definition, involves an infinite number of parameters.

Traditionally, a physical theory within its own range of validity is taken to be independent of scales, and its parameters should be constants correspondingly. Then why are so many physicists willing to assume that there is a continuous scale dependence of parameters in a field theoretical system? To my knowledge, the only justification in this context is given by appealing to the idea of the screening effect of a charge in a polarized medium, as first suggested by Dirac.

[9] L. P. Kadanoff (1966): 'Scaling laws for Ising models near T_c', *Physics* **2**: 263–72; L. P. Kadanoff *et al.* (1967): 'Static phenomena near critical points: theory and experiment', *Rev. Mod. Phys.* **39**: 395–431. For Wilson, see note 2.

Here I wish to register my first puzzlement with the idea of scale dependence, and thus the idea of renormalization group. It seems to me that the screening effect is imaginable only in terms of particles, both in terms of real test particles and virtual particles resulting from the polarization of the medium. Yet the concept of particle makes sense only when its Compton wavelength is smaller than the distance between particles. That is, it makes sense only for large distances or low energy phenomena. Thus, if the scale dependence of parameters in quantum field theory is solely caused by the screening effect, or the anti-screening effect in the case of non-Abelian gauge theories due to their nonlinearity, then, although in understanding the derivation of effective theories from a fundamental theory through integrating out high energy degrees of freedom, this heuristic mechanism is very helpful, it would not work all the way through, as suggested by the formal argument of renormalization group transformations, to justify the idea of asymptotic freedom, which is essentially a statement about the very high energy behavior of local fields. Moreover, it is very difficult to understand the scale dependence of mass parameter in terms of screening effect, although this difficulty can be sidestepped by assuming that every parameter that appears in a Lagrangian is a coupling constant, thus is directly or indirectly related with interactions which are susceptible to the screening effect.

Physical justification aside, the assumption of a continuous scale dependence justifies the group property of invertible transformations which is adopted in deriving the results of asymptotic freedom from low energy to high energy scales and of confinement or effective theories from high energy to low energy scales.

There is another more sophisticated argument for scale dependence, which is based on the idea of anomalous breakdown of scale invariance. However, this argument assumes the necessity of renormalization, which in itself assumes the scale dependence. A circularity in this argument has spoiled its being a justification of scale dependence.

Fisher: I think you must say that slower again, because I may say I only agreed to chair this session because Cao said it would be informal, and we could have some to and fro. I mean I do feel that in nearly all the remarks you're making in which you characterize renormalization group, you are looking on rather superficial characteristics that the expositors of the technique emphasize, because those are the ones they value. And you're assuming that because the others haven't been advertised, they are not there.

So as I was arguing before on the question of whether you can go backwards or forwards, I think you have a deep misunderstanding. I think here, as I said on my first slide yesterday, that scale dependence is just a shorthand term. Let's assume that if in doing measurement we disagree in two labs. Then you will say, 'You measured it on a very short scale, that's not what I really mean, it's defined on long scale.' And I say, 'Yes, I agree with you, I'll go back to my lab.' Surprisingly enough, when I do it on the longer scale, I agree with you. The picture there is that on macroscopic scales, there's a unique answer.

What David Nelson showed here, and which you can then check in a number of cases experimentally, that you do the experiment, namely measure an elastic constant (leaving aside the theory now, I'm talking about the facts of the physical world), then you find when you do the measurement, which you thought was the same measurement, and you do it on a different scale, you get a different answer.

That's a physical fact. The renormalization group does not depend on whether that fact is true or untrue. So you're making, to my mind, a number of non sequiturs. Now I'll let you go on, but I think I have to put you on notice that I feel that many of the things where you're characterizing the renormalization group, you're characterizing it in one or two situations. It works differently in different ones. So let me let you go on, but again I explain to the audience that I only agreed to accept the chair—

Cao: We have agreed that he has the freedom to intervene. May I make a comment on your comment? First, what you have said about doing experiments at different scales and getting different answers is quite different from what I have been talking about, the scale dependence of parameters that characterize a theoretical model, which underlies the idea of renormalization group. In fact, your scale dependence of parameters, as you said, is a shorthand term of very complicated physical processes of comparing experimental results, usually involves change of theoretical model or change of physics, thus has undermined the group nature of operations involved and the related invertibility of these operations. Second, I'm interested in physical situations in general, but not in detail. That's why I'm a philosopher, not a physicist. But if, as David Nelson agrees, basically renormalization group in the context of statistical mechanics is a semi-group, as you yourself also just said a few minutes ago, then I think the essential point is here. You said you can map back a little bit or map back a lot. But if you can get new things from running forward, and nothing new from mapping backward, then my argument is there. Can you get something new?

Fisher: Well, I think that will depend on—

Cao: You see, field theorists can get something new. They can use the renormalization group, and go back to very high energies. They get something new. There's asymptotic freedom as suggested by David Gross, and there are ultraparticles and ultrasymmetries as argued convincingly by Detlev Buchholz.[10] But you cannot. The difference in two cases comes from the group property or the invertibility of the renormalization transformations in one case and its lack in another.

Fisher: The difference there is merely that we always accept the cutoff, he would like to have not a cutoff.

Cao: Yes, that's my point: there is a difference in different theories, or in different physical systems described by different theories. Your system and field system, they are different. Thus what are usually put under the same heading, the renormalization group, are in fact two quite different things.

Fisher: It's not a difference in the different systems, they are of course different, and the way the renormalization group will come out, as again David illustrated this morning, looks quite different in the different systems. But you see, where David Gross was using the running backwards in a really physical way, is to say, if I take the different running coupling constants and I run them back, and Weinberg made the same point, I finish up—

Cao: Much earlier, Low and Gell-Mann, they made the same argument and go back to higher and higher energy scale.

Nelson: No, but I will disagree with you in the sense that Gell-Mann–Low didn't predict the existence of quarks by running backwards.

[10] D. Buchholz (1995): 'Quarks, gluons, colour: Facts or fictions?' hep-th/9511002.

Cao: No, no. That's the limit of their formulation of the renormalization group. But with much more sophisticated ideas and techniques, mainly developed in the context of algebraic quantum field theory, it becomes possible to predict the mathematical possibility although not necessarily the physical reality of quarks and gluons.

Nelson: But for them as well as us, there is no difference. David Gross can't predict whether there are preons inside quarks by running it backwards. He takes the theory and runs it forwards and backwards, great. We can do the same, as Michael stressed. We are both subject to the limitation of not being able to deduce the next layer of complexity merely by running the renormalization group backwards.

Cao: No single concept is powerful enough to get everything. Let me finish my paper because I have some comments, in fact I have some suspicions about their work too.

In general, a physical parameter summarizes the physics governing the smaller scales and is measurable. Then why should parameters at different scales, which summarize physics governing different scales, be taken as the manifestations of the same parameter at different scales? In other words, the very idea of scale dependence seems to suggest some mysterious effects that are caused by physical processes at infinitely many length scales shorter than the scale at which a parameter is defined, and these effects are so coherently cooperative that they give rise to one effective scale-dependent parameter. For explaining this cooperative effect, or using mathematical jargon, to justify the self-similarity transformations, we have to assume that the physical system describable by quantum field theory, or for that matter describable by renormalization group transformations, must be essentially homogeneous.

Here I wish to register my second puzzlement with the idea of scale dependence: how we can reconcile the idea of homogeneity and another basic assumption about the short distance behavior of a quantum field system, namely that it is full of extremely violent fluctuations. This puzzlement is greatly reinforced by another fact revealed by renormalization group itself, that is that renormalization group transformations generally will lead to an infinite number of parameters that are constrained only by symmetry. This suggests that the underlying property of the system is not so homogeneous.

My third puzzlement with the idea of renormalization group is more complicated, involving our understanding of causal connections in quantum field systems. As is well known, certain physical interactions can be defined only at certain energy levels, which is characterized by a specific set of entities and certain specific symmetry. For example, the color interactions can only be defined at high energy levels, which are characterized by quarks and gluons and color symmetry. At low energy, quarks and gluons disappear, in their places are hadrons, which can only exist at low energy levels, but not at high energy levels, and their interactions are characterized by the approximate yet broken chiral symmetry.

This fact is apparently different from, yet not incompatible with, another fact: that renormalization group transformations, carrying out the changes of parameterization or the changes in boundary conditions of the solutions of a theory, are continuous transformations of the solutions of a theory constrained by the symmetry of solutions, which ensures that the change in boundary conditions has no effect on the form of the solutions. The latter fact, consistent with the idea of homogeneity, was articulated by Wilson in the following statement: the system describable by the renormalization group approach is characterized by the absence of characteristic scales. This explains why the physical processes at high energy scales can have renormalization effects upon low energy physics, which have revealed deep causal connections between field fluctuations at different length scales.

Now we know that based on the idea and technique of renormalization group, there are important works by Symanzik on decouplings and on effective field theories,[11] which are logically connected with the concept of spontaneous symmetry breakings. All these would cause no problem if the decoupled domains were really decoupled. But this is not the case. The physical processes at high energy levels involving heavy particles and a higher symmetry, according to the understanding based on the idea of renormalization group, can produce renormalization effects and other corrections upon the low energy physics which is characterized by different entities and a lower symmetry. These effects are only suppressed by a power of the experimental energy divided by the heavy mass. We can take these effects and corrections as an indication that there is a hierarchical ordering of couplings between quantum fluctuations at different energy scales: while some couplings between fluctuations at higher and lower energy scales exist universally and are manifested in the renormalization effects in low energy physics, others are suppressed and reveal no observable clues in low energy physics.

Although we have no problem with such a complicated picture of causal connections in the microscopic world, the physical process is not clear to me. My uninformed question here to particle physicists is this. Can physicists calculate the renormalization effects of quarks and gluons upon pion–nucleon physics? If you can, then it is fine. Otherwise, the whole conceptual structure smacks of artificiality. The problem for me here is that quarks and gluons are not represented in the Lagrangian of nucleon–pion interactions.

These puzzlements have raised a question: should we take the scale dependence, or running parameters, as a representation of physical reality, or just mental manipulations for the sake of convenience? In the traditional conception of renormalization, since only the final unique result is important and real, there is no necessity to take all these intermediate steps as physically real. Only in the discourse of effective field theories, in which all finite changes of parameters due to the change of scale are measurable, has the question of the physical reality of scale dependence been forced upon us.

Yet at this stage, there seems to be no unanimous consensus upon this issue among physicists. Let me just give two examples. First, about four years ago, after I visited Michael Fisher and Ken Wilson and Leo Kadanoff, I met another expert in statistical mechanics, who tried very hard to convince me that the whole idea of renormalization group is only a paradigm in the Kuhnian sense. That is, this is only a means for some young physicists to get quick success because the paradigm has dominated the discipline. When it was successful, he said to me, it was not necessary because all its successes can be explained in other terms. Besides, he said, the paradigm had many conceptual difficulties. He gave me a long list of difficulties, but I cannot remember. I am not an expert. Of course I think all of you know all the difficulties, because I

[11] K. Symanzik (1973): 'Infrared singularities and small-distance-behavior analysis', *Comm. Math. Phys.* **34**: 7–36; (1983): 'Continuum limit and improved action in lattice theories', *Nucl. Phys.* **B226**: 187–227. See also T. Appelquist and J. Carazzone (1975): 'Infrared singularities and massive fields', *Phys. Rev.* **D11**: 2856–61.

asked Leo Kadanoff, I asked Wilson, they said, yes, of course, there were. But I want your comments on that.

Perhaps some of you will argue that this is only a comment made by someone who is not in the mainstream. Then let me give you another example which involves a physicist nobody can doubt is a central figure in the physics community. His name is Steven Weinberg. Steven Weinberg in his recent book *The Quantum Theory of Fields* published a few months ago claims, I'll give you a quote, 'the difference between the conventional and the Wilson approach is one of mathematical convenience rather than of physical interpretation' because the 'coupling constants of the Wilson approach must ultimately be expressed in terms of observable masses and charges, and when this is done the results are, of course, the same as those obtained by conventional means.'[12]

My personal reading is that this quotation smacks of an instrumentalist interpretation of the scale dependence. Of course, I may be wrong on this point, and on many points I have touched upon in this talk. The hope is that physicists here will be willing to correct my mistakes.

Discussions

Fisher: There was a question addressed to the particle physicists, that can one say, through the renormalization group, or perhaps through the renormalization effects, or whatever you want, something about quarks and gluons on nuclear physics and the hadrons? Is there a particle physicist here who would like to address that issue?

Low: Yeah, the answer is 'no'.

Audience: I want to make a comment on this. The answer might be 'no' in the pure sense. But still it has an impact if you look at lattice field theory. I'm recently much interested in constructing fixed points from the lattice field theory point of view. The fixed point helps us a lot to find points in the renormalized trajectory of Wilson, which represent points that are free of lattice artifacts. Those would enable computer simulations, even on small lattices or small computers, to give reliable information about QCD, and would ultimately solve QCD at the right spectrum, explain maybe confinement and everything which is needed. And in this indirect practical sense, it has a strong impact even on the entire system.

Low: I think a lot of this argument is about words. When Murray and I started working on what is now called the renormalization group, we didn't call it that. We were interested in the consequences of renormalization and the scale invariance that was involved for the structure of high energy processes. We studied that question and found a result. A quite different seeming procedure was discovered by Kadanoff, Wilson, Michael, others, which involved integrating out certain degrees of freedom and combining them, and then moving to a macroscopic answer. I think it's very unfortunate that this method was called the renormalization group. It must be Ken Wilson who did it, because the only person outside of the Soviet Union and Japan who read our paper was Ken Wilson. [laughter] I think it's an unfortunate nomenclature. Although maybe Steve is right, and in some way the renormalizability of a quantum field theory is like integrating out degrees of freedom that happen in condensed matter physics. It's not obvious. They really look like two very

[12] S. Weinberg (1995): *The Quantum Theory of Fields* (Cambridge University Press).

different things. But the fact that any of these things can assist you in a numerical calculation is interesting. If you're doing numerical calculations, naturally you use asymptotic forms as controls on what happens in the region you're integrating in. So good, I have no quarrel with that.

Fisher: But there's a more specific case in which I wanted to say what I think Steven Weinberg was saying and maintains. If we want to ask why hadrons are as they are, then we have to say that entails looking inside them. And inside them we find the quarks and the gluons. And nuclear physics has gone on happily, some of us feel almost too happily, for a long time without quarks and gluons. But the reason why we don't easily see the effects of quarks and gluons is because the theory is a very difficult nonlinear theory. There are hopes, in which I think one can rationally pin quite a lot of faith, that through the program of lattice gauge theory, one will be able to see specific aspects of the hadron–hadron interactions as we see them in nuclear physics, which reflect the presence of the quarks and the gluons, and calculate from the knowledge of quarks and gluons parameters that are otherwise fitting parameters in nuclear physics. I mean that is the general program, and it is in that sense that I go along entirely with Steven Weinberg, that things are reductionist.

I think, I'm going to ask you to respond, but I feel you make a mistake in judging the philosophy that at this stage we can't do certain things, therefore we shouldn't argue as if we can do it. Because physics goes a lot on faith. I mean when a many body wave function was first written down, the only ones we could do were those where they were independent particles. And then when we could do things that were near independent particles, someone like Eliot Lieb, who's really a superb applied mathematician, and not a physicist, despite the fact that he has his degree at Birmingham from Jerry Brown and from Peierls, can sidestep the need for the approximate methods; they can solve, and have solved, beautiful problems. But as I emphasized with the Onsager solution, you have a beautiful solution, you don't understand it. The same is true of a number of things that Eliot has solved. The Kosh–Lestalitz transition, where this magic of the flow equations was demonstrated, in fact turns out to have been solved by Eliot Lieb many decades before. But nobody understood what it meant or what it applied to, including Eliot Lieb.

The fact that the same model could apply to many systems in a certain limit, came entirely from the renormalization group. The hope is always that you can find at the lower level things of importance. Nuclear physics is by and large said to be irrelevant to atomic physics and chemistry, yet the chemists rely on certain properties of nuclei, their statistics and their spin, and in certain more delicate cases their size, which are outside the subject. So though by and large you can do chemistry with just knowing about atoms and electrons, occasionally you want to know about positrons and nuclear physics. So I feel that you were saying, well, there are certain things we can't do now and therefore they can't be done.

Cao: No, I never had such intention. And of course I agree with Francis Low: what we are debating here is only about words, or I would say about concepts. Several times you asked me, where were the philosophers, what's the concern of the philosophers? I say that now this is an example. Philosophers are concerned with ideas, how to put these ideas together to have a conceptual framework. This conceptual framework should be consistent. Now we examine what is meant by renormalization group. Now I look at the group. What is a group? Mathematicians say for a group, you can have invertible operations. Is it possible, I try to look very carefully at all these operations of averaging out, these Kadanoff transformations. I find that it's very difficult to imagine an invertible operation. You can average it. Once you average it, how can you get it back? Now I have difficulties in understanding why these operations should be called renormalization group transformations because they don't form a group and they are obviously different from those transformations in quantum field theory. So in the most profound sense, I also agree with Francis Low that the renormalization group procedure usually adopted in condensed matter physics and the renormalization group procedure used in quantum field theory are two very different things, although in some cases, such as turbulence, condensed matter physicists also use field theorists' procedure.

Fisher: There are limited theorems worth knowing. In probability theory, we normally look at the mean and maybe the breadth of the distribution, the moment. But there are very powerful mathematical theorems that say that if you know all the moments, and all the moments are averages, then you know the probability distribution. There are interesting exceptions to it. Usually the physicists don't want all that information. So I think where I'm personally worried, is that I think most of the practitioners have a fairly clear conceptual framework. I'd like to let Shankar talk to it, because he comes from a different perspective also.

Shankar: I would like to respond to one topic, because that has to do with the seeming possibility in relativistic field theories to run the flow backwards to very short distances. That's a problem I really do understand, so I want to comment on that. In a quantum field theory when David Gross runs to shorter and shorter distances, and he knows how my coupling is flowing, what he really means is, if you guarantee to me that the only term in the Lagrangian is the Yang–Mills term, what coupling should it have at that distance to reproduce the numbers at much greater distances, then he finds a number, that's his running coupling constant.

If you do this in condensed matter theory, why do you get 15 more coupling constants? Because in condensed matter theory, you want to calculate correlation functions right up to the cutoff. If the cutoff is finite and you're interested, you have to put all these coupling constants. And even in a field theory, if you wanted to compute things right up to the cutoff, it's not enough to vary Gross's coupling constant. You will have to put the same irrelevant terms that we don't care about. So it's a question of what you want to compute. Even in a field theory, if you're obsessed with physics at the cutoff, you cannot just deal with just one or two couplings. And you will have to put all the ugly terms: because you ask a stupid question, you get a stupid answer.

Cao: I agree with you completely. In fact one of the suggestions I have made here is that the renormalization group, if we take it as a mathematical property, is a property of certain transformations adopted in quantum field theory but not a property of averaging out operations adopted in condensed matter physics. The reason for this, as you rightly suggested, is that for having a group character, we have to make dramatic simplifications, and we have to forget many many parameters occurred during the process of renormalization transformation. Also, as you indicated, this is possible only in certain cases of quantum field theoretical problems, but not in most problems of condensed matter physics.

Shankar: Everybody agrees it's a semigroup, from the very first day. I don't think anybody—

Fisher: We also have to agree with Francis Low that it was an unfortunate name. But the inventor of great things usually gets the privilege of naming them. Occasionally they get superseded. It was a tribute, I think, to the inspiration that Ken found in that work by you and Murray Gell-Mann. And I was not about to argue with him, although I might have done at the time, Christ's sake, call it something different, because I already mentioned in fact that De Castro and Jona-Lasinio said it's important, and the way they looked at it, Ken agreed with dismissal. But at any rate, we're stuck with the name. Everybody agrees it's not a group, it's a semigroup. But remember that semigroups also can come backwards. And the degree of invertibility of a semigroup is something to debate. And by and large, in the renormalization groups we can run them as far backwards as we want to if we take the trouble. David has a comment.

Nelson: To some extent I had the same gut feeling as Francis has, that some of what we're arguing about is words and definitions. But there are aspects of the renormalization group, about which very bright people, practising physicists as well as philosophers, feel passionately. And I think Eliot Lieb is a good example. Brilliant applied mathematician. And I remember as a young man hearing that Lieb had said that Ken Wilson was the Lysenko, and that he was leading a whole generation of young people like myself down.

But Freeman Dyson will tell you that in general, God is in the detail. And in a sense there are mysteries of smaller and smaller length scales, which I think all of us in this room would grant. I agree with Michael that we can run the renormalization group back, as far as we need to, in many specific circumstances. But as Shankar points out, if we start putting this multitude of couplings that are there, probably at the cutoff in particle physics, we run things backwards, then there'll be divergent flows in all directions. What was a bunch of irrelevant variables for us condensed matter physicists will all become relevant, will be extremely sensitive to our long wavelength initial conditions when we run things backwards, and we won't know what's inside a quark or more profound questions. Fortunately, there's a set of things we can calculate at long distances. I think there are things about which people can feel passionately, and not all justifiable in my opinion.

Stachel: I want to preface my comment by admitting I know nothing about renormalization theory. But I think this discussion illustrates very well the difficulty of communication between philosophers and physicists, in particular, the point I tried to make yesterday: only the physicist knows where the shoe pinches. In developing a physical theory, a premature demand for consistency can be very dangerous. Obviously, at a certain point, it's very important to demand logical coherence. But a physicist will generally have a feeling about when such a consistency requirement is important and when it's not important at some point in time.

Of course, a *completely* consistent theory is a *dead* theory. It's only when a theory has been superseded, and you already know its limitations that you can give a complete and consistent account of it, axiomatize it, and so forth. But while a theory is still developing, to call for closure and demand logical consistency is extremely dangerous and risks damaging its development. A creative physicist should not be simply paralyzed. It's extremely dangerous for a philosopher who does not have this 'einfuhlung' for the physics to try prematurely to demand consistency. Of course, I am not saying that a philosopher of science cannot develop 'einfuhlung'.

Fisher: Well, I think that's correct. In the defense of the philosophers again, I think they can and have, certainly for some of the physicists, I count myself as one, been helpful by pulling out the principles by which in fact we work. And this is again a role that the axiomatic field theorists did. The advantage of having people set down axioms is that you can see which one you want to get rid of, when you want some different answer. And that's a method which one has to use time and time again in physics. If there are certain no-go theorems, then you have to find some way around or you have to say, you know, this particular fundamental principle I'm going to do without. I think there's another comment, please.

Schnitzer: I think looking at axiomatic structures and finding axioms one wants to get rid of involve an *a posteriori* vision of what goes on. Let's look at the attack on locality that string theory presents as just a paradigm of what we're talking about. One did not sit down and say let's get rid of locality. Rather, there were physical issues on the agenda, and the issue of locality arose as a result of analyzing those issues. Then *a posteriori*, one said we can eliminate locality and still keep consistency. If one is looking at the historical development, one should not adopt an *a posteriori* reinterpretation.

Fisher: I was incorrect. I misexpressed myself. I said, you see the facts, and you want a certain theory which you're told is ruled out by some axioms. Then we're always prepared to go ahead and say, well, we don't need that axiom. It's sometimes nice to know that this is, if you like, what you violate that people would have imposed upon you. Please, Roman.

Jackiw: Another example is the development of supersymmetry. It was a longstanding desire in particle physics to combine internal symmetries with space-time symmetries in a nontrivial fashion. It never worked. There was the Coleman–Mandula theorem, which showed you couldn't do it. But then, remarkably, it was done. And it was done by people who weren't paying attention to the previous development at all, but came at the problem from something like string theory, and they were forced, in order to get a result, to use anticommutators rather than commutators in setting up a Lie algebra, and thereby completely circumvented the no-go theorem. In retrospect you can say, the hypothesis in the theorem was that you should use commutators and we'll do away with that. But actually people's brains don't work that way. You don't look at a set of hypotheses in a theorem, and say I'll cross out this one hypothesis, and now I'll make progress.

Fisher: But in the defense of the philosophers, it was not inappropriate from them. And this was sort of a metatheorem to say, well, with these assumptions you can't do this. The usual mistake that I think both the mathematicians and the physicists make is to assume that the conditions of the theorems are just small technicalities. Often the whole guts is in there. And in this case, as you demonstrated, that was the crucial error, if you like, of the theorem.

Jackiw: It's not so much an error, nor is it a technicality, it's just that nature chooses something else.

Cao: Another point that I feel very interesting is that there's an underlying assumption about the scale dependence of parameters. That is that the underlying properties of the system described by the renormalization group should be homogeneous.

Fisher: What do you mean by homogeneous?

Cao: Homogeneous means that the parameters have smooth and continuous dependence on the scale so that you can have smooth and continuous transformations.

Fisher: I still don't understand it. I mean we all accept—

Nelson: Do you mean a differential equation as opposed to—

Cao: Yes. The homogeneity is implicitly assumed by differential renormalization group equations, which exploit the scale dependence; but it is not assumed by operations of averaging out, in which case, the scale dependence, as Michael said, is only a shorthand term of a complicated physical process of comparing experiments.

Nelson: He means differential renormalization group.

Fisher: But I mean we're happy with discrete renormalization group. The original Wilsonian conception was always a discrete one. For most of us, it's a technicality.

Cao: No. It is much more than a technicality. It involves a fundamental difficulty in understanding renormalization group. In my talk, I first said that the same mathematical concept, in this case the renormalization group, can have two different interpretations in different theoretical contexts. Then I said a few words about statistical mechanics. My focus was on quantum field theory. In quantum field theory, we use the differential equations with an underlying assumption that the basic properties of the system should be homogeneous.

Fisher: I think the only assumption that the particle physicists would claim they were making is what David Gross put very forcefully, there are no grounds for quantizing space and time, they seem to be continuous, why not take advantage of that because differential formulations are often simpler to write down.

Cao: Yes, I agree. But my point is that there's a conceptual difficulty in it. Why? When David Gross tried to use the renormalization group equations to prove the asymptotic freedom, what is required is more than continuity but also the smoothness of the substratum, which is not space-time, but the field system itself. But this subtle difference doesn't matter too much here. However, this smoothness or homogeneity requirement cannot be met in any local quantum field system. Why? Because as you get shorter and shorter distances, the quantum fluctuations will inevitably become more and more violent according to the basic assumption of quantum mechanics, the Heisenberg uncertainty principle. This in fact is the argument I took over from John Wheeler, because when he tried to compromise his geometrodynamics with quantum principle, he found that there's no way to have a differential geometry because of the quantum effect.

Shankar: What large fluctuations are you talking about in Yang–Mills theory? There are no large fluctuations. That's exactly the point of asymptotic freedom. Fluctuations in geometry have no place in field theory with no gravity in it. So he can do his calculation within his field theory.

Cao: I don't think it involves any calculation. It's just a metaphysical image of the physical reality described by a local field. Of course, if you assume asymptotic freedom, then by definition there should be no large fluctuations. But how do you know that there is asymptotic freedom? You have to argue for it. And the most effective way, perhaps up till now the only convincing way or plausible way to argue for it is by using renormalization group equations, as David Gross and some others did in the 1970s. But then the question arises: how can we justify the use of renormalization group equations? The only acceptable justification is provided by the homogeneity or smoothness assumption. But this assumption, as

I try to argue, runs counter to Heisenberg's uncertainty principle. Every quantum field theorist would agree that with the distance getting shorter and shorter, the quantum fluctuations will become more and more violent. Thus there's a fundamental conflict between the homogeneity assumption underlying the renormalization group and the basic assumption underlying any local quantum field theory. Essentially this is the same argument that had forced John Wheeler to give up his dear geometrodynamics.

Fisher: Well, the argument is an argument.

Stachel: Try to localize the particle too much, you can create energy fluctuations that exceed mc^2, and you don't have one particle anymore.

Cao: Yes, that's why I argue that if you try, David Gross really tried to use renormalization group to argue for the possibility of asymptotic freedom, it wouldn't work. Because he had to assume that this group can work all the way to the last point, to the shortest distance. Of course, formally, mathematically, it worked as shown by Gross. But he presumes the antiscreening effect, which, as I discussed moments ago, runs afoul of the Compton wavelength argument.

Fisher: He works that it goes down to the reasonable distances given by the energy scales.

Audience: Tian, this is a bit broader, more philosophical question, if I might, about trying to get at your more philosophical concern. You seem to be worried that there's a certain instrumentalism creeping into quantum field theory here, okay.

Cao: Yeah, that's another important point.

Audience: But it seems to me that quantum theories in general are instrumentalist by nature.

Cao: Do you think they agree?

Audience: I don't, I mean I'm just raising this for discussion. Lacking an account of measurement, quantum theory is a great tool for making predictions. It's a great instrumental theory. It doesn't seem to me that it's clear that there is any real ontology there other than what you can churn out as far as predictions. Now do you want to address that?

Cao: Oh yes. I'm eager to say that—

Low: Before you do that, could you define ontology and instrumentalist?

Fisher: They refused to define ontology before. But I'm going to force them to do it now. Define ontology for us.

Low: And instrumentalist.

Cao: The instrumentalists claim that all theoretical terms are only instruments for calculations and predictions, don't care about what's physical reality is reflected or described by these terms. That is, the theoretical term has nothing to do with reality. The only thing which matters is its effectiveness. That's instrumentalism. But there's a big problem with instrumentalism: how can we explain its effectiveness if a theory has nothing to do with physical reality? I think no physicist would accept that 'what I'm doing has nothing to do with reality; it is very effective only because there's some magic making it so.'

Fisher: Does that answer it, Francis, for you? Now ontology.

Cao: By ontology we mean basic entities assumed in a physical theory. If you look at Newtonian mechanics, then you find particles are assumed to be the basic entities to be treated with. If you look at quantum field theory, then you have local fluctuating fields. That's just the basic entities described by a theory. If we take a purely instrumentalist position, then all these discussions are meaningless. However, at least some physicists would argue that something we described has something to do with reality. Now the difficult point is that you cannot take everything as a true replica of reality. There must be a dividing line. Some concepts and structures are reflections of the reality, others are pure conventions for manipulations. Then the question is, where is the dividing line? Should we take the scale dependence as a reflection of the physical reality? Or maybe we should take it as a conceptual tool for manipulations. Now I'm not quite sure which one is the case. I just laid down the good arguments for its realistic interpretation. But I also put doubts about this realistic interpretation, because some things you cannot take them very seriously.

Audience: Where do experiments come into this?

Cao: That's my weakness. I'm not familiar with experiments.

Fisher: But I'm telling you the experiments. David showed you that there are cases where nobody but a fool would challenge the scale dependence of the parameter. If you don't, you can't get out of this by saying you're not an experimentalist. You have to take our word for this fact.

Cao: Yes, I take your word. But even though I'm not familiar with the experiments, I know that in the effective field theory discourse when the structure of the theory remains the same, the scale dependence is something real because you can measure it. In the context of condensed matter physics it's also real. Of course there is a subtle difference between the scale dependence used in quantum field theory as part of a renormalization group invariant structure and the scale dependence as Michael defined as a shorthand term for comparing experiments in condensed matter physics. Subtleties aside, the question becomes, as Weinberg argued, if everything in intermediate steps must ultimately be expressed in terms of observable renormalized quantities, then it doesn't look very real.

Nelson: But you can observe it at different scales.

Cao: That is true. But what Weinberg discussed was the traditional renormalization. This means you have to take the cutoff to infinity and define your parameters at the energy level of your laboratory. In this case, you can only get one unique observable parameter and the scale dependence, as Ken Wilson says, will be totally invisible.

Rovelli: I would like to express a sense of unease. I am a physicist, I work in a physics department and publish in *Phys. Rev.* D. I tried once to publish in a philosophy journal, and my paper was rejected. So, there is no doubt to which camp I belong. I came to this conference invited by philosophers, and understood this invitation as a request of dialog. I may be wrong, but what I have mainly heard from the physicists' side in this conference is an attitude of lecturing: 'you do not understand this, we'll explain to you that...' Several requests of discussion have been dismissed with an attitude that sometimes sounded to me almost closed-minded arrogance. At the evening panel, for example, Simon Saunders asked a question (interesting, I thought) and was dismissed with no answer. There have been moments in which, as a physicist, I was a bit embarrassed. Today, some discussion has emerged after Cao's lecture. But an immediate reaction against terms such as 'instrumentalism' or 'ontology' has followed. I think that in order to have a dialog, we have simply to learn the meaning of words used by the other side. Do my fellow physicists really think that only their difficult words have meaning?

Fisher: It's a term that we don't normally use and so we find it not useful.

Rovelli: That's exactly the point. There may be terms that we don't normally use, but may still be useful for communicating with people outside our community. There are two different languages here, and we need an effort to understand language and subtleties of the other camp. Philosophy is subtle; if we do not understand, it is not necessarily because 'they' are confused. We need an attitude of reciprocal respect—

Fisher: No! You cannot ask respect for people who are not prepared to define their terms! Any physics term that comes up here, there's not a physicist here who would not attempt to define it, and the philosophers have the same responsibility. We go to another question, Roman.

Rovelli: No, let me—

Fisher: If you have something concrete, then please say it.

Rovelli. Okay. Thanks. About today's topic, the meaning of the renormalization group, I have heard in the past any possible disagreement between physicists; in condensed matter and much more in field theory. It is or it is not fundamental. In spite of the empirical success of today's theories, which lead to Nobel prizes and all that, physicists are confused about the meaning of some aspects of these theories. More often than not, different physicists hold contradictory views on the actual meaning of the renormalization group. I would simply like to correct the impression that might have emerged from yesterday and today's discussion, that physicists know exactly what is going on in their theories, or that everybody agrees. And I want to tell physicists: if we are interested in knowledge, wouldn't it be more productive to fully display our confusion and our disagreements?

Part Eight. Non-Abelian gauge theory

20. Gauge fields, gravity and Bohm's theory

NICK HUGGETT AND ROBERT WEINGARD[†]

Introduction

Recently, the following question appeared in the *American Journal of Physics* (Andereck, 1995): how can gravity be unified with the other fundamental forces when it is not a force at all, but an aspect of the geometry of space-time? In general relativity (GR) the gravitational field is the space-time metric, but in the standard model of quantum field theory the electro-weak and strong forces are manifestations of massless vector particles – of gauge fields. If our most complete contemporary theories treat forces so differently how can they be unified?

Roughly, there are two kinds of answers; either all four fundamental forces are aspects of the geometry of space-time, so the vector fields are also absorbed into the metric. Or, gravity is a force field like the other fundamental forces, so GR is formulated somehow as a gauge theory with curved space-time as an effective low energy approximation. Both proposals put all the forces on a similar footing.

But the primary interest of this paper is not unification, in any of its senses. Rather we want to consider the other questions raised here. What is gravity? What are forces more generally? How does gravity relate to the other forces? These questions are as old as physics, and have been reposed and reinterpreted after every revolution in physical thought. In the present century, remarkable progress has been made in finding potentially definitive anwers to them, and this paper is intended as an account of this work. There are two crucial points that we aim to bring out: first, that running through the physics of gravity are deep connections and analogies between GR and non-Abelian gauge theory; and second, Bohm's theory can give a powerful interpretative insight into the formal theories.

1 Geometrizing gauge fields

The first proposal for unifying GR and gauge theory is almost as old as GR itself, made by Kaluza in 1921 and Klein in 1926 (see Chyba, 1985, for a full introduction). It is based on the fact that gauge groups can be realized by the isometries of a compact physical space; in Kaluza's case, that O(1) (rotations around a compactified circle) and U(1) (the Maxwell field) are locally isomorphic. The original Kaluza–Klein theory then postulated a five-dimensional space-time, $M^4 \times S^1$ (the direct product

---

[†] A few months after presenting this paper, Robert Weingard died suddenly of heart failure, costing philosophy of physics one of its most knowledgeable and insightful thinkers. It will be a fitting memorial if others are stimulated by the papers in this volume to investigate the topics Bob found so fascinating.

of four-dimensional Minkowski space and circle with a radius so small as to be undetectable). In certain coordinate systems the components of the metric are:

$$g_{\alpha\beta} = \begin{pmatrix} g_{\mu\nu} + e^2\kappa^2\phi A_\mu A_\nu & e\kappa\phi A_\mu \\ e\kappa\phi A_\mu & \phi \end{pmatrix}$$

$$= \begin{pmatrix} E & & & & & & M \\ & I & & & & & A \\ & & N & & & & X \\ & & & ST & & & W \\ & & & & E & & E \\ & & & & & I & L \\ & & & & & & N \quad L \\ M & A & X & W & E & L & L \end{pmatrix},$$

where α, β run through 0 to 4 and μ, ν 0 to 3 for the desired 5×5 matrix. Then the U(1) gauge transformation is taken to be a coordinate transformation in the compactified dimension x_5. The gauge field, here the electromagnetic four-vector potential, A_μ, is composed of the $g_{\alpha 5}$ components of the metric.

The approach also can cover non-Abelian gauge fields by the addition of more compact dimensions. U(1) × SU(2) × SU(3) of the standard model requires at least seven such dimensions, for a total of 11. Kerner (1968) was the first to correctly state the form of the metric. But there are apparently insuperable problems with carrying out this extension in that it is known that parity is not conserved in weak interactions, but there is no way of incorporating suitable 'chiral' fermions in Kaluza–Klein theory. There is then something very remarkable about the way that the theory links gravity and gauge theory, but it is not clear how deep the parallel runs.

2 Quantum gravity

But this does bring us to the next topic, for the key idea was that the dynamics both for four-dimensional gravity and for the gauge fields is to be derived from the dynamics of a big space-time metric; i.e., from $4 + N$ dimensional GR. And this implies that a quantum Kaluza–Klein theory is just quantized GR, albeit in higher dimensions. In a standard approach this means quantizing the Einstein field equations for $g_{\mu\nu}$, the metric field.

There are three problems with quantizing GR that we would like to highlight. The first is the nest of problems called the 'problem of time', an aspect of which is the following (see Callender and Weingard, 1995). The classical GR Hamiltonian is identically zero, which leads when quantizing in the Schrödinger picture to the equation of motion $\hat{H}\Psi = 0$, so we have no quantum dynamics. In particular, if R is the radius of the universe, then $d\langle\Psi|\hat{R}|\Psi\rangle/dt = 0$, which seems to imply that the universe is necessarily static, contrary to observation. Second, we have the apparent non-renormalizability of quantized GR in $3 + 1$ and higher dimensions, which is unacceptable if GR is a fundamental theory, valid at all length scales.

And third, there is the measurement problem, which affects all quantum theories. According to a standard interpretation of quantum mechanics, a property P of a system has a definite value if and only if the state of the system is an eigenvector of $\hat{P}$ (the Hermitian operator representing P). This is a consequence of the so called 'eigenvalue–eigenvector' link. Almost certainly, however, the state of the universe is not an eigenstate of the operators representing properties such as the radius of the universe, the number of galaxies, or the number of earth-like planets. Indeed, more than likely, $|\langle 0|\Psi\rangle| \neq 0$, where $|0\rangle$ is the vacuum. Since there are galaxies and so on, we have a conflict.

There are several distinct solutions to this problem, among them the many minds, the many worlds, the continuous collapse theory of Ghirardi, Rimini, Weber and Pearle, Penrose's gravitational approach, and Bohm's interpretation. Of these we find Bohm the most plausible, in part because it not only solves our third problem but also helps with the first (again, see Callender and Weingard, 1995).

This is because Bohm theory abandons the eigenvalue–eigenvector link. In the case in hand, the metric field is taken to have definite values at all times, whether or not the quantum state is an eigenvector of the operator, $\hat{g}$, representing it. Therefore, we have two equations of motion: the Schrödinger equation for the quantum state $\Psi[g, \phi]$, and the Bohm equation for the motion of the metric, g, and the matter fields, ϕ. In minisuperspace models, this dynamics is just $\partial R/\partial t \sim \partial S/\partial R$, where S is the phase of the universal wavefunction, and R the radius of the universe. In effect, something like an absolute time is defined by $\partial S/\partial R$.

3 Gauge field gravity

However, we still have our second problem, that of non-renormalizability, and Bohm cannot help us with that because it takes the wavefunction as input for the Bohm equation of motion. In this regard, if a quantum theory is not defined, then neither is the corresponding Bohm theory. However, non-Abelian gauge theories are renormalizable. This suggests that rather than absorbing the other forces into gravity, we try the converse; we explore the second of the answers mentioned at the start of the paper, and formulate GR in analogy with gauge field theory, rather than by directly quantizing the field equations for the metric.

The natural way to do this invokes the analogy between gauge invariance and Lorentz invariance:

$$\begin{array}{ccc} \text{Global gauge invariance} & \Longleftrightarrow & \text{Global Lorentz invariance} \\ \downarrow & & \downarrow \\ \text{Local gauge invariance} & \Longleftrightarrow & \text{Local Lorentz invariance} \end{array}$$

That is, as we move from global to local gauge invariance in field theory, we should formulate our 'gaugey' GR by demanding local Lorentz invariance. To make the transition from global to local Lorentz invariance we introduce (in four space-time) four tetrad fields $e^{m}{}_{\mu}(x)$, which, at each point x, define a local inertial (or Lorentz) coordinate system. Since these tetrads relate the GR metric to the local Lorentz metric, they allow us to represent all physical quantities in Lorentz coordinates; i.e. in terms of the indices, m, of the local Minkowski metric, η_{mn}. Local Lorentz invariance then means that the action is invariant under arbitrary (independent) Lorentz transformations at each point; in particular that we can define a covariant derivative. A covariant derivative in gauge theory requires the introduction of a matrix valued gauge connection $\hat{A}_{\mu}$, which is the inner product of the generators of the gauge transformation and the gauge fields, $\tau_i A^i{}_\mu$. So in GR it requires a connection $\omega^\mu{}_{mn}$ – the so called 'spin connection'. (Why do we carry out this for Lorentz invariance but not general covariance? Because the latter is continuous and hence not independent at each point.)

We can then treat the tetrads and spin connection as independent fields and formulate the Einstein action in terms of them – this is the 'Palatini action'. In vacuum GR in three-dimensional space-time this approach produces an interesting result. The Palatini action is $\sim \int e \wedge (d\omega + \omega \wedge \omega)$ which looks very similar to the Chern–Simons three-form $\text{Tr}(\hat{A} \wedge d\hat{A} + \frac{2}{3}\hat{A} \wedge \hat{A} \wedge \hat{A})$ that gives rise to a gauge invariant action (under infinitesimal transformations). And indeed, in a classic paper, Witten (1988) showed that the two are equivalent. Furthermore, while $2+1$ dimensional gravity appear to be non-renormalizable by power counting when formulated in terms of the metric g, Witten showed that the quantum theory based on Chern–Simons action is renormalizable! In a sense this is not surprising, since in three dimensions matter-free space-time is flat and quantum states are just superpositions of distinct flat topologies, but it is instructive none the less. It shows that a theory can appear non-renormalizable when formulated one way, and yet when formulated in another be renormalizable (basically, the new formulation allows a perturbative expansion about a different vacuum).

Might the same be true in $3+1$ dimensional GR? Not in this way it seems; even in the absence of matter the Palatini action is $\sim \int e \wedge e \wedge (d\omega + \omega \wedge \omega)$ and as Witten points out, the corresponding Chern–Simons form $\text{Tr}(\hat{A} \wedge \hat{A} \wedge d\hat{A} + \hat{A} \wedge \hat{A} \wedge \hat{A} \wedge \hat{A})$ does not lead to a sensible gauge theory. It doesn't seem that this approach helps with renormalizability.

That leaves open the possibility of slicing space-time up into three-dimensional space and one-dimensional time and giving a Hamiltonian formulation of GR, now based on the Palatini action. But if that is all we do, then we recover standard GR, and again we have made no progress with our problem. However, Ashtekar discovered that if we require the spin connection to be self-dual with respect to the Lorentz indices m, n (i.e. the 'internal' space) then ω is complex and the $3+1$ decomposition of the Palatini action is very similar to that of an SU(2) (complex) Yang–Mills theory. Remarkably, the Hamiltonian constraint, $H = 0$, is much simpler than in the standard approach; it is just a polynomial in the basic variables. In particular, in the canonically quantized standard formulation, no general solution of $\hat{H}\Psi[g] = 0$ has ever been found (in full superspace that is). But such a solution has been found in the Ashtekar approach. Note, however, that being a solution to the Hamiltonian constraint is not enough. There are other constraints to be satisfied too.

4 Strings[‡]

Consequently, we would like to speculate in a different direction, namely towards string theory, and in particular towards string field theory. First quantized string theory cannot be the whole story for several reasons, two of which are (i) that the string perturbation expansion probably diverges (it certainly does for the oriented

---

[‡] The ideas in this section are due to Weingard, and in preparing the manuscript for publication my contribution here has been that of editing – NH.

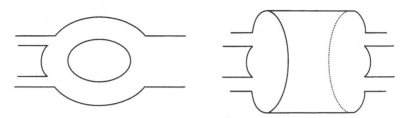

Fig. 1. The emergence of a closed string from the scattering of open strings.

closed bosonic string) and (ii) the rules for constructing the perturbation expansion don't have a principled derivation. We need a non-perturbative formulation from which the rules follow, and a strong field would provide just such a formulation, as QED does for the interaction of electrons and photons.

Our interest in string theory springs from the hope that it provides a consistent quantum theory of gravity, which is bolstered by the fact that the closed string contains a spin two massless mode – a graviton perhaps. Indeed, even the open interacting string automatically contains closed strings (and hence gravitons), though only as intermediate states. For example, the diagram on the left of the figure, in which open strings scatter via open strings, is topologically equivalent to that on the right in which the intermediate state is a closed string.

Of course, in a full quantum theory of gravity we would like gravitons to appear as full fledged particles, not just virtually, and hence we require a closed string field theory. However, for the purposes of this paper we will explore open bosonic string field theory because it can also be formulated in analogy with non-Abelian gauge theory – most closely in Witten's (1986) formulation. The closed string on the other hand does not seem to be formulable in the same way: it has non-polynomial action, not all the constraints follow from the action, and there are problems with BRS invariance.

To see how Witten's open bosonic string field theory works, we formulate Yang–Mills theory in terms of differential forms:

$$\hat{A} = A^i{}_\mu \tau_i$$

$$F = \mathrm{d}\hat{A} + \hat{A} \wedge \hat{A}$$

$$S = \int \mathrm{Tr}(F_{\mu\nu}F^{\mu\nu})\,\mathrm{d}x = \int \mathrm{Tr}(\hat{F} \wedge {}^*\hat{F}) \text{ and}$$

$$\delta_\Lambda(\hat{A}) = \hat{A}\Lambda - \Lambda\hat{A} - \mathrm{d}\hat{A}$$

Here, ${}^*\hat{F}$ is the dual, and Λ an infinitesimal gauge transformation.

In comparison, Witten proposed a string field one-form A and string field strength F such that

$$F = QA + A * A,$$

(where Q is the BRS operator) and

$$\delta_\varepsilon(A) = Q\varepsilon + A * \varepsilon - \varepsilon * A$$

under infinitesimal gauge transformations, ε. By analogy we have the following dictionary:

String operations		Yang–Mills operations
$*$	$\Longleftrightarrow$	$\wedge$
Q	$\Longleftrightarrow$	d
$\int$	$\Longleftrightarrow$	$\int \mathrm{Tr}$

All that's missing in the analogy is the action for the string field. That is because it cannot be the Yang–Mills action; that depends on the dual of F, in $\hat{F} \wedge {}^{*}\hat{F}$ (or $F_{\mu\nu}F^{\mu\nu}$), but there is no analogue of these operations in the string axioms for $*$ and $\int$. The closest thing available for the string field is $\int \hat{F} * \hat{F}$, but $\delta_A \int \hat{F} * \hat{F} = 0$ identically, so we would not obtain any equations of motion by minimizing this action.

Instead, an acceptable action, invariant under infinitesimal gauge transformations, is just the stringy Chern–Simons three-form,

$$S = \int \hat{A} * Q\hat{A} + \tfrac{2}{3}\hat{A} * \hat{A} * \hat{A},$$

which leads to the equation of motion $F = 0$, rather than the Yang–Mills $DF = 0$. It seems then that the analogy is imperfect, but there is one more point of comparison. When Yang–Mills theory is formulated in terms of loop variables, the equation of motion is that the loop space curvature is zero (Hong-Mo *et al.*, 1986).

5 Beables for quantum string theory

Earlier we noted that Bohm theory provides a solution to several problems facing quantized GR. At least one of them, the measurement problem, remains for a quantized version of string field theory. Does that mean that we will want a Bohmized covariant string field? The answer is 'no'.

The reason is the ghost fields that couple to the string field via the BRS charge, Q (at least). According to our Bohmian treatment, the string field will be the 'beable' – an element of reality that always has definite values no matter what the wavefunction. Since it couples to the ghost fields, the ghosts would therefore appear in the Bohm equation of motion and so be beables themselves! To avoid this we turn instead to non-covariant lightcone gauge string field theory. That doesn't mean, from a Bohm point of view, that the covariant theory is unimportant; it is crucial, for it determines the wavefunctional, which must be covariant. However, the Bohm equation of motion for the strings (as opposed to the Schrödinger equation for the wavefunction) is not itself covariant. There is a unique preferred frame in which they hold: this will never conflict with the observed covariance of nature, because that follows from the covariant quantum dynamics.

Something similar occurs in the Bohm theory of Yang–Mills theory. Here too, the ghosts couple to the gauge fields $A^i_{\ \mu}$ in the usual covariant gauges. However, in axial gauge the ghosts decouple from the gauge fields just as in lightcone gauge string theory.

The question then is, can Bohm theory be applied to lightcone gauge string field theory? Well, notice first that there is no problem in developing a Bohm theory of first quantized (free) strings, since in lightcone gauge it consists of $D-2$ scalar fields in the two-dimensional space-time of the string's worldsheet (Weingard

1996). However, when we try to apply Bohm theory to lightcone gauge string field theory, a problem arises.

To obtain the theory we start with Schrödinger wavefunction, $\psi[x(\sigma), \tau]$, for the space-time coordinates, x, of points parameterized by τ and σ on the string worldsheet. Next we second quantize by treating ψ as a classical string field, and finding its wavefunction, $\Psi[\psi(x(\sigma)), \tau]$. The Hamiltonian is

$$\hat{H} = \int \mathcal{D}x(\sigma)\psi^+[x(\sigma)]H\psi[x(\sigma)]$$

where H is the quantized Hamiltonian. Quantization is now straightforward by imposing equal (string) time commutators. Since the canonical momentum, $\pi = i\psi^+$, they are (ignoring certain subtleties),

$$[\psi(x(\sigma), \tau), \psi(x'(\sigma), \tau)] = \pi\delta(x(\sigma) - x'(\sigma))$$

This gives us the free open bosonic string field theory in lightcone gauge. When we now try to give it a Bohmian interpretation we run into trouble. The problem is simple and derives from the fact that the equation of motion of the string field, $\psi[x(\sigma), \tau]$, is first order in time, and therefore has canonical momentum $\pi = i\psi^+[x(\sigma), \tau]$. Now, to obtain the Bohm equation of motion we rewrite the full wavefunction, $\Psi = \mathrm{Re}^{iS}$, which leads to a Hamiltonian–Jacobi equation in which $\partial S/\partial \psi$ and $\partial S/\partial \psi^+$ play the role of canonical momenta. Therefore we identify π with $\partial S/\partial \psi$, and hope this yields the velocity, as it would for a non-relativistic particle. But since here $\pi = i\psi^+$, and not $\partial \psi/\partial t$, rather than giving us the velocity of the field, this procedure just gives us a constraint on it.

A similar problem arises when we try to give a Bohm interpretation of the Dirac field. The field equation (i.e. the Dirac equation) is also first order in time, with canonical momentum $\pi = i\phi^+$. However, here the problem can be solved by writing the Dirac equation in the second order form known as the van der Waarden equation (Sakuri, 1967),

$$\partial^2\chi/\partial t^2 - (\vec{\sigma} \cdot \vec{\nabla})\chi + m^2\chi = 0$$

where χ is a two-component spinor, $\vec{\sigma}$ the Pauli matrices and $\vec{\nabla}$ the covariant derivative. This is used by Valentini (1992, see also Huggett and Weingard, 1994) to give a Bohmian formulation of the Dirac field.

Whether the analogous problem in lightcone string theory can be solved analogously remains to be seen. There are other approaches to it as well, such as that proposed by Bell (1987) also for the Fermi field. But that requires discretizing space and bounding all observables. (Vink, 1993, considers the important issue of the continuum limit of this approach.) The point here is that there is not a fixed recipe for applying Bohm to any quantum theory. As new theories such a string theory come along, it is important to ask whether a Bohm interpretation is possible.

A final point. Although the covariant closed string interaction is non-polynomial, as we noted above, in lightcone gauge it becomes polynomial. Further, if we can apply Bohm theory to the free lightcone gauge string field, then we can apply it to the interacting theory since the interaction terms don't contain derivatives. Thus, supposing we can overcome the difficulty discussed above, and thus achieve a Bohm theory of open and closed (though only) bosonic string field theory, does that mean we have a satisfactory interpretation of the theory? One that answers our questions about gravity?

The answer is again 'no', because the lightcone theory assumes a space-time background. The theory does allow one to compute graviton–graviton scattering (in 26 dimensions of course), but only against a fixed background space-time. According to string theory, however, the graviton is an excitation of the closed string, so the curvature itself must arise from the string, not be presupposed by it. Since the BRS charge, Q, as usually formulated does require such a metric, the covariant string field theory, even that of the open string, is not completely satisfactory.

Horowitz *et al.* (1986) proposed, for the open string, the cubic action $\int A * A * A$, from which they could obtain, in a formal manner, the Witten open string faction. It has been shown that the field equation derived from this action, $A * A = 0$, has solutions which do not contain the space-time metric (Horowitz *et al.*, 1988). However, even if this were satisfactory, it still leaves the closed string with its non-polynomial interaction term.

Whatever the solution to these problems may be (assuming they have a solution!), there have been many indications from topological QFT, from recent work in the Ashtekar program, and as we have seen, from string field theory, that our picture of space-time will have to be radically revised. It will be interesting to see if Bohm theory – or something in its spirit – can survive these revisions and still provide what we think is the best available solution to the problems of the interpretation of quantum mechanics.

References

Andereck, B.S. (1995), 'Is there a gravitational force or not?', *American Journal of Physics* **63**: 583.

Bell, J.S. (1987), 'Beables for quantum field theory', in *Speakable and Unspeakable in Quantum Mechanics*, Cambridge, UK: Cambridge University Press.

Callender, C. and Weingard, R. (1994), 'The Bohmian model of quantum cosmology', in D. Hull, M. Forbes and R.M. Burian (eds.) *PSA 1994: Proceedings of the Biennial Meeting of the Philosophy of Science Association*, East Lansing, MI: Philosophy of Science Association, 218–27.

Chyba, C.F. (1985), 'Kaluza–Klein unified field theory and apparent four-dimensional space-time', *American Journal of Physics* **53**: 863–72.

Hong-Mo, C., Scherbach, P. and Tson, S.T. (1986), 'On loop space formulation of gauge theories', *Annals of Physics (NT)* **166**: 396–421.

Horowitz, G.T., Lykken, J., Rohm, R. and Strominger, A. (1986), 'Purely cubic action for string field theory', *Physical Review Letters* **57**: 283–6.

Horowitz, G.T., Morrow-Jones, J., Martin, S.P. and Woodard, R.P. (1988), 'New exact solutions for the purely cubic bosonic string field theory', *Physical Review Letters* **60**: 261–4.

Huggett, N. and Weingard, R. (1994), 'Interpretations of quantum field theory', *Philosophy of Science* **61**: 370–88.

Kerner, R. (1968), 'Generalization of the Kaluza–Klein theory for an arbitrary non-abelian gauge group', *Ann. Inst. H. Poincaré* **9**: 143.

Sakuri, J.J. (1967), *Advanced Quantum Mechanics*, Addison-Wesley.

Valentini, A. (1992), *On the pilot wave theory of classical, quantum and sub-quantum physics*: PhD Thesis. Trieste, Italy: International School for Advanced Studies.

Vink, J.C. (1993), 'Quantum mechanics in terms of discrete beables', *Physical Review* **A48**: 1808–18.

Weingard, R. (1996), 'Exotic (quixotic?) applications of Bohm theory', in R. Clifton (ed.), *Perspectives on Quantum Reality*, Kluwer.

Witten, E. (1986), 'Non-commutative geometry and string field theory', *Nuclear Physics* **B268**: 253–94.

Witten, E. (1988), '2 + 1 dimensional gravity as an exactly soluble system', *Nuclear Physics* **B311**: 46–78.

Discussions

Jackiw: I want to make some comments first. I think it's very nice to separate the problems of general relativity: the conceptual problems of quantum gravity and the practical ones dealing with actual computation, which are frustrated by the infinities. I suggest that you could examine at least the first set of problems in lower-dimensional gravity. You can consider gravity theory not in four space-time dimensions, but in three space-time dimensions, with some imagination even in two space-time dimensions. The advantage of those lower-dimensional theories is they are every bit as diffeomorphic as the four-dimensional theory. Consequently they have exactly the same problems of time and the same problems of interpretation as the four-dimensional theory, but they have no propagating gravitons. And because they have no propagating gravitons, you don't have any perturbative divergences. So this whole bane of infinities disappears, and you can just address questions, deep questions of principle. And what's nice about it is that the theories are actually sufficiently simple that you can let the mathematics lead you and you are led to answers. And once you are presented with the answers, you can then think about what they mean. So there's a lot you can do about that. On the last point, I know nothing about Bohm theory, but I gather, from your little presentation, that it is tied to the Schrödinger realization of a quantum theory.

Weingard: It's tied to a configuration, taking the configuration space whether it's the configuration space of particles or the configuration space of fields, absolutely.

Jackiw: But it's a Schrödinger representation. Now the difficulty with the Schrödinger presentation for a field theory is perhaps not appreciated by you. You can talk about functionals of fields, but those functionals don't lie in a Hilbert space, or the space is not separable, the probability interpretation becomes really odd. You don't know what you mean when you take the modulus squared of a functional, and of course nothing is normalizable. So although there is a formalism for a Schrödinger type of presentation of quantum field theory, even for Fermi fields, I think that the Bohm theory, with its attempts to have a physical meaning associated with those wave functions and so forth, what can they mean? Do they mean the probability of finding a field localized in some space in field space? Maybe. But then if the field itself is a fermionic field, in other words, a Grassmann variable, do you mean localization in Grassmann space? It all becomes very odd.

Weingard: I agree with you, those would be problems. I'm not saying they're all solved. If you're going to take the Bohm theory seriously, you have to address those kinds of difficult questions, absolutely, bite the bullet.

Jackiw: Right. I think you can do gravity without Bohm theory, you can do the first portion of gravity, actually the conceptually difficult portion of gravity. The infinities are not conceptually difficult. They're just infinities.

Weingard: I don't think you can forget about three even if you're working at lower dimensions. At least part of the point I want to make, or points of view I want to express, is that the measurement problem has to be always addressed. Otherwise you do not have a coherent physical theory. You have got to take some stance, either you opt for many worlds, or you think there's a collapse postulate, or you think there's no collapse and there's Bohm theory, whatever. You've got to address that question, you can't forget about it.

Jackiw: But you do that at the stage when you have something in hand, let's not say a theory, but you have a structure in hand, which you can examine. The measurement problem of quantum mechanics arose after quantum mechanics existed, after people calculated with it, then they started to think, what does it all mean? But in gravity we don't even have the calculations yet, or at least we're only beginning to have them and only in lower dimensions. Once lower-dimensional theories are solved, just kind of by machine, not by intelligence, then you can start thinking about the meaning.

Weingard: I take your point, but it's also true that how you might solve the interpretation problem might shed some light on a problem in the physical theory. And an example to me is the fact that in a Bohm interpretation, you can get a dynamics for the radius of the wave function even if the quantum dynamics is stationary.

DeWitt: First of all, there is no time problem in quantum gravity. *Time is what a clock measures.* If you put into your system a clock and, if you want, also an elastic medium, i.e., a physical coordinate system to determine where you are, you can compute amplitudes for, say, the curvature as projected onto this intrinsic coordinate frame at a certain time, when the clock reads such and such. The projection is gauge invariant, i.e., completely diffeomorphism invariant, and can be computed, and has been. It was computed in connection with what you might call the measurement problem. If you claim the measurement problem has not been solved, then you have to say that it wasn't solved by Bohr either. Bohr and Rosenfeld analyzed the measurability of space-time averages of the electromagnetic field, and showed that the accuracy of measurement agreed exactly with the limitations imposed by the Heisenberg uncertainly relations. The same has been done for the gravitational field, as long as you stay beyond the Planck range where conventional, old fashioned quantum gravity is valid. One can make measurements of space-time averages of the curvature as projected on physical coordinate systems with an accuracy equal to that allowed by the commutation relations. One can say nothing in the Planck range.

Weingard: I have a number of things to say. First, not any clock will work. You have to have a suitable material system which is going to behave appropriately. But more importantly, clocks measure time. Now maybe all time is nothing more than certain changes in physical systems—

DeWitt: Time is what a clock measures even in classical general relativity.

Weingard: That's certainly a view you can have and we could argue about.

Ashtekar: Actually this is in a way a continuation of Roman's and Bryce's comments. I think I agree with Bryce completely that there's an issue of time, but not a problem in the sense that time is defined operationally. But there is still an issue: in a given model, how would we actually set up this formalism and say that something is evolving with respect to something else? Technically, this can be hard. But there are models which have been worked out in great detail, including, in particular, the mini-superspaces, but also some models with midi-superspaces which have infinite number of degrees of freedom. They have been worked out in the configuration representation. And there is a Hilbert space; there is no problem with the measure, everything is well defined in these cases.

So such models exist. I understand the mini-superspace may be complicated, but I was wondering if, in fact, (a) you had looked at any of these models? In these you can implement Carlo's time dependent observable idea, which is what Bryce was saying; how do correlation between two physical quantities change? Does Bohm theory shed any new light, or does it give us any new information? And (b), maybe this model has been not looked at, but perhaps along the lines that Roman was saying, you can just look at three (i.e. $(2 + 1)$) dimensional gravity, where again the issue of time which has been looked at by Carlip, for example, and various things have been done. So does Bohm theory shed any new light on any of these examples?

Weingard: I can't really answer most of your questions. They raise very interesting things for me to look at. I wasn't trying to say that Bohm theory was the way to solve the problem of time. What I wanted to remark was, it struck me as an example in trying to solve the interpretation problem, it also provides a solution to what I think of as the problem of time. Now it may be that you can also give an account in terms of material fields or clocks. But I don't think that's incompatible necessarily with what I'm saying at all, because it would just mean that the Bohm time and that time agree. My point was just that from the problem of the interpretation, it looked like this interesting thing follows.

21. Is the Aharonov–Bohm effect local?

RICHARD HEALEY

1 Introduction

Aharonov and Bohm (1959) drew attention to the quantum mechanical prediction that an interference pattern due to a beam of charged particles could be produced or altered by the presence of a constant magnetic field in a region from which the particles were excluded. This effect was first experimentally detected by Chambers (1960), and since then has been repeatedly and more convincingly demonstrated in a series of experiments including the elegant experiments of Tonomura *et al.* (1986).

At first sight, the Aharonov–Bohm effect seems to manifest nonlocality. It seems clear that the (electro)magnetic field acts on the particles since it affects the interference pattern they produce; and this must be action at a distance since the particles pass through a region from which that field is absent. Now it is commonly believed that this appearance of nonlocality can be removed by taking it to be the electromagnetic potential A^μ rather than the field $F^{\mu\nu}$ that acts locally on the particles: indeed, Bohm and Aharonov themselves took the effect to demonstrate the independent reality of the (electro)magnetic potential. But the nonlocality is real, not apparent, and cannot be removed simply by invoking the electromagnetic potential. While there may indeed be more to electromagnetism than can be represented just by the values of $F^{\mu\nu}$ at all space-time points, acknowledging this fact still does not permit a completely local account of the Aharonov–Bohm effect. Rather, it shows that, though unavoidable, quantum non-locality may be of radically different varieties depending on how one interprets quantum mechanics. In this way consideration of the Aharonov–Bohm effect reinforces a lesson one can learn by reflecting on violations of Bell-type inequalities. For any reasonable account of either phenomenon violates at least one of two distinct locality conditions – the principle of local action and the condition of separability.

2 The Aharonov–Bohm effect

An example of the Aharonov–Bohm effect is illustrated in Fig. 1.

If no current flows through the solenoid behind the two slits, then the familiar two-slit interference pattern will be detected on the screen. But if a current passes through the solenoid, generating a constant magnetic field B *confined to its interior* in the z-direction parallel to the two slits, the whole two-slit interference pattern is shifted by an amount

$$\Delta x = \frac{l\lambda}{2\pi d}\frac{e}{\hbar}\Phi \qquad (1)$$

298

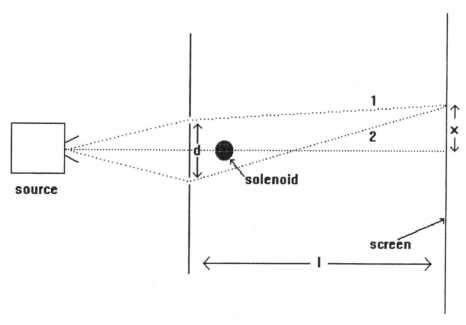

Fig. 1

where d is the slit separation, l is the distance to the screen, λ is the de Broglie wavelength of the electrons in the beam, and Φ is the magnetic flux through the solenoid.

One can account for this shift in the interference pattern quantum mechanically by representing the effects of electromagnetism by a classical magnetic vector potential A, and inserting this into the Schrödinger equation. The vector potential A is no longer zero either inside *or outside* the solenoid because of the current flowing through it, even though $B = \text{curl}\, A = 0$ everywhere outside the solenoid. The solution to the modified Schrödinger equation predicts the fringe shifts given by Eq. 1. However, this shows that the A–B effect is local only if A may be regarded as a physically real field, capable of acting on the electrons directly. But there is reason to doubt that the magnetic vector potential is a physically real field, since A is not gauge-invariant, unlike the magnetic field B and the phase-shift between paths 1 and 2. That is to say, both A and

$$A' = A + \nabla \chi \qquad (2)$$

are to be regarded as specifying the same physical state, for an arbitrary (but suitably differentiable) function χ.

If one nevertheless maintains that in some way A represents a physically real field, the following argument appears to establish that its gauge-dependence excludes a local account of the A–B effect.

With no current flowing, A is zero everywhere outside the solenoid; or more precisely, there exists a function χ such that A can be set at zero everywhere outside the solenoid by the transformation $A \rightarrow A'$ defined by Eq. 2. But even *with* a current flowing, this transformation permits one to set A equal to zero over a very wide region outside the solenoid! By a suitable choice of χ one can, for example, set A equal to zero outside the solenoid everywhere except within a segment, of arbitrarily small angle α, of a solid cylinder of infinite thickness whose inner radius coincides with the outside of the solenoid. One could thus set A equal to zero everywhere along path 1, or everywhere along path 2 (but not both at once).[1] Now for the shift in the interference pattern to be produced by a direct local interaction between each individual electron and the magnetic vector potential A outside the solenoid, that interaction would have to be different when a current is passing through the solenoid. However, the potential is defined only up to a gauge-transformation; and for any continuous path from source to screen that does not enclose the solenoid there is a gauge-transformation that equates the value of A at every point on that path when a current is flowing to its value when no current is flowing. The shift in the interference pattern cannot therefore be produced by a direct local interaction between individual electrons following such continuous paths and the magnetic vector potential A outside the solenoid. Thus accepting the physical reality of the vector potential fails to render the A–B effect local; while denying its physical reality leaves one without any local explanation of the effect.

3 The Dirac phase factor

At this point, one might naturally appeal to the analysis of Wu and Yang (1975), for they showed how to give a gauge-independent description of electromagnetism which could still account for the A–B effect. Following their analysis, it has become common to consider electromagnetism to be completely and nonredundantly described in all instances neither by the electromagnetic field, nor by its generating potential, but rather by the so-called *Dirac phase factor*

$$\exp\left[-(ie/\hbar)\oint_C A^\mu(x^\mu) \cdot dx^\mu\right]$$

where A^μ is the electromagnetic potential at space-time point x^μ, and the integral is taken over each closed loop C in space-time. Applied to the present instance of the Aharonov–Bohm effect, this means that the constant magnetic field in the solenoid is accompanied by an association of a phase factor $S(C)$ with all closed curves C in space, where $S(C)$ is defined by

$$\exp\left[-(ie/\hbar)\oint_C A(r) \cdot dr\right]$$

This approach has the advantage that since $S(C)$ is gauge-invariant, it may readily be considered a physically real quantity. Moreover, the effects of electromagnetism outside the solenoid may be attributed to the fact that $S(C)$ is nonvanishing for those closed curves C that enclose the solenoid whenever a current is passing through it. But it is significant that, unlike the magnetic field and its potential, $S(C)$ is not defined at each point of space at each moment of time. There is an important sense in which it therefore fails to give a local representation of electromagnetism in the A–B effect or elsewhere.

---

[1] Indeed, if one generalizes the concept of a choice of gauge along the lines of Wu and Yang (1975), it is even possible to choose a global gauge according to which A is zero *everywhere* outside the solenoid (though this will not be a global gauge in which there is a single value of A in each region within the solenoid).

4 Locality

I have claimed that the usual quantum mechanical account of the Aharonov–Bohm effect fails to be local, whether it represents electromagnetism by the (electro)magnetic field $F^{\mu\nu}$, by the (electro)magnetic potential A^{μ}, or by the Dirac phase factor $S(C)$. But to substantiate this claim it will be necessary to be clearer both about what is required of a local account of the effect and also about how the accounts provided by various different interpretations of quantum mechanics differ among themselves. I first explain two distinct concepts of locality which are at stake here (as well as in the case of violations of Bell-type inequalities[2]). The next section shows why at least one of these is violated by each of several rival interpretations of quantum mechanics as applied to the A–B effect.

Let me begin with the following general account of locality formulated by Einstein (1948).

> ... it appears to be essential for [the] arrangement of the things introduced in physics that, at a specific time, these things claim an existence independent of one another, insofar as these things 'lie in different parts of space' ....
>
> Field theory has carried this principle to the extreme, in that it localizes within infinitely small (four-dimensional) space elements the elementary things existing independently of one another that it takes as basic, as well as the elementary laws it postulates for them. For the relative independence of spatially distant things (A and B), this idea is characteristic: an external influence on A has no immediate effect on B; this is known as the 'principle of local action', which is applied consistently only in field theory.
>
> (pp. 322–3)

As has now been widely recognized,[3] one can find two distinct ideas in this and similar passages from Einstein's writings. I shall call these the principle of *Local Action* and the principle of *Separability*, and state them as follows.

Local Action

If A and B are spatially distant things, then an external influence on A has no immediate effect on B.

Separability

Any physical process occurring in space-time region R is supervenient upon an assignment of qualitative intrinsic physical properties at space-time points in R.

The idea behind *Local Action* is that if an external influence on A is to have any effect on B, that effect must *propagate* from A to B via some continuous physical process. Any such mediation between A and B must occur via some (invariantly) temporally ordered and continuous sequence of stages of this process. Nonrelativistically, such mediation could not be instantaneous, and so an effect on B could not occur at the same time as the external influence on A. Thus although in the nonrelativistic case the term 'immediate' in *Local Action* may be read as ambiguous between 'unmediated' and 'instantaneous', that ambiguity seems relatively harmless, insofar as any instantaneous effect would have to be unmediated.

---

[2] The companion paper Healey (1997) gives a more thorough discussion of *Local Action* and *Separability*, and is particularly concerned to point out the parallel ways in which the A–B effect and violations of Bell-type inequalities challenge these two distinct locality conditions.

[3] See, for example, Howard (1985), Redhead (1987), Healey (1991, 1994).

Applied to the Bell case, *Local Action* entails that a measurement on a particle *A* in one wing of an Aspect-type device has no immediate effect either on a particle *B* on which a measurement is performed in a different wing of the device, or on the apparatus which performs that measurement. Given *Local Action*, a measurement on particle *A* in one wing of an Aspect-type device can affect either particle *B* or the apparatus which performs a measurement on it only if some continuous process mediates that effect. But the experimental conditions are designed precisely so as to rule out the possibility that any process could mediate between the two measurement events.[4] This supports the conclusion that the results of the two measurements are causally independent. It is this consequent condition of causal independence that is taken (explicitly or implicitly) to justify more specific 'locality' conditions appealed to in derivations of Bell-type inequalities.

Applied now to the Aharonov–Bohm case, *Local Action* entails that a change in the current passing through the solenoid has no immediate effect either on the behavior of any electron outside the solenoid, or on any real field or potential associated with electromagnetism and located away from the solenoid. Here the force of the term 'immediate' is to require that any effect of the field inside the solenoid on the behavior of electrons or electromagnetism outside the solenoid be mediated by some influence propagating from the solenoid and then acting where these are – somewhere outside the solenoid.

The principle of *Separability* requires some explanation, especially because it differs from formulations offered by other authors.

On one common understanding, any 'entangled' quantum systems are nonseparable in so far as they must be described quantum mechanically by a tensor-product state-vector which does not factorize into a vector in the Hilbert space of each individual system

$$\Psi_{12...n} \neq \Psi_1 \otimes \Psi_2 \otimes \ldots \otimes \Psi_n \qquad (3)$$

This is related to a more general understanding of nonseparability, according to which any two spatially separated systems possess their own separate real states. For if the state vector of a quantum system gives its real state, then any spatially separated quantum systems described by the 'entangled' state vector of Eq. 3 will count as nonseparable, on this more general understanding.[5] But even if spatially separated systems do possess their own separate real states, the system they compose may still fail to be separable if its real state does not supervene on theirs. This suggests the following formulation of separability:

Spatial Separability
The qualitative intrinsic physical properties of a compound system are supervenient on those of its spatially separated component systems together with the spatial relations among these component systems.

---

[4] In fact, these conditions at most exclude the possibility of mediation via a *separable* process. The measurement events may still be connected by a nonseparable process (as in Healey (1994)), in which case the question of causal dependence is reopened. Healey (1992) argues that our concept of causation may then not be sufficiently univocal to permit this question to be decisively answered.

[5] Note that this conclusion would not follow if, like Einstein, one were to *deny* that a quantum system's real state is given by its state vector. Indeed, as Howard (1985) and others have pointed out, a number of Einstein's reasons for this denial assumed that the real state (unlike the quantum state) must be separable in the sense just noted.

Here I take the real state of a system to be given by its qualitative intrinsic physical properties.

Now while the condition of *Spatial Separability* is naturally applied to an *n*-particle system that figures in an Aspect-type experiment, it is less clear how it is relevant to an Aharonov–Bohm experiment. But, as shown in Healey (1991), *Spatial Separability* is itself a consequence of a yet more general principle (there called *Spatiotemporal Separability*) which is immediately applicable to the Aharonov–Bohm case. Here is a condensed statement of that principle.

Separability

Any physical process occurring in space-time region *R* is supervenient upon an assignment of qualitative intrinsic physical properties at space-time points in *R*.

According to this principle, whether a process is nonseparable depends on what qualitative, intrinsic properties there are. Deciding this involves both conceptual and scientific difficulties. The conceptual difficulty is to say just what it means for a property to be qualitative and intrinsic.

Intuitively, a property of an object is *intrinsic* just in case the object has that property in and of itself, and without regard to any other thing. This contrasts with *extrinsic* properties, which an object has in virtue of its relations, or lack of relations, to other things. Jupiter is intrinsically a body of mass 1.899×10^{27} kilograms, but only extrinsically the heaviest planet in the solar system.[6] Unfortunately, philosophers have been unable to agree on any precise definition of the distinction between intrinsic and extrinsic properties, or even on whether such a definition can or should be given.[7] This is true also of the distinction between qualitative and individual properties, where a property is *qualitative* (as opposed to *individual*) if it does not depend on the existence of any particular individual. Having a mass of 1.899×10^{27} kilograms is a qualitative property of Jupiter, while both the property of being Jupiter and the property of orbiting our sun are individual properties of Jupiter.

After such an inconclusive resolution of the conceptual difficulty, it may seem premature to consider the scientific difficulty of discovering what qualitative, intrinsic properties there in fact are. But this is not so. Whatever a qualitative, intrinsic property is *in general*, it seems clear that science, and in particular physics, is very much in the business of finding just such properties.

What is meant by a process being supervenient upon an assignment of qualitative intrinsic physical properties at space-time points in a space-time region *R*? The idea is familiar. It is that there is no difference in that process without some difference in the assignment of qualitative intrinsic physical properties at space-time points in *R*. I take the geometric structure of *R* itself to be already fully specified by means of the spatiotemporal properties of and relations between its constituent points.[8] The

---

[6] Note that I follow philosophers' usage rather than physicists' here. I take Jupiter's mass to be intrinsic to it even though Jupiter's mass may vary, or indeed might have always been different, from 1.899×10^{27} kilograms. Physicists tend to use the term 'intrinsic' differently, to refer only to unchanging, or even essential, properties (where an essential property is one which an object could not have lacked while remaining that very object).

[7] David Lewis (1986, pp. 61–9), for example, tentatively offers two possible definitions but argues that the distinction is both possible and necessary even if it cannot be defined in terms of anything more basic.

[8] If *R* is closed, it may be necessary to add information on how points in *R* are related to points just outside *R*.

supervenience claim is that if one adds to this geometric structure an assignment of qualitative intrinsic physical properties at space-time points in R, then there is physically only one way in which that process can occur.

Spatial Separability (and hence also *Separability* itself) is in question in Aspect-type experiments to the extent that the intrinsic properties of an 'entangled' compound quantum system fail to supervene on those of its components. A violation of *Separability* of this kind would be associated with a kind of physical holism (see Healey (1991)). Note that since *Local Action* does not presuppose *Separability*, a nonseparable account of these experiments may be given which conforms to *Local Action* (see Healey (1994)).

The Aharonov–Bohm effect challenges *Separability* in a different way. In this case what is at issue is whether either the process by which each electron passes through the region outside the solenoid, or the (electro)magnetism present there throughout the time of its passage, supervenes on qualitative intrinsic physical properties of (objects at) points in that region at moments during that time.

5 Nonlocality in the A–B effect

A completely local account of the A–B effect would conform both to *Local Action* and to *Separability*. It would provide a separable description both of the electrons' passage through the apparatus and also of electromagnetism in the apparatus. Moreover, it would give an account, conforming to *Local Action*, of the effects of electromagnetism on the electrons in the apparatus, and also of the effects on electromagnetism of altering the current through the solenoid. No reasonable interpretation of quantum mechanics provides such an account.

Consider first the view that it is meaningless to ascribe any position to individual electrons passing through the apparatus. This view provides no description of the electrons' passage through the apparatus, and certainly no separable description. And for the same reason it precludes the possibility of giving a local account of the effects of electromagnetism on the electrons.

Next consider a view according to which electrons passing through the apparatus have a position, but that position is imprecise – perhaps it corresponds to the region in which the amplitude of the electrons' wave-function differs 'significantly' from zero. On a view like this, the region occupied by each electron simultaneously traces out paths like 1 and like 2 in Fig. 1, and the electron thereby actually traverses closed curves C enclosing the solenoid. But the gauge-invariant phase factor $S(C)$ is defined on precisely such closed curves. Hence, on this interpretation, the spatially nonlocalized electron interacts directly with electromagnetism in conformity to *Local Action*.

On a Bohmian interpretation, each electron has a precisely defined position at all times, so that in the A–B effect some electrons follow a continuous path above the solenoid, while others follow a continuous path below it. At first sight, the electromagnetic interaction with these electrons appears to violate *Local Action*, since each electron is affected by $S(C)$ for closed curves C enclosing the solenoid, even though its trajectory does not trace out any such curve. But on the Bohmian interpretation the effects of electromagnetism on an electron are mediated by its wave-function, which is taken to represent a physically real field: and the wave-function is nonzero everywhere along both path 1 and path 2, hence everywhere on C. The

Bohmian interpretation can therefore try to offer an account of the A–B effect that is in conformity to *Local Action*.

Here is such an account. Since the single-particle wave-function describing the electrons is not gauge-invariant, it must be some gauge-invariant function of it that directly represents the locally mediating influence on the electrons. There are essentially two candidates for this role: the velocity field, $v = 1/m(\nabla S - eA)$ and the so-called 'quantum potential' $Q = -(\hbar^2/2m)\nabla^2 R/R$, where the wave-function has been written as $\Psi = Re^{iS/\hbar}$. Holland (1993) and Bohm and Hiley (1993) seek to preserve conformity to *Local Action* by taking electromagnetism to act indirectly on an electron by first modifying the quantum potential throughout the region surrounding the solenoid: the electron's trajectory through the region is then altered, since its velocity at each point x is now given by the resulting $v(x) = 1/m(\nabla S(x) - eA(x))$. On this view, the quantum potential generates a 'quantum force' which acts in a quasi-classical manner on the electron, whose trajectory is consequently modified. Changing the current through the solenoid changes the magnitude of this force, and consequently alters the electron's trajectory. Alternatively (Dürr, Goldstein and Zanghi (1992)), one may appeal directly to the velocity field $v = 1/m(\nabla S - eA)$ as an independently postulated physical field associated with the wave-function, thus avoiding any mention of the quantum potential. In that case, changing the current through the solenoid simply changes this (gauge-independent) velocity field, with the same consequent effect on the electron's motion.

But now the second challenge presents itself. If electromagnetism outside the solenoid is not representable by a field defined at each point of that region, then how can its action on either the quantum potential or the velocity field, both of which are so defined, conform to *Local Action*? This would not be such a problem if the electromagnetic interaction could be taken to occur with $S(C)$ for an arbitrarily small curve C surrounding the instantaneous location of the Bohmian electron. But of course $S(C)$ is zero for any such curve: $S(C)$ differs from zero only for curves C enclosing the solenoid.

Now appeals to the Hodge decomposition theorem may appear to offer a way of meeting this challenge. As can be shown, that theorem entails that the magnetic vector potential A [represented as a one-form (or covector) rather than a (contra)vector] may always be expressed uniquely as the sum of three components, one of which vanishes just in case the magnetic field strength is zero. Hence in the field-free region outside the solenoid, the magnetic potential at a point may be represented as the sum of just two one-forms (the so-called exact and harmonic components). Significantly, only the former of these one-forms is gauge-dependent. Hence the value, at each point outside the solenoid, of the latter, harmonic one-form ω_h may be thought to provide a local, gauge-invariant representation of electromagnetism in the Aharonov–Bohm effect. Moreover, when the current is constant, these values of ω_h are uniquely determined by the flux through the solenoid.

A Bohmian may appeal to this apparently purely local representation of electromagnetism in support of the claim that quantum mechanics, on this interpretation, does indeed provide an account of the Aharonov–Bohm effect that conforms to both *Local Action* and *Separability*. Qualitatively, the account now goes as follows. Passing a current through the solenoid produces a flux through the solenoid. The flux gives rise to a corresponding harmonic one-form field. This field is a genuine physical field, insofar as it constitutes a unique, gauge-invariant, representation of the (electro)magnetic potential at each point outside the solenoid. The value of ω_h at a point affects either the quantum potential or the velocity field at that point, and consequently affects the motion of any electron whose trajectory takes it through that point. Both electromagnetism and the motion of the electrons conform to *Separability*, and the action of electromagnetism on the electrons conforms to *Local Action*.

But although this account conforms to *Separability*, it must still explain how changing the current through the solenoid affects electromagnetism outside the solenoid. The account takes electromagnetism at a point outside the solenoid to be represented by ω_h when the (electro)magnetic field strength is zero at that point. A larger constant current through the solenoid will correspond to a larger value of ω_h. But can one represent electromagnetism at a point outside the solenoid by intermediate values of ω_h during the transition period between a state of lower constant current and a subsequent state of higher constant current? Clearly not. Since changing the current through the solenoid will induce varying electromagnetic fields both inside and outside it, a representation corresponding to zero electric and magnetic field at a point outside the solenoid will not be adequate during the transition period. An adequate representation is provided only by the full electromagnetic potential A^μ, but this, of course, is gauge-dependent.

One might wish, analogously, to uniquely express the full electromagnetic potential A^μ as the sum of three components: one of these would correspond merely to a choice of gauge, a second would vanish in a field-free region, and the third could be held responsible for the Aharonov–Bohm effect (as a four-dimensional generalization of ω_h it would be gauge-invariant and need not vanish even in a field-free region). Granted this wish, one could hope to show that changing the field through the solenoid produced changes in this third component that propagated continuously from the solenoid in conformity to *Local Action*, until a new stable state was reached in which the third component could once again be given a three-dimensional representation just by a larger value of ω_h.

Unfortunately, this wish would be frustrated by the failure of the Hodge decomposition theorem to generalize to manifolds (like Minkowski space-time) whose metric is not Riemannian but pseudo-Riemannian. When the magnetic (or electromagnetic) potential in the A–B effect is represented by a one-form on a compact manifold, it is the topology (not the metric) that allows it to be closed but not exact (in the field-free case). This is the significance here of de Rham's results on cohomology groups. But only if the metric is Riemannian does Hodge's theorem yield a unique harmonic one-form representative of each cohomology class. Thus while ω_h is uniquely defined in some compact region of three-dimensional Euclidean space, there is no analogous unique one-form in a compact region of four-dimensional Minkowski space-time which could be taken to represent the underlying, gauge-invariant, physical field that is responsible for the Aharonov–Bohm effect in both the steady-state and transitional regimes.[9]

If one supposes that harmonic one-form field ω_h represents a real physical field defined at each point outside the solenoid when a constant current is passing through

---

[9] I wish to thank Professors Robert Kotiuga of Boston University and David Malament of the University of Chicago for helping to educate me on the mathematics of Hodge decomposition; and Professor Robert Geroch of the University of Chicago for providing a simple counterexample to the assertion that this decomposition is unique also on manifolds with Minkowski metric.

it, then one can therefore give no account in conformity to *Local Action* of how turning on the current in the solenoid gives rise to this field. To say that changes in the current instantaneously affect the values of ω_h is to admit that *Local Action* is violated in the A–B effect, while if one is prepared to offer no account at all of how the values of ω_h are affected by changes in the current, one is essentially abandoning the claim that ω_h represents a real physical field. And indeed I believe that claim should be abandoned. No empirical evidence supports the existence of such a field in the Aharonov–Bohm case or anywhere else, since the effects of electromagnetism here and elsewhere may be accounted for purely in terms of the (nonseparable) gauge-invariant Dirac phase factor.

To summarize, attempts to secure conformity to *Local Action* in a Bohmian account of the A–B effect founder because of the impossibility of giving a purely local account of an interaction between electromagnetism (a nonseparable process) and the electron which is subject to it (and whose trajectory constitutes a separable process). This contrasts with the earlier result, that if the motion of the electrons through the apparatus is a nonseparable process, then it is possible to account for the A–B effect in terms of a purely local interaction between (nonseparable) electromagnetism and this process.

6 Quantum field theory

Einstein believed that a field theory would naturally provide a local account of its subject matter, as the earlier quote made clear. And while he doubtless had in mind a classical field theory, one might suppose that quantum field theory contains the resources to provide a completely local account of the A–B effect. For that theory represents the electrons as well as electromagnetism by means of quantized fields defined at space-time points, and postulates that these fields interact via an interaction term $e\bar{\psi}(x)\gamma^\mu A_\mu(x)\psi(x)$ in the total Lagrangian that couples them at each space-time point x. This secures overall gauge invariance, since while neither $\psi(x)$ nor $A_\mu(x)$ is individually gauge-invariant, their interaction is: a gauge transformation

$$\psi(x) \rightarrow \exp(i\Lambda(x))\psi(x): A_\mu(x) \rightarrow A_\mu(x) - \partial_\mu\Lambda(x)$$

preserves the total Lagrangian, and hence leaves the dynamics unaltered. It may appear that this gives us a gauge-invariant, separable, account of electromagnetism and electrons, plus an account in conformity to *Local Action* of how these interact in general, and so in particular in the A–B effect. But this appearance proves illusory.

First there is the general problem faced by any interpretation of a quantum field theory. In this case this involves understanding the relation between the quantized electron field and the electrons which are its quanta on the one hand, and the relation between the quantized and classical electromagnetic fields on the other. Without some account of the ontology of quantum fields one can give no description of either electromagnetism outside the solenoid or the passage of the electrons through the apparatus, still less a separable description. And one is therefore in no position to show that interactions between these two processes conform to *Local Action*.

Suppose one were to represent the passage of quasi-localized electrons through the apparatus by wave-packets formed by superposing positive-energy particle solutions to the Dirac equation, with non-negligible amplitudes only for momenta corresponding to trajectories through the top slit that go over the solenoid and trajectories through the bottom slit that go under the solenoid. Then *given a choice of gauge* one could think of electromagnetism's effect as that of locally altering the phase of each pair of the overall wave-packet so as to change the relative phases of the different elements of the superposition. But actually neither the electromagnetic potential nor the local phases are well-defined, since each is gauge-dependent. In fact, the interaction is between electromagnetism (represented by the Dirac phase factor) and the entire wave-packet. This changes the overall wave-packet's amplitude at each point by altering the phase difference around curves enclosing the solenoid. But there is no localized interaction between a quantized representative of electromagnetism (such as a quantized ω_h field) and the component of the electron wave-packet superposition with non-negligible amplitude at that point.

In the absence of an agreed interpretation of the ontology of quantum field theories we have no clear quantum field-theoretic account of the A–B effect. But we have good reason to believe that any account that is forthcoming will be nonseparable, and in that sense nonlocal, irrespective of whether or not it could be made to conform to *Local Action*.

7 Non-Abelian gauge theories

There has been much theoretical discussion of analogs to the A–B effect that occur in certain non-Abelian gauge theories. The prospects for giving a completely local account of such effects appear even more bleak than in the original A–B case. For the non-Abelian analog to the (electro)magnetic potential is a generalized gauge potential represented by a smooth one-form, defined not on some region of physical space(time) but rather on a principal fiber bundle over such a region. A fiber bundle representation of the (electro)magnetic potential is also possible, but in that case the bundle is always trivial (in the absence of magnetic monopoles), so the one-form defined on it can be simply projected down onto the base space. But in the general non-Abelian case, the fiber bundle on which the gauge potential one-form is defined is not trivial, and so it is not even possible to define a smooth potential one-form over a space(time) region appropriate to the generalized A–B effect, still less to decompose such a form uniquely into gauge-dependent and gauge-independent parts via an analog to the Hodge theorem. The point is that in the general non-Abelian case, even in the 'field-free' region the homogeneous non-Abelian field equations do not take the form $dG = 0$ for the gauge field G: hence there need not be any gauge potential one-form W defined on that region such that $G = dW$. Without such a one-form W defined on the region 'outside the solenoid', we can't apply the Hodge decomposition to get any gauge-invariant locally defined harmonic one-form generalizing ω_h of the original (magnetic) A–B effect.

8 Conclusion

The Aharonov–Bohm effect arises from a quantum mechanical interaction between electrons and (electro)magnetism in a region that would classically be regarded as free of (electro)magnetic fields. It was predicted by nonrelativistic quantum mechanics and has been confirmed by experiment. But on no interpretation does quantum mechanics give a completely local explanation of the effect. On some interpretations, the quantum mechanical explanation violates a principle of local action, while on others it violates a principle of separability. There is no completely local quantum field-theoretic explanation, despite the fact that both electrons and electromagnetism are there represented by local quantized fields that interact at each space-time point. Nor is there any local explanation of generalizations of the A–B effect to non-Abelian gauge theories. While the interpretation of quantum mechanics remains controversial, and even our best confirmed theories remain conjectural, it seems reasonable to conclude that it is the Aharonov–Bohm effect itself, and not merely our current theoretical accounts of it, that is nonlocal.

Acknowledgements

Much of this material was based upon work supported by the National Science Foundation under grant no. SBER94-22185.

References

Aharonov, Y. and Bohm, D. (1959), 'Significance of electromagnetic potentials in the quantum theory', *Physical Review* **115**: 485–91.
Bohm, D. and Hiley, B.J. (1993), *The Undivided Universe* (New York: Routledge).
Chambers, R.G. (1960), 'Shift of an electron interference pattern by encoded magnetic flux', *Physical Review Letters* **5**: 3–5.
Dürr, D., Goldstein, S. and Zanghi, N. (1992), 'Quantum equilibrium and the origin of absolute uncertainty', *Journal of Statistical Physics* **67**: 843–907.
Einstein, A. (1948), 'Quantum mechanics and reality', *Dialectica* **2**: 320–4. (This translation from the original German by Howard in Howard (1989), pp. 233–4.)
Healey, R. (1991), 'Holism and nonseparability', *Journal of Philosophy* **88**: 393–421.
Healey, R. (1992), 'Chasing quantum causes: How wild is the goose?', *Philosophical Topics* **20**: 181–204.
Healey, R. (1994), 'Nonseparability and causal explanation', *Studies in History and Philosophy of the Physical Sciences* **25**: 337–74.
Healey, R. (1997), 'Nonlocality and the Aharonov–Bohm effect', *Philosophy of Science* **64**: 18–41.
Holland, P.R. (1993), *The Quantum Theory of Motion* (Cambridge: Cambridge University Press).
Howard, D. (1985), 'Einstein on locality and separability', *Studies in History and Philosophy of Science* **16**: 171–201.
Howard, D. (1989), 'Holism, separability and the metaphysical implications of the Bell experiments', in Cushing, J. and McMullin, E., eds. *Philosophical Consequences of Quantum Theory: Reflections on Bell's Theorem* (Notre Dame, Indiana: University of Notre Dame Press).
Lewis, D. (1986), *On the Plurality of Worlds* (Oxford: Blackwell).
Redhead, M.L.G. (1987), *Incompleteness, Nonlocality and Realism* (Oxford: Clarendon Press).
Tonomura, A. *et al.* (1986), *Physical Review Letters* **56**: 792.
Wu, T.T. and Yang, C.N. (1975), 'Concept of nonintegrable phase factors and global formulation of gauge fields', *Physical Review* **D12**: 3845.

Session discussions

DeWitt: I get a little bit worried. Philosophers seem to want to use words in a very precise sense, and I don't believe you can here. You used the term intrinsic, and let me take you as an example, as a macroscopic object. It's impossible that you be intrinsic, or your properties be intrinsic, except in a phenomenological sense. Because you are there, what makes your wave packet continually collapse is the medium around you, the photons, the molecules in the air, you cannot escape that. You're not intrinsic, the whole universe is part of you.

Healey: Philosophers do try to use words like intrinsic as precisely as they can. A number of philosophers have got worried about the fact that they haven't got a good definition of this, and there're actually a large number of journal articles trying to address the question, what do we mean by intrinsic? Some people think that one can give a definition in clearer terms, others think it can't be defined, it has to be taken as a primitive term which does a lot of work in various philosophical theories. So this is something we've worried about. Now as for me, remember that intrinsic is a characteristic not of objects but of properties, so to say that I am intrinsic is confusing.

DeWitt: Your properties are not even intrinsic.

Healey: Now, I think physicists often use the word in a way that differs from the way philosophers use it. I think they think of intrinsic as being somehow unchanging or essential. Now I have the properties that I have, the intrinsic properties of my sense that I have, by virtue of all kinds of other things. Like I have an intrinsic property corresponding to my mass, which would have been different if I'd had different things for breakfast. And I might have a completely different mass from the mass I in fact have, but that doesn't mean that my mass is not an intrinsic property of me. It just means it's not an essential and unchanging property of me. So, that all these photons are changing my intrinsic properties, they're not implying that I have no intrinsic properties.

Jackiw: I want to make one last comment for mathematically minded Aharonov–Bohm aficionados, there's a new chapter in the story, and it's a field theoretic description in terms of a Chen–Simons interaction, which is completely local, the way a field theory is, and it might be of interest.

DeWitt: I wanted to make a couple of comments about history. When Feynman gave the talk at Warsaw, at which I was present, he said (you know Feynman), 'Ah, the gravity stuff, that's a bit complicated. If you don't mind, I'll do it with Yang–Mills; it suffers from the same disease.' So he actually presented it in terms of Yang–Mills. He pointed out at that time that he was only able to apply his arguments to one-loop order.

In '64, I discovered a way to get the rules for two loops, which in principle would allow you to get them for any number of loops. I said to myself, there must be an easier way. And in late '65 I discovered it. It didn't get published until '67. I showed that one has to drag in the inverse of the dam ghost determinant. The only way I knew of to express it as a path integral was to put the ghost integration inside the integrand, and do the ghost integral first and then the other. That's the way you would have to do it on a computer.

There are jokes nowadays that somebody has developed a computer chip to do Berezin integration. It would certainly be lovely for QCD people. It was essential to have the Berezin scheme in order to discover BRST invariance. That's why Tyutin discovered it early, being a Russian. People like Schwinger knew of the superdeterminant, and knew how to manipulate anticommuting numbers, but they never discovered Berezin integration, and that was crucial.

I just want to ask one question. I have never understood the term 'gauge fixing'. The term should be 'gauge breaking' because one doesn't stick to a particular gauge. The gauges get smeared out by a Gaussian integral. Gauge invariance is broken. What you start with is a local group and what you wind up with is only a global group and an anticommuting parameter. What's the reason for the terminology?

Jackiw: I can't answer your question about nomenclature, but I too wanted to make a kind of ghost busting statement. The operative reason for the ghosts in physical theory is Feynman rules. It is because you want to use Feynman rules that you get into this whole ghost story. You remember when Weinberg was speaking, he was talking about the S-matrix, and he said that there was something involving the interaction Hamiltonian. And that is truly what happens in quantum mechanics. It is the interaction Hamiltonian that determines the S-matrix. However, Feynman rules, first of all they are Lorentz covariant, but the Hamiltonian is not a scalar. Second of all they come from the Lagrangian. So what you would like to have in the formalism is the interaction Lagrangian rather than the interaction Hamiltonian. Now you remember Weinberg said that mostly they are the same, except sometimes there are complications, as with the Coulomb interactions. Well, mostly these are just small complications. In gauge theories, however, there are big complications. In order to go from the Hamiltonian to the Lagrangian, you have to put in some further nonlocal quantities, and if you want to localize them you introduce the ghosts. However, it's the desire to use Feynman rules and Lorentz covariant expressions that leads you into this channel of using ghosts.

Actually a precursor to all of this arises even if you're not doing a gauge theory. If you're doing a theory with a lot of velocity dependent forces, then the interaction Hamiltonian is not the same as the interaction Lagrangian. So as soon as you have velocity dependent forces, again these nonlocal terms arise, and again you have to do something about them. And indeed in my opinion the first people to have realized this were Lee and Yang who wrote a paper on the harmonic oscillator in ordinary quantum mechanics, a harmonic oscillator with high velocity forces which necessitated an addition of a nonlocal term. There is also a paper by Steve Weinberg and co-authors, of whom I am one, which does this for a nonlinear sigma model, again not a gauge theory, but with a lot of velocity dependent forces, which require a nonlocal modification to go from the Hamiltonian to the Lagrangian. This nonlocal modification can be left as a nonlocal modification, or it can be represented in terms of ghost propagation. Now, whether that makes the ghosts physical or not is for the future to decide. But that's my point of view about quantum ghosts.

Audience: I agree completely that in most physical theories like Yang–Mills you start without ghosts. You can at least formally write down the path integral and that has nothing to do with ghosts. It's just for certain calculations we find it useful at some point to introduce them. Although we might want to take them more seriously. The topological field theories are examples where you start with an action zero and the entire theory is given by the ghost. So at least you can use that as an example to show that maybe we should keep an open mind. This leads me into the question of trying to get the main point of what you, as a philosopher, might be telling the physicist. It seems to be to think more carefully about foundations, and was that an example of what you mean?

Ronald Anderson (Boston College): I think one thing that comes up from the story of ghosts is the issue about the meaning of the formalism. Now I'm not certain this is partly of issue for philosophers or partly an issue for the practising physicist. But how do you understand the formalism, what's the significance of the terms in the formalism? So there is an issue of interpreting the formalism, and also an issue related with what choices are built into the formalism that influence that decision.

Jackiw: I think from a physicist's point of view the story of ghosts is very much like a replay, or may become like a replay, of the story of the vector potential, which also began as an attempt to localize nonlocal structures in electrodynamics, by introducing the vector potential. I was taught that it is unphysical and one should never think that any physics resides in the vector potential; I was taught this by classical physicists, teaching classical electrodynamics. Then slowly in quantum mechanics the vector potential acquired significance, first in the Aharonov–Bohm effect, and then it actually pops out in the Berry phase. Now in the Berry phase you have a vector potential without electrodynamics, it's just there. And finally in gauge theories, non-Abelian gauge theories I mean, it becomes essential even in the classical formulation. All those other stories, Berry phase, Aharanov–Bohm effect, all those stories are quantum stories. But in Yang–Mills theory, even on the classical level, you can't write down the equations, you can't state the equations, without introducing the vector potential. So it's acquired a life of its own even though it still has redundant aspects to it. So maybe ghosts will also acquire a life of their own and there will be a ghost effect.

Stachel: Ron asked what I thought about this comment by Gockler and Schucker on how Einstein's formulation of general relativity in a holonomic frame masked the fact that it's a gauge theory. Holonomic frame just means coordinate system, so I don't think that this is the crucial point. Everything meaningful that can be said, can be said in a coordinate system, even though it's obviously much better to use the coordinate-free formulation of differential geometry, which was developed many years after Einstein formulated general relativity.

The crucial point is that a basic concept was missing when Einstein formulated general relativity, namely the concept of parallel transport. Without the concepts of parallel transport and affine connection, how could one formulate the theory as a gauge theory? If you have that concept, you can formulate Newton's theory of gravitation as a gauge theory; without it you can't formulate Einstein's theory in that way. It's no accident that Weyl, the person who introduced the concept of generalized parallel transport, was the first person to introduce the concept of gauge transformations.

Rohrlich: Roman, maybe you want to explain why it is necessary to go from the Hamiltonian to the Lagrangian, why you can't stay with the Hamiltonian formulation.

Jackiw: I'm happy to stay with the Hamiltonian. Ugly rules, it's more complicated.

Rohrlich: But then you avoid the problem of all the ghosts.

Jackiw: You can stay with the Hamiltonian, but then you can't use Feynman rules. You use essentially the old fashioned perturbation rules. You don't have the elegance and the simplicity and the transparency of the Feynman rules. But it is also true that the Feynman rules are not so elegant, nor are they so transparent when there is high derivative coupling.

Rohrlich: Exactly, so you don't save much by going, you pay.

Jackiw: You do, for conceptual purposes it's better to go and use the ghosts than to try to work with the physical degrees of freedom exclusively.

Rohrlich: Well, my point is essentially this, since it is not necessary to do that, you could in principle also work with the Hamiltonian, so it indicates that the whole story with ghosts is really a matter of formalism rather than a matter of basic fundamental significance.

Jackiw: Well, that's what they said of the vector potential too.

Rohrlich: It's true.

Jackiw: Now it's moved on to a more central role in science, and so maybe that is what will happen to ghosts as well.

Stachel: Roman, since you brought up Aharonov–Bohm now, I'd like to point out that there's a classical gravitational analog of the Aharonov–Bohm effect, which was actually first noted – without really understanding it – by Marder. He pointed out that the external gravitational field of an infinite rotating massive cylinder is locally static. I showed that this is a sign that one should distinguish between globally static gravitational fields and fields that are locally static but globally stationary. Indeed, if you use the Kaluza–Klein approach you can unify the gravitational and the electromagnetic Aharonov–Bohm effects, view them both as just two sides of the same coin. And a *classical* test is possible in the gravitational one, using a classical light wave divided and passed around both sides of the rotating cylinder. You can tell whether the cylinder is rotating or not by measuring the phase shift between the two waves when reunited after passing around the cylinder.

Part Nine. The ontology of particles or fields

22. The ineliminable classical face of quantum field theory[1]

PAUL TELLER

This paper has two independent parts. Part 2 examines the interpretive ramifications of the fact that quantum field theory employs a classical, point valued, space-time parameter. But the editor, eager for the present volume to provide information about the present state of the art as well as new research, has very kindly encouraged me to do a little sketch of the prior work I have done on interpreting quantum field theory. Doing so, in Part 1, will also enable readers to see in some detail how the examination in Part 2 fits into the larger project. Part 1 can be skipped by those not in need of the survey material.

Part 1

An Interpretive Introduction to Quantum Field Theory (Teller 1995; hereafter IIQFT)[2] supersedes most of my prior work on quantum field theory. The gossip mill has described this book as a popularization of the most elementary parts of Bjorken and Drell (1965), which into the 1980s was the most widely used quantum field theory text. As with any good caricature, there is a great deal of truth in this comparison. Like Bjorken and Drell, who published in 1965, IIQFT presents the theory largely as it existed in the 1950s. But in order to see aspects of structure and interpretation more clearly, IIQFT presents the theory stripped of all the details needed for application. IIQFT also does not treat contemporary methods, such as the functional approach, and important recent developments, especially gauge theories and the renormalization group. Nonetheless it is hoped that by laying out the structure of the theory's original form in the 1950s, much of which survives in contemporary versions, and by developing a range of ways of thinking about that theory physically, one does essential ground work for a thorough understanding of what we have today.[3]

---

[1] The subject of this paper was inspired by remarks in Schwinger (1966), called to my attention by Gordon Fleming. This paper has benefitted in many ways from detailed contents by Gordon Fleming, Andrew Wayne, and Martin Jones, as well as many others.

[2] I still regard the metaphorical 'harmonic oscillator' interpretation of Teller (1990) as a viable way of thinking about quantum field theory, but recommend the field theoretic treatment in IIQFT as better capturing the valid aspects of the prior treatment. Teller (1990) does have a few bits of analysis, such as how to think about cloud chamber tracks field-theoretically, which did not resurface in IIQFT. Teller (1989) is reprinted almost without change as chapter 7 of IIQFT, and Teller (1988) does add some considerations about renormalization not discussed in the book.

[3] An objective in writing IIQFT was also to make the theory more accessible than it is from the physics texts, while not misrepresenting the theory.

Why do we use the term 'quantum *field* theory'? A good fraction of the work done in IIQFT aims to clarify the appropriateness, accurate development, and limitations of the application of the epithet 'field', as well as examination of alternatives. While called a 'field theory', quantum field theory (QFT) is also taken to be our most basic theory of 'elementary particles'. Sometimes the 'particles' are described as 'quantized field excitations', or alternatively, the 'particle aspect' is associated with number representations and the 'field aspect' with complementary representations.

Both of these last two suggestions prove to be problematic. Instead, IIQFT begins by developing the little studied idea of dropping from our accustomed ways of conceptualizing 'particles' that particles are 'things' to which identity applies. Instead, I think of the subject matter of QFT as comprising 'quanta', stuff which can be 'measured' only in discrete units but which cannot be 'counted', insofar as counting suggests that order matters, and cannot be switched or exchanged in thought or in description. For quanta, as opposed to particles, the puzzles about Bose statistics never arise. In addition IIQFT presents a methodological argument for the quantal construal, turning on the fact that description in terms of 'particles' results in uninstantiated or uninterpreted theoretical structure.[4] Starting with the conventional quantum mechanical description of quantum systems, IIQFT shows how one can easily build up the usual Fock space representation of QFT, including both multiple quanta states and states with indefinite numbers of quanta.[5]

What then is the connection with anything that might appropriately be called a 'field theory'? We re-express the theory as one with the form of a field theory by assuming that the individual quanta in the Fock space description are quanta with definite momenta and then re-express in the position basis, using the transformation also borrowed from the one quantum theory. IIQFT traces this redescriptive process, reversing the subject's history, and showing the interplay of first, second and field quantization. There is a pedagogically novel treatment of antiquanta as 'negative energy states' and a brief survey of the puzzles of exact localization in any relativistic treatment.[6]

Some readers have jumped to the conclusion that, because IIQFT starts with the quantal description and then re-expresses in a field theoretic formalism, I think the former is 'right' and the latter 'wrong'. Not at all. IIQFT indicates (e.g., p. 61, where the point should have been made more forcefully) that both perspectives provide useful ways of thinking about the theory physically. Indeed, full understanding of QFT requires inclusion of both perspectives.

Field and second quantization, as well as transformation from the quantal representation, all provide something with the *form* of a field theory. But in what way is QFT a genuine field theory? The question gets bite from the circumstances that QFT is often described as presenting an 'operator valued field'. But operators are much more like (representations of) determinables or physical quantities than the

---

[4] Further technical work connected with this argument is presented in Redhead and Teller (1991). IIQFT attempts to present quanta by stripping our ordinary conception of quanta of the philosophically traditional idea of 'haeccities' or 'primitive thisness', thought of as that which confers identity on an object, or that in which an object's identity consists. Since publication of IIQFT I have come to see that the point is put more clearly by simply denying the applicability of strict identity, instead of bringing in 'haeccities' or 'primitive thisness', both in danger of misleading and irrelevant reification. This further development in thinking about quanta is presented, in two stages, in Teller, To Appear (*a* and *b*).

[5] This presentation closely follows Merzbacher (1970).

[6] This material briefly introduces Fleming's thinking in terms of 'hyperplane dependence'.

values of these quantities. IIQFT closely examines the 'operator valued field' conception and argues that the field theoretic content is much less misleadingly put in terms of field configuration descriptions of specific quantum states. Along the way IIQFT suggests resolutions of puzzles surrounding vacuum expectation values and Rindler quanta.[7]

Included also are an elementary description of how interactions are treated with the S-matrix expansion, a brief description of the puzzles surrounding Haag's theorem, and a careful discussion of the interpretation of Feynman diagrams. Here the main point is that individual Feynman diagrams represent elements of a superposition, what may be called 'analytic parts', and not ordinary, or 'mereological' parts, such as the parts of a chair. This treatment dispels the impulse to reify so called 'virtual quanta'.

Finally, the old renormalization procedure from the 1950s is laid bare and demystified. By stripping away the technicalities, IIQFT makes clear (what has been generally clear for a long time to most practitioners) that the old renormalization procedure involves no mathematical sleight of hand, but constitutes a consistent procedure for incorporating experimentally observed values of parameters to cover for certain bits of theoretical ignorance. It should be stressed that this last chapter does not at all treat the more general recent insights to be gained from the renormalization group.

IIQFT also leaves a great many other problems untouched.[8] The work below, on the interpretive ramifications of QFT's appeal to a classical space-time parameter, enables me along the way to tie up just one loose end. In IIQFT I presented field theories from a substantivalist's point of view, which takes space-time points to be 'things' to which values of field quantities can be attributed. But in my (1991) article I reject substantivalism in favor of a form of relationalism according to which the space-time facts are exhausted by (actual and possible) space-time relations between physical bodies. By the end of Part 2 we will see how I propose to resolve this conflict.

Part 2

1 Introduction

Quantum field theory is formulated in terms of a space-time parameter. It is a too little noticed fact that this parameter refers to a *classical* conception, so that the theory appears to have a classical element built in right from the start. I plan here to examine this appearance and its repercussions. Conclusions vary with the interpretation taken for the nature of space-time, and I will argue that on the most plausible interpretation quantum field theory cannot be viewed as a conceptually fundamental theory of physical matter.

I had better start with a word on what I intend by the epithet 'classical'. While this is not the occasion for a thorough analysis, I shall assume that a classical system of physical objects or fields is to be characterized in terms of simultaneous exact values for conjugate quantities, such as position and momentum in the familiar phase space representation for particles with exact space-time trajectories. When it

---

[7] An independent exposition of the interplay of field and quantal aspects of QFT appears in Teller (1996).

[8] Some of these are briefly described in tongue in cheek 'problem' sections. This is a resource for graduate students looking for thesis topics, but beware: some are much too hard!

comes to space-time and the space-time parameter, 'classical' will just mean 'point valued', with no thought of further analysis. Though alternatives are occasionally proposed, we hardly have any viable alternative to describing space-time as point valued.[9]

I had better also say right away what I intend by a 'conceptually fundamental theory of physical matter'. Quantum theories are intended as accounts of physical matter, in the extended sense of physical objects and physical fields. No one should see a problem if such a theory presupposes an independent conception of space-time, however 'classical' that conception might be. Quantum theories are about physical matter, and if space-time is an entirely different sort of entity there is no reason why quantum theories should not help themselves to facts about this different sort of entity. If space-time and physical matter are two entirely different sorts of things, there need be no strain if, within the same conceptual framework, the first is described in terms of exact values and the second is not so described.

But some might be troubled if quantum theories, as accounts of physical objects and fields characterized as without exact values for conjugate quantities, were to presuppose a contrary conception of such things *with* exact values. This is the circumstance which I want to summarize with the slogan, 'quantum field theory is not a conceptually fundamental theory of physical matter'. I do not myself find this circumstance problematic. But, if it occurs, it is a conclusion about the conceptual structure of our theories worth noting, as is the fact that this conclusion depends on the presupposed conception of space-time.

2 Quantal vs. classical space-time properties

Our eye will be sharpened for the possibility of repercussions of the classical status of the space-time parameter if we pay attention to the very non-classical characterization of location given by all quantum theories. On the description given by non-relativistic quantum theories, no actual physical objects have exact locations. Precise location would require complete indefiniteness in momentum, which in turn would require an unbounded amount of energy. While we use delta function descriptions in manipulating the theory, it is understood that these are idealizations which, strictly speaking, do not correspond to states recognized by the theory. (Comparable comments apply for exact momenta.)[10] Relativistic formulations of quantum theories involve additional barriers to description of exact location. The states arising from application of the usual field raising operators are parameterized by the space-time variable, but these states are not orthonormal and so not interpretable as exact location states. One can define exact location states, the so called Newton–Wigner states, but they give the desired description only in one frame of reference and have very unpleasant transformation properties. (See IIQFT pp. 85–91 for a summary of the relevant facts.) In sum, all quantum theories describe physical objects as having only imprecise, or spread out locations in space-time. (Teller (1979) and (1984, pp. 387–93) discuss the idea of spread out quantal inexact locations.)

---

[9] Of course, one can start with space-time volumes as primitive, and recapture the exact valued points as limits of nested spheres. For present purposes we may include such an approach to space-time as point valued by extension.

[10] See Teller (1979) for a more detailed exposition of these arguments.

On the other hand, the space-time parameter is itself point valued. Although quantum theories never describe actual objects as having precise trajectories, the space-time parameter can be used to describe exact trajectories in space-time. And it is entirely in terms of the space-time parameter that the constraints of relativistic Lorentz invariance are imposed on the theory. (Indeed, alternative imposition of Lorentz invariance, perhaps statistically, would require radical restructuring of the theoretical framework, if possible at all.) But Lorentz invariance is just uniqueness of formulation over alternative inertial reference frames, and reference frames are traditionally understood in terms of idealized, *non-quantal* rods and clocks. On the face of it, the whole relativistic space-time scaffolding on which quantum field theory is erected appears to be understood in terms of a decidedly classical conception of physical objects with exact space-time trajectories.

In addition, we must note that the quantal characterization of location is parasitic on the classical. The quantal notion is of indefinite or spread out location, where the spread is characterized in terms of the presupposed point valued classical idea of space-time. (Again, analogous comments go for momentum.) For example in conventional quantum mechanics, spread is given by the wave function's probability amplitude at various spatial points, classically conceived, and in quantum field theory, 'particle density' – really the expectation for particle detection – is again parameterized by the classical space-time variable.

From Bohr's point of view this is all exactly as it should be. Bohr insisted that quantum formalisms only function to give statistical predictions of experimental outcomes, classically described (see Teller (1981), pp. 226–7). Applying Bohr's dictum to quantum field theory, this means that location observations are observations of the space-time quantity, conceived classically. When descriptions are read from the point of view of a literal and realist construal of the quantum field theoretic formalism, actual objects never have exact space-time locations, so that the classical point values are idealizations. But, according to Bohr, the epistemological facts of life dictate that, however idealized these exact values might be from the quantal point of view, from *our* point of view they are the only ones which we can legitimately call 'real'.[11] Bohr could only take satisfaction in seeing the marks of his interpretive attitude built right into the formalism.

I don't see that one has to share Bohr's attitude. One can take states that involve physically realizable, slightly 'spread out' positions (and momenta) as the ones that describe actual measurement outcomes. These can be used in the overlap functions to calculate experimental distributions. Since we get only negligible differences in both probability and physical spread when we calculate in this way as opposed to using the delta function idealizations, the practice of using the latter can be seen as a harmless calculational idealization rather than as a conceptual imperative.[12]

Bohr's insistence on classical concepts provides one way of understanding the dual quantal/classical role of location in quantum field theory: only classical location is to be accepted as real. Quantal locations function merely as a formal calculational tool for predicting location (and momentum) outcomes, classically conceived. But those

---

[11] This loose description of Bohr's attitude passes over his rejection of a literal construal as 'meaningless', appropriately I would maintain, since apart from unexplicated references to 'the need for unambiguous communication' he never provides us with a theory of meaning.

[12] Since, in relativistic theories, position delta functions are not 'position eigenstates', their use actually idealizes doubly in passing over the complexities of a proper position operator and its eigenstates. But this further distortion also makes no difference in most practical applications.

inclined to take a more realistic stance towards quantum field theory need to inquire further how to understand the role of the classical space-time parameter in the theory. To do so, one must take a stand on how space-time, classically conceived, should be understood.

3 Substantivalism

Most practitioners seem to hold a substantivalist view, according to which there are individual space-time points, vaguely thought of as 'things', at which ordinary physical objects can be located and which can have various field properties. On this view quantum field theory is a field theory in the sense that the theory ascribes properties, such as probability amplitudes and expectation values, to individual space-time points or volumes.[13]

Substantivalism seems easily to coordinate the dual classical/quantal role of location. According to substantivalism, space-time is an objectively existing thing, the collection of classical space-time points. Description of this object is incorporated into quantum field theory which characterizes various properties which hold of space-time points, including indefinite or spread out locations of objects as a property, not of individual points, but of collections of points.[14] On this conception we have the incorporation of one conceptually fundamental theory in another.

One might worry whether this attitude is compromised when we take into account application of the theory. Application must always be in terms of some concrete reference frame, and a reference frame must be characterized in terms of physical bodies. And if the reference frame is classical, corresponding to the classical space-time parameter, then, it might seem, the characterizing physical bodies must also be described as classical; and the theory would seem to presuppose not only classically conceived space-time but classically conceived physical bodies as well.

I go up and down about this worry, but, finally, suspect that it is misguided. In principle, one ought to be able to describe the bodies which serve to determine a reference frame as large but quantum theoretically described physical objects. As such they are subject to indefiniteness of position and momentum, and thus can describe a reference frame only approximately. But, from a substantivalist point of view, this approximate description is of an exact, objectively existing entity – space-time – and the approximation is plenty good enough for practical applications.

However, this kind of worry becomes much more acute for alternatives to substantivalism.

4 Relationalism

I reject substantivalism as an adequate account of space-time. Let's recall the Leibnizian indiscernibility argument. For a reductio, assume a substantivalist account of space-time. Then one ought to have alternative possible cases described as exactly like the real world, but with everything uniformly 'moved over' in space-time. The

---

[13] Traditionally quantum field theory is said to describe an 'operator valued field'. Teller (1995), p. 93 ff., explains why this is a misleading description.

[14] In a wide range of physically realizable cases, well behaved and reasonably well localized states will have non-zero amplitude throughout the whole of space, since no matter how well concentrated such states are in their spatial distribution they will have infinitely extended, though exceedingly small, tails throughout space.

Paul Teller

fact that such alternative descriptions do not seem like descriptions of real alternatives gives one pause about the substantivalist assumption. A new souped-up version
of this argument – the Einstein–Earman–Norton hole argument – works similarly,
but takes advantage of the contingency of the space-time metric in a general relativistic setting. The hole argument describes alternative arrangements in space-time by
moving everything – including the metric – around, but only in a localized volume.
To give the intuitive idea of this argument, consider an alternative world in which
all the masses doubled overnight (with correlative changes in other quantities in
which mass figures as a dimension). Such a world would be indistinguishable from
our own. The hole argument works similarly, reconstruing a passive coordinate transformation restricted to a localized space-time volume as an active transformation.[15]

The hole argument persuades me to reject substantivalism in favor of what I call
liberalized relationalism.[16] On this view space-time facts are all facts about space-
time relations between actual and possible bodies, including not only the actually
exemplified space-time relations but possible ones as well.[17] A quick way to see the
plausibility of liberalized relationalism is to consider that any reference to where
things are or might be requires use of a coordinate system, which in turn characterizes
location in terms of actual or possible space-time relations to the bodies used in
setting up the coordinate system.

Liberalized relationalism requires a recharacterization of what we mean by a field
theory, classical or quantal. Instead of attributing values of the field quantity to
independently existing space-time points, one says that values of the field quantity[18]
simply occur with the requisite space-time relations to the bodies in terms of which
some coordinate system has been set up.

This recharacterization puts quantum field theory on a very different footing than it
had in a substantivalist framework. If the space-time framework of liberalized relationalism is taken to be classical, classical bodies are tacitly being presupposed.
Saying, for example, that a certain expectation value holds at a given space-time
point must be reinterpreted as saying that the expectation value occurs with specific
space-time relations to various actually existing bodies. If the space-time framework

---

[15] The hole argument appeals to the conceptual apparatus of the general theory of relativity, and one might
wonder about the appropriateness of applying such an argument to a question about a framework – that
of quantum field theory – which involves only flat Minkowski space-time. The tacit logic of the application goes like this. We appeal to the conceptual apparatus of the general theory of relativity to reach a
conclusion about the nature of space-time. Then, armed with our best account of the nature of space-
time, we want to see how this best account is to be integrated into a distinct conceptual framework
which involves only the special case of flat space-time.

[16] The argument is not conclusive, as there are several ways of trying to reconcile substantivalism with the
hole argument. See Teller (1991) for detailed exposition of the hole argument, a summary of possible
responses, and a more detailed characterization of liberalized relationalism. Teller (To Appear, *b*)
explains my reservations with Maudlin's and Butterfield's substantivalist responses to the hole
argument.

[17] This statement suffices for a statement of liberalized relationalism concerning spatial facts, as discussed
in the Leibniz–Clark correspondence, where temporal facts are taken for granted. But in a general
relativistic setting, and if bodies are taken to persist over time, the actual and possible relations really
need to be taken to hold between bodies as characterized by varying complexes of properties which
do the work of clocks, uniquely individuating bodies at proper (relative) times. This will be understood
in the following references to bodies serving to set a space-time frame. I am grateful to Martin Jones for
pointing out the need for this correction to the statement of liberalized relationalism in my 1991 paper.

[18] A more general discussion would examine when to understand these values as occurrent properties
(though not obtaining *of* anything!) and when to understand them in terms of dispositions for the
behavior of test particles or of 'measuring instruments'. Probably further options should also be
considered.

is, literally, taken to be classical, so too must the space-time relations and the pre-supposed bodies in terms of which the framework is set up. If the space-time relations and framework setting bodies are not both taken to be classical, it is hard to see how an exact valued space-time scaffolding will result.

Why the difference in dependence on presupposed classical notions when one assumes a relationalist as opposed to a substantivalist understanding of space-time? On the substantivalist conception, classical space-time is assumed as a freestanding entity, not itself dependent on bodies. Quantum field theory then puts itself forward as a theory of matter. Quantum field theory assumes the prior classical conception of space-time, but since the prior theory pretends to be a theory of independently under-stood space-time, not of matter, there is no presumption of classically conceived bodies involved in presenting quantum field theory. As I explained, a quantum theoretic account of bodies looks to be adequate for characterizing the way we approximately identify the needed space-time referents in applying the theory.

But from the point of view of liberalized relationalism, there is no prior conception of space-time independent of physical matter. All space-time facts are facts about actual or potential *space-time relations* between *physical bodies*. The presupposed space-time relations are classical, exact valued; so the presupposed physical bodies between which the relations do or would have to be taken, in this respect, are classical too.

Here is a suggestion for circumventing the relationalists' apparent dilemma.[19] In the context of a classical relationalist theory one could take a sufficiently variegated classical field to play the role of the physical bodies which serve to set the space-time frame. But if a classical field can so serve, could a liberalized relationalist avoid dependence on classical bodies-or-fields by using the *quantum* field to serve to set the frame? The idea would be that, typically, the quantum field includes patterns of specific structured amplitudes, which could then be characterized as bearing various space-time relations to one another; and if this pattern is sufficiently rich, it could provide the needed reference frame.

I don't think that this suggestion will work. One must ask: What will be the intended content of asserting that two 'bits' of structured amplitude bear some space-time relation to each other? What, for example, would be the difference between characterizing a first 'bit' as three, as opposed to four, feet from a second 'bit'? We certainly don't want to impose a standard of operational definition. But such attribu-tions of relative position will have to be connectible with measurement procedures for them to have any useful application. The straightforward way of doing so would be to appeal to amplitudes for joint manifestation of physical objects – or measurement outcomes – at specified relative spatial, or space-time locations. And now we must ask, what kind of physical objects or measurement outcomes? If these are construed in terms of quantum mechanically described physical objects, we do not look to be any farther ahead in the project of getting an interpretation of the point valued space-time parameter. And if the presupposed physical objects are thought of classi-cally, the point at issue has already been conceded.

I have not shown that there is no less straightforward way of appealing to space-time frame setting complexes of quantum theoretical amplitudes. But the challenge is to find a way to make this work without reverting to an entirely different theory.

---

[19] Ingeniously suggested by Mathias Frisch.

Liberalized relationalists might also attempt to circumvent the presupposition of classical bodies with an idealization strategy. Perhaps one might say that the presupposed network of bodies that identify the reference frame, as well as the space-time relations between these bodies, could, in principle, be described in terms of quantum theoretical analogs, with the classical description functioning as a practical idealization only.

I can offer no conclusive reason why this could not be done. But it hardly seems straightforward. The imprecision of location of quantum objects is stated in terms of the exact valued space-time parameter. Thus any straightforward statement of the idealization in terms of the limiting case of indefinitely heavy bodies would already require presupposition of the exact space-time framework for its very statement.

How about the relation of relative space-time location needed on a liberalized relationalist account? Perhaps, one might suggest, at least this could be assumed as an exact valued primitive, much as substantivalists assume space-time points as a primitive. But relative space-time location is a relation. What is it a relation between? If substantivalism and an assumption of classical bodies are both to be avoided, the relata will have to be quantal objects, characterized by imprecise location. But the only way we know of characterizing imprecise location is in terms of a distribution over possible precise locations, which was just the conception to be analyzed.

None of this proves that the idealization strategy can't be carried out. But the difficulties suggest that, if it could be carried out at all, the result would provide such a different way of describing quantum phenomena that it would fairly count as a new theory.

Short of possible radical innovations which would plausibly count as new theory, relationalists seem unable to view any quantum field theory as a conceptually fundamental theory of physical matter. As things stand, relationalists must take any quantum field theory as a model which very accurately simulates the behavior of matter in the small, but which is conceptually parasitic on a prior, classical understanding of the matter of ordinary experience.

Substantivalists may then argue from the premise that quantum field theory is obviously a conceptually fundamental theory of physical matter to the conclusion that we have in these considerations just one more nail in the relationalists' coffin. I am very happy to bite the bullet and reject the premise: I believe that there are no fundamental theories, and that science only presents us with models, each of which has a limited domain of application and accuracy. It is interesting that in this case the accurate model of matter in the small must appeal conceptually to a prior understanding of matter in the large. This fact provides partial vindication for Bohr's insistence that we must start with classical concepts.[20] While I see no reason to believe this claim in its full generality, in this case it appears to be built into the structure of our theories.

[20] As far as I know, the closest Bohr came to offering an explicit argument for the primacy of classical concepts was to cite the need for unambiguous communication as a requirement for objective knowledge. The idea appears to have been that without a classical framework within which all applicable quantities could be assumed to have a value, there would be indeterminateness in what quantities were 'well defined', and indeed in what object was in question. We have in the present considerations a different approach to the issue, which Bohr appears never to have mentioned. In particular I find no trace of it in Bohr and Rosenfeld (1933), which appears to follow physicists' usual presumption of a substantivalist's understanding of space-time.

As mentioned in Part 1, the presentation in IIQFT proceeds entirely from the point of view of a substantivalist understanding of space-time, and readers may fairly wonder how I square this with my commitment in Teller (1991) to liberalized relationalism. All the pieces have been prepared for a simple resolution, at which the last paragraph but one hints. Quantum field theory is a (collection of) model(s) which appeal(s) to the space-time frame setting concepts of classical models. Although classical and quantal models would provide inconsistent descriptions if simply conjoined, no outright inconsistency occurs in the present application. In the quantum field theoretic models (tacit) appeal is made to classical bodies *only* in their space-time frame characterizing role, and not as objects 'in' the model which could interact with quantal objects.

Once a space-time framework has been presupposed, characterizable in terms of coordinate systems, one can harmlessly use the substantivalists' way of speaking.

References

Bjorken, J.D., and Drell, S.D. (1965). *Relativistic Quantum Fields*. New York: McGraw-Hill.

Bohr, Niels, and Rosenfeld, Leon (1933). 'On the question of the measurability of electromagnetic field quantities,' originally published in *Mat.-fys. Medd. Dan. Vid. Selsk.* **12**, no. 8. Translation by Aaga Peterson (1979) in *Selected Papers of Leon Rosenfeld*, R.S. Cohen and J. Stachel, eds. Dordrecht, pp. 357–400; reprinted (1983) in *Quantum Theory and Measurement*, J.A. Wheeler and W.H. Zurek, eds. Princeton: Princeton University Press, pp. 479–522.

Merzbacher, Eugen (1970). *Quantum Mechanics, Second Edn.* New York: John Wiley and Sons.

Redhead, M.L.G., and Teller, Paul (1991). 'Particles, particle labels, and quanta: the toll of unacknowledged metaphysics.' *Foundations of Physics* **18**: 43–62.

Schwinger, Julian (1966). 'Relativistic quantum field theory.' *Physics Today* **19**(6): 27–43.

Teller, Paul (1979). 'Quantum mechanics and the nature of continuous physical quantities.' *Journal of Philosophy* **76**: 345–62.

Teller, Paul (1981). 'The projection postulate and Bohr's interpretation of quantum mechanics.' *PSA 1980: Proceedings of the 1980 Philosophy of Science Association Meetings, Vol. II*, R. Giere and P. Asquith, eds. East Lansing: The Philosophy of Science Association.

Teller, Paul (1984). 'The projection postulate: a new perspective.' *Philosophy of Science* **51**: 369–95.

Teller, Paul (1988). 'Three problems of renormalization', in *Philosophical Foundations of Quantum Field Theory*, H.R. Brown and R. Harre, eds. Oxford: Clarendon Press.

Teller, Paul (1989). 'Infinite renormalization.' *Philosophy of Science* **56**: 238–57.

Teller, Paul (1990). 'A prolegomenon to the proper interpretation of quantum field theory.' *Philosophy of Science* **57**: 595–618.

Teller, Paul (1991). 'Substance, relations, and arguments about the nature of space-time.' *The Philosophical Review* **100**: 363–97.

Teller, Paul (1995). *An Interpretive Introduction to Quantum Field Theory*. Princeton: Princeton University Press.

Teller, Paul (1996). 'Wave and particle concepts in quantum field theory,' in *Perspectives on Quantum Reality*, R. Clifton, ed. Dordrecht: Kluwer.

Teller, Paul (To Appear, *a*). 'Haeccities and quantum mechanics,' to appear in *Physical Objects. Identity Individuation, and Constitution of Objects in Physics*, Elena Castellani, ed. Princeton: Princeton University Press.

Teller, Paul (To Appear, *b*). 'The ins and outs of counterfactual switching.'

23. The logic of quanta*

STEVEN FRENCH[†] AND DÉCIO KRAUSE[‡]

Are there quantum objects? What are they if they are neither given nor resemble anything familiar? To answer these questions we have to abstract from the substantive features of familiar things, delineate the pure logical forms by which we acknowledge objects and show how the forms are fulfilled in quantum theories. We have to explicate, in general terms and without resorting to the given, what we mean by objects. The clear criteria will enable us to affirm the objectivity of quantum theories.

Auyang 1995, p. 5

1 Introduction

Cantor, famously, defined a set as '... collections into a whole of definite, *distinct* objects of our intuition or of our thought' (Cantor 1955, p. 85, our emphasis). On this basis the standard formulations of set theory, and consequently much of contemporary mathematics, are erected. Reflecting upon this definition, and the underlying ontological presuppositions, the question immediately arises, how are we to treat, mathematically, collections of objects which are not distinct individuals? This question becomes particularly acute in the quantum mechanical context, of course. As Yu. I. Manin noted at the 1974 American Mathematical Society Congress on the Hilbert Problems,

We should consider possibilities of developing a totally new language to speak about infinity [that is, axioms for set theory]. Classical critics of Cantor (Brouwer *et al.*) argue that, say, the general choice is an illicit extrapolation of the finite case. (...) I would like to point out that it is rather an extrapolation of commonplace physics where we can distinguish things, count them, put them in some order, etc. New quantum physics has shown us models with quite different behavior. Even *sets* of photons in a looking-glass box, or of electrons in a nickel piece are much less Cantorian than the *sets* of grains of sand.

(Manin 1976)

As is by now well known, consideration of the Indistinguishability Postulate which lies at the heart of quantum statistics suggests two metaphysical possibilities: either

---

* Earlier versions of this paper were presented at the Senior Seminar of the Department of Philosophy, University of Leeds and the workshop on *The Ontology of Particles and Fields*, Boston Colloquium for the Philosophy of Science, Boston University. We are grateful to Tian Yu Cao for inviting us to the latter Conference and to Richard Healey, James Ladyman, George Mackey, Anna Maidens, Abner Shimony, Peter Simons and Paul Teller for comments.
† E-mail: s.r.d.french@leeds.ac.uk
‡ Visiting Scholar in the Department of Philosophy at the University of Leeds, supported by CNPq (Brazil). Permanent address: Department of Mathematics, Federal University of Paraná, 81531-990, Curitiba, PR, Brazil. E-mail: dkrause@mat.ufpr.br

quantum 'particles', for want of a more neutral term, can be regarded as individuals subject to accessibility constraints on the sets of states they can occupy, or, alternatively, they can be regarded as non-individuals in some sense (French and Redhead 1988; French 1989; van Fraassen 1991). Surprisingly, perhaps, *both* options suggest revisions of the classical apparatus of set theory.

In the first case, quantum particles are still regarded as distinct individuals, so this would suggest no revision is necessary. However, there is the issue of entangled states and superpositions in general to consider. One way of viewing the situation in which we have particles, so regarded as individuals, in an entangled state is to describe the 'entanglement' in terms of 'non-supervenient' relations (Teller 1986; 1989); that is, relations that are neither determined by nor dependent on the monadic properties of the relata, in this case the particles (French 1989a). This leads to an interesting kind of ontic 'vagueness' in which we have individuals but we cannot in principle say which one is which (Lowe 1994; French and Krause 1995; French, Krause and Maidens, preprint). It is in this sense that quantum particles are 'strongly' indistinguishable in a sense which goes beyond the classical notion of merely having all intrinsic or state independent properties in common. Classical set theories like ZFC are inadequate for capturing collections of strongly indistinguishable particles which are 'veiled' by these non-supervenient relations in the above sense. One suggestion is to introduce Dalla Chiara and Toraldo di Francia's notion of a 'quaset', which is a collection of individual, strongly indistinguishable objects (1993, 1995). The elements of a quaset may still be regarded as individuals since the equality relation continues to hold among them, unlike the case of quasi-sets to be considered below. Significantly, because of this non-classical indistinguishability, such 'sets' have cardinality but may not have ordinality; that is, we can give the number of objects in such a 'set' but we cannot order them. We shall return to this distinction shortly, but for now we leave the introduction of quasets into this particular metaphysical context as merely a suggestion.

The more well-known option is to regard the quantum 'particles' as non-individuals in some sense, a proposal which was made by Born (1943, pp. 27–29), Weyl (1949, Appendix B), Schrödinger (1957) and Post (1963). The idea of entities which are not individuals is of course metaphysically rather mysterious, particularly given the inclination to regard such entities as particles (an inclination fostered by elements of experimental practice, for sure), and we are inclined to press the point that if such a view is to be anything more than the object of a lot of philosophical hand waving, the 'sense' of the above claim needs to be explicated. Of course, given the basic metaphysical level at which we are operating here it is no surprise that such explication proves to be difficult.

One way of approaching the issue is to go back to the distinction between distinguishability and individuality, a distinction which greatly exercised Scholastic minds for example. Thus Suarez argued that whereas the notion of distinguishability involves more than one object, individuality pertains to the object considered on its own (Gracia 1988, pp. 34–35). Thus Post (1963) spoke of 'Transcendental Individuality' in this context, in the sense of the particles' individuality 'transcending' their (intrinsic) properties. The individuality is then typically taken to be conferred either in spatio-temporal terms, which raises obvious problems, particularly in this context, or by some notion of 'haeccity' or 'primitive thisness' (French 1989a; Teller 1995, pp. 16ff.) situated within the philosophical framework which regards individuals as 'bare particulars' possessing sets of properties (Gracia *op. cit.*, pp. 86–90). However, equating non-individuality with a lack of primitive thisness really doesn't get us very far when it comes to having a formal grip on the former notion. What we want is some way of employing the usual logical structure of terms and quantifiers but in a manner which allows us to 'refer' in some general sense to entities which are non-individuals (we shall briefly return to the notion of reference in the conclusion). Returning to the properties, broadly construed, which an object can have, there is one which it and it alone can possess and that, of course, is the property of being identical to itself. A possible way of understanding non-individuality, then, is in terms of a lack of self-identity and indeed this is precisely the manner in which it has been expressed in the quantum context by Schrödinger and Hesse, for example (Schrödinger 1952, pp. 16–18; Hesse 1970, pp. 48–50).[1] Schrödinger (*loc. cit.*), for instance, stressed that

When you observe a particle of a certain type, say an electron, now and here, this is to be regarded in principle as an *isolated event*. Even if you observe a similar particle a very short time later at a spot very near to the first, and even if you have every reason to assume a *causal connection* between the first and the second observation, there is no true, unambiguous meaning in the assertion that it is the *same* particle you have observed in the two cases. The circumstances may be such that they render it highly convenient and desirable to express oneself so [that is, in saying that one and the same particle was observed twice at different instants of time], but it is only an abbreviation of speech; for there are other cases where the 'sameness' becomes entirely meaningless; and there is no sharp boundary, no clear-cut distinction between them, there is a gradual transition over intermediate cases. And I beg to emphasize this and I beg you to believe it: It is not a question of our being able to ascertain the identity in some instances and not being able to do so in others. *It is beyond doubt that the question of 'sameness', of identity, really and truly has no meaning.* [Last emphasis ours]

Schrödinger is not talking here about the impossibility of re-identifying an electron once observed, which is related to the old philosophical problem concerning *genidentity*, but rather he is expressing a much more profound insight, namely, the fact that the concept of identity simply cannot be applied in general to elementary particles.[2]

Returning to the 'Manin problem' of mathematically characterising collections of such objects, regarding non-individuality in this way allows for the development of formal systems, termed 'Schrödinger logics' and also 'quasi-set theories', in which identity does not hold in general (da Costa and Krause 1994, Krause 1992, preprint, Dalla Chiara, Giuntini and Krause forthcoming). These are (in a certain sense) many-sorted logics in that for a certain category of objects the expression $x = y$ simply is not a well-formed formula (and likewise for the negation $x \neq y$). For all other entities classical logic is maintained. The semantics of Schrödinger logics can then be given in terms of 'quasi-sets' as suggested by da Costa (1980, p. 119).

2 The mathematics of non-individuals

Mathematical frameworks for dealing with collections of objects such as elementary particles have two independent but related origins. The first lies with da Costa's

---

[1] Auyang (op. cit., p. 125): ' ... identity is constitutive of the general concept of the individual ...'.

[2] Cf. Mackey 1963, p. 110, 'The accepted physical interpretation of such a state of affairs [in quantum statistics] is that the particles are rather far from being like billiard balls in that thay have no individual identities ...'.

suggestion of a more general framework to provide semantics for Schrödinger logics, and a theory of *quasi-sets* was first developed in Krause (1990) (see Krause 1992; the terminology *quasi-sets* was proposed by da Costa (loc. cit.)). However, in 1983 Dalla Chiara and Toraldo di Francia independently proposed their *quaset theory* mentioned above (see their 1993 paper). According to this approach, standard set theories are not adequate for representing microphysical phenomena, since the ontology of physics apparently does not reduce to that expressed in standard set theory (also cf. Dalla Chiara and Toraldo di Francia 1995). As they have also remarked, one of the basic motivations underlying such a supposition is that collections of objects like elementary particles cannot obey the axioms of set theories like Zermelo–Fraenkel owing to their indistinguishability. In addition, these authors have suggested that identity questions in the quantum context demand a kind of *intensional semantics*, for which quaset theory (and also quasi-set theory) may provide the (meta)mathematical framework.

The basic idea underlying *quaset* theory is that of a collection of objects which have a well-defined cardinal, but there is no way to tell (with certainty) which are the elements that belong to the quaset. This is achieved by distinguishing between the primitive predicates $\in$ and $\notin$ (which is not the negation of the former), meaning 'certainly belongs to' and 'certainly does not belong to' respectively. The postulates imply that $z \in y$ entails $\neg z \notin y$, but not the converse. So, it may be the case that it is false that z certainly does not belong to y, but this does not entail that z (certainly) belongs to y. The elements z to which it may be said that 'it is false that they certainly do not belong to y' might be said to be members *in potentia* of y. Since the cardinal of the quaset is fixed, there is a kind of 'epistemic' indeterminacy with respect to its elements. Convenient postulates provide the basis for the whole theory, and a semantical analysis of microphysics has also been sketched.[3]

The 'epistemic' point just mentioned is fundamental for understanding the distinction between quasets and quasi-sets. The main characteristics of quaset theory concerns the distinction between a quaset and its *extension*, namely, the collection of those objects that 'really belong' to the quaset. This feature notwithstanding, a theory of identity continues to hold in its underlying logic (which should be regarded as being the first-order predicate calculus with identity). This is precisely what we mean by saying that in a quaset we have an 'epistemic' indeterminacy with respect to its elements; the elements *are* individuals (if we understand this as meaning that there is a reasonable theory of identity which holds among them), but they are *epistemically indeterminate* since we cannot define precisely which individuals belong to a quaset.

Despite the similarities, quasi-set theory differs substantially from the above on this point. In quasi-set theory,[4] the presence of two sorts of atoms (*Urelemente*) termed *m*-atoms and *M*-atoms is allowed, but the concept of identity (on the standard grounds) is restricted to the *M*-atoms only. Concerning the *m*-atoms, a weaker 'relation of indistinguishability'[5] is used instead of identity. Since the latter (that is, the predicate of equality) cannot be applied to the *m*-atoms, there is a precise sense in saying that

---

[3] Other papers of related interest are Dalla Chiara 1987, 1987a, Dalla Chiara and Toraldo di Francia preprint.

[4] We will make reference to the theory $\mathfrak{Q}$ presented in Krause preprint, which constitutes a distinct theory from that of Krause 1992.

[5] Denoted by the symbol $\equiv$. It is postulated that $\equiv$ has the properties of an equivalence relation only.

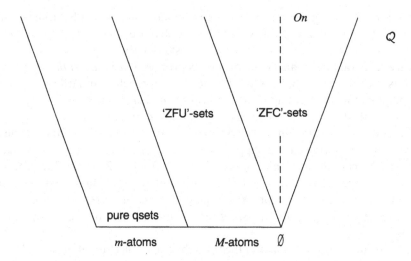

Fig. 1. The quasi-set universe.

they can be indistinguishable without being identical. As a consequence, Leibniz's Law is violated. So, contrary to the case of quasets, the lack of sense in applying the concept of identity to the *m*-atoms produces in quasi-set theory a kind of 'ontic' indeterminacy.[6]

The universe of $\mathfrak{Q}$ is then composed of *m*-atoms, *M*-atoms and collections of them, termed *quasi-sets*. The axiomatics is adapted from that of ZFU (Zermelo–Fraenkel with *Urelemente*), and when we restrict ourselves to that part of the theory which does not consider *m*-atoms, quasi-set theory is essentially equivalent to ZFU, and the corresponding quasi-sets can be named 'ZFU-sets' (similarly, if also the *M*-atoms are ruled out, the theory collapses into ZFC). The figure explains intuitively the Cumulative Hierarchy of quasi-sets, which can be defined in an obvious way (see Krause preprint). 'Pure' quasi-sets are quasi-sets whose elements are *m*-atoms only.

In order to preserve the concept of identity for the 'well-behaved' objects, an *Extensional Equality* is introduced for those entities which are not *m*-atoms on the following grounds: for all x and y, if they are not *m*-atoms, then

$$x =_E y := Q(x) \wedge Q(y) \wedge \forall z(z \in x \leftrightarrow z \in y) \vee (M(x) \wedge M(y) \wedge x \equiv y)$$

In $\mathfrak{Q}$ it is possible to prove that $=_E$ has all the properties of classical identity. Among the very specific axioms of $\mathfrak{Q}$, there are a few which perhaps deserve further explanation here.

For instance, in order to form certain elementary quasi-sets, such as that containing 'two' objects, we cannot use something like the usual 'pair axiom', since its standard formulation presupposes identity. Thus, we use the weak relation of indistinguishability instead:

---

[6] The use of quasi-set theories in the discussion of 'vagueness' in quantum mechanics is presented in French and Krause 1995 and French, Krause and Maidens preprint. For further discussion of the distinction between quasets and quasi-sets, see Dalla Chiara, Giuntini and Krause op. cit.

[The 'Weak-Pair' Axiom] For all x and y, there exists a quasi-set whose elements are the indistinguishable objects from either x or y. In symbols,

$$\forall x \forall y \exists_Q z \forall t (t \in z \leftrightarrow t \equiv x \vee t \equiv y)$$

Such a quasi-set is denoted by $[x, y]$ and, when $x \equiv y$, we have $[x]$ by definition. This quasi-set *cannot* be regarded as the 'singleton' of x, since it contains *all* the indistinguishable elements from x, so its 'cardinality' (see below) may be greater than 1. A concept of *strong singleton*, which plays an important rôle in the applications of quasi-set theory, may be defined on distinct grounds, as we shall mention below.

Clearly, the concept of *function* also cannot be defined in the standard way, since usually a function distinguishes among its arguments and values, and this cannot happen if there are indistinguishable m-atoms involved. Thus, we have introduced a weaker concept of *quasi-function*, which maps collections of indistinguishable objects into collections of indistinguishable objects; when there are no m-atoms involved, the concept reduces to that of function as usually understood. Relations, however, can be defined in the usual way, although no order relation can be defined on a quasi-set of indistinguishable m-atoms. Roughly speaking, the reasons are as follows: partial and total orders require antisymmetry, and this property cannot be stated without identity. Asymmetry also cannot be supposed, for if $x \equiv y$, then for every relation R such that $\langle x, y \rangle \in R$, it follows that $\langle x, y \rangle =_E [[x]] =_E \langle y, x \rangle \in R$, by the axioms of $\mathfrak{Q}$.

It is important to note that it is possible to define a translation from the language of ZFU into the language of $\mathfrak{Q}$ in such a way that there is a 'copy' of ZFU in $\mathfrak{Q}$, as the figure exemplifies. In this copy, all the usual mathematical concepts can be defined, and the 'sets' (in reality, the '$\mathfrak{Q}$-sets' which are 'copies' of the ZFU-sets) turn out to be those quasi-sets whose transitive closure (this concept is like the usual one) does not contain m-atoms.

In standard set theories, although the concept of cardinal can be defined independently of that of ordinal, according to Weyl 'the concept of ordinal is the primary one' (Weyl 1963, pp. 34–35). However, as we have seen, quantum mechanics seems to present strong arguments for questioning this thesis, and the idea of presenting collections which have a cardinal but not an ordinal is one of the most basic presuppositions of quasi-set theories, as in the case of quasets already mentioned.

The concept of *quasi-cardinal* is taken as primitive in $\mathfrak{Q}$, subject to certain axioms that permit us to operate with quasi-cardinals in a way similar to that of cardinals in standard set theories. Among the axioms for quasi-cardinality, we mention the following:[7]

[Quasi-cardinality] Every qset has a unique quasi-cardinal which is a cardinal (as defined in the 'classical' part of the theory, which resembles ZFU) and, if the quasi-set is in particular a set, then this quasi-cardinal is its cardinal stricto sensu:[8]

$$\forall_Q x \exists! y (Cd(y) \wedge y =_E qc(x) \wedge (Z(x) \rightarrow y =_E card(x)))$$

---

[7] Here, '$qc(x)$' stands for the 'quasi-cardinal' of the quasi-set x, while $Z(x)$ says that x is a *set* (in $\mathfrak{Q}$). Furthermore, $\forall_Q$ and $\exists_Q$ are the quantifiers relativised to quasi-sets, while $Cd(x)$ and $card(x)$ mean 'x is a cardinal' and 'the cardinal of x' respectively, defined as usual in the 'copy' of ZFU.

[8] Then, every quasi-cardinal is a cardinal and the above expression 'there is a unique' makes sense. Furthermore, from the fact that $\oslash$ is a set, it follows that its quasi-cardinal is 0.

There is also an axiom which says that if the quasi-cardinal of a quasi-set x is α, then for every quasi-cardinal $\beta \leq \alpha$, there is a subquasi-set of x whose quasi-cardinal is β, where the concept of *subquasi-set* is like the usual one. In symbols,

[The quasi-cardinals of subquasi-sets]

$$\forall_Q x(qc(x) =_E \alpha \rightarrow \forall \beta(\beta \leq_E \alpha \rightarrow \exists_Q y(y \subseteq x \wedge qc(y) =_E \beta)))$$

Another axiom states that

[The quasi-cardinal of the power quasi-set]

$$\forall_Q x(qc(\mathcal{P}(x)) =_E 2^{qc(x)})$$

where $2^{qc(x)}$ has its usual meaning.

As remarked above, in $\mathfrak{Q}$ there may exist quasi-sets whose elements are m-atoms only, called 'pure' quasi-sets. Furthermore, it may be the case that the m-atoms of a pure quasi-set x are all indistinguishable, in the sense of sharing the indistinguishability relation $\equiv$. In this case, the axiomatics provides the grounds for saying that nothing in the theory can distinguish the elements of x from one another. But, in this case, one could ask what it is that sustains the idea that there is more than one entity in x. The answer is obtained through the above mentioned axioms (among others, of course). Since the quasi-cardinal of the power quasi-set of x has quasi-cardinal $2^{qc(x)}$, then if $qc(x) = \alpha$, for every quasi-cardinal $\beta \leq \alpha$ there exists a subquasi-set $y \subseteq x$ such that $qc(y) = \beta$, according to the axiom about the quasi-cardinality of the subquasi-sets. Thus, if $qc(x) = \alpha \neq 0$, the axiomatics does not forbid the existence of α subquasi-sets of x which can be regarded as 'singletons'. Of course the theory cannot prove that these 'unitary' subquasi-sets (supposing now that $qc(x) \geq 2$) are distinct, since we have no way of 'identifying' their elements, but quasi-set theory is compatible with this idea. That is, in other words, it is consistent with $\mathfrak{Q}$ to maintain that x has α elements, which may be regarded as absolutely indistinguishable objects. Since they may share an equivalence relation (precisely $\equiv$), these elements may be further understood as belonging to a same 'equivalence class' (being all indistinguishable electrons, say) but in such a way that we cannot assert either that they are identical or that they are distinct (i.e., they act as 'identical electrons' in the physicist's jargon).[9]

It is also worth mentioning the 'weak' axiom of extensionality used in $\mathfrak{Q}$ which is stated as follows. First of all, we define x and y as *Similar* quasi-sets (in symbols, $Sim(x,y)$) if the elements of one of them are indistinguishable from the elements of the other, that is, $Sim(x,y) \Leftrightarrow \forall z \forall t(z \in x \wedge t \in y \rightarrow z \equiv t)$. Furthermore, x and y are *Q-Similar*($QSim(x,y)$) if they are similar and have the same quasi-cardinality. Then, since the quotient quasi-set $x/_\equiv$ may be regarded as a collection of equivalence classes of indistinguishable objects, then the 'weak' axiom of extensionality is:

---

[9] The application of this formalism to the concept of non-individual quantum particles has been proposed in Krause and French (1995).

[Weak Extensionality][10]

$$\forall_Q x \forall_Q y (\forall z (z \in x/_\equiv \rightarrow \exists t (t \in y/_\equiv \wedge QSim(z, t)) \wedge$$

$$\forall t (t \in y/_\equiv \rightarrow \exists z (z \in x/_\equiv \wedge QSim(t, z)) \rightarrow x \equiv y))$$

In other words, the axiom says that those quasi-sets that have 'the same quantity of elements of the same sort'[11] are indistinguishable.

Finally, let us remark that quasi-set theories have been found to be equiconsistent with standard set theories (like ZFC) (see Krause 1995, da Costa and Krause preprint a, Krause preprint for proofs of this fact for the different quasi-set theories).

2.1 The quasi-set version of the Indistinguishability Postulate

As we have seen, quasi-set theory permits us to formalise the notion of a collection of indistinguishable non-individual objects, and this is achieved in a rather precise sense of the words 'indistinguishability', 'identity' and 'collection'. But a number of other related consequences can also be established (see Krause and French 1995), mainly regarding the 'set-theoretical' meaning of the Indistinguishability Postulate mentioned in the beginning.

Intuitively speaking, the Indistinguishability Postulate asserts that permutations of indistinguishable elementary particles cannot be regarded as observable. In order to provide an interpretation of this fact in $\mathfrak{Q}$, let us introduce the following definition:

Definition 1 A strong singleton of an object x is a quasi-set x' which satisfies the following property $St(x)$:

$$St(x) \Leftrightarrow x' \subseteq [x] \wedge qc(x') =_E 1$$

That is, x' is a subquasi-set of $[x]$ (the collection of *all* indistinguishable from x) that has just 'one element'. This definition makes sense since the quasi-cardinals are cardinals, due to the 'Quasi-cardinality' axiom. Furthermore, it is straightforward to prove in $\mathfrak{Q}$ that for every x there exists a strong singleton of x (cf. Krause *op. cit.*).

Then, by recalling further that the quasi-set operations of difference, intersection and union acquire intuitive meanings as in the standard set theories, we can state the following theorem:

Theorem 1 [Unobservability of Permutations] *Let x be a qset such that x is distinct from z, and z an m-atom such that $z \in x$. If $w \equiv z$ and z does not belong to x, then there exists w' such that*

$$(x - z') \cup w' \equiv x$$

The proof is an immediate consequence of (mainly) the axiom of Weak Extensionality. By recalling that z' (respect. w') denotes the strong singleton of z (respect. of w), then when $w \notin x$ we may interpret the theorem as saying that we have 'exchanged' an element of x by an indistinguishable one, and that the resulting fact is that 'nothing

---

[10] This formulation differs from that presented in other versions of quasi-set theory.
[11] In the sense that they belong to the same equivalence classes of indistinguishable objects.

has occurred at all'. In other words, the resulting quasi-set is indistinguishable from the original one in the sense of the axiom of Weak Extensionality.

3 Models and the Fock space formalism

Returning to the two metaphysical options with which we began this paper, the undetermination is broken by appealing to quantum field theory (QFT): the particles-as-non-individuals view is to be preferred to the particles-as-individuals plus state accessibility restrictions package because the former meshes better with QFT whereas the latter does not. This claim has been developed by Redhead and Teller (1991, 1992) and recently by Teller himself (1995) in terms of 'surplus structure', or elements of the formalism which do not correspond to anything in the real world. Thus the view of quantum particles as individuals leaves a residue of inaccessible non-symmetric states as surplus formal structure arising from the attribution of particle labels, whereas the non-individuals package avoids this.[12] In particular, an alternative formalism can be developed incorporating this latter package – the Fock space formalism of QFT.

In quantum mechanics labels are assigned to the particles, these particle labels are then permuted and on the basis of the Indistinguishability Postulate we may conclude that the particles are actually non-individuals. This is what Post meant by Schrödinger 'getting off on the wrong foot' (Post op. cit., p. 19). The Fock space formalism of QFT avoids particle labels from the word go, or so it is claimed, and the particles-as-non-individuals get expressed by what Teller calls 'quanta'. Thus,

> ... things with primitive thisness can be *counted*: that is, we can think of the particles as being counted out, the first one, the second one, the third, and so on, with there being a difference in principle in the order in which they are counted, a difference that does not depend on which particle has which properties. By way of contrast quanta can only be *aggregated*; that is, we can only heap them up in different quantities with a total measure of one, or two, or three, and so on, but in these aggregations there is no difference in principle about which one has which properties. The difference between countability and susceptibility to being merely aggregated is like the difference between pennies in a piggy bank and money in a modern bank account.

(Teller op. cit., p. 12)

The money-in-the-bank analogy is one that has been given numerous times before, as Teller acknowledges (and he returns to it in greater detail on p. 29), notably by Schrödinger (1957) and Hesse (1970), again. The latter's discussion of this analogy is particularly interesting from our perspective as she explicitly invokes it to illustrate what she means by quantum particles lacking self-identity: '[w]e are unable to identify individual electrons, hence it is meaningless to speak of the self-identity of electrons ...' (*ibid.*, p. 50). Carrying this over into the domain of QFT, as Redhead and Teller do, suggests a way of responding to the Manin problem, namely by regarding aggregates of quanta in terms of quasi-sets.

---

[12] Doubts concerning such claims have been indicated in French (forthcoming) and more explicitly in Huggett (1995, pp. 74–5). In particular, if the Indistinguishability Postulate is taken to be part of the theory, as it should, then 'surplus structure' simply does not arise in any meaningful sense. The point is driven home if the Postulate is conceived of group theoretically.

By doing so we can represent and bring out, philosophically, the distinction emphasised by Teller between 'countability' and 'susceptibility to being aggregated'.[13] There are fundamental philosophical issues here. Thus Lowe, for example, in his discussion of primitive substances, situates individuality within the framework of differentiation or distinctness and stresses that the latter must be achieved in such a manner that the individuals can be regarded as countable. A necessary condition of countability is that

... the items to be counted should possess determinate identity conditions, since each should be counted just once and this presupposes that it be determinately distinct from any other item that is to be included in the count.

(Lowe 1994)

Thus, if a plurality is countable, the entities of which it is composed must possess self-identity, according to Lowe. However, the notion of 'countability' that is invoked here is not at all clear. In particular we need to distinguish *cardinality* from *ordinality* and this is precisely achieved in quasi-set theory (and also in quaset theory as we have indicated above). With regard to aggregates of quanta one can have good theoretical and experimental reasons for ascribing a determinate number of quanta to such a collection but without being able to count them, in the sense of putting them in a series and establishing an ordering. As we have seen, a quasi-set of quantum particles possess a cardinal but not an ordinal in exactly this sense – the elements of a quasi-set are countable, *in the sense of possessing (quasi-)cardinality but not in the sense of possessing ordinality*. It is in this manner, we believe, that Teller's distinction can be formally represented.

What we are attempting to do here is to suggest a formal framework for understanding the notion of quanta and it is our firm conviction that such a framework is absolutely necessary if something more than mere lip service is to be paid to this notion and its underlying metaphysics.[14] We wholeheartedly agree with Auyang, who says,

A proper interpretation of quantum theories calls for a general concept of the objective world that adequately accommodates both classical and quantum objects.

(op. cit., p. 6)

where we would extend the notion of a 'proper interpretation' into metaphysics.

However, focusing on the notion of 'non-individuality' and a set theory suitable for capturing it in this way highlights a two-fold tension with regard to the Fock space approach to QFT, advocated by Redhead and Teller, and the model-theoretic approach to the latter, advocated by Teller in his book. We can reveal this tension by first considering what it is to 'axiomatise' a scientific theory. This is a fundamental issue not only in the foundations of physics but in philosophy of science as a whole

---

[13] Barcan Marcus notes that in the foundations of set theory '... there is an ambiguity in the notion of "gathering together" and "conceiving a totality" that is reflected in the distinction between an itemization of elements and a statement of conditions for membership' (Barcan Marcus 1993, p. 93). It seems to us that this broadly corresponds to the distinction between 'countability' and 'susceptibility to being aggregated' and correspondingly to that between ordinality and (quasi-)cardinality captured by quasi-(and qua)set theory. As Barcan Marcus goes on to note, '[t]he obliteration of the difference by the logic of set theory does not generate too many perplexities for mathematics ...' (ibid.) but it is a different matter when it comes to physical objects.

[14] There is obviously a crucial issue as to what 'understanding' consists in here. At the very least we are sympathetic to Carnap's feeling that '[w]hen I considered a concept or a proposition occurring in a scientific or philosophical discussion, I thought that I understood it clearly only if I felt that I could express it, if I wanted to, in a symbolic language' (Carnap 1963, p. 11).

and we certainly do not claim to adequately address it here. However, on first iteration we can say something like the following: the axiomatic basis of a theory basically
encompasses three 'levels' of axioms: (1) the axioms of the underlying logic of the
theory, say first-order logic with identity; (2) the axioms of the mathematical apparatus, say Zermelo–Fraenkel set theory; and (3) the specific axioms of the theory itself.

   According to the model-theoretic or 'semantic' approach to theories these axioms
delineate the class of models, or more generally, mathematical structures specified by
the theory. As Teller puts it, these models are viewed as '... abstract objects embody-
ing just those properties and relations which are of interest to us in the physical
systems to be modeled' (op. cit., p. 4). This is an approach to which we are very sympathetic; however, for these 'abstract objects' to be models, in the strict sense, they
must be 'satisfiers' of some sort. According to Suppes, famously, what they satisfy
is a set-theoretic predicate and to axiomatise a theory is to define just such a predicate
(Suppes 1957, 1967). Thus, if axiomatisation delineates a class of models, then of
course it also involves axiomatisation at levels (1) and (2), at least implicitly, since
these axioms 'affect' a theory's theorems. (In this regard it is perhaps worth recalling
Church's remark (made with respect to mathematics) that

... for the precise syntactical definition of the particular branch of mathematics it is necessary
to state not only the specific mathematical postulates [what we would call axioms of level 3] but
also a formulation of the underlying logic, since the class of theorems belonging to the branch
of mathematics in question is determined by both the postulates and the underlying logic and
could be changed by a change of either.

                                                                    (1956, p. 317))

   This brings us to the Fock space approach, advocated by Redhead and Teller as
being the framework in which non-individuality can be understood in terms of a
'label-free formalism'. However, if this is to be characterised in model-theoretic
terms, as Teller intends, then axiomatisation via a set-theoretic predicate must be
assumed, again at least implicitly and, in this particular case, such axiomatisation
cannot be left as 'an exercise for the reader', as it were, precisely because of the metaphysical issues driving the move to this approach. Let us be specific: standard presentations of the Fock space formalism, such as given by Teller (op. cit.), Geroch 1985,
Landau and Lifschitz 1959, for example, are elaborated within the framework of
'naïve' (Cantorian) set theory which is precisely based upon ontological presuppositions which this formalism supposedly denies![15] In other words, at levels (1) and (2),
classical logic and mathematics have been left intact as the necessary formal underpinning; only the axioms at level (3) have been exchanged from those of the first-
quantised approach to Fock spaces. The problem, then, is to consider the adequacy
of the class of models these axioms describe and the intuitive motivation which was
used for the Fock space formalism, namely the rejection of 'primitive thisness' and
individual elementary particles.

[15]   In the paper in which he introduces the first axiomatic set theory, Zermelo says that Cantor's 'definition'
of the concept of set certainly requires some limitation, '... although no one has yet succeeded in replacing it by another definition, equally simple that is not exposed to any such doubt'. Then he proceeds in
applying the axiomatic method by 'starting from the historically existing "theory of sets" [i.e., Cantor's
naïve theory], to seek out the principles which are required as a basis for this mathematical discipline'
(Zermelo 1908). Then, taking into account that all other axiomatic set theories derive from Zermelo's
in one way or another, we may say that no such theory has destroyed Cantor's intuition, and it is
precisely on these grounds that we say that the ontology of set theories and hence of all classical mathematics is essentially an ontology of individuals (distinguishable objects).

3.1 Suppes predicate for QFT

Let us define a *Fock structure* as an ordered triple:

$$\mathfrak{F} = \langle F, a_i^\dagger, a_j \rangle_{i,j \in I}$$

where

1. F is a Fock space (understood as a direct sum of tensor products of Hilbert spaces in the usual sense – see Geroch, *op. cit.*).
2. $a_i^\dagger$ and a_j are operators on F that satisfy the following conditions:
   - for every $i \in I$, $a_i^\dagger$ and a_i are Hermitean conjugates of each other;
   - these operators obey the standard commutator and anticommutator relations (Merzbacher 1970, p. 515).

Given such a definition, we can construct an appropriate 'Suppes predicate', since all models of non-relativistic QFT can be obtained in this way (cf. van Fraassen 1991, p. 448).[16] So, we may say that X is a *Fock space quantum field theory* if X satisfies the following set-theoretical predicate:

$$FSQFT(X) \Leftrightarrow (\exists F)(\exists a_1^\dagger, a_2^\dagger, \ldots)(\exists a_1, a_2, \ldots)(X = \langle F, a_i^\dagger, a_j \rangle \wedge (1)\text{–}(2) \text{ as above})$$

Let us consider briefly the nature of the models of this predicate. We take $\mathcal{H}$ to be a complex Hilbert space with orthonormal basis $\{|v_i\rangle\}$ ($i = 1, 2, 3, \ldots$), where the v_i are eigenvectors of some maximal (non-degenerate) Hermitian operator A defined on $\mathcal{H}$. Then we define F to be the direct sum

$$F = \bigoplus_{n=0}^{\infty} \mathcal{H}^n$$

where $\mathcal{H}^0 = \mathcal{C}$ (the complex Hilbert space), $\mathcal{H}^1 = \mathcal{H}, \mathcal{H}^2 = \mathcal{H} \otimes \mathcal{H}$ and so on (Geroch, *op. cit.*). In this Fock space vectors such as $\alpha_1 = v_1 \otimes v_2$ and $\alpha_2 = v_2 \otimes v_1$, $(v_1 \neq v_2)$ appear, since they are both elements of $\mathcal{H}^2$. But this introduces 'surplus formal structure' and for that reason the model is to be rejected for the present discussion (Redhead and Teller 1991, 1992; Teller 1995).

More appropriate models can of course be constructed, such as the *boson model*: here we take $\mathcal{H}^n$ as above but define on it the function $\sigma^n : \mathcal{H}^n \to \mathcal{H}^n$ ($n = 1, 2, \ldots$) by

$$\sigma^n(v_1 \otimes \cdots \otimes v_n) = (1/n!)\Sigma_P P(v_1 \otimes \cdots \otimes v_n)$$

where P is an element of the permutation group of $\mathcal{H}^n$. We now define

$$\mathcal{H}_\sigma^n = \{\sigma^n(v) : v \in \mathcal{H}^n\}$$

and

$$F = \bigoplus_{n=0}^{\infty} \mathcal{H}_\sigma^n$$

where, once more, $\mathcal{H}_\sigma^0 = \mathcal{C}$. Clearly $\mathcal{H}_\sigma^n$ is the symmetric subspace of $\mathcal{H}^n$ (van Fraassen, op. cit., p. 437). The *fermion model* can be obtained in a similar way by defining functions $\tau^n : \mathcal{H}^n \to \mathcal{H}^n$ such that

$$\tau^n(v_1 \otimes \cdots \otimes v_n) = (1/n!)\Sigma_P s^P P(v_1 \otimes \cdots \otimes v_n)$$

---

[16] As exemplified for instance in Mattuck 1967, Chap. 7 (see also Landau and Lifschitz, op. cit.), with the help of 'raising' and 'lowering' operators all of second-quantised quantum mechanics can be erected.

where s^p is the *signature* of P, that is, s^p is $+1$ if P is even and -1 if P is odd (van Fraassen op. cit., p. 386). Then, in analogy to the symmetric case, we define the anti-symmetric subspace of $\mathcal{H}^n$ as

$$\mathcal{H}_\tau^n = \{\tau^n(v) \colon v \in \mathcal{H}^n\}$$

and

$$F = \bigoplus_{n=0}^\infty \mathcal{H}_\tau^n$$

The problem now is to define the operators $a_i^\dagger$ and a_j. Let us define, for each $\alpha \in \mathcal{H}^n$, $\alpha = v_{i_1} \otimes \cdots \otimes v_{i_n}$, arithmetic functions n_α, taking values on the set $\{1, 2, \ldots, n\}$ and ascribing to each $k \in \{1, 2, \ldots, n\}$ the quantity of subindices among the i_j which equal k. The numbers $n_\alpha(k)$ are called the *occupation numbers of v_{i_j} in α*. It can be shown that $n_\alpha(k) = n_\beta(k)$ iff $\sigma^n(\alpha) = \sigma^n(\beta)$ $(i = 1, \ldots, n)$ (ibid.). Then we can pay exclusive attention to the occupation numbers and write them as an n-tuple $\langle n(1), n(2), \ldots, n(n) \rangle$ (where the subindices have been omitted and $\Sigma_n(i) = n$) for every basis vector of $\mathcal{H}^n$ (and similarly for the fermion model). Thus it has been claimed that in this manner we arrive at an 'individual-label-free notation', since the vectors are mentioned by their occupation numbers only (ibid., p. 438; in the fermion case of course these are either 0 or 1).

It is worth noting that the basis vectors of the Fock space vectors may be represented by sequences whose $(n+1)th$ element is precisely $\langle n(1), n(2), \ldots, n(n) \rangle$ and whose other elements are zero vectors. Then, by paying attention exclusively to their occupation numbers as noted above, we need not make reference to the full sequence but only to this $(n+1)th$ term. Hence the basis vectors of the symmetric subspace can be represented as usual by:

$$|n\rangle = |n(1)n(2)\ldots n(n)\ldots\rangle$$

We insist that this notation *stands* for a certain vector in the Fock space, which is uniquely characterised, as remarked by van Fraassen (op. cit., pp. 440–41) jointly by the occupation numbers $n(i)$, whose sum equals n (this n indicates to which $\mathcal{H}^n$ the vector belongs) and by the values of the $n(i)$ themselves, which show to which vectors in the basic Hilbert space $\mathcal{H}$ the 'true' vector is linked. In this way, we may say that $|20700\ldots\rangle$ represents a state (a vector) with two quanta with eigenvalue v_3 and seven quanta with eigenvalue v_7 with respect to the observable A from which we have picked up the discrete orthonormal basis $\{|v_i\rangle\}$ of $\mathcal{H}$ (Teller 1995, p. 37ff.).

We are now in a position to define, for each $i = 1, 2, \ldots$

$$\alpha_i^\dagger |n(1)n(2)\ldots n(i)\ldots\rangle = \sqrt{n(i)+1}\, |n(1)n(2)\ldots (n(i)+1)\ldots\rangle$$

$$\alpha_i |n(1)n(2)\ldots n(i)\ldots\rangle = \sqrt{n(i)}\, |n(1)n(2)\ldots (n(i)-1)\ldots\rangle$$

In the usual way it can be shown that the $a_i^\dagger$ and a_i have the properties of the 'raising' and 'lowering' operators respectively (ibid.) and as was remarked above, all the second-quantised formalism can be derived.

A potential problem now arises: the 'label-free notation' mentioned above is achieved by taking the symmetric Fock space (and likewise for the antisymmetric case, of course). But we began with a Hilbert space $\mathcal{H}$ with basis $\{|v_i\rangle\}$, as van Fraassen, for example, has emphasised and the question is whether this procedure of constructing Fock space from a labelled Hilbert space avoids the criticism directed by Redhead and Teller to the 'labelled-tensor-product-Hilbert-space-formalism' of the first-quantised approach.[17]

The right way of going about this, as both van Fraassen and Teller have suggested, is in fact to begin with an 'abstract' Fock space in which only the occupation numbers are mentioned. That is, we should construct the Fock space by taking the set of vectors as being the sequences $|n(1)n(2)\ldots\rangle$. This is precisely what Teller claims by saying that such vectors 'generate' the elements of the direct sum in the definition of the Fock space (op. cit., p. 39). However, from a mathematical perspective this is rather odd, since we cannot define vector space operations on n-tuples of natural numbers. In other words, we might understand the claim that $|1\,1\,0\ldots\rangle$, $|2\,0\,0\ldots\rangle,\ldots$ for example, *generate* a Hilbert space only by regarding these n-tuples to be the vectors they represent in $\mathcal{H}^n$, but then we arrive again at the Hilbert space $\mathcal{H}$ and conclude, on the same grounds of Redhead and Teller's criticisms of the first-quantised approach, that we are still not free from primitive thisness.

Of course, the direction in which to proceed has been indicated by van Fraassen and Teller, but the procedure followed is essentially that of Weyl in Appendix B of his (1963) work. Here, in discussing 'aggregates of individuals' (some of which are intended to act as collections of quanta), Weyl simply takes a *set S* (whose cardinal is n, for example), together with an equivalence relation $\sim$ on S and considers the equivalence classes of the quotient set $S/\sim$. Then, by 'forgetting' the 'nature' of the elements of S (recall what we said at the beginning of this paper concerning the fact that in standard set theories a set is a collection of distinct objects) and paying attention exclusively to the cardinality $n(i)$ $(i = 1,\ldots,k)$ of the equivalence classes, he obtains the 'ordered decomposition' $n(1) + \cdots + n(k) = n$ (which notably bear a certain resemblance to the occupation numbers) which, as Weyl emphasises, is what is considered in quantum mechanics. However, as one of us has noted elsewhere (Krause 1991) this procedure effectively 'masks' the individuality of the elements of the set S, and cannot be said to be a mathematical treatment of indistinguishable objects. With respect to the models described above, we suggest that something similar is occurring when we take the symmetric and antisymmetric subspaces by 'forgetting' the initially attached labels.

Taking into account that the notation $|n\rangle = |n(1)n(2)\ldots\rangle$ is only representing a vector in the Fock space, then, paraphrasing Teller when he asks for a 'rationale' for the labels of the Hilbert space formalism, typical of first quantisation (cf. Teller 1995, p. 25), we could ask: how are the labels of these vectors (in the Fock space) to be understood? Recall once more that every one of these occupation numbers is related to a pure state in the original Hilbert space $\mathcal{H}$ (cf. van Fraassen, op. cit., p. 437). In our opinion, there have yet to be constructed models of *FSQFT* in which the non-individuality of the quanta is ascribed 'right at the start', to use Post's phrase (op. cit.).

4 Conclusion: the objectivity of quanta

Given the above and, specifically, the ontological considerations driving the shift to the Fock space approach, the question, then, is how we are to read the n-tuple

---

[17] Cf. Huggett op. cit., pp. 73–4.

$|n(1)n(2)\ldots n(n)\rangle$. A clue perhaps is given by Auyang who writes

To say the field is in a state $|n(1)n(2)\ldots n(n)\rangle$ is not to say that it is composed of $n(k_1)$ quanta in mode k_1 and so on but rather $n(k_1)$ quanta show up in an appropriate measurement.

(Auyang 1995, p. 159)

On these grounds, we suggest that perhaps we can provide a semantics in terms of quasi-set theory, by reading the 'crucial' vector $|n(1)n(2)\ldots n(n)\rangle$ as representing a quasi-set with quasi-cardinality n whose elements are equivalence classes of indistinguishable m-atoms with quasi-cardinalities respectively $n(i)$ $(i = 1,\ldots,n)$. This interpretation in terms of 'collections' of something is implicit in the metaphysical discussion of QFT; even if adopting a broadly realist stance according to which the 'essential reality' is a set of fields (Weinberg 1977), yet still we cannot dismiss their 'particle grin', as Redhead puts it (1988). Granted that quanta are not well defined in realistic situations, where we have interaction, it is the number of quanta in an 'aggregate' (that is, the cardinality of the appropriate quasi-set) that is typically measured in a scattering experiment, for example. As Auyang puts it, the eigenvalues $n(k)$ of the number operator $N_i(k)$ (defined in the above formalism as usual: $a_i^\dagger a_i = N_i$) are 'partial manifestations' of the properties of a free field in experiments (op. cit., p. 160). Questions as to the 'reality' of such quanta are sidestepped in favour of the claim that they are, at least, 'objective', a move which meshes nicely with recent 'ontic' forms of structural realism (Ladyman preprint).

But objective what? 'Are field quanta entities?' asks Auyang; 'If an entity is a this-something, then field quanta are not entities; they lack numerical identities' (ibid., p. 159). However, an entity does not have to be a 'this-something', or possess 'primitive thisness'; as Barcan Marcus has put it:

... all terms may 'refer' to objects, but ... not all objects are things, where a thing is at least that about which it is appropriate to assert the identity relation. (...) If one wishes, one could say that object-reference (in terms of quantification) is a wider notion than thing-reference, the latter being also bound up with identity and perhaps with other restrictions as well, such as spatiotemporal location.

(Barcan Marcus 1993, p. 25)

Thus, just as the Scholastics separated distinguishability from individuality, we insist on a further conceptual distinction between entities and individuals. Indeed, it is hard to see what sense can be made of the notion of 'quanta' without such a distinction. And of course it maps onto that made by Barcan Marcus, between 'object' and 'thing'-reference. As she goes on to note, if the distinction is denied at this level then it pops up again at the level of the denotation of names (ibid.). If reference is understood in this wider fashion, then it simply makes no sense to assert that '$x = y$' where 'x' and 'y' 'refer' to non-individual entities, and of course this result is precisely captured by quasi-set theory.[18] In our terms, 'entity' is a wider notion than 'individual' and field quanta are precisely examples of non-individual entities which are objective in the sense of being partially manifested in experiments.

We insist: the 'particle grin' remains (indeed, an empiricist might be inclined to propose this as the 'essential reality').[19] How then are we to understand its peculiar nature? A step to such understanding can be achieved by situating it in an appropriate

---

[18] We are grateful to Paul Teller for pressing us on this point.

[19] 'The representation of QFT has a structure compatible with atomism, and that remains true however you decide to express the states' (Huggett, op. cit., p. 74).

formal framework and it is the beginnings of such a framework that we hope to have delineated here. We leave the final word to Auyang:

Other people reject quantum objects because they are different but all their argument shows is that there is nothing like classical objects in the quantum realm, not that there is no quantum object.

(op. cit., p. 5)

References

[1] Auyang, S. Y., *How is Quantum Field Theory Possible?*, Oxford, Oxford Univ. Press, 1995.

[2] Barcan Marcus, R., *Modalities: philosophical essays*, New York and Oxford, Oxford Univ. Press, 1993.

[3] Born, M., *Experiment and Theory in Physics*, Cambridge Univ. Press, 1943.

[4] Butterfield, J., 'Interpretation and Identity in Quantum Theory', *Studies in History and Philosophy of Science* **24**, 1993, 443–76.

[5] Cantor, G., *Contributions to the Founding of the Theory of Transfinite Numbers*, New York, Dover, 1955.

[6] Carnap, R., 'Intellectual biography', in Schillip, P. A. (ed.), *The Philosophy of Rudolf Carnap*, Open Court, La Salle, 1963.

[7] Church, A., *Introduction to Mathematical Logic*, Vol. I, Princeton, Princeton Univ. Press, 1956.

[8] da Costa, N. C. A., *Ensaio sobre os fundamentos da lógica*, São Paulo, Hucitec, 1980 (2nd edn, 1994).

[9] da Costa, N. C. A. and Krause, D., 'Schrödinger logics', *Studia Logica* **53**, 1994, 533–50.

[10] da Costa, N. C. A. and Krause, D. (preprint a), 'Set-theoretical models for quantum systems'. Abstract in the *Volume of Abstracts* of the Xth International Congress of Logic, Methodology and Philosophy of Science, Florence, August 19–25, 1995, p. 470.

[11] da Costa, N. C. A. and Krause, D., 'An intensional Schrödinger logic', preprint b.

[12] Dalla Chiara, M. L., 'An approach to intensional semantics', *Synthese* **73**, 1987, 479–96.

[13] Dalla Chiara, M. L., 'Some foundational problems in mathematics suggested by physics', *Synthese* **62**, 1987a, 303–15.

[14] Dalla Chiara, M. L. and Toraldo di Francia, G., 'Individuals, kinds and names in physics', in Corsi, G. *et al.* (eds.), *Bridging the Gap: philosophy, mathematics, physics*, Dordrecht, Kluwer Acad. Press, 1993, 261–83.

[15] Dalla Chiara, M. L. and Toraldo di Francia, G., 'Identity questions from quantum theory', in Gavroglu *et al.* (eds.), *Physics, Philosophy and the Scientific Community*, Dordrecht, Kluwer, 1995, 39–46.

[16] Dalla Chiara, M. L. and Toraldo di Francia, G., 'Quine on physical objects', preprint.

[17] Dalla Chiara, M. L., Giuntini, R. and Krause, D., 'Quasiset theories for microobjects: a comparison', in Castellani, E. (ed.), *Quantum Objects*, forthcoming.

[18] French, S., 'Identity and individuality in classical and quantum physics', *Australasian Journal of Philosophy* **67**, 1989, 432–46.

[19] French, S., Individuality, supervenience and Bell's theorem', *Philosophical Studies* **55**, 1989a, 1–22.

[20] French, S., 'On the withering away of quantum objects', in Castellani, E. (ed.), *Quantum Objects*, forthcoming.

[21] French, S. and Redhead, M., 'Quantum physics and the identity of indiscernibles', *British Journal for the Philosophy of Science* **39**, 1988, 233–46.

[22] French, S. and Krause, D., 'Vague identity and quantum non-individuality', *Analysis* **55**, 1995, 20–26.

[23] French, S. and Krause, D., 'Opaque predicates and their logic', preprint.

[24] French, S., Krause, D. and Maidens, A., 'Quantum vagueness', preprint.

[25] Geroch, R., *Mathematical Physics*, Chicago Univ. Press, 1985.

[26] Gracia, J. J. E., *Individuality*, Suny Press, 1988.

[27] Haag, R., *Local Quantum Theory*, Springer-Verlag, 1992.

[28] Hesse, M., *Models and Analogies in Science*, Univ. Notre Dame Press, 1970.

[29] Huggett, N., 'What are quanta and does it matter?', *PSA 1994*, Vol. 2, PSA 1995, pp. 69–76.

[30] Krause, D., *Non-reflexivity, Indistinguishability and Weyl's Aggregates* (in Portuguese), Thesis, Univ. São Paulo, 1990.

[31] Krause, D., 'Multi-sets, quasi-sets and Weyl's aggregates', *Journal of Non-Classical Logic* **8**, 1991, 9–39.

[32] Krause, D., 'On a quasi-set theory', *Notre Dame Journal of Formal Logic* **33**, 1992, 402–11.

[33] Krause, D., 'Non-reflexive logics and the foundations of physics', in Cellucci, C., Di Maio, M. C. and Roncaglia, G. (eds.), *Logica e Filosofia della Scienza: problemi e prospettive*, Atti del Congresso Triennale della Societa Italiana di Logica e Filosofia delle Scienze, Lucca, Gennaio 1993, Ed. ETS, Pisa, 1994, 393–405.

[34] Krause, D., 'The theory of quasi-sets and ZFC are equiconsistent', in Carnielli, W. A. and Pereira, L. C. (eds), *Logic, sets and information*, Proceedings of the Xth Brazilian Conference on Mathematical Logic, Itatiaia, 1993, UNICAMP, Col. CLE Vol. 14, 1995, 145–55.

[35] Krause, D., *O Problema de Manin: Elementos para uma Análise Lógica dos Quanta*, Universidade Federal do Paraná, Curitiba, 1995.

[36] Krause, D., 'Axioms for collections of indistinguishable objects', *Logique et Analyse* **153–154**, 1996, 69–93.

[37] Krause, D. and French, S., 'A formal framework for quantum non-individuality', *Synthese* **102**, 1995, 195–214.

[38] Krause, D. and Béziau, J.-Y., 'Relativizations of the concept of identity', *Bulletin of the IGPL* **5**(3), 1997, 327–38.

[39] Ladyman, J., 'Structural realism and the model-theoretic approach to scientific theories', preprint.

[40] Landau, E. and Lifschitz, E. M., *Quantum Mechanics*, Vol. 3 of *Course of Theoretical Physics*, London, Pergamon Press, 2nd edn, 1959.

[41] Lowe, E. J., 'Vague identity and quantum indeterminacy', *Analysis* **54**, 1994, 110–14.

[42] Mackey, G., *Mathematical Foundations of Quantum Mechanics*, New York, Benjamin, 1963.

[43] Manin, Yu. I., 'Problems of present day mathematics: I (Foundations)', (in) Browder, F. E. (ed.) *Proceedings of Symposia in Pure Mathematics* **28** American Mathematical Society, Providence, 1976, p. 36.

[44] Merzbacher, E., *Quantum Mechanics*, New York, Wiley and Sons, 1970.

[45] Mattuck, R. D., *A Guide to Feynman Diagrams in the Many-Body Problem*, London, McGraw-Hill, 1967.

[46] Post, H., 'Individuality and physics', *The Listener* **70**, 1963, 534–7; reprinted in *Vedanta for East and West* **32**, 1973, 14–22.

[47] Redhead, M., 'A philosopher looks at quantum field theory', in Brown, H. R. and Harré, R. (eds), *Philosophical Foundations of Quantum Field Theory*, Oxford, Clarendon Press, 1988, 9–23.

[48] Redhead, M. and Teller, P., 'Particles, particle labels, and quanta: the toll of unacknowledged metaphysics', *Foundations of Physics* **21**, 1991, 43–62.

[49] Redhead, M. and Teller, P., 'Particle labels and the theory of indistinguishable particles in quantum mechanics', *British Journal for the Philosophy of Science* **43**, 1992, 201–18.

[50] Schrödinger, E., *Science and Humanism*, Cambridge Univ. Press, Cambridge, 1952.

[51] Schrödinger, E., 'What is matter?', *Scientific American*, September 1953, 52–7.

[52] Schrödinger, E., *Science Theory and Man*, Allen and Unwin, London, 1957.

[53] Suppes, P., *Introduction to Logic*, North-Holland, 1957.

[54] Suppes, P., 'Set-Theoretical Models in Science', Stanford University, 1967.

[55] Teller, P., 'Relational holism and quantum mechanics', *British Journal for the Philosophy of Science* **37**, 1986, 71–81.

[56] Teller, P., 'Relational holism', in Cushing, J. and McMullin, E. (eds.), *Philosophical Consequences of Quantum Theory*, Univ. Notre Dame Press, 1989, 208–23.

[57] Teller, P., *An Interpretive Introduction to Quantum Field Theory*, Princeton, Princeton Univ. Press, 1995.

[58] Terricabras, J.-M. and Trillas, E., 'Some remarks on vague predicates', *Theoria – Segunda Época* **10**, 1989, 1–12.

[59] Toraldo di Francia, G., 'What is a physical object?', *Scientia* **113**, 1978, 57–65.

[60] Toraldo di Francia, G., *The Investigation of the Physical World*, Cambridge, Cambridge Univ. Press, 1981.

[61] Toraldo di Francia, G., 'Connotation and denotation in microphysics', in Mittelstaed, P. and Stachow, E. W. (eds.), *Recent Developments in Question Logics*, Mannheim, 1985, 203–14.

[62] van Fraassen, B., *Quantum Mechanics: an empiricist view*, Oxford, Clarendon Press, 1991.

[63] Weinberg, S., 'The search for unity: notes for a history of quantum field theory', *Daedalus* **106**, 1977, 17–35.

[64] Weyl, H., *Philosophy of Mathematics and Natural Science*, New York, Atheneum, 1963.

[65] Zermelo, E., 'Untersuchungen über die Grundlagen der Mengenlehre, I', *Math. Ann.* **65**, 1908, 261–81.

Discussions

Shimony: One of the things that seems to me to be missing in the things that you've said is the way in which individuality comes into and goes out of being. I have a particular recent instance. Since the discovery of down conversion pairs of photons, one has a nice answer to an old question raised by Otto Frisch about how do you 'take' a photon. Well, once one has the pairs and the pairs arrive rather infrequently, if one catches one of the partners, that one is absorbed and gone, but then one finally can take the other photon, and it has some individuality. So individuality and lack of individuality seem very much to be context dependent. And that's hard to catch in a logical scheme.

French: I think Michael Redhead at one time referred to this notion of ephemeral, and then later on said, I'm going to give that up because he was pressed by Bob Weingard on the field theoretic nature of quantum field theory. But Michael still insists, well, there's still this particle grin, you know, that you get in scattering experiments and so on. And I guess all we're trying to do is catch the grin of the Cheshire cat here.

Healey: I'd like to follow up on Abner's question, because there does seem to be something of a problem about the observability of these quasi-sets. I mean that's a strange sort of problem, one doesn't expect sets to be observable in any direct sense.

French: Maddy thinks so.

Healey: Yes, I know. But one can imagine having a two-photon state, and then having a photon detector, and detecting first a photon and then another photon. Now it seems that one can refer to photon detection events as individual events. Can't one use those events, just referred to, to tag the photons which figured in them, and then say, at least in observation we have a set of two photons? Now perhaps that's not enough to have observed the quasi-set, one could say, there's a quasi-set of observed photons before, and now there's an observed set of two photons, not an observed quasi-set, but an observed set. But that's going to mean that the quasi-sets cannot be observed.

French: Sure, I don't know if this answers it. I mean the driving force behind this whole discussion of individuality and non-individuality is quantum statistics and the nature of the indistinguishability postulate: when the permutation operator acts, what exactly is going on when you're permuting the labels? Are you just somehow considering the particles as individuals, with the Hilbert space dividing up into these noncombining subspaces, or as these authors have claimed, are you losing the individuality there? So when you consider the statistics, the claim is that it becomes awfully hard to maintain that these things are individuals in the sense of having this primitive thisness. They're indistinguishable in a much stronger sense than classical electrons. That's the whole point. It's an attempt to come up with a sort of an axiom for, or axioms of, strongly indistinguishable objects in that sense.

The issue here is whether we can take the photon detection events as unequivocally 'tagging' the photons *as individuals*. Regarding quantum particles as such is certainly one metaphysical package that can be adopted, as I indicated. However they can also be taken to be 'non-individuals' in some sense, a view which Heinz Post always claimed to mesh well with the foundations of QFT. From the latter perspective – which of course is the one we are concerned with here – these photon detection events are Redhead's 'particle grin'. And, on this view, what the grin 'tags', if that is the right word, is not an individual per se, but a quantum, the logic and metaphysics of which Professor Krause and I are concerned with in our paper.

Mackey: I should identify myself, I'm actually a mathematician, and I was worried about some of these issues many years ago. And about 33 years ago I published a book, which was an attempt to explain quantum mechanics to mathematicians in their own language. And I brought this very question, and I think there isn't time to give you my solution, but I recommend that you might be somewhat satisfied by what I say about this in my book.

French: Yes, we've read your book [laughter]. But now I'm going to have to admit I found it incredibly difficult.

Mackey: But I spend two or three pages describing this, and, well, I'll give you a rough idea. If you think of a vibrating string with certain hickeys in it, how do you interchange two of those? Well, I discuss that at some length, and it gave me a very satisfying answer to the question you've raised.

Jackiw: So the question is what is the individuality of.

French: Well, in a sense, I guess part of it is language. If you're just going to talk about it entirely field theoretically, then you might want to say, well look, don't even talk about quanta, you've just got these fields. Well, then there's a further issue, Auyang, I think, goes into this, that you still have issues of individuality, when you're talking about the fields. And she approaches it from a sort of fiber bundle perspective. I'm just saying that if some people are using this language of quanta, and are talking about them as non-individuals, then we can at least give some kind of logical mathematical framework for grasping that. But maybe what we should be doing is just looking at the concept of a field as Paul Teller has tried to do.

24. Do Feynman diagrams endorse a particle ontology? The roles of Feynman diagrams in *S*-matrix theory

DAVID KAISER

1 Introduction

Feynman diagrams have long been a staple tool for theoretical physicists. The line drawings were originally developed in the late 1940s by the American physicist Richard Feynman for the perturbative study of quantum electrodynamics, and are taught today to students of quantum field theory in much the same manner as Feynman originally designed them. Yet for a time during the 1950s and 1960s, Feynman diagrams became an equally indispensable tool in the hands of quantum field theory's rivals – particle physicists following the lead of Lev Landau and Geoffrey Chew, who aimed to construct a fully autonomous *S*-matrix theory built around a scaffolding of Feynman diagrams, but conspicuously lacking all field-theoretic notions of quantum fields or Hamiltonians. By the early 1960s, Chew and his collaborators had commandeered the field-theoretic diagrams while denouncing quantum field theory as being both bankrupt and dead. In the process, Feynman diagrams were re-invested with distinct meanings by the *S*-matrix theorists, and were taken to represent a fundamentally different basis for studying the constitution and interactions of particles.

There was a series of steps by which *S*-matrix theorists crafted an autonomous diagrammatic method, some of which we consider here.[1] First came Geoffrey Chew's 1958 particle–pole conjecture. This work was soon extended by Chew in collaboration with Francis Low in their 1959 work on scattering off of unstable targets. In that same year, one of the earliest and most strident calls for an autonomy from quantum field theory was sounded by Lev Landau, who codified the new usage for Feynman diagrams. Richard Cutkosky's work from 1960 extended Landau's diagrammatic methods to treat Stanley Mandelstam's double dispersion relations, which had first been published in 1958. Out of this work emerged the twin notions of 'nuclear democracy' and a hadron bootstrap. In this new arena, what was 'pictured' in Feynman diagrams became starkly re-interpreted, now tied to a distinct ontological approach from that of quantum field theory. It is this change in the symbolic function of Feynman diagrams that we examine here.

---

[1] This is certainly not meant to be an exhaustive survey of the history of the *S*-matrix. The fullest historical account of this program can be found in James Cushing's study, *Theory Construction and Selection in Modern Physics: The S-Matrix* (New York: Cambridge University Press, 1990). See also Tian Yu Cao, 'The reggeization program 1962–1982: attempts at reconciling quantum field theory with *S*-matrix theory,' *Archive for the History of the Exact Sciences* **41** (1991): 239–83. The dispersion relations program of the 1950s is treated in Andrew Pickering, 'From field theory to phenomenology: the history of dispersion relations,' in *Pions to Quarks: Particle Physics in the 1950s*, edited by Laurie Brown, Max Dresden, and Lillian Hoddeson (New York: Cambridge University Press, 1989), pp. 579–99.

343

The essence of the shift between the quantum field theorists' use of Feynman diagrams and the new roles given to the same diagrams by S-matrix theorists can be summarized in the statement: for quantum field theorists, the diagrams served as a bookkeeping device to keep track of less and less important contributions within an infinite series, whereas for the S-matrix theorists, Feynman diagrams served as 'pole-finders' to locate the single most important contributions (nearby singularities) to the total S-matrix. As discussed in section 3, this shift entailed a change of emphasis from 'entity' physics to 'process' physics, with Feynman-like diagrams now characterized principally by their singularity structure instead of their particular particle content.

After reviewing in section 2 some of the steps during the late 1950s and early 1960s that led to this new dynamical outlook, we will examine the new ontological role of Feynman diagrams for the S-matrix theorists in section 3. Some concluding thoughts follow in section 4.

2 From diagrammatic polology to the Mandelstam representation

By the mid-1950s, the so-called 'Feynman rules' for quantum field theory were well-codified. Students learned from textbooks such as the 1955 *Mesons and Fields*, by Silvan Schweber, Hans Bethe, and Frederic de Hoffmann, how to draw diagrams for a given process as part of a perturbative solution for the transition matrix, S, and how to relate, in an unambiguous way, each feature of these diagrams to elements of the mathematical scattering amplitude. By the early 1960s, readers of several other popular quantum field theory textbooks, including Schweber's mammoth *An Introduction to Relativistic Quantum Field Theory* (1961) and James Bjorken and Sidney Drell's *Relativistic Quantum Fields* (1965), encountered essentially the same treatment of Feynman diagrams and the Feynman rules.[2] In particular, students learned to integrate over all possible momenta for internal (virtual) particles, and to add higher- and higher-order diagrams to increase the accuracy of the calculated amplitude for a given process.[3]

In the early 1950s, however, particle theorists hit trouble with the strong interactions between protons, neutrons, and pions. Although interaction Hamiltonians

---

[2] See Silvan Schweber, Hans Bethe, and Frederic de Hoffmann, *Mesons and Fields*, two volumes (White Plains: Row, Peterson, 1955), volume 1, pp. 192–6; Silvan S. Schweber, *An Introduction to Relativistic Quantum Field Theory* (Elmsford, NY: Row, Peterson, 1961), especially pp. 415–506; and James Bjorken and Sidney Drell, *Relativistic Quantum Fields* (New York: McGraw-Hill, 1965), especially pp. 173–208. Note in particular that none of these textbooks proceeded by means of the generating functionals Z or W, which are characteristic of more recent pedagogical approaches to quantum field theory. Cf., for example, Claude Itzykson and Jean-Bernard Zuber, *Quantum Field Theory* (New York: McGraw-Hill, 1980), and Lewis Ryder, *Quantum Field Theory* (New York: Cambridge University Press, 1985).

[3] In his recent textbook on quantum field theory, *Diagrammatica: The Path to Feynman Diagrams* (New York: Cambridge University Press, 1994), Martinus Veltman remarks on Feynman diagrams' curiously successful role in quantum field theory: 'the resulting machinery [of Feynman diagrams and the perturbative study of quantum field theory] is far better than the originating theory. There are formalisms that in the end produce the Feynman rules starting from the basic idea of quantum mechanics. However, these formalisms have flaws and defects, and no derivation exists that can be called satisfactory' (p. xi). Neither the usual operator formalism nor Feynman's path integral 'derivations' of Feynman diagrams satisfy Veltman, who instead sees the unexplained success of Feynman-diagram-based perturbative quantum field theory calculations as *itself* the support for these otherwise poorly grounded approaches. Veltman's characterization of the place of Feynman diagrams in ordinary quantum field theory is thus quite similar to the case presented here, of the originating role the diagrams played for the autonomous S-matrix approach.

could be specified for these particles' interactions, much as they could be specified for electrons and photons, the coupling constant between the nucleons and pions was much larger than one. That is, the nucleons and pions were coupled with a strong-interactions charge $g \gg e$. In fact, where quantum field theorists could exploit the smallness of $e^2 \approx 1/137$ for their perturbative bookkeeping within quantum electro-dynamics, the strong interaction seemed to be coupled with $g^2 \approx 15$ (in units where $\hbar = c = 1$).[4] Now, if physicists tried to treat pion–nucleon scattering in the same way as they treated photon–electron scattering, with a long series of more and more complicated Feynman diagrams, each higher-order diagram would include extra factors of g^2, and hence would overwhelm the lowest-order contributions. Such a scheme, therefore, could never be applied for the infinite series of terms involved in the usual perturbative approach to quantum field theory. Perturbation theory had broken down.

In the face of this failure, some particle theorists began to work with a new way to represent particles' scatterings, in the form of dispersion relations. The scattering amplitude for an arbitrary process could be written as an analytic function of the center-of-momentum energy. Beginning in 1954, Marvin Goldberger, Murray Gell-Mann, Geoffrey Chew, Francis Low, Yoichiro Nambu and others began to work in terms of these dispersion relations.[5] Although most of these theorists agreed in the mid-1950s that dispersion relations were merely a convenient representation within quantum field theory, the direct ties with field theory were not always sought out or proven.[6]

The key to the dispersion-relations approach was that the amplitudes were *analytic* functions of the center-of-momentum energy: once the functions' values were known in some small region, the functions could (in principle, at least) be determined every-where within their domains of analyticity. Geoffrey Chew sought to capitalize on this feature in 1958, when he outlined a possible procedure for treating the pion–nucleon interactions without getting mired in the breakdown of perturbation theory. He began by postulating an interaction Hamiltonian for nucleons and pions in the usual way, $H_{\mathrm{I}} = g\bar{\psi}\gamma_5\psi\phi$. With this form for H_{I}, the lowest-order Feynman diagram for the scattering of nucleons off of nucleons could be written down, as in Chew's figure (reproduced here in Figure 1): two incoming nucleons of momenta p_1 and p_2 could exchange a pion, scattering into the new momenta p_1' and p_2'. Defining the invariant momentum transfer for this process, $\Delta^2 = (p_1 - p_1')^2 = (p_2 - p_2')^2$, the

[4] The large magnitude of g^2, and what this magnitude portended for perturbative quantum field-theoretic treatments of the strong interactions, dominated discussion at the 1954 Rochester conference. See the transcript of the 'Theoretical Section' in *Proceedings of the Fourth Annual Rochester Conference on High Energy Nuclear Physics*, edited by H. P. Noyes, E. M. Hafner, J. Klarmann, and A. E. Woodruff (Rochester: University of Rochester, 1954), pp. 26–40. For more on the Rochester conferences, including summaries of the main content of each meeting, see John Polkinghorne, *Rochester Roundabout: The Story of High Energy Physics* (New York: W. H. Freeman, 1989).

[5] Two useful introductions to these dispersion relations, both of which contain brief reviews of the history of dispersion relations, may be found in the 1960 summer school lectures by J. D. Jackson and Marvin Gold-berger: see Jackson, 'Introduction to dispersion relation techniques,' in *Dispersion Relations: Scottish Universities' Summer School, 1960*, edited by G. R. Screaton (New York: Interscience, 1961), pp. 1–63; and Goldberger, 'Introduction to the theory and applications of dispersion relations,' in *Relations de Dispersion et Particules Élémentaires*, edited by C. DeWitt and R. Omnès (Proceedings of the 1960 Les Houches Summer School) (New York: John Wiley & Sons, 1960), pp. 15–157.

[6] Cushing documents carefully throughout his study that many of the dispersion-relations theorists, and even some of the S-matrix theorists, treated these new developments as branches within quantum field theory; only a few physicists, such as Chew and Landau, later advocated a more complete break from quantum field theory. See Cushing, *Theory Construction*, pp. 116–17, 119–22, 129–33, 142–5, and 187.

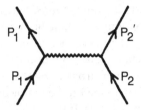

Fig. 1. Reprinted with permission from Geoffrey Chew, 'Proposal for determining the pion–nucleon coupling constant from the angular distribution for nucleon–nucleon scattering,' *Physical Review* **112** (1958): 1380–83, on page 1381.

integrand for the corresponding lowest-order Feynman integral would be

$$g^2 \frac{(\bar{u}_{p_1'} \gamma_5 u_{p_1})(\bar{u}_{p_2'} \gamma_5 u_{p_2})}{\Delta^2 + \mu^2},\tag{1}$$

where μ is the mass of the pion.[7] This contribution would dominate all other possible contributions to nucleon–nucleon scattering if the exchanged pion were put 'on mass shell', that is, if its momentum-squared were set equal to its mass-squared, $k^2 = \mu^2$. From equation (1), this would mean setting $\Delta^2 = -\mu^2$. Because none of the higher-order diagrams became singular at the point $\Delta^2 = -\mu^2$, all of the remaining infinite series of possible diagrams could be neglected, regardless of the fact that $g^2 \gg 1$. By setting the exchanged pion on mass shell, in other words, Chew had avoided the pitfall of perturbation theory.

However, such a straightforward assignment of momentum could not be undertaken lightly: evaluating the momentum transfer in the center-of-momentum frame yielded $\Delta^2 = 2k^2(1 - \cos\theta)$, where θ is the angle of scattering. If Δ^2 were set equal to $-\mu^2$, then $\cos\theta > 1$, which is impossible for any physical angle θ. Here is where Chew exploited the analytic nature of the amplitude: the amplitude could be analytically continued from the physical region, $-1 < \cos\theta < +1$, to the unphysical, yet dominant region, $\Delta^2 \approx -\mu^2$. Once the value of the amplitude was determined in this interesting, unphysical region, it could be determined anywhere else within the domain of analyticity. In addition, by extrapolating to the point $\Delta^2 \to -\mu^2$, near the *singularity* or *pole* of the amplitude, the residue of the pole would simply be proportional to the physical pion–nucleon coupling constant, g^2, rather than to any 'bare' coupling constants which would appear in a quantum field-theoretic Lagrangian.

This 1958 paper by Chew contained one of the first speculations on what was later termed the 'particle–pole' conjecture: any scattering amplitude would have a singularity at just the place where the virtual particle was assigned the mass it would

[7] Geoffrey Chew, 'Proposal for determining the pion–nucleon coupling constant from the angular distribution for nucleon–nucleon scattering,' *Physical Review* **112** (1958): 1380–3. Here we have written the pseudoscalar nucleon–pion coupling simply as γ_5, rather than as Chew did ($\tau_\alpha \gamma_5$), in keeping with the treatment in Schweber, Bethe, and Hoffmann, *Mesons and Fields*, volume 1, p. 225. Cf. Cushing, *Theory Construction*, pp. 109–11. Here Cushing rewrites Chew's equation (1) in a way which makes the extrapolation more clear: the denominator is $(t - \mu^2)$, with $t = -\Delta^2$ the invariant momentum transfer for the t-channel. Thus, assuming that this momentum-squared t is carried by the single exchanged pion, then the amplitude becomes singular when the pion is put on mass shell, $t = +\mu^2$, though, given $t = -2k^2(1 - \cos\theta)$, the pion pole lies outside of the physical region.

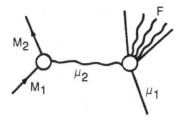

Fig. 2. Reprinted with permission from Geoffrey Chew and Francis Low,
'Unstable particles as targets in scattering experiments,' *Physical Review* **113**
(1959): 1640–48, on page 1648.

carry if it were a free, physical particle. At a conference at CERN in July 1958, Chew
stated his hope that every pole of the *S*-matrix would correspond in this way to a physical particle, with no other singularities occurring anywhere in the *S*-matrix.[8] This
would allow theorists to continue to avoid perturbation theory, by restricting attention to the poles of the lowest-order Feynman diagrams, and by extrapolating
between physical and unphysical regions of the energy or momentum-transfer
variables.

This approach was broadened several months later by Chew in collaboration with
Francis Low. The two theorists presented a means of studying the effects of particle
scattering which could not actually be measured in the laboratory. For example,
experimentalists were unable to make beams of pions for use in scattering experiments. Chew and Low showed that information about $\pi + \pi \rightarrow \pi + \pi$ scattering
could be extracted from laboratory results for $N + \pi \rightarrow N + 2\pi$ scattering, by exploit-
ing the analytic continuation of a lowest-order Feynman diagram to the dominant
pole.[9] (See Figure 2.) Again, the key was to evaluate the residue of the amplitude's
pole near the exchanged pion's mass shell, $\Delta^2 = -\mu^2$. In this way, by analytically continuing the momentum-transfer variable to unphysical regions, Chew and Low could
evaluate the cross section for pure, though unobserved, pion–pion scattering. The
particle–pole conjecture and these methods from 1958–9 for singling out the most
singular contributions to scattering amplitudes were soon labeled 'polology'.[10]

In the hands of Lev Landau, Chew and Low's pole-finding methods were generalized and codified for any arbitrary Feynman diagram. Landau entered the fray in

[8] Geoffrey Chew, 'The nucleon and its interactions with pions, photons, nucleons and antinucleons,' in
Proceedings of the 1958 Annual International Conference on High Energy Physics at CERN, edited by
B. Ferretti (Geneva: CERN, 1958), pp. 93–113, especially on pp. 94–5.

[9] Geoffrey Chew and Francis Low, 'Unstable particles as targets in scattering experiments,' *Physical
Review* **113** (1959): 1640–8, on p. 1640. This paper was received on 3 November 1958. For more on
some of Low's other work from this same period, see the essays collected in the *Festschrift* for his sixtieth
birthday, *Asymptotic Realms of Physics: Essays in Honor of Francis E. Low*, edited by Alan Guth, Kerson
Huang, and Robert Jaffe (Cambridge: MIT Press, 1983).

[10] The Chew–Low polology received immediate attention from Berkeley experimentalists, as well as from
other theorists such as Michael Moravcsik, Sidney Drell, and R. H. Dalitz. See J. Ashkin, 'Experimental
information on the strong interactions of pions and nucleons,' in *Proceedings of the 1960 Annual
International Conference on High Energy Physics at Rochester*, edited by E. C. G. Sudarshan *et al.*
(Rochester: University of Rochester, 1960), pp. 634–6; Moravcsik, in *Dispersion Relations*, edited by
G. R. Screaton; Drell, 'Peripheral contributions to high-energy interaction processes,' *Reviews of
Modern Physics* **33** (1961): 458–66; Dalitz, *Strange Particles and Strong Interactions* (New York:
Oxford University Press, 1962), pp. 103–5, 156, 164–9.

1959 with his paper 'On analytic properties of vertex parts in quantum field theory'.[11] Unlike Chew and Low, however, Landau argued not that the diagrammatic pololology offered a trick for bypassing quantum field theory's perturbative expansions, but rather that quantum field theory itself was incorrect for interacting particles. Only for the case of non-interacting particles (that is, when $H_I = 0$) could the quantum fields' equations of motion be solved exactly, and thus, Landau stated strongly, quantum field theory was not the correct way to proceed when studying the non-trivial cases of interacting particles. With quantum field theory so unhesitatingly pronounced dead, Landau outlined how best to sculpt a new theory of the strong interactions.

The key lay for Landau in a theory which treated only the relations between observable, free particles; and this led him to Chew's particle–pole conjecture. The opening paragraph of Landau's *Nuclear Physics* article alluded to the Chew and Chew–Low type of pololology:

As has become clear recently, a *direct study of graphs* is the most effective method of investigating the location and nature of the singularities of vertex parts.... By posing the problem of analytic properties of quantum field values, we actually go beyond the framework of the current theory. An assumption is thereby automatically made that there exists a non-vanishing theory in which ψ-operators and Hamiltonians are not employed, yet *graph techniques are retained*. In evolving dispersion relations, therefore, the employment of the graph technique is, indeed, solely consistent, since the problem becomes meaningless if the graph technique is rejected.[12]

Landau proposed to replace quantum field theory, and yet to 'retain graph techniques', continuing to rely on Feynman diagrams for studying the strong interactions. More forcefully than Chew and Low's diagrammatic pololology, then, Landau here drove a wedge between Feynman's diagrams and their field-theoretic birthplace. The Feynman diagrams were to act as a heuristic scaffolding to help in the construction of the new theory.[13]

Landau began by writing the Feynman integral corresponding to an arbitrary nth-order Feynman diagram, and concluded, after a short series of steps, that the

[11] Lev Landau, 'On analytic properties of vertex parts in quantum field theory,' *Nuclear Physics* **13** (1959): 181–92. For more on Landau's renouncement of quantum field theory in this 1959 work, cf. Cushing, *Theory Construction*, pp. 129–33, 167–72.

[12] Landau, 'On analytic properties,' p. 181; emphasis added. Landau's article contained only three references, none of them to the work by Chew or Chew and Low. However, Landau was well versed in Chew's pololology by the time of the 1959 Kiev conference. As Chew recalls in his interview with Capra, Landau approached him at this conference, and told Chew, 'You know, you have discovered an absolutely crucial point, which is that particles correspond to poles, that the *S*-matrix is an analytic function, and that the poles of the *S*-matrix are the particles.' ('Bootstrap physics: A conversation with Geoffrey Chew,' in *A Passion for Physics: Essays in Honor of Geoffrey Chew*, edited by Carleton DeTar, J. Finkelstein, and Chung-I Tan (Singapore: World Scientific, 1985), pp. 247–86, on p. 256).

[13] Landau invented an additional type of 'graph technique' in this 1959 paper, which was eventually labeled the 'dual diagram'. These diagrams were featured in several pedagogical settings in the early 1960s, and elaborated upon in the 1966 textbook, *The Analytic S-Matrix*, by R. J. Eden, P. V. Landshoff, D. I. Olive, and J. C. Polkinghorne (Cambridge: Cambridge University Press, 1966), pp. 60–2, 73–6, 104–8. (See also J. C. Taylor, 'Analytic properties of perturbation expansions,' *Physical Review* **117** (1960): 261–5; L. B. Okun and A. P. Rudik, 'On a method of finding singularities of Feynman graphs,' *Nuclear Physics* **15** (1960): 261–88; J. C. Polkinghorne, 'The analytic properties of perturbation theory,' in *Dispersion Relations: Scottish Universities' Summer School 1960*, edited by G. R. Screaton (New York: Interscience, 1961), pp. 65–96.) The dual diagrams offered a purely geometrical means of finding and characterizing singularities and thresholds for generic vertices, and could be analyzed in direct analogy to Kirchhoff's rules for electric circuits. Despite this early attention, these alternative diagrams never really caught on. They are historically significant, however, because they did *not* look anything like Feynman diagrams; thus it was not predetermined in 1959 that the *S*-matrix theorists should continue to rely so heavily on Feynman diagrams – a contrasting diagrammatic technique was put forward, but not exploited.

all-important poles of the scattering amplitude occurred *only* when the exchanged virtual particles were put on mass shell. Much as Chew had conjectured for his audience at CERN the previous summer, then, Landau had demonstrated that the only poles of the scattering matrix occurred for physical particles.[14] Feynman diagrams with internal lines satisfying $q_i^2 = m_i^2$, which dominated any given scattering process, corresponded to freely propagating, physical particles. By studying the analytic properties of arbitrary scattering diagrams, therefore, Landau hoped to construct a theory which related the scattering amplitudes of free particles, while abandoning all unobservable virtual particles and the bankrupt formalism of Hamiltonians.

Thus, by 1960, there had emerged two distinct sets of practices for translating between Feynman diagrams and scattering amplitudes. On the quantum field theory side, theorists taught their students to proceed along four steps:

1. Specify an interaction Hamiltonian in terms of quantum fields.
2. Derive propagators for these fields from the fields' equations of motion and commutation relations.
3. Integrate over all possible momenta for the internal, virtual particle lines.
4. Add more and more complicated Feynman diagrams within a perturbative expansion to improve the accuracy of the scattering amplitude.

Against each of these steps, dispersion-relations specialists countered:

1. Isolate the single most dominant Feynman diagram for any given interaction, based on the singularity structure of the corresponding amplitude, thereby avoiding all perturbative expansions.
2. Evaluate the amplitude near the pole, with all particles on the mass shell.
3. Use analytic continuation to extrapolate between physical and unphysical regions of the amplitude.
4. Avoid all reference to (unobserved) virtual particles.

The very same diagrams were bandied about by different groups of people and interpreted to mean different things.[15]

Although Chew-styled diagrammatic polology continued to dominate strong interaction physics into the mid-1960s, Chew himself never returned to this work (and only rarely cited it) after 1959. Instead, he was captivated by the prospects of producing an autonomous *S*-matrix theory. For this quest, Chew was most influenced by Stanley Mandelstam's 1958 work on double dispersion relations, and by Landau's 1959 attempts at total independence from quantum field theory. Within this next phase, Feynman diagrams, still further re-interpreted, continued to play a central role.

---

[14] A few years later, a team of British theorists demonstrated that in fact a 'new class' of singularities, not treated by Landau, existed for these same complex amplitudes, corresponding to 'solutions of the Landau equations with infinite internal momenta'. These new singularities arose from the intricacies of analytic continuation and the distortions of integration contours into the complex plane. See D. B. Fairlie, P. V. Landshoff, J. Nuttall, and J. C. Polkinghorne, 'Singularities of the second type,' *Journal of Mathematical Physics* **3** (1962): 594–602. Cf. Cushing, *Theory Construction*, p. 131.

[15] Actually, these groups were not always so cleanly separated. It is not often recalled, for example, that the classic treatise on quantum field theory by Bjorken and Drell, *Relativistic Quantum Fields* (1965), contains long excursions into dispersion relations and Landau's rules (pp. 210–82). We have seen above that Drell's 1961 review article featured the Chew–Low technique; similarly, Bjorken's 1959 dissertation was on the Landau rules (see p. 232). This is in keeping with Stanley Mandelstam's 1960 characterization of the dispersion relations work as simply mass-shell quantum field theory (see Cushing, *Theory Construction*, pp. 131–2).

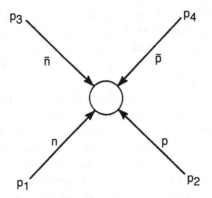

Fig. 3. Reprinted with permission from Geoffrey Chew, *S-Matrix Theory of Strong Interactions: A Lecture Note and Reprint Volume* (New York: W. A. Benjamin, 1961), p. 12.

Two months before Chew's 1958 CERN talk and his paper on the pion–nucleon coupling constant, Stanley Mandelstam, a young post-doctoral fellow at Columbia, presented a new kind of dispersion relation at the Washington, D.C. meeting of the American Physical Society, in May 1958. Whereas the earlier dispersion relations had been written as functions of only one complex variable, Mandelstam constructed a representation in which the scattering amplitude was a function of two complex variables, the center-of-momentum energy *and* the momentum transfer.[16] Mandelstam considered the case of general two-body elastic scattering, and wrote down the scattering amplitude as the sum of a series of terms:

$$A(s_1, s_2, s_3) = \frac{1}{\pi} \sum_{i=1}^{3} \int ds'_i \, \frac{\rho(s'_i)}{(s'_i - s_i)} + \frac{1}{\pi^2} \sum_{i \neq j} \int ds'_i \int ds'_j \, \frac{\rho(s'_i, s'_j)}{(s'_i - s_i)(s'_j - s_j)}. \qquad (2)$$

Each of the invariants, s_i, referred to the center-of-momentum energy for the $2 \to 2$ scattering in one of the three possible channels, and simultaneously to the momentum transfer in one of the 'crossed' channels.[17] (See Figure 3.) As had been done in the Chew, Chew–Low, and Landau work, the analytic nature of the amplitude A could next be exploited: each of the density functions $\rho(s_i, s_j)$ would be non-zero only in one of three non-overlapping regions, and the amplitude could be evaluated in any of these three regions, or physical channels, by means of analytic continuation from one physical region to another.

Soon Feynman diagrams were put to use as a means of determining the boundaries of the physical regions for these $\rho(s_i, s_j)$ functions. Building explicitly upon Landau's diagrammatic method of isolating singularities for the single-variable dispersion relations, Richard Cutkosky analyzed the new double-dispersion relations with the

---

[16] The abstract for Mandelstam's talk is published in the meeting's program: *Bulletin of the American Physical Society* **3** (1958): 216. He soon published fuller discussions of this work in Mandelstam, 'Determination of the pion–nucleon scattering amplitude from dispersion relations and unitarity. General theory,' *Physical Review* **112** (1958): 1344–60; *idem.*, 'Analytic properties of transition amplitudes in perturbation theory,' *Physical Review* **115** (1959): 1741–51; and *idem.*, 'Construction of the perturbation series for transition amplitudes from their analyticity and unitarity properties,' *Physical Review* **115** (1959): 1752–62.

[17] The Mandelstam kinematic variables were often written as (s, t, u), though Chew consistently wrote them in the even more symmetric form, (s_1, s_2, s_3). See, e.g., Geoffrey Chew, *S-Matrix Theory of Strong Interactions: A Lecture Note and Reprint Volume* (New York: W. A. Benjamin, 1961), pp. 10–22.

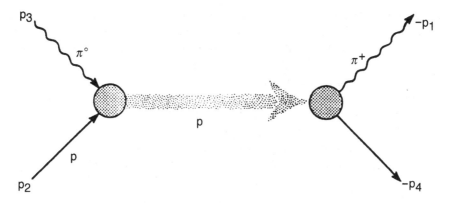

Fig. 4. Reprinted with permission from Geoffrey Chew, *S-Matrix Theory of
Strong Interactions: A Lecture Note and Reprint Volume* (New York: W. A.
Benjamin, 1961), p. 16.

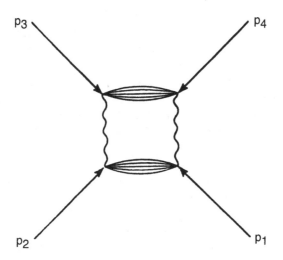

Fig. 5. Reprinted with permission from Geoffrey Chew, *S-Matrix Theory of
Strong Interactions: A Lecture Note and Reprint Volume* (New York: W. A.
Benjamin, 1961), p. 26.

help of Feynman's ubiquitous diagrams in 1960.[18] Cutkosky revealed that the density
functions $\rho(s_i, s_j)$ could be evaluated most simply by replacing all internal propagators by mass-shell delta functions: $(p^2 - m^2)^{-1} \rightarrow i\pi\delta(p^2 - m^2)$. The lowest-order
Feynman diagrams to contribute to the double-variable density functions, $\rho(s_i, s_j)$,
were fourth-order 'box' diagrams. Again, low-order Feynman diagrams, now suitably
re-interpreted with the propagator-to-delta-function prescription, served as a
heuristic device in the construction and calculation of the new theory.

These relations between Landau and Cutkosky rules and particular Feynman
diagrams were quickly summarized by constructing 'Landau graphs' and 'Cutkosky
graphs' to stand in for entire classes of individual Feynman diagrams, each member of
these classes sharing the same *singularity structure*. (See Figures 4 and 5.) Whereas in

---

[18] R. Cutkosky, 'Singularities and discontinuities of Feynman amplitudes,' *Journal of Mathematical
Physics* **1** (1960): 429–33. Cf. Chew, *S-Matrix Theory of Strong Interactions*, pp. 23–6.

quantum electrodynamics one pictured an in-principle infinite series of more and more complicated diagrams, obtained both from the permutation of specific particles in a given diagram and from the inclusion of higher-order loop corrections at each vertex, in the emerging *S*-matrix work the single most relevant feature of the Feynman-like diagrams was the presence of one, two, or *n* exchanged particles *between* vertices. It was these exchanged lines in the diagrams which gave rise to the poles and branch cuts of the amplitude, once the exchanged particles were put on mass shell; complications from perturbatively added loops at vertices now were simply irrelevant, and the isolation and location of singularities with diagrammatic means became central. As we will see in the following section, this shift in the use of Feynman diagrams came tied with a deep shift in how particles and their interactions were conceived ontologically.

3 'Nuclear democracy' and a hadron bootstrap

For Chew, the Landau–Cutkosky rules signified more than convenient appropriations of Feynman diagrams for use in various dispersion relations calculations. Instead, their shared restriction of all interesting physics to physical, on-mass-shell particles, seemed to portend a further break from quantum field theory. As he explained in his 1961 lecture note volume, even through the Landau–Cutkosky rules 'completely determine the dynamics, they contain not the slightest hint of a criterion for distinguishing elementary particles [from composite ones].... If one can calculate the *S*-matrix without distinguishing elementary particles, why introduce such a notion?'[19] In quantum field theory, the only lines to appear within Feynman diagrams corresponded to the *elementary* quantum fields appearing in the interaction Hamiltonian; composite particles were excluded. Yet, if the Landau–Cutkosky diagrammatic recipe were treated independently from this field-theory proviso, then particles of all sorts could be treated on an equal footing, subject only to their being on mass shell. Deuterons, for example, could be treated symmetrically with neutrons and protons. Borrowing the phrase from Murray Gell-Mann, Chew soon dubbed this notion by the colorful term, 'nuclear democracy'.[20]

Further consideration throughout 1960–61 of the Mandelstam representation, and its treatment by the Landau–Cutkosky diagrammatic machinery, led Chew to the essential point of his autonomous *S*-matrix program: not only were composite particles to be treated equivalently to elementary ones, but all (strongly interacting) particles could in fact be seen as composites *of each other*. This 'bootstrap' of all hadrons arose for Chew from diagrams such as Figure 3, combined with the crossing

[19] Chew, *S-Matrix Theory of Strong Interactions*, p. 5. Chew hinted at this as early as his 1958 CERN rapporteur talk: 'The nucleon and its interactions,' pp. 94, 95, 100. Landau had mentioned this same possibility in 1959: see his 'On analytic properties,' pp. 181, 191; and his talk of the same title at the 1959 Kiev conference, in *Ninth International Annual Conference on High Energy Physics* (Moscow: International Union of Pure and Applied Physics), volume 2, pp. 95–101, on p. 99.

[20] Chew attributes the term to Gell-Mann in his 1964 lecture notes, 'Nuclear democracy and bootstrap dynamics,' in *Strong-Interaction Physics: A Lecture Note Volume*, by Maurice Jacob and Geoffrey Chew (New York: W. A. Benjamin, 1964), pp. 103–52, on p. 105.

relations between the three Mandelstam channels.[21] As he explained in a series of lectures published in 1964, the notion of a hadron bootstrap sprang from the assumption of nuclear democracy:

The bootstrap concept is tightly bound up with the notion of a democracy governed by dynamics. Each nuclear particle is conjectured to be a bound state of those S-matrix channels with which it communicates, arising from forces associated with the exchange of particles that communicate with 'crossed' channels. . . . Each of these latter particles in turn owes *its* existence to a set of forces to which the original particle makes a contribution. In other words, each particle helps to generate other particles which in turn generate it.[22]

As he wrote about it in 1964, then, the combination of the bootstrap concept with a democracy amongst all hadrons bespoke not merely the lack of a *formal* distinction between supposedly elementary or composite particles, but rather the lack of any *ontological* distinction. It was not just that the Landau–Cutkosky rules happened to treat elementary and composite particles in the same way, but that the field-theoretic division of the world into elementary and composite particles had been a mistaken one.

The role of Feynman diagrams in effecting these new philosophical views of the microcosmos cannot be too strongly emphasized here: the diagrams were divorced from their specific, unambiguously defined meanings in terms of field-theoretic amplitudes; following Landau, the same pictorial scheme was then used to build the new theory. The theorists were now free to 'read' the 'content' of the diagrams as they saw fit: if the *diagrams* simply tracked particles' propagation and scattering, and if various rotations of these diagrams showed mutually interchanging exchanges between various channels, then 'particles' no longer had to bear any relation to the descriptions previously forced upon them by quantum field theorists. By 'retaining graph techniques' but jettisoning fields and Hamiltonians, S-matrix theorists' *philosophical* picture shifted radically.

The democratic-bootstrap notion of the mid-1960s fixed upon a fundamentally different ontological vision of particles and their interactions from that of quantum field theory. In S-matrix physics, 'dynamics' took all precedence over 'entities'.[23] The

---

[21] Chew developed this bootstrap idea in a pair of papers with Stanley Mandelstam in 1960–61, and in work with Steven Frautschi. See Geoffrey Chew and Stanley Mandelstam, 'Theory of low-energy pion–pion interaction,' *Physical Review* **119** (1960): 467–77; *idem.*, 'Theory of low-energy pion–pion interaction, II,' *Il Nuovo Cimento* **19** (1961): 752–76; Geoffrey Chew and Steven Frautschi, 'Principle of equivalence for all strongly-interacting particles within the S-matrix framework,' *Physical Review Letters* **7** (1961): 394–7. See also Chew's exuberant invited paper before the New York meeting of the American Physical Society, January 1962, published as 'S-matrix theory of strong interactions without elementary particles,' *Reviews of Modern Physics* **34** (1962): 394–401; Chew, 'Nuclear democracy and bootstrap dynamics'; and Geoffrey Chew, Murray Gell-Mann, and Arthur Rosenfeld, 'Strongly interacting particles,' *Scientific American* **210** (February 1964): 74–93. For more on the bootstrap idea, see Cushing, *Theory Construction*, chapter 6; *idem.*, 'Is there just one possible world? Contingency vs. the bootstrap,' *Studies in History and Philosophy of Science* **16** (1985): 31–48; and Yehudah Freundlich, 'Theory evaluation and the bootstrap hypothesis,' *Studies in History and Philosophy of Science* **11** (1980): 267–77.

[22] Chew, 'Nuclear democracy and bootstrap dynamics,' p. 106. Chew emphasized that the principal difference between his 1961 and 1964 lectures was that the latter rested on 'the unequivocal adoption of nuclear democracy as a guiding principle.' (Ibid., p. 104.)

[23] This point is made by Fritjof Capra in his popular exposition of modern physics, *The Tao of Physics*, 2nd Edn. (New York: Bantam Books, 1983 [1975]), p. 252. Capra worked as a physicist with Chew in Berkeley, and Chew is included among the people to whom Capra dedicated *The Tao of Physics*. See also Chew's popular articles on the subject: '"Bootstrap": A scientific idea?,' *Science* **161** (23 August 1968): 762–5; 'Hadron bootstrap: triumph or frustration?,' *Physics Today* **23** (October 1970): 23–8; and 'Impasse for the elementary-particle concept,' in *The Great Ideas Today*, edited by Robert M. Hutchins and Mortimer J. Adler (Chicago: Encyclopedia Britannica, 1974), pp. 92–125.

particles entering into reactions were themselves merely *derivative* of the 'violently nonlinear' crossed-channel bootstrapped dynamics.[24] As Chew wrote, 'Since any particle may play the role of intermediary in a two-step reaction, the concept of "particle" is seen to be interlocked with the concept of "reaction".'[25] Furthermore, '[a]ccording to the bootstrap hypothesis, no nuclear particle is intrinsically more significant than any other.'[26] Thus, the focus for the *S*-matrix theorists remained not at the level of equally insignificant democratic particles, but rather at the level of the underlying dynamical *reactions* which gave rise to the appearance of these particles within various reaction channels – and these reaction channels were to be studied in terms of low-order Feynman-like diagrams (Figures 4 and 5) and their rotations.[27]

For Chew, here lay a crucial contrast with the ontological vision of quantum field theory. Where quantum field theorists focused on a small set of fundamental, elementary entities (such as photons or electrons, or, increasingly throughout the 1960s, quarks), Chew urged that the truly important subjects of study were the Yukawa-like *exchanges* between particles, rather than the end-state particles themselves. Thus, he confessed in 1970,

I would find it a crushing disappointment if in 1980 all of hadron physics could be explained in terms of a few arbitrary entities. We should then be in essentially the same posture as in 1930, when it seemed that neutrons and protons were the basic building blocks of nuclear matter. To have learned so little in half a century would to me be the ultimate frustration.[28]

Quantum field theory, with its focus on elementary entities, like its postulation of virtual particles and quantum fields, simply failed for Chew to capture the real essence of the strong interactions. The hadronic world was to him and his collaborators an endlessly shifting, mutually bootstrapping world in which 'particles' were the mere epiphenomena of dynamical interactions. To unravel this singly self-consistent network, Chew urged an attention to processes over particular entities. And this attention to processes sprang from and was undergirded by a new way of handling Feynman diagrams.

4 Conclusions: the circular role of Feynman diagrams in *S*-matrix theory

Much like the bootstrapping particles eventually 'seen' in them, Feynman diagrams played a circular role in the emerging *S*-matrix theory ontology. From the dispersion relations and pololology work in the late 1950s, theorists learned to restrict attention to lowest-order diagrams when treating the strong interactions, and to only consider these diagrams with all particles put on mass shell. By serving as 'pole-finders', the diagrams became a bookkeeping device of a different sort from their original function in quantum field theory. Now a single Landau or Cutkosky graph organized infinite

---

[24] Chew, 'Nuclear democracy and bootstrap dynamics,' p. 106.

[25] Chew, 'Impasse for the elementary-particle concept,' p. 104.

[26] Ibid., p. 119.

[27] As is clear from Figures 4 and 5, these Feynman-like diagrams were *drawn* with the same pictorial form as Feynman diagrams. This could engender confusion, as Gabriel Barton feared. In his 1965 textbook, *Introduction to Dispersion Techniques in Field Theory* (New York: W. A. Benjamin, 1965), he explicitly cautioned his readers about how to 'read' such new Feynman-like diagrams: 'Such diagrams must be carefully distinguished from Feynman diagrams; their interpretation is as follows: on the *left* we symbolize the *process* in question: the source *j* vomits out two particles' (p. 149).

[28] Chew, 'Hadron bootstrap: triumph or frustration?,' p. 25.

numbers of Feynman diagrams according to their (singularity-generating) exchanges between vertices. After these generalizations by Landau and Cutkosky, it became clear that no formal distinction could then be made between elementary and composite particles in the diagrams themselves. At least, no distinctions appeared to be inherent in the line drawings themselves, from which the new autonomous theory was to be constructed.

As the Feynman diagrams were further peeled away from their original quantum field theory package, this lack of a formal distinction was elevated to a guiding principle, and a full democracy amongst all hadrons was proclaimed. By further tinkering with these low-order diagrams, the close interconnectedness of the various Mandelstam channels could be highlighted: rotations of the diagrams allowed the (s_1, s_2, s_3) channels to emerge on an equal footing, with 'incoming' and 'outgoing' reactants in one channel simultaneously becoming exchanged 'force-carrying' particles in a crossed channel. These diagrammatic rotations now linked the democratic particles in a hadronic bootstrap; and in so doing, the diagrams ceased to depict stable, elementary entities, revealing instead a self-supporting menagerie of reactions and exchanges.

Thus, from the re-interpretation of Feynman's diagrammatic machinery by Landau, Mandelstam, and Cutkosky arose the S-matrix focus upon bootstrapped dynamics. And from the S-matrix focus upon such bootstrapped dynamics arose a further re-interpretation of the physical content pictured in Feynman's line drawings. Whereas these diagrams had once been held, in the hands of quantum field theorists, to depict 'aristocratic' scatterings amongst foundational, elementary entities, they were now taken to portray the wild dances of exchanges from which emerged a spectrum of 'democratic' particles. While these simple lines were arranged the same way pictorially by Feynman in 1948 and by Chew in 1964, the theoretical significance read into these lines had undertaken a sea-change.

In response to the question posed in the title of this chapter, then, the answer appears to be 'no', or at least 'not uniquely'. Between 1958 and 1964, the diagrams' contortions inspired, for some theorists, a new ontology for the strong interactions, which in turn contributed to a re-interpretation of the physical content of the diagrams themselves. When particle experimentalists earlier in the century had struggled with the vagaries of 'artificial clouds and real particles',[29] some theorists at mid-century saw even these 'particles' dissolve in a sea of real lines.

Discussions

Mackey: I'm surprised that you didn't mention a somewhat earlier use of Feynman diagrams in the many body problem. In 1957, I think that was the key year, there were several different approaches to doing this. But this is all written out for the first time in a book that came out by David Pines [*The Many Body Problem*, a lecture note and reprint volume (W. A. Benjamin, New York, 1961)]. That was even earlier. To say another word, the main point is that Feynman diagrams occur whenever you want to do the perturbation theory at a high level. And that is its reason for persistence, in a sense.

[29] This phrase is borrowed from the title of Peter Galison and Alexi Assmus's study of C. T. R. Wilson's cloud chamber: 'Artificial clouds, real particles,' in *The Uses of Experiment*, edited by David Gooding, Trevor Pinch, and Simon Schaffer (New York: Cambridge University Press, 1989), pp. 225–74.

Kaiser: You're certainly right. There are other areas, condensed matter theory being a very important one, where these diagrams were being used in still a distinct way. And so they get drawn the same way and used differently.

Mackey: There's always the question of having some problem you can approach by perturbation and using it to higher and higher orders, and you use the Feynman diagrams to keep track of the bookkeeping.

Kaiser: Yes, except it seems at least in the double dispersion relations, as far as I understand them, the goal was not to have a perturbative theory.

25. On the ontology of QFT

FRITZ ROHRLICH

1 Philosophical introduction

The following seemingly trivial story raises important questions which lead to the philosophical issues essential for our discussion.

My cat is alive and well. I can describe her by size, color, behavior, etc. I can watch her, touch her, play with her, and feed her. I know this animal is really there. But my biologist friend came to visit and informed me that my cat is really nothing but a very complicated collection of different kinds of cells in mutual interaction, and he can prove it to me by having me look through his microscope, and, if I would only let him, by dissecting my cat. But the biochemist next door has a quite different opinion: he claims that my cat is *really* just a very complex assemblage of organic compounds, proteins, nucleic acids, etc., etc., all in mutual interaction. He, too, can prove his claim.

Of course, physicists know better: a cat, like all matter living or dead, is in last analysis *really nothing but* atoms bonded together to form the complicated molecules my biochemist friend is talking about. And they can prove that too. But, finally, one cannot ignore the elementary particle people who tell us that atoms are just composed of quarks, gluons, and electrons, so that a cat is *really, really* just a very complicated assembly of those few types of elementary particles in mutual interaction.

The story ends here. But only temporarily. Weinberg's final theory may be just around the corner, and then we shall know what my cat is *really, really, really*, like.

This story raises a number of questions about our knowledge of the world. Is one science reducible to another so that the only science we really need is the final theory and everything else follows from it? Or, if this is not the case, does the world have many faces? And are they consistent with one another? What then becomes of scientific truth? Of the unity of nature? Of ontology?

Many different answers have been given to these and closely related questions. A fuller discussion clearly lies outside the present workshop. But I can state here my own views formed over the years and based on the work of many others too numerous to list. The beliefs I shall present, I have justified (Rohrlich and Hardin 1983; Rohrlich 1988, 1994, 1995), and I shall use them to answer some of the problems of quantum field theory (QFT). Very briefly, they are as follows.

(1) Our scientific knowledge falls naturally into a number of '*cognitive levels*' on which we can perceive the world. These differ in scale of length, time, and energy. Each level requires for its exploration different tools of research and different methodologies. On each level one acquires knowledge in a different way and forms

images of the world that are different. They differ in their claims of what components the world is made of (cells, molecules, atoms, etc.), their properties, and their (inter)actions.

(2) It is not possible to claim that the description on one level is 'better' (simpliciter) than that on another level. Each level offers aspects not fully implied by any other level, so that levels *complement* one another. Thus, atoms have no color, are not rigid, etc. These properties cannot be derived from lower levels but are nevertheless necessary for the description on higher levels.

There existed for a long time the claim that theories on deeper levels ('finer graining' or more fundamental) imply the theories on higher ones ('coarser graining' or less fundamental). That claim, called *'reductionism'*, has been repeatedly proven false in recent years. Higher levels contain concepts (and associated words) that have no meaning on lower levels; they cannot be derived by logical arguments from lower levels (incommensurability of concepts): they must be 'invented' (emergence of concepts). But since the higher level theory is already known (being developed historically earlier), this invention involves just a recall of concepts incommensurable with lower level concepts.

(3) The reason that so many physicists continue to claim the validity of reductionism is primarily due to the following two erroneous beliefs: (a) that a mathematical derivation of the equations of theory S from the equations of a deeper theory T suffices to prove reduction, and (b) that theory T is both needed and able to explain all the problems that arise in theory S; S is therefore superfluous and can be replaced by T. Point (a) ignores the incommensurability of concepts, and point (b) confuses *reductive explanation* with theory reduction.

As to the latter: there are in fact many questions on the level of theory S that *can be* explained satisfactorily by S and for which T is in fact utterly useless; and there also are other questions that indeed *cannot* be explained by S and that do require T. But none of this implies that the whole theory S follows deductively from T: reductive explanation does not imply theory deduction.

(4) There exists a certain *autonomy* of scientific theories. First, there is the concept of validity domain; it is characterized by the errors made in using S for a given set of physical variables instead of T. Given the validity domain, the theory remains correct within those errors even after a new theory, T, on a deeper level has been developed. Therefore, it cannot be 'overthrown' and replaced in a scientific revolution by any empirical evidence that falls within those limits. This is of course well known to physicists who use Newtonian gravitation theory 'where it applies' rather than general relativity. The latter theory obviously did not run Newtonian gravitation theory out of business.

A consequence of the above is that even though a new, deeper level of theory may be essential for the explanation of *some* of the unanswered questions on a higher level, it does *not* by any means make the higher level theory superfluous, make research on those higher levels no longer 'relevant', or make it 'less important' than research on that new deeper level. Just look at the present blooming of condensed matter physics which involves a theory on a much higher level than the 'more fundamental' theory of quarks and leptons!

The view expressed by the above claims can be dubbed *'scientific pluralism'*. It has important implications for *scientific truth*: to assert something to be true without any means of confirmation is meaningless in science; but if one does have a means of confirming or refuting a truth claim, one thereby also has a specification of the level on which this confirmation or refutation is to be carried out. Thus follows that *scientific truth is not necessarily meaningful outside its cognitive level.*

It also follows that the *ontology* implied by a given theory is *tied to the cognitive level on which the theory functions.* Applied to the theory of interest here, QFT, the ontology of the next deeper level theory (string theory? quantum gravity with a 'grainy' space-time?) can be expected to be entirely different from the ontology of QFT.

The contextual necessity implied by the validity domain of a theory is of course intuitively well-known to us as physicists. But we often neglect that necessity out of convenience. Thus, we simply *know* when to use classical mechanics, and when to use quantum mechanics; but we also talk carelessly about 'classical point particles' when we know very well that classical physics does not apply below a Compton wave length or so, and that both the mass as source of a gravitational field and the charge as source of a Coulomb field yield divergent energies in the classical point limit.

2 Fields and particles in classical physics

Classical particles are characterized by space-time localization, a continuous world-line, and occasional additional properties (angular momentum, charge, etc.). They also have individuality and extension (since the classical description fails for lengths below a certain size).

Classical fields come in two very distinct varieties: linear and nonlinear, depending on their defining equations. The former permit superposition and the latter do not. There is no superposition in the velocity field of a fluid satisfying the Navier–Stokes equation; but there sure is superposition for the linear Maxwell field.

Fields are functions of space and time. They carry identity which they preserve throughout their continuous evolution: the dynamics of a small volume of fluid can be followed in its time development, and so can an electromagnetic pulse.

The common classical *linear* fields (the electromagnetic and the gravitational fields) are emitted and absorbed by particles (including macroscopic objects like stars and planets) which act as their sources and sinks. The fact that in Newtonian gravitation theory these fields move with infinite speed is irrelevant to this property. But the interaction between sources and sinks via these fields makes them mediators of 'forces'. So the linear fields lead to a dichotomy of entities: *fields* as mediators of forces, and *particles* as emitters and absorbers of these forces.

At this point one observes that while the detection of particles or of volume elements of nonlinear fields (fluids) raises little controversy, the detection of classical linear fields is less trivial: a linear field can only be detected when it exerts a force on a particle. But this does not imply that the electric and the magnetic fields do not have ontological status: their presence can be ascertained by suitable experiments.

3 Fields and particles in quantum mechanics

Any interpretation of quantum mechanics (QM) must start with a formulation of the theory in its own right and not as 'quantized classical mechanics'. The notion of 'quantization' indicates that one has not been freed of the heuristic level on which the theory has been developed originally.

In QM particles are treated as nonrelativistic while electromagnetic fields are intrinsically relativistic. The theory is therefore strictly speaking inconsistent from the space-time symmetry point of view: it is neither Galilean nor Lorentz invariant.

Also, contrary to classical mechanics, QM still has an unsolved fundamental problem. That unsolved problem is of course the issue of time development: are there two qualitatively different time developments as first suggested by von Neumann, a deterministic continuous time development specified by the Schrödinger equation, and another indeterministic and discontinuous 'collapse of the wave packet'? Or, is the latter only apparently so, and is the collapse actually due to a physical cause: environmental influence (Zurek 1982), dynamical reduction expressed by an (*ad hoc*) stochastic term in the Schrödinger equation (Ghirardi *et al.* 1986, Pearle 1989), or other causes? Until that is resolved, our interpretation must remain tentative.

While QM is not the topic here, I cannot leave it without pointing out one important ontological difference due to its different cognitive level: the words 'particle' and 'field' have a meaning in QM entirely different from their meaning in classical physics. The so-called wave–particle conundrum is largely due to a *metaphoric* use of these words. This comes about because the classical dichotomy between fields and particles becomes confused in QM: the particle is described by a wave function which depends on space and time and is therefore a (linear!) field; and the linear electromagnetic field of classical mechanics takes on particle properties, the photons of QM. Both 'particle' and 'field' have a new and different meaning. One must therefore distinguish the classical meaning of these words from 'quantum particle' and 'quantum field'.

Now, the experimental physicist *observes only particles*, i.e. localized manifestations of them; for example, the electron is detected by a scintillation in a well-defined location on a screen. He does not observe its wave function (which might be thought of as a 'field'). The ontological status of the wave function is resolved in a Hilbert space formulation of QM where it is seen to be a representation of the state vector; it is only one of several alternative representations (the configuration space representation). It is therefore not a field in the usual sense, and it is certainly not an observable. Only its absolute square represents an observable (corresponding to the density operator) under the usual kinds of measurement. I shall not discuss 'protective measurements' here (Aharonov *et al.*, 1993).

In contradistinction, in order to detect a photon it must interact with a massive *particle* whether by collision (Compton effect) or by absorption in an atom. Thus, even though the photon has particle properties, when it comes to detection, it is seen only in the way classical fields are seen: in their effect on a charged (and therefore massive) particle. In a sense, one could say that, at least in QM, the observer sees only massive particles, and only the theorist deals primarily with fields some of which are unobservable. But this does not mean that photons are not particles; few physicists would deny the photon ontological status.

4 Fields and particles in quantum field theory

Relativistic QFT has a consistent space-time symmetry (in contradistinction to QM): both particles and fields are treated in a Lorentz invariant way.

What I said about the incompleteness of QM is even more applicable to QFT: the theory is still not complete even mathematically. Its interpretation can therefore be only tentative. No mathematically rigorous results exist as yet for QFT in four dimensions especially in the presence of zero mass particles. But if the results on two- and three-dimensional models are an indication, there is good reason to believe that the four-dimensional theory can also be put on a sound mathematical foundation. Meanwhile, many calculations in QFT are still done on a heuristic level.

Of course, a theory can be given meaning only when it is formulated in its own right: a formulation based on a 'second quantization' (which is of course only a first quantization for the electromagnetic field) is just as inadequate as a formulation of QM based on 'quantization' of classical mechanics. Like QM, relativistic QFT must be treated in its own right.

But the mathematical questions are not the only reasons for my hesitation at a philosophical interpretation of QFT. Perhaps even more important for considering QFT as not yet ready is the fact that QFT is not yet an *established* theory (in the sense of Rohrlich and Hardin 1983). This means that QFT has so far not been 'overthrown by a scientific revolution' – the next level theory is not yet here. Only then can one know the exact validity limits of QFT; and only then can one make the corrections that the development of the next level theory necessitates.

The following general comments are made with this in mind. I emphasized the changed meaning of 'particle' and 'field' in the transition from the classical to the QM level. A further change takes place when one goes from the QM level to the QFT level. In the operator formulation, the field practically takes over and one must make some effort to extract a particle description. Let me explain.

What the quantum field theorist computes are primarily expectation values of (products of) field operators (more precisely, of operator valued distributions). Of course, he or she also uses operators that have a more direct significance such as operators for spin, isospin, charge, etc. But the theory is formulated in terms of fields and their interactions; these are the primary quantities of interest to the theorist.

But now consider the experiments. What are detected at SLAC, Fermilab, CERN, etc. are *particles* and their properties, not fields. People build particle detectors, not field detectors; they measure particle tracks, particle counts, particle collisions and particle decay modes. The results of the large experimental groups are stated in terms of particles, their properties, and their decay modes. They don't look for an intermediate vector boson *field* but for intermediate vector bosons, not for a Higgs field but for a Higgs particle.

Is there a contradiction? No, because the quantum field theorist knows how to extract particles from fields: the fields can be expressed in terms of creation and annihilation operators which are identified as referring to particles even though very often only to 'bare' ones rather than to the 'physical' ones: the Fock representation saves the day. But there is a mathematical difficulty: the Fock representation is valid usually only for free fields. A rigorous treatment shows that nontrivial interactions require representations that are *unitarily inequivalent* to the Fock representation. One must then attempt to get to the Fock representation at least asymptotically assuming the particles to be asymptotically free. Even that is not always possible rigorously. For example, in the case of a Coulomb field there is no Fock representation. But such difficulties are usually fudged.

What I am saying is not that one cannot get from fields to particles on the theoretical level, but that it is much less trivial than might be thought if one wants to go beyond crude approximations. And while the ontological status of elementary particles and their properties is in little doubt, being accessible to empirical evidence, just exactly what else in the theory has ontological status is far less obvious.

Note the vague similarity to QM: the central concepts of the latter are the operators which *represent* the observables and the wave functions (more generally, the rays of Hilbert space) which *represent* the states of the system; neither of these are by themselves observable. Instead, the observable outcomes of the theory are given by probabilities. These are the squares of the magnitudes (positive numbers) of inner products in Hilbert space. Typically, they refer to the interactions between *particles*.

The situation in QFT is seen in a different light when it is formulated by means of path integrals as was first suggested by Feynman. Here, the particle concept enters *ab initio*. And the theory is from the beginning much closer to the empirical situation. In the same spirit, Feynman objected to the use of creation and annihilation operators because they permit the creation or annihilation of a single charged particle, thus violating charge conservation.

But the difference between the operator formulation and the path integral formulation is enormous; when Feynman first presented the path integral formulation of QM at the Shelter Island conference in 1947, Bohr, then considered to have deeper insight into QM than anyone else, did not recognize it as describing QM. Needless to say, the equivalence seems to have been obvious to Feynman all along. And he then proceeded to generalize the path integral formulation to QFT. A heuristic proof of the path integral and the then conventional formulations of QFT was given for QED soon thereafter by Dyson (1949). But it took almost a quarter century till the mathematical equivalence of the operator formulation and the path integral formulation could actually be proven rigorously (Osterwalder and Schrader 1972, 1973, 1975).

The lesson for philosophers from this story is that one should not attempt an ontology of a theory without full knowledge of its various mathematically equivalent formulations.

5 Some specific issues in QFT

The following is only a small selection of the questions of philosophical interest.

(a) What is the most important difference between QM and QFT?

Perhaps the most important difference is the number of degrees of freedom: finite for QM and infinite for QFT. It has crucial mathematical consequences which bear on the particle concept.

For a finite number of degrees of freedom, a theorem by von Neumann states that all representations of the canonical commutation relations are unitarily equivalent. But for an infinite number of degrees of freedom as is the case in QFT, there exist nondenumerably many unitarily inequivalent such representations (Gårding and Wightman 1954). Of all these only one equivalence class, the one involving the Fock representation, permits a number operator. It has a vacuum which is characterized by the eigenvalue zero of that number operator. Free massive fields typically have such a representation. But interacting fields in general do not have Fock representations: the interaction determines which representation of the canonical commutation relations must be used (Streater and Wightman 1978). Thus, for a particle interpretation it is necessary to have at least the asymptotic states, the in- and out-states, span a Fock space.

If the system has no energy gap, i.e. when there is a continuum of energy states down to the vacuum, there is no Fock representation and therefore no number operator. This is the case for interaction with a field of zero mass particles for which therefore no particle number can be specified at low energies. Other representations must then be used; the coherent state representation is a common example. (For higher energies one can approximate the zero mass field by a field of particles with energies above a lower bound so that the particle number becomes finite.)

(b) What is an elementary particle?

Fundamental particles may be composites; for example, the proton consists of quarks. Massive composite particles have spatial extension. But those that are not composites have no structure and are therefore *point* particles such as the electron or the muon (as far as is known to date). Contrary to classical physics, noncomposite particles in QFT *can* decay into other particles; the muon is an example. In contrast to massive particles, zero mass particles are not composites; they are all elementary particles. But they cannot be localized, and in that sense they are not point particles.

(c) Are virtual particles real?

A virtual particle has all the properties of a real particle except that it is not confined to a mass shell. Such particles arise naturally in perturbation expansions in terms of free particles, and they are most tangibly present in Feynman diagrams. It is tempting to give them literal interpretation and assign them ontological status. However, as has been argued by several authors (Weingard 1982 and 1988, Redhead 1983 and 1988), this would be incorrect. One argument against accepting them as real is the fact that a mathematically equivalent and empirically identical representation of QFT (the path integral formulation) does not involve such particles. Virtual particles arise in perturbation expansion, i.e. in a particular approximation method. Other methods of solving the problem (for example of finding the scattering amplitude) do not necessarily involve them. Another argument recalls that the exact results of calculations are sums of Feynman diagrams, and only those sums can be given physical significance. It follows that virtual particles are an artifact of the perturbation expansion into free particle states.

(d) How does QFT distinguish matter and force?

It would seem that source particles such as protons, electrons, quarks, etc. are all fermions while interaction mediating particles such as photons, gluons, and gravitons are all bosons. And one may therefore be tempted to conclude that fermions describe what is classically called 'matter', while bosons describe the interaction, the 'force'. Unfortunately, this holds at best in a crude approximation because we know of interactions for which this simplistic rule does not hold. Thus, photons interact with one another via closed electron loops.

That the simplistic correspondence, fermions/matter and bosons/force, cannot be taken too seriously also follows because many bosons have mass. On a more sophisticated level the failure of the correspondence fails because of supersymmetric theories. There, fermions and bosons can convert into one another so that the whole matter/force distinction based on a fermion/boson distinction fails.

6 The ontological commitment

Once a theory has become mature and well supported by empirical evidence (become empirically adequate), the first requirement for its acceptance is almost met. But there are others: it should be reductively explanatory, answering questions not understood on the next higher level of its competence, and it should be coherent with other theories. The latter requirement refers both to coherence with the higher level theory and to coherence with neighboring theories of the same cognitive level. For example, QFT should be consistent with QM as well as with cosmology; and when applied to the latter it should give results that agree with what is known from other astrophysical considerations.

Thus, the scientific community accepts a theory if it can be sufficiently *justified*. And when that is the case, the community is *committed* to the ontology of that theory just as much as it is committed to its mathematics, its technical terms, its methodology, etc. Such a commitment is therefore based on a lot more than pure pragmatism. Of course, and I keep emphasizing this point, a theory and its ontology are necessarily meaningful only relative to a certain cognitive level. An ontic commitment must refer to entities (particles, fields) and their predicates (properties, behavior) that can be empirically confirmed (directly or indirectly). Only then does it become an essential truth claim of science.

Some day we shall have a theory of quantum gravity which may perhaps be part of a TOE (theory of everything). And in that case, we shall have reached a new and probably quite different cognitive level. That level may be as different from QFT as QFT is from classical physics, or at least as different as QFT is from QM. Let me just quote one well-known conjecturer: 'Are fields and particles foreign entities immersed in geometry, or are they nothing but geometry?' (Wheeler 1962).

It is therefore no more meaningful to claim that the world is *really* what QFT claims it to be, than it is to claim that it is what QM claims it to be. QFT is just another level of knowing and it is not the last one – if there is a last one.

References

Aharonov, Y., Anandan, J., and Vaidman, L. (1993), *Phys. Rev. A* **47**, 4616.

Asquith, P.D. and Nickles, T. (1982, 1983), *PSA 1982*, vol. 1; vol. 2. East Lansing: Philosophy of Science Association.

Brown, H.R. and Harré, R. (1988), *Philosophical Foundations of Quantum Field Theory*, Oxford: Clarendon Press.

Dyson, F.J. (1949), 'The radiation theories of Tomonaga, Schwinger, and Feynman', *Phys. Rev.* **75**, 486–582.

Gårding, L. and Wightman, A.S. (1954), *Proc. Nat. Acad.* **40**, 617 and 622.

Ghirardi, G.C., Rimini, A., and Weber, T. (1986), *Phys. Rev.* **D34**, 470.

MacKinnon, E. (1982), 'The truth of scientific claims', *Philosophy of Science* **49**, 437–62.

Osterwalder, K. and Schrader, R. (1972), 'Feynman–Kac formula for Euclidean Fermi and Bose fields', *Phys. Rev. Lett.* **29**, 1423–5.

Osterwalder, K. and Schrader, R. (1973, 1975), 'Axioms for Euclidean Green's functions', *Commun. Math. Phys.* **31**, 83–112, and **42**, 218–308, respectively.

Pearle, P. (1989), *Phys. Rev.* A **39**, 2277.

Redhead, M.L.G. (1983), 'Quantum Field Theory for Philosophers', in P.D. Asquith and T. Nickles (1983), pp. 57–99.

Redhead, M.L.G. (1988), 'A philospher looks at quantum field theory', in H. R. Brown and R. Harré (1988), pp. 9–23.

Rohrlich, F. and Hardin, L. (1983), 'Established theories', *Philosophy of Science* **50**, 603–17.

Rohrlich, F. (1988), 'Pluralistic ontology and theory reduction in the physical sciences', *British Journal on Philosophy of Science* **39**, 295–312.

Rohrlich, F. (1994), 'Scientific explanation: from covering law to covering theory', pp. 69–77 in *PSA 1994* (D. Hull *et al.*, eds.), East Lansing: Philosophy of Science Association.

Rohrlich, F. (1995), 'Why scientific theories are largely autonomous', *Intl. Congress on Logic, Methodology, and Philosophy of Science*, Florence 1995.

Streater, R.F. and Wightman, A.S. (1978), *PCT, Spin and Statistics, and All That*, 2nd printing with additions and corrections, Reading: Benjamin/Cummings Publishing Co., p. 168.

Weingard, R. (1982), 'Do virtual particles exist', in P.D. Asquith and T. Nickles (1983), vol. 1, pp. 235–42.

Weingard, R. (1988), 'Virtual particles and the interpretation of quantum field theory', in H.R. Brown and R. Harré (1988), pp. 43–58.

Wheeler, J.A. (1962), 'Curved empty spacetime as the building material of the physical world', *LMPS 1960* (E. Nagel, ed.), Stanford: Stanford University Press, pp. 361–74.

Zurek, W.H. (1982), *Phys. Rev.* D**26**, 1862.

Discussions

Brown: You stressed that you need an S-matrix and asymptotic states and so on, but experimentalists think they observe things like quarks and gluons, and doesn't this pose a bit of a problem? You have to regard the composite things as more fundamental than the things they are made of. Please explain the ontological basis of quarks and gluons.

Rohrlich: The problem of defining what you mean by fundamental is not a trivial problem. One way to say it is that 'anything that does not have structure, and presumably therefore must be a point, is not extended, could be regarded as fundamental.' On the level of quantum field theory, electrons are presumably elementary, they have no structure. The same is true for massless particles. On the level of nuclear physics, protons have no structure. On the level of quantum field theory, standard theory, they do have structure. So it depends on what level you're talking about.

Healey: I'd like to follow up on the last question a little, because I'm still worried about the ontology of quantum field theory. I think theoreticians would happily say, at least in their introductory lectures, that a proton is made up of three quarks. But that is a situation in which one is not dealing with objects where they're, as it were, nearly free. Yet they talk about the existence of such particles in strongly interacting states. On the other hand, one theoretician, Paul Davies, wrote a provocatively entitled paper, 'Why particles don't exist in quantum field theory', and gave various arguments as to why, even in the free field case, it was a bad idea in general to give a particle ontology. So I think the situation is really rather more confused than you'd have us believe. At least I'm confused.

Rohrlich: There is a question of whether or not every particle has to have a free state. That's an open question, I don't think the last word has been said. At the present time, people seem to tend to give quantic reality to quarks even though they do not have free states. They are permanently bound in nucleons or mesons. But this is not a completely settled question.

Anandan: I don't know whether this is what Richard Healey's referring to, but in the Unruh effect it's possible for an accelerated detector to detect a particle even though a stationary detector will not. And you were mentioning that experimentalists talk about a particle, and here's an experiment that you can do, and depending on how you do it, you see a particle or you don't see a particle. So I don't think that the concept of a particle is very meaningful. So I would think that the ontology of field theory is that the field is the basic object that exists, and that you can talk about the states of the field and so on, but the concept of particle is not really a very good concept.

Rohrlich: When the particle has a mass, its existence is fairly unique. But when the particle does not have a mass it becomes a little more tricky. If you have a Coulomb field, for example, then you would say that you do not have a particle, you just have a field called a Coulomb field. If you take a detector and accelerate it past the Coulomb field, that detector will detect photons. This is a little tricky, but that is a fact. So there is a question about under what conditions one can see particles. Perhaps one should refer to observations by inertial observers only.

Teller: A couple of you have mentioned the Unruh effect, and of course in that situation we have a situation in which the world is responding to us in ways which are in part particle-like and in part field-like. To my way of thinking, what needs doing in this situation is to go back and untangle the classical conceptions of field and particle, analyze them, observe the way that in quantum theories some of those notions must be rejected, and you can take the remainder and combine them in a kind of consistent comprehensive package, which shows the way in which this very strange sort of quantum phenomenon is in some respects field-like and in some respects particle-like.

There's an inclination to see science as giving us the full truth of exactly the way things are. Another way to think of what science does in helping us to understand the world is, as I put it at the end of my presentation, that it gives us a plethora of models, each one of which gives us a window on certain aspects of the way the world looks, but none of which gives a whole panorama. The fact that one window gives us a limited view, another one gives a limited different view, and that the concepts that we use in those views may even be in conflict needn't be seen as any reason for thinking any one of the views is incorrect. Rather, each view is limited, and the understanding, the growing understanding that we have, of the way the world is that science gives us, is by ever increasing these sometimes scattered, often overlapping, but never in one picture, comprehensive pictures, of how we should think about the world. That's a restatement of at least part of your message, or is it different?

Rohrlich: Yes this is part of my message. But you point to something that I left out in my talk, and that is the following. When we talk about what there is, we must not forget that we must relate it to what we can observe, what we can find out by empirical methods with observation or experiments. In other words, the issue of measurement comes into the picture. And every measurement has a finite error. So when you talk about measuring photons, if the photons are having longer and longer wavelengths, the apparatus can no longer detect them. So you measure only a certain number of them, and the others are below threshold of the measurement. No matter how accurate your apparatus, there are always photons that are softer, that cannot be measured. So even though you have an infinity of photons, you can always measure only a finite number. This is the type of situation where the issue of how many particles are there becomes fuzzy, you see, because it has to do with the finite accuracy of the measurement. That is the problem of the infrared divergence.

Part Ten

26. Panel discussion

STANLEY DESER (MODERATOR), SIDNEY COLEMAN, SHELDON LEE GLASHOW,
DAVID GROSS, STEVEN WEINBERG AND ARTHUR WIGHTMAN

Deser: We are about to see what is to me at least a unique 'five fork' round table here. That is to say we have five people who are amongst the most important creators and makers of quantum field theory. Sidney Coleman is of course the conscience of modern field theory and the greatest teacher in the field. He's done a few nifty things himself besides that. I'm going alphabetically here of course. Shelly Glashow, as we have been told yesterday, is a self-confessed consumer of quantum field theory, but don't let that fool you. The *faux naive* attitude hides a very profound knowledge of the field as well. David Gross is of course Mr QCD, and unlike Moses, he's also been able to make the transition over to the promised string land, so that we have it both ways with him. Steve Weinberg has really made much of quantum field theory, of what is modern quantum field theory over the decades, and he has used it, in spades. He has now just published a classic textbook on the subject. Finally, in alphabetical order, is Arthur Wightman, who is the father of rigorous quantum field theory amongst many other things, for example things which have now passed into the vernacular, the use of Euclidean techniques to understand quantum field theory and the wisdom of understanding many of the things as you heard in talk, understanding of how quantum field theory was formed. Now when I was asked to chair this session, to replace Mr Cushing, I was promised each of them had 15 minute position papers, which they would read, and then after that we would go around. Instead, I've just discovered, there's no such thing. They want, like oysters, to be irritated into action. They will then tell us all sorts of deep things. I will try to make sure those deep things include topics of interest not only to us physicists, but in terms of history, foundations and philosophy. So I will start off the questions, and let us start just at the further end from me. The first specific question: when you die and go to heaven, what will you ask God, Arthur?

Wightman: I have a remark about something that irritated me. Here you have five wizards and a retiree with an attitude. And one of the things that irritated me was all the abuse which renormalization theory took because it was done by all these ugly different methods, and it seemed to me that what was overlooked and not mentioned in the session, was the fact that, at least for perturbative renormalization, Hepp tells you a simple set of conditions, under which all of them have to yield the same answer up to a finite renormalization. So the answer, when you write it down, is some kind of thing that you build out of an operator-product expansion. And what's so ugly about that? And all those other things are auxiliary constructions. Most people in mathematical analysis don't complain because constructions are ugly, provided the final product is something really good. And so, I'm irritated with that.

Weinberg: I share Arthur's irritation actually. I hope I wasn't one of the ones you were irritated at. I didn't think I said anything bad about renormalization. But I was always irritated at Dirac for complaining that renormalization was sweeping the infinities under the rug. In fact you have to renormalize whether the theory is finite or infinite. It's simply a matter of recognizing what it is that you observe. The m that you put in the equation is not the same m in the table of constants published by the Lawrence Berkeley lab. And you simply have to recognize that, whether the renormalization is finite or infinite.

I think that is a deep question. In fact maybe this is a good thing to talk about. That is, should we be bothered by the fact that the correction is infinite? I'm not saying that I'm worried, in the sense of actually doing the calculations, because I agree there are well-defined ways of doing the calculations. It's not a practical problem. But is it a signpost toward the future? For example, the fact that in a sense string theory is a finite theory, does that really matter, as compared to the fact that quantum field theory is not finite? But even if it's nonrenormalizable, you can still renormalize it in the sense I was describing this afternoon. I don't know whether that's important or whether I care whether a theory is finite or infinite, and I'd like to know what other people think about it.

Coleman: What's important is how far you can trust the theory, right? If the non-renormalizable interactions in the improved standard model, including the non-renormalizable things, are really all of gravitational strength, then you are not forced to consider new physics until you are up to the Planck scale, whereupon in the old Fermi theory of the weak interaction, you had to consider new physics, what was it? Gribov bound? Shelly, do you remember?

Weinberg: Well, anyway, it was 300 GeV.

Coleman: Yeah, but trying to remember my own history in the subject, I think, I really did think, when I first started studying the subject, that there was a really big difference between renormalizable and nonrenormalizable theories. It was a while before I caught on that renormalizable really meant insensitive to new physics at short distances, and nonrenormalizable meant sensitive to that physics. But I can't quite place the moment of my enlightenment. I think it must have been something that either gradually dawned on me or you told me when I first met you or something.

Weinberg: No, it wasn't when I first met you because it took me a while too. But you're not answering my question which is—

Coleman: What irritated me? [laughter]

Weinberg: No, that was not my question. My question was: assuming renormalization works and that it does not matter whether the theory is renormalizable or not, which is not a practical problem, should we still care that there was ever an infinity there at all? Is that a problem for a theory? Would it be nice if the theory was really finite?

Coleman: Well, it has to be really finite if it has to be really a theory that goes down to arbitrarily small distances.

Gross: QCD is fine.

Coleman: No, I'm not saying that every theory that is really finite is extendible to arbitrarily small distances.

Weinberg: No, but you're saying the opposite, which QCD is the counter-example of.

Coleman: No, I'm saying—

Gross: It is finite and extendible, if I remember correctly.

Weinberg: It is not finite.

Gross: Yes it is.

Coleman: It's a question of whether you multiply or divide by zero.

Fisher: It's much easier to sell if you don't have to divide by zero or subtract infinity!

Coleman: It's hard to formulate the theory in terms of bare coupling constants without a cutoff.

Gross: It's the only way we actually know how to calculate precisely in the theory.

Coleman: No, I know, but you can't say g_0 is zero and then there are these radiative effects that are infinity and the product of zero.

Gross: That's why you encounter infinities when you do perturbation theory. But, in fact, the only rigorous way we have of either formulating the theory or doing numerical calculations never encounters any infinity.

Coleman: I agree, but we could also do quantum electrodynamics by doing the subtractions in the integrand rather than—

Gross: And if we formulated that theory in the same nonperturbative way, we would encounter exactly the infinities in physical quantities as when we went to the continuum limit.

Weinberg: Well, certainly many kinds—

Coleman: No, I don't understand that, David, could you expand?

Gross: That's the reason, in fact, people believe that theories like QED don't have a nonperturbative existence. The infinities in QED are the infinities that you encounter in physical objects, in two-point functions, or in other Green's functions—

Coleman: Are you talking about the Landau pole?

Gross: Yeah.

Coleman: Oh, oh those infinities.

Weinberg: I'm not talking about those infinities, I'm talking about just the ordinary perturbative infinities.

Gross: But there is a way of physically measuring the bare coupling, which is after all the coupling that you measure at infinitely high energies. So there is a physical meaning to these infinities. In the standard model, with certain kinds of Higgs, in some range of mass, the energy at which these infinities show up could be an accessible energy.

Coleman: And that would be an internal sign that there has to be new physics coming out of that energy.

Gross: So these infinities are bad, but infinities of the type you encounter in QCD I think are just artifacts of approximations done in certain ways.

Weinberg: Well, if I can restate this, what I was asking was whether it was important that you encounter infinities when you do perturbation theory, and you're telling me it's not important what happens when you do perturbation theory.

Gross: Well, no. What I'm saying is that you can do perturbation theory in such a way that you never encounter any infinities.

Coleman: For QCD.

Gross: Yeah, for QCD.

Weinberg: Yeah, well, I don't know how to do that, except by putting in a cutoff, and then taking it away.

Gross: No, you can do it *à la* Hepp. You just subtract and then you never see—

Coleman: No, no, Hepp has a cutoff. You mean *à la* Zimmermann.

Gross: *À la* BHPZ. But Steve asked an interesting question which I wanted to give a different answer to. Is finiteness good? We actually know two kinds of finite theories, not just string theory. There are field theories that are finite. There is, for example, $n = 4$ supersymmetric Yang–Mills theory is perfectly finite.

Weinberg: Yeah, but I thought you were just saying QCD was a finite theory. That was what was confusing me.

Gross: $N = 4$ supersymmetric Yang–Mills is a finite theory in the sense you were talking about.

Weinberg: No.

Gross: $N = 4$ supersymmetric Yang–Mills is a finite theory in the sense that there are no renormalizations at all.

Coleman: It may be free of them.

Gross: Right.

Weinberg: Okay.

Gross: The beta function is zero. That makes it a very nice theory in some sense, but not as a physical theory, since such a theory is scale invariant. String theory is finite partly by having a dimensional parameter. So it really has the possibility of being finite and predictive, of predicting all other dimensional parameters. So that is an advantage, I think, for such a finite theory.

Deser: But the real payoff is having all the parameters with dimensions remain finite in terms of the single one. That's of course the holy grail.

Coleman: No, but I think it's a quantitative question rather than a qualitative one. The appearance of divergences or Landau ghosts or other pathologies are clues as to where the theory is inapplicable. The fact that quantum electrodynamics has problems at e^{137} times the mass of the electron, or maybe it's even $e^{63.5}$, I don't know what the constants are. But at such a huge energy—

Weinberg: It's in the book.

Coleman: The fact is useful only if you believe QED presents you with a contradiction, only if you believe QED describes the world on that distance scale. If you think QED is just good on a much smaller energy scale, that fact is irrelevant.

Weinberg: I never understood Landau in his rejection of quantum field theory because of his—

Coleman: Well, he thought the behavior was generic. He thought no matter what theory he replaced it with, it would have parallel problems.

Gross: Right. And it wasn't an argument against QED so much as an argument against attempts to construct strong interactions where the problem occurs at a GeV.

Coleman: In fact the remarkable fact is that the standard model, including QCD, is renormalizable. That is to say, there is no trace of nonrenormalizable interactions in current experiments, which means something that the string theorists would like you to believe.

Gross: We understand that.

Coleman: That's right. But that means there is no new physics, or at least it is a significant possibility that there is no new physics up to the Planck scale.

Gross: No, that by itself is not proof. There could be new physics at any scale that we haven't yet observed.

Weinberg: Yeah, but there would have to be some special properties to keep the neutrino mass zero and keep proton decay slow. You'd have to—

Gross: In fact there is, called supersymmetry. It's right around the corner.

Weinberg: Well, that has problems with avoiding baryon and lepton violation and there are ways of dealing with it, but it involves violent hand-waving.

Gross: Supersymmetry?

Weinberg: Yeah, I mean—

Gross: Low energy supersymmetry.

Weinberg: Low energy supersymmetry makes it more natural than not to have baryon and lepton violation. I know it can be avoided, but it's not trivial. I don't think it's trivial.

Jackiw: I'd like to make a comment. It's really curious to sit here and listen to all these people saying that they don't really care whether the theory is finite or not finite, that it makes no difference. Now I think that's just an example of human psychology. We're given a bad deal, so everyone is trying to make the best of it. Now tell me, suppose Schrödinger's theory, which everybody always knew was incomplete, suppose it were to predict nonsense at some stage, for example the norm of a state is equal to zero, or something else like that; suppose it were to make self-contradictory predictions, as opposed to basically inaccurate ones; you wouldn't have been as happy with it as people were with Schrödinger's theory. Maybe you would have renormalized it, and then you would have developed this colossal evasion, 'well, it doesn't make any difference'; 'isn't it lucky that there are infinities because they show us that the theory is incomplete?' I believe that occurrence of infinities is a defect. And I think probably, eventually, we'll fix that defect. But to keep saying over and over again that it isn't a defect, that it doesn't exist, that's what David Gross said: 'It's not even infinite.'

Gross: In just QCD.

Jackiw: That's self-delusion. Others say that, well, you wouldn't trust it in the place where the infinity makes a difference. Well, I think it'd be nicer, even if you didn't trust it at high energies, if it made sense at high energies.

Coleman: Well, it would be nice to have a complete theory. Firstly, it seems to me I'm being criticized for learning from experience. And if so, I accept the criticism. But if we have an ultimate complete theory of everything, it should certainly be free of pathologies. But if you have a theory—

Jackiw: If you have an incomplete theory, it still is nice if it's free of pathologies.

Coleman: Indeed it is, but it's not necessarily a disaster if it has pathologies as long as they don't propagate into the domain of validity of the theory.

Jackiw: It is not a disaster, but it is a little odd to say it makes no difference.

Coleman: Well, we have a universal rug. I mean even if you don't believe in string theory, you can say obviously when fluctuations in the metric become of the size of the metric itself, the whole concept of a theory in space-time must break down, whatever it is, without any specific statements about the theory. And then we can say – well, things you heard a dozen times here – you know, if, at some energy scale below that Planckian energy scale, things do look like a field theory with all the coupling constants of order one, as that's consistent with what we— Well, I don't want to finish the paragraph, you've heard it before.

Gross: The typical case in physics, in our previous physical theories – classical mechanics, especially classical electrodynamics – was that they were incomplete at short distances.

Fisher: But it did not show up in the theory in the same way.

Gross: Yes it did. Of course it did.

Coleman: If you put charged particles in.

Gross: Of course it showed up in the theory, people knew about it and were concerned about it.

Fisher: Do you mean in classical *mechanics*?

Gross: Classical electrodynamics.

Fisher: Yes, classical *electrodynamics*! But classical thermodynamics and classical mechanics displayed the miracle of being quite consistent.

Deser: I wouldn't want to ask for the Newtonian self-energy of a particle, either.

Weinberg: Classical thermodynamics had a few problems of its own.

Gross: Thermodynamics has a problem, distinguishability.

Jackiw: When those theories had problems, people said, they have problems, they didn't say, 'these are no problems', or 'it makes no difference'. Think of electron self-energy debates.

Gross: Actually, they most often noticed that they had problems after they had solved the problems. They went back and said, 'ah ha, well, we used to have problems'. And it's very rare the problems are really recognized until they're solved.

Deser: No, but I think, Roman, you want recognition of the fact that we are not yet at a stage that we can do without the props of the low energy theory? Or do you just feel that until and unless we have a theory without infinities, that we shouldn't be self-satisfied? That's certainly true, but—

Jackiw: No, I just think you should be honest, I don't think you should be satisfied or dissatisfied. I think you should just admit that there is a defect; you should admit that there are infinities.

Gross: Actually, I wanted to answer your question about what question I would ask God. Because it is an interesting question and you gave me time to think about it since you asked it to me five minutes before. So, there are two answers I would give. The first is kind of a joke. I'd paraphrase Gertrude Stein and ask, 'What's the question?' Because, in some sense, that's really the hardest thing in physics, to ask the right question. In explaining to me the question, He would have to explain to me all the answers I'd like to know.

Deser: He's subtle. We don't know what He would say.

Gross: The other question is one that from the point of view of particle physics seems the most mysterious, to all of us, that is why is the cosmological constant zero or so small.

Weinberg: I think we all always keep asking the same question, which is why things are the way they are.

Deser: Sure, but there are some things that are more 'the way they are' than others. And it's not clear to me why the cosmological constant is picked on specifically as being—

Weinberg: It's a lot of orders of magnitude.

Deser: No, I understand that.

Coleman: 120 orders of magnitude difference between the rough order of magnitude estimate and the experimental observation.

Deser: That should tell you that it can't be the fundamental problem. But anyway, that would take longer to discuss. Shelly, what would you like to ask God, or do you know all the answers?

Glashow: I'm working at that, I'm working. One question, just one question: why is the top quark so heavy?

Deser: That's a good question. You don't care about the tau lepton?

Weinberg: Yeah, I would have said, you know that's the easy one, because that's the mass you would expect.

Coleman: Why are the others so light?

Gross: Why is the neutrino so light?

Glashow: Yeah, well, you start off with the easy questions, see.

Weinberg: Why is the electron so light? That's really hard. I mean the electron is the mysterious particle, not the top quark.

Deser: And why is this difference of emphasis important?

Weinberg: Well, it does direct the way you think. I mean, some people think you have to give the top quark new kinds of interactions which are different from the other particles. And other people think you have to invent new symmetries that keep the light quarks light. I'm not sure, obviously, which is right, but there is a difference.

Glashow: Steve, we're asking the same question: why is there this little, curious factor of 10^5 between the mass of the lightest particle and the mass of the heaviest particle?

Weinberg: Yeah, that's the right one to ask. And it really is an amazing thing.

Deser: And you have no leptonic questions?

Weinberg: Same question.

Deser: Same question, okay. Does the audience have any questions of the five gods here? Bryce?

DeWitt: This goes back a little bit to the insistence that Roman had given in his talk about the usefulness of the infinities. And I would like to direct the question to David Gross, who keeps insisting that QCD is a finite theory. Well, why is it that in deriving the beta function for students, you look at the coefficients of the divergent pole terms? You're sure making use of the divergences there. Is there another way to derive the beta function, for students?

Gross: Sure. As I said, you do encounter infinities in certain ways of dealing with the theory. The theory is characterized, in the continuum formulation, by a coupling at arbitrarily short distances, an idealized concept, which vanishes. And you're trying to express finite quantities in terms of zero. So you encounter infinite quantities. But there is nothing physical in the theory that diverges.

Weinberg: Well of course not. That's not—

Gross: But there are divergences in other kinds of quantum field theories, and these give rise to the suspicion that these infinities are real and dangerous. That's the distinction I wanted to make. And there are other formulations of QCD, for example, the one we would use if we wanted to follow the program that Arthur Jaffe was talking about, of rigorously formulating the theory, in which infinities would actually never arise. People who put QCD on a lattice, which is a way of defining the theory nonperturbatively, in fact the only way we know how to define the theory nonperturbatively and do numerical calculations, never encounter infinities, or subtractions, or anything like that.

Weinberg: I'm sorry, they do encounter subtractions. In fact lattice calculations shown by Lepage are tremendously improved if you make subtractions so that the coupling constant that you work with is not the Wilson coupling constant, but a renormalized coupling constant.

Gross: No, no, no, as you just said yourself, renormalization is something you do in a finite theory.

Weinberg: But it took a long time for the lattice people to discover that they really had to work with a renormalized coupling, not with a bare coupling, which depends on the cutoff.

Gross: They are working with a bare coupling, but they're just making a wiser choice of that coupling. They still have to carefully take the continuum limit. They just make a wiser choice of how to approach it, so that the effective parameter is smaller.

Weinberg: Well, the wiser choice amounts to including renormalization effects in the—

Gross: Finite.

Weinberg: I mean it's not. David, you must be saying something more than if you have a cutoff then theory is finite. That's what a lattice theory is, a theory with a cutoff.

Gross: If you were to do the same thing for a nonrenormalizable, non-asymptotically free theory, you'd really run into infinities. If you were trying to set up such a theory on the lattice in shrinking the cutoff to zero, you would have to adjust the parameters in your computer program to approach infinity. You couldn't avoid that. So at some point you'd have in your computer program arbitrarily large numbers as your accuracy got better and better. In QCD, you will never encounter numbers less than any bound that you want. That's what I mean, that's the difference between finite and infinite. There is a difference. So I think it's a matter of confusion which goes back to the old way of thinking about divergences in field theory. I want to really make this point provocatively and strongly. In QCD there are no infinities. I'll defend that.

Huggett: First of all, a comment about this debate. I mean one of the things that seems to have been established in the other talks is there's an important difference between exact theories and the perturbative theories. And I'm not quite clear here whether you're talking about infinities in the perturbative theories or infinities in the exact theories. And it seems, maybe, there's an equivocation there that's leading to confusions. I'd like a comment on that. Can I ask two other questions about the standard model? One is related to this. We heard that ϕ^4 theory doesn't exist in the exact formulation. And yet the Higgs field is supposed to be a ϕ^4 theory. How does that work? And second, what is the mechanism, the dynamics for spontaneous symmetry breaking supposed to be? Are there answers to these questions?

Coleman: That's choosing between many possibilities, Steve. I mean, do you want just one possibility?

Weinberg: Well, we still think it's very likely that it's a scalar field, and the fact that some people say the theory doesn't exist has very little effect. What they mean by saying that it doesn't exist is that they are unable to define it as a theory which is applicable at all energies, but we already knew it wasn't applicable at all energies. So it doesn't bother us. The energies at which it breaks down are high enough so that it's not a problem. Although if the Higgs mass is actually very heavy, it becomes a problem, because then the energy at which the theory breaks down is not so very high.

Deser: And nobody wants to ask God, for example, why we are in an effectively four-dimensional universe, given that in any generalized Kaluza–Klein level, or ten dimensions, say, that we still have no clue whatsoever as to why four is a stable point. You're not interested, Shelly?

Weinberg: You know there is maybe a deep point to be made here, and it is in a way the one David made by saying we have to ask what the question is. At various times, physicists become fixated on something as being the important thing. It was one time when people thought the value of the fine structure constant was important. Now of course it's still important, of course, as a practical matter, but we now know that the value it has is a function, that in any fundamental theory you derive the fine structure constant as a function of all sorts of mass ratios and so on, and it's not really that fundamental. The idea of finding some master formula for the fine structure constant now exists only among the crackpots. It may be four dimensions also. I'm not saying that it's not an integer, but that it may be that you get to the number four at the end of a long chain of inference, and that it's very far removed from the fundamental principles. We don't know.

Deser: That's harder to believe, given that string theory, at least as we know it now, or as we knew it until recently, certainly gave a specific prediction, at least at the ten-dimensional level, right?

Gross: No, no, no, no, no.

Weinberg: No, string theory can be formulated in four dimensions.

Deser: It can be formulated in four, but ten does play a magic role. And then there are many ways for it to come down to four. Or any other number.

Weinberg: A complicated dynamical thing which is not of the highest importance. Except to us.

Deser: And to the renormalizability of our problem.

Gross: Well, it's usually the case that the most symmetric formulation of a theory isn't necessarily the ground state.

Coleman: The person who asked the question, was that a satisfactory answer?

Huggett: Yes, the other was the dynamics of spontaneous symmetry breaking.

Weinberg: It's scalar fields.

Huggett: Can it be unitary?

Coleman: No, it could be scalar fields, but it's not at all uncommon for the Lagrangian of a theory to possess a symmetry that is not possessed by the ground state. It is not true that all nuclei, all atomic nuclei, have $j = 0$, despite the fact that the laws of nuclear forces are rotationally invariant. And this is a triviality for a nucleus, but it has far reaching consequences when it applies to the vacuum state. But, you know, there are lots of kinds of dynamics that make a nucleus not have $j = 0$, and there are lots of kinds of dynamics that can make a vacuum state be asymmetric. Is that a satisfactory answer?

Huggett: My worry is there's supposed to be a transition from an unbroken symmetry to the state.

Coleman: With what, with temperature? A transition with what, with changing the fundamental parameters of the theory or with—

Huggett: Right. I mean isn't this a dynamic evolution, something that happens in the history of the universe?

Coleman: Oh, it happens with temperature, yeah. Typically at high temperature you're very far from the ground state but the density matrix or whatever has the symmetry. Have I got it right, Steve? You were one of the first to work this out.

Weinberg: Yeah, it doesn't always happen, but it usually happens.

Coleman: Yes, typically at high temperatures the density matrix has a symmetry which then disappears as the temperature gets lower. But that's also true for ordinary material objects. What's that famous dialogue, Fitzgerald and Hemingway, 'The very rich are very different from you and me.' 'Yeah, they have more money.' Okay, it's the same thing. The difference between the vacuum and every other quantum mechanical system is that it's bigger. And that's from this viewpoint the only difference. If you understand what happens to a ferromagnet when you heat it up above the Curie temperature, you're a long way towards understanding one of the possible ways it can happen to the vacuum state.

Fisher: I had a question for Steve, who mentioned Jean Zinn-Justin. Maybe Steve has seen the recent review of his book by Zinn-Justin in *Physics World* (January 1996 p. 46)?

Weinberg: Oh yes! The reviews have been complimentary.

Fisher: However, Zinn-Justin had a critical remark with respect to your general program and I thought you might comment on its validity. He emphasized your approach regarding the physical requirements of any relativistic theory of quantum particles explaining why you should expect quantum field theory. And he writes: 'However, it may give a less careful reader the erroneous impression that quantum field theory in its present form is unavoidable.' And he goes on: 'This is not necessarily so. For instance, microcausality is really an additional assumption.' I would appreciate your comment on that.

Weinberg: Yeah, I disagreed with that. I mean, obviously if you make that assumption, it simplifies proving all sorts of things. What I was suggesting in my talk was that I didn't see how you could have a Lorentz invariant theory that satisfied the cluster decomposition principle which also didn't have some kind of microcausality in it. I mean I realize there are such things as the Coulomb force which is then canceled by the noncovariant part of the propagator. But still in some generalized sense, microcausality is satisfied even in electrodynamics. And I believe that it is a consequence of Lorentz invariance and quantum mechanics and the cluster decomposition principle and should not be taken as a separate principle. And hence, as I said, analyticity should not be taken as an independent principle. But maybe, actually I think this is maybe a good question for Arthur, because you've been writing papers for years in which you start out with an axiom of causality.

Wightman: I'm in favor of causality.

Fisher: Microcausality as well? I think it was the question.

Wightman: Yes, yes, I mean microcausality. And there are general results using the analyticity that follows from the spectral condition which implies that if you could find a tiny little place at space-like separations, where the commutator vanished, it would vanish all the way into the standard microcausality.

Weinberg: You mean just the vacuum expectation value of the commutator or the whole commutator?

Wightman: I was thinking of the commutator itself, but imbedded in a vacuum expectation value. But, I mean, it would already be something if you began with a two-point function, and it's true there. The other thing is that if it falls off too fast, falls off faster than a Gaussian, it vanishes also. So it's really a tight situation. My attitude was, you might just as well assume it exactly from the beginning instead of focusing on these weaker hypotheses; these theorems are nice things for the communications in mathematical physics, but...

Coleman: But it's a trivial argument that if you can transmit information over an arbitrarily small space-like separation, which of course you can do by interfering measurements if you don't have microcausality, then you can also transmit information backwards in time by an arbitrarily large amount if you have Lorentz invariance. The drawing is like this; you go off, staying very close to the light cone to a point a million light years plus ε away, a million years in the past. Then you do the same thing in the other direction. And even though the space-like separation is only of order ε, you've communicated information. So, you know having a little violation of microcausality is like being a little bit pregnant.

Weinberg: I think there's an assumption when you recoil in horror from this prospect. It seems to me part of your horror is your assumption that the fields are observable.

Coleman: Oh, no. But from a Wightman-type viewpoint, that's the answer to your question about electrodynamics. It's only the observable quantities that obey microcausality.

Weinberg: Yeah.

Gross: Yeah, but look, in the end your argument boils down to the Hamiltonian view.

Coleman: You're talking to him, not me, right?

Gross: Right.

Weinberg: No, I mean I'm the last one to claim any degree of rigor for my argument. But the argument was that you get the S-matrix from a sum of time ordered products of the Hamiltonian, and the important thing is not that the Hamiltonian is observable, but just that it's what you calculate the S-matrix from. The S-matrix is observable. I think it's a challenge. I mean I don't think it's the crucial issue facing physics, but I think it will be an interesting problem for a bright graduate student to try to prove that within perturbation theory, it's the only context in which I can imagine proving anything, that these axioms about Lorentz invariance and cluster decomposition and so on imply microcausality.

Deser: I wouldn't necessarily advise any graduate student to jump onto that offer.

Audience: I'm asking this question because I believe that the world is ultimately geometrical. The question is this: gauge fields have a beautiful geometric interpretation. They are interpreted as connections, as Bryce DeWitt mentioned today. Also the matter fields which couple to gauge fields are interpreted as cross sections on vector bundles. So they are actually treated asymmetrically, classically. And yet when they are quantized by introducing canonical commutation relations they are treated similarly, and I think Steven Weinberg mentioned today that the photon and the electron are treated similarly in quantum field theory.

However, classically, the photon corresponds to a connection, whereas the electron corresponds to a cross section of a vector bundle. So my question is why is it that when they are quantized all the geometry is lost and they are treated similarly. And so, even prior to encountering renormalization problems and infinities, I would be unhappy about the way gauge fields are quantized.

Coleman: The canonical quantization procedure is of wide applicability, and one should be happy we have a procedure of wide applicability that can be applied with equal success to objects as different as gauge fields and matter fields. Certainly in the quantum theory, the gauge fields end up being very different than the matter fields. We have low energy theorems for the gauge, well let's take electrodynamics, a prototypical gauge theory. We have low energy theorems for soft photons, not for soft electrons. So there is a difference. And just the fact that the method is very general, you know, it's like the US mail. You can send a bill, you can send a love letter, the fact that you use the same mail system to send them doesn't mean that you are destroying the distinction between a letter from your wife and one from American Express.

Audience: What bothered me as a physicist is the question why in the standard model are there so many phenomenological parameters? I mean all these coupling constants, all these masses. I think it should also bother you. So the question is why we are so optimistic that we are on a good track anyway.

Weinberg: Welcome to the club. Shelly and I and others and Sidney have been saying for years that the standard model is clearly not the final answer. It has 18 free parameters—

Glashow: I have 19.

Weinberg: He has 19, all right. I think the point is a similar point. And we know that that means it's not a final theory. But it's pretty good. I mean, it accounts for an awful lot of nature. It's true it's more parameters than appeared in Newton's theory of gravitation. But look how much more of nature it accounts for. We know what it is, it's a stopping point; it's a waystation on the way to a better theory. We know that.

Fisher: But do you believe that there is any chance that, for example in string theory, you will be able to calculate at least some of these parameters that they will appear somehow?

Weinberg: Yes, David.

Gross: Yeah, there's a hope. I mean, Shelly said, and I said the same thing about the standard model. It's more than just that there are 19 parameters, it's the feeling that the theory, as it is now, and with any minor modification of it, and probably any improved field theory, is not going to explain those parameters, or is going to have more parameters. There are beautiful attempts, for example, Georgi and Glashow's attempt at $SU(5)$ unified theory, which answered some of the questions—

Glashow: And introduced many more parameters.

Gross: —by doubling, no, tripling the number of parameters. So the general feeling, my feeling, and Shelly, although he's not a proponent of string theory, agrees that just going on with quantum field theory *à la* standard model, is not going to deal with the parameter question.

Coleman: Well, the Technicolor people feel it will.

Gross: No, of course not. They're in an even worse situation. Are you kidding? They have to introduce—

Glashow: They add more and more architecture and none of the architecture is constrained—

Coleman: You're right. I was thinking of their first hopes.

Gross: But if you go to unified theories that produce Technicolor theories, you have to have more and more parameters.

Glashow: But David, it's very important to remind the audience, because they don't know it as well as we do, that despite the fact that there are 19 parameters, most of which play very little role in anything often observed. More like four or five parameters have to do with all of physics, including nuclear physics. We get a perfect explanation of the physical world up to our ability to calculate it, with this theory. The damn theory works. And that is the greatest frustration to non-string theory people, because how can we be asked to do things any better when we're doing things as well as we could possibly do? The string theory people want to come in and say, well, your methodology cannot possibly do any better, try ours. Our response to that is, well, maybe yours is the right way, but maybe there's a third way.

Gross: Well, they're not in contradiction.

Glashow: No.

Gross: The only point I wanted to make again about string theory is that one of the claims of string theory is that, as far as we know, it is a theory in which all dynamical parameters, all parameters in the theory, all dimensionless parameters in the theory, are dynamical.

All coupling constants appear in string theory as the vacuum expectation value of certain dynamical objects. Thus, as far as we know, all parameters in this theory, unlike field theories, are determined by the dynamics.

Therefore, the theory has the potentiality of calculating all those parameters. So you ask, why haven't we calculated anything?

Roughly speaking, the situation is that our knowledge of string theory is very primitive and will probably change, and we don't yet probably have a handle on the theory as a whole. At the moment, it appears that these dynamical parameters are sitting in a potential well which is flat. And the dynamics just says that they can sit anywhere in that potential well. And every place they sit is equally possible with our control of the dynamics, and at every place they sit, there's a different value for their expectation value of those dynamical objects, therefore a different coupling constant or a different mass for the electron and so on. We don't yet have control over the dynamics which we would hope, as in all other quantum mechanical systems, lifts this degeneracy and picks a unique state and would fix all those parameters.

Now that might never transpire, and there might still be ambiguity left even after we know what picks the correct vacuum, the correct solution of this theory. But at least string theory has the possibility of doing that. And that's a new feature of string theory, one which is not so often discussed as the fact that it includes quantum gravity, but for me this is one of its most attractive features—

Glashow: David, can I ask you a question?

Gross: —it offers the possibility that quantum field theory doesn't, of setting—

Glashow: No. You have a dream. You and your colleagues who do strings have a dream that you will have a unique theory, that the 19 parameters will be one or zero. But have you ever had the following dream, that 75 years pass, and you're quite an old man, and finally the string theory has been worked out, and you've figured out how it all goes, and you've got all these numbers, and the fine structure constant comes out to be 17? [laughter]

Gross: It's called a nightmare.

Weinberg: I'd like to comment on the exchange between Shelly and David in two respects. One is kind of a sedative remark, and that is that there are obvious tactical differences among the people at the table. I think this remark is really directed to the philosophers and the historians here, rather than to the physicists, who I think already know what I would say. There are tactical differences here. Shelly is skeptical about the future of string theory. David is very enthusiastic. I tend to agree more with David on this, because I've taken the trouble to teach courses in string theory, although I haven't contributed anything to it. But you know, I don't think there's any ideological differences, as compared to tactical differences. That is, if a string theory tomorrow, based on fundamental principles, without *ad hoc* assumptions, turned out to predict a low energy theory which was the standard model, and gave the right values for the parameters, everyone at the table would regard that as a tremendous triumph. And then if Shelly's nightmare took place, David would probably give up. We know, my point is, and again, I don't think this is something that's any surprise to the physicists, but may be some surprise to philosophers of science who think, in terms of externals and social settings and so on, putting us in a historical context, we are ideologically different. We really score theories the same way. You know we really are the same kind of physicist.

Deser: Some better than others. But all of them are opportunistic, when it comes to theories.

Weinberg: Opportunistic, yes, we're all opportunistic, that's right.

Fisher: I want to say, 'Hear, hear' to that, because I think it is a very important point that all serious theoretical physicists share the same values.

Deser: Resolved.

Weinberg: But believe me, a lot of philosophers don't agree with that, don't understand it, and think we are some kind of tribal island society, where we're working out a kind of social evolution toward a hierarchical pecking order.

Deser: And Steve has thought very much about this problem, has written some very good things on it in case you want to follow up. I think it's a very important remark to the nonphysics side of the audience.

Weinberg: Yeah, by the way, with regard to string theory, could I also make the other remark I wanted to make? This is more to the point. And that is that with regards to the flat potentials David was mentioning – this I guess is on Shelly's side of the argument – but for a long time we heard that the reason you can't find the curvature of these potentials and find where the minimum is is because that necessarily involves nonperturbative effects. And I remember Seiberg made that argument back in '86, I think. And now the string theorists are learning how to deal with nonperturbative effects. There's been tremendous progress in the past year. And they still can't find the minimum of the potential, even though now they have some ability, wherever the nonperturbative effects are that produce the minimum in the potential, they're not where they're discovering them now.

Deser: And it's true that lest there's any misunderstanding, that still there's perhaps more than one 'string theory' and still it isn't just the 20—

Weinberg: Well—

Deser: Wait, wait! And there's not just the 19 numbers. It's not as if we know the nature, the matter content, it's just a matter of degree as it were.

Gross: No, there's only one string theory.

Deser: But there's no $SU(3) \times SU(2) \times U(1)$ coming out of that one string theory yet.

Gross: That's one solution.

Deser: That's one solution.

Saunders: I just wanted to make a comment about something that Steve brought up this afternoon concerning Dirac and the distinction he drew between the photon and the electron in the early '30s. And it's true that in that time he did want to say that the electromagnetic field had a rather special role, that it was indeed the means by which we obtain information about the matter fields, in particular the electron. But the point about that view was that this was Dirac's way of not fitting in with Bohr's Copenhagen interpretation. This was Dirac's way of trying to address the problem of measurement.

Weinberg: It was silly when Dirac said it and it's silly now.

Saunders: Well, that brings me to my question. Michael Fisher very kindly asked philosophers in the audience to try to say what philosophers of physics were interested in and what they wanted to learn from quantum field theory. And I think one thing I feel rather embarrassed about, and I think most of the philosophers of physics here tonight feel embarrassed about, is that it's very simple: we're interested in the problem of measurement. But we appreciate that this is sort of like making a bad smell or something, it's not something that one ought to do in front of physicists. And we've learnt that through many years of interacting with physicists, that physicists don't like to talk about it. And that's very interesting too, to philosophers, and something that we don't understand very well. But now I want to put my question—

Deser: Before you do, I just want to say that at dinner time I was treated to a very interesting discussion by Prof. Coleman on precisely the question of measurement in quantum mechanics. So don't assume that it's—

Saunders: I do apologize—

Coleman: I do say, you came to my office and I displayed willingness to discuss the problem at great length, and you never returned. [laughter]

Deser: Ah, you have forfeited your question.

Saunders: No, Sidney, please.

Coleman: Let him finish his question.

Saunders: May I finish my question, please. Sidney, I do apologize, and I have very great sympathy with your point of view—

Deser: Let's get to the question.

Saunders: —and indeed Bryce DeWitt's. No, the question is this, and it's one that's already been asked, and I'm restating it because it hasn't been answered. Namely, how we are to understand spontaneous symmetry breaking and specifically the dynamics of spontaneous symmetry breaking. The problem is somewhat related to the problem of measurement, insofar as in measurement processes we have what appears to be a violation of a unitary dynamics. It seems we cannot implement measurement processes unitarily. Now, it seems, correct me if I'm wrong, that equally we cannot implement a dynamical representation of the process of spontaneous symmetry breaking, unitarily. Now, my thought is that people are not particularly interested in addressing that question, and now this is my question: is the disinterest the same as the disinterest in the problem of measurement?

Coleman: Well, you phrase questions, may I?

Deser: By all means.

Coleman: There is a position on measurement theory which grossomodo is shared by at least Bryce and me and probably other people, which eventually goes back to Hugh Everett and which denies the existence of non-unitary process of measurement, and says that is just a process that can be eliminated. That's easy to prove but hard to believe. And I don't want to go into it now, although I would be happy to go into it privately. And during your stay at Harvard I gave a public lecture on this topic. Now spontaneous symmetry breaking. I'm not sure what I'd mean. You mean the action of the symmetry group cannot be realized unitarily? Is that what you mean?

Saunders: I mean the process by which the symmetry is spontaneously broken.

Coleman: That's ordinary unitary time evolution. In order to have it happen, you have to imagine adjusting some of the control parameters of the theory as a function of time. So you could change the coupling constants, or in cosmology change the temperature.

Saunders: So we don't have unitarily unequivocal representations associated with different values of the spontaneously broken symmetry, different ground states?

Coleman: Well, it's tricky what you mean. If you take a two-dimensional, if you take an ordinary free field theory and change the mass of the particle as a function of time, then in a technical sense – and here I'm very glad to have Arthur Wightman as part of the same panel – as I recall, it is not a unitary transformation. However, it's as good as a unitary transformation if you use the language of local observables and physical equivalents and all that odd stuff. And I think, my feeling, although I will withdraw this sentence in a moment if Arthur so much as winks at me, my feeling is that the same situation is true in the situation of changing the parameters so spontaneous symmetry breakdown occurs. The problem in defining the unitary operator is basically an infrared one. It wouldn't exist in a box, it does exist in infinite space. And I imagine it can be cured by the same means, although I must say I haven't thought hard about it. You might have thought hard about it.

Wightman: No, I haven't.

Coleman: Is there anything wrong with what I've said?

Wightman: No, no.

Fisher: I want to make a remark which perhaps embodies a misunderstanding but which might be a useful reinterpretation of the question. If I take a system which at high temperatures is completely symmetric and I cool it down, then we know from real physics that it breaks up into at least two phases – certainly in the simplest cases. So it would follow cosmologically that there are either domain walls or there are parts of the universe where the phase is different.

Coleman: That is a real problem, a real phenomenon in cosmology. It represents a significant constraint on cosmological models.

Fisher: Well, I do not know whether this issue was partly in the mind of the questioner. But one does see that one does not just 'break the symmetry of the ground state' and find that the whole universe happens to be in one broken-symmetry phase. As we cool down, there is a mechanism which, if the high-temperature state was really symmetric originally, tells us something about what we should find now on a very large scale.

Coleman: Well, eventually, one domain will take over. No, I shouldn't say that. Anyway, from Dr Saunders's viewpoint that doesn't matter. The state is still not invariant under the action of the symmetry group even if the order parameter has a nonzero value at only one place.

Weissman: The question about reduction wasn't satisfactorily handled earlier, so I want to put a simple example. Consider three line segments. Now, configure them as a triangle. Instantly, we get new properties; angularity, and a new law;

the Pythagorean theorem. Relations explain the difference. Instead of aggregation, we get emergent properties and laws. Now, I'd like to ask Mr Weinberg, why not?

Weinberg: I'd like to ask you, why did you think I would say not?

Weissman: Well, I think the implication of your—

Weinberg: No, that's not the implication. I mean, I know perfectly well that triangles are different than line segments. And that they have different properties. I said this earlier. I don't know why I keep saying this again and again. Of course triangles are different than line segments; entropy is different from particles; supply and demand is different from individual consumers. The question you have to ask always is why? Why are these things the way they are?

Now in the case of line segments, it's not because of the properties of line segments, it's because of the rules of Euclidean geometry, I mean, that's what the reduction is. Reduction isn't always reduction to something smaller or to the parts of the whole. When Einstein explained Newton's laws in terms of the curvature of space and time, it wasn't in terms of something smaller. That's what I call petty reductionism. And what I'm talking about is grand reductionism. I don't commit petty reductionism.

Weissman: So the relations are there in that—

Weinberg: The relation is one of explanation. You have to ask, why is it the way it is? And it isn't always in terms of something smaller, but it is in terms of something deeper.

Weissman: But it isn't always then just aggregation? Thank you.

Weinberg: No, no.

Tisza: We should say what we have is a non-uniform convergence in the transition from relativity to Newtonian mechanics. And this means that we have a continuity for reduction, at the same time the non-uniform convergence leads to very different concepts, so there is a discontinuity of concepts. So in a way we can have both. It's very interesting that this subtle concept is so important in physics. Now of course Newton didn't know about it. In his third rule of reasoning he said when we have a continuous transition between macroscopic objects and atoms, therefore the mechanics must be the same. And that was of course wrong. But when Einstein said, 'Is the moon there when we don't observe it?', of course it's again nothing to do with it because it's non-uniform convergence. Why don't we use this term more often, because it would clarify things?

Weinberg: Okay, I don't disagree with what you say. There are times when it is the right way to look at it.

Zinkernagel and Rugh: Our problem is about the cosmological constant, I mean the vacuum energy density in the empty space. Could one establish a normal ordered form of quantum field theory that would cure some subset of the problem about having energy density in the vacuum? And what is your interpretation of the energy–time uncertainty relation?

Coleman: I'll answer the first one, but not the second, because I've never studied the energy–time uncertainty principle. So, one could always by establishing a proper convention make the cosmological constant vanish. In free field theory, this would be normal ordering, and in an interacting theory it would be much more complex, and this may be a reason why people did not think seriously about the problem until the rise of spontaneous symmetry breakdown. With spontaneous symmetry breakdown there were several choices in a typical model for the ground state of a quantum field theory for the vacuum. And in general they had different energy densities. And you can't reorder the Hamiltonian so that the ground state energy density is the same in all of them. Furthermore, the energy density, the actual form of what is called the effective potential, is dynamically determined. It's something you have to compute, order by order in perturbation theory, you get higher and higher order corrections. Then you begin to worry, why is it exactly zero in the one we ended up in, rather than being zero in one of the other possible vacuums that happens to have slightly higher energy density, and some unbelievably large negative number in the one we ended up in. One can never prove that one can't just add a constant to the Hamiltonian to make the cosmological constant zero, but it seems to be unnatural in a technical sense.

Deser: What was the second?

Coleman: Time–energy uncertainty relation.

Weinberg: Well, if there is an uncertainty relation that's relevant, presumably the time is some cosmological time, which would make the energy very small, and that's not the problem. It's not a very small energy, it's a very large energy we don't understand. I mean energy is really not that well defined because the universe is expanding. But that's a very tiny effect.

Gross: As Sidney said, you could, in ordinary quantum field theory, unnaturally subtract away the vacuum energy. But we feel very uncomfortable having to do that over 130, 140 orders of magnitude in energy and through lots of phase transitions. But in string theory the reason we are having difficulty in solving this problem is because in string theory you can't subtract away. As I said string theory is a theory which is not arbitrary. You cannot subtract away the vacuum energy.

Weinberg: But I'd like to point out another possible explanation, which is that in fact it's not determined and that the wave function of the universe has many terms, in most of which it's very large, and we're living in a term in the wave function of the universe in which it's small, and the reason for that is because it's the only term in the wave function that allows for the existence of people. This is close to an argument that Sidney gave some years ago when your paper, following up Hawking's work when you talked about the wave function of the universe. You tried to show that the wave function of the universe was infinitely peaked at zero cosmological constant.

Coleman: That paper I remember, right.

Weinberg: But my point is, even if it isn't infinitely peaked, as long as it's smooth around zero cosmological constant, we would naturally be in a term in the wave function in which in fact it is very small.

Coleman: This is the modified limited anthropic—

Deser: Anthropic principle, right. Another subject which I will not permit questions on.

Weinberg: This is the thinking man's anthropic principle.

Zinkernagel and Rugh: Did we understand Coleman's answer in the way that had it not been for the implementation of mass giving the Higgs field, then actually one could establish normal ordered language so that the cosmological constant problem would be...

Coleman: I would say before the rise of spontaneous symmetry breakdown, which certainly has nothing necessarily to do with the Higgs field (there is no Higgs field in most condensed matter problems, no fundamental scalar field), one could always say the vacuum was the vacuum, this ridiculous vacuum energy is just nothing, I'll just add a counter-term that's a constant, add it to the Lagrangian and cancel it out. Now that we learn, by studying spontaneous symmetry breakdown, that vacuum energy differences are dynamic, one can consider many possible vacuum states in a typical theory. And vacuum energy differences, which have nothing to do with, which are insensitive to, adding a constant to the Lagrangian, are dynamical quantities that have to be computed, order by order in perturbation theory, and the answer to the computation has definite physical consequences.

Then you become much less sanguine about just getting rid of the problem by defining it away, although, as long as you don't consider gravity, you can get the same Feynman diagrams. It's only gravity that tells you absolute energy. It's like making the mass of a particle zero by fine-tuning the parameters when there's no symmetry reason for it to be zero. You can always do that, you can always make the mass zero. But you're typically unhappy in doing that when you say if I introduce some new interaction very very weak, it will make a large effect on the mass that'll have to cancel out. This is a naturalness criterion. Likewise, you know, the cosmological constant is the mass of a box of empty space. You can always fine-tune it to zero. And nobody will say you can't do it, but nobody will applaud you when you do it, either.

Gross: I'd like to emphasize one point here. What Sidney said is the way you think about a cosmological constant in the absence of theory of gravity. Once you have a theory of gravity, you're much more constrained. Not just string theory, but say supersymmetric gravity, supergravity. Because the extra symmetries involve space-time, the cosmological constant term you can otherwise put in to subtract away the vacuum energy is not always allowed. In fact it's not. And so there you don't have this ability. You either get that the cosmological constant equals zero or is small, or not. In string theory even more so. You get whatever you get. And that's why in theories like that in finding the vacuum and deciding what the dynamics is that picks the vacuum, one is going to have to solve this problem. One cannot fudge it as in previous theories.

Deser: Well, there is one field theory, namely $d = 11$ supergravity in which you cannot put in a cosmological constant term, period. The lower-dimensional ones you have that choice. The hour is getting a little bit late, and we have exploited my five distinguished colleagues, or tortured them, quite considerably. Let us thank everybody for their contributions. [applause]

Name index

Abrikosov, A. A., 269
Adler, Stephen, 12, 18, 27, 158
Aharonov, Y., 298, 309, 360, 364
Aharony, 102, 106–7, 123, 126–7, 129–30
Amit, Daniel J., 94, 126, 129
Anderson, Philip W., 129, 252, 255, 258, 260
Anderson, Ronald, 312
Appelquist, T., 277
Araki, Huzihiro, 44
Ashtekar, Abhay, 174, 187–206, 223–4, 229–30, 236, 240, 290, 294–7
Atiyah, 222
Auyang, S. Y., 324, 326, 338–9, 342

Baker, Jr. George A., 91, 95, 98, 103, 113–14, 126, 129
Balaban,Tedeusz, 146, 165
Barbour, J., 212, 229
Baxter, Rodney J., 111, 113, 121, 129
Becchi, C., 251
Bell, John S., 28, 293–4
Bergman, D. J., 126, 189
Bergman, O., 159, 174
Bergmann, Peter, 211
Bernoulli, Daniel, 78
Bervillier, C., 90, 98, 129
Bethe, Hans A., 38, 344, 346
Birkhoff, G., 122, 132, 134
Bjorken, Jame D., 314, 323, 344, 349
Blokhintsev, Dmitri Ivanovich, 168–9
Bogoliubov, N. N., 117, 129, 135
Bohm, David, 28, 298, 309
Bohr, Niels, 6–7, 28, 70, 84–5, 92, 167, 169, 296, 318, 322–3
Boltzmann, Ludwig, 93, 117
Bombelli, L., 203
Borchers, 43
Born, Max, 241, 325
Brézin, E., 95, 122, 126, 129
Bridgeman, Percy W., 76, 82
Bronstein, Matvei Petrovich, 170
Brown, H. R., 323, 340, 364
Brown, Jerry, 279
Brown, Laurie, 13, 252–6, 259–60, 263, 343
Brush, Stephen G., 128
Buchholz, Detlev, 14, 45–6
Buckingham, M. J., 102, 111, 118, 129
Burford, Robert J., 97–8, 115, 130
Burkhardt, T. W., 94, 111, 126, 129

Butterfield, J., 320, 339

Cantor, G., 324, 334, 339
Cao, Tian Yu, 1–33, 38–40, 75, 90, 120, 127–9, 268–86, 324, 343
Capra, Fritjof, 348, 353
Carazzone, J., 277
Carnap, Rudolf, 30, 75, 333, 339
Chambers, R. G., 298, 309
Chew, Geoffrey, 47–8, 252, 255, 343, 345–54
Cohen, Robert S., 168, 233–4, 239–40, 323
Coleman, Sidney, 32, 54, 76, 185, 368–86
Connes, Alain, 20, 144, 146, 229
Cushing, James, 27, 254, 258, 309, 343, 345–6, 349
Cutkosky, Richard, 343, 350–5

Da Costa, 326, 331, 339
Dalitz, R. H., 347
De Castro, 281
De Rham, 306
Debye, P., 111
Derrida, 34
Descartes, René, 212, 216, 219, 225, 227–9
Deser, Stanley, 70, 158, 174, 189, 368–86
DeTar, Carleton, 348
DeWitt, Bryce, 10, 30, 70, 173–4, 176–86, 202, 205, 211, 226, 296–7, 310, 378, 382
DeWitt, C., 345
Di Castro, C., 117, 126, 129
Dimock, John, 45, 140
Dirac, Paul A. M., 13, 35, 40, 54–5, 77, 84, 139, 155, 158, 189, 200, 216, 231, 241–3, 250, 253–4, 257, 259–60, 269, 272–3, 369, 381–2
Domb, Cyril, 91–2, 94, 98, 102–4, 111, 113, 115, 127–30
Doplicher, Sergio, 14, 146
Drell, Sidney D., 314, 323, 344, 347, 349
Dresden, Max, 343
Drouffe, J.-M., 94–5, 122, 126, 131
Dürr, D., 305, 309
Dyson, Freeman J., 14, 38, 57, 137, 146, 243, 247, 250, 255–6, 258–9, 268–9, 272, 281, 362, 364

Earman, J., 19, 22, 211–12, 215, 229–30
Eden, Richard J., 255, 348
Einstein, Albert, 5, 11, 17, 19–20, 28–9, 58, 72, 75, 83–5, 166–9, 212, 215–16, 225, 230, 233, 237, 239–40, 247, 290, 301–2, 309, 312, 384
Essam, J. W., 91–2, 102, 111, 130

Euler, H., 242, 250
Everett, Hugh, 382

Faraday, L. D., 11, 176, 209, 216
Feinberg, Gary, 76
Feldman, Joel, 144, 146
Fermi, Enrico, 45, 170, 259, 263, 265, 293
Ferretti, B., 347
Feshbach, H., 92
Feynman, Richard P., 38, 57, 66, 93, 143, 146, 173, 180, 209, 216, 241, 247–8, 250, 258–9, 310, 343, 362
Field, H., 22
Fierz, Marcus, 169, 173
Fine, Arthur, 3, 27
Finklestein, David, 88, 233
Fisher, Michael E., 13, 15, 48, 72, 86–135, 164–5, 231, 249–50, 264, 266, 270–2, 274–84, 286, 377, 382–3
Fleming, Gordon, 314–15
Frank, Philipp, 75, 82
Frautschi, Steven, 353
Fredenhagen, Klaus, 40, 46, 146
French, Steven, 5, 9, 324–42
Frenkel, Victor Yakov, 166–7, 169–73
Furry, 242, 254, 259, 263

Gallavotti, Giovanni, 94, 96, 117, 129
Gårding, L., 362, 364
Gaunt, D. S., 121, 130
Gawedski, K., 94, 131
Gell-Mann, Murray, 3, 14, 115, 117–8, 131, 181, 268–9, 281, 345, 352–3
George, H., 82, 379
Geroch, Robert, 306, 334–5, 340
Ghirardi, G. C., 37, 40, 289, 360, 364
Gibbs, Josiah Willard, 93–4, 108, 117
Ginzburg, V. L., 91–2, 96
Glashow, Sheldon Lee, 18, 26–7, 74–88, 253, 368–86
Glimm, James, 146
Glymore, C., 19
Goldstone, Jeffrey, 151
Gomis, Joaquim, 163–4, 247, 251
Gorelic, Gennady E., 166–7, 169–72, 175
Gorishny, S. G., 123, 131
Gracia, J. J. E., 325–6, 340
Grassi, R., 37, 40
Green, Melville S., 95, 111, 114, 119, 126, 128–31
Griffiths, R. B., 102, 126, 131
Grisaru, M., 158
Gross, David, 31–2, 56–67, 69–73, 83, 86–8, 90, 128, 131, 138, 159, 165, 245, 251, 261, 267, 275–6, 280, 283, 368–86
Grünbaum, Adolf, 19, 225, 229
Gunton, J. D., 102, 114, 119, 128–9, 131–2
Gupta, A., 131
Guth, Alan, 347

Haag, Rudolph, 30, 43–6, 340
Hall, A. R., 229
Hall, M. B., 229
Halperin, B. I., 126–7, 131
Harada, M., 247, 251
Hardin, L., 208, 361, 365
Harré, R., 76, 247, 251, 323, 340, 364–5

Hartle, 181
Hawking, Stephen W., 222, 251
Healey, Richard, 298–313, 324, 341, 365–6
Heisenberg, Werner, 7, 64, 70, 84, 151, 168–9, 241–2, 250, 254–5
Hepp, 371
Hermann, A., 84
Hess, M., 326, 332, 340
Hiley, B. J., 309
Hohenberg, P. C., 111, 126–7, 131
't Hooft, Gerard, 76, 241, 247, 250–1
Horgan, John, 78, 82
Horowitz, G. T., 294
Howard, D., 5, 230, 301, 309
Huang, Kerson, 347
Huggett, Nick, 287–97, 332, 337, 340, 376
Hunter, D. L., 92, 111, 129
Huygens, Christian, 227

Iliopoulos, John, 76
Isham, Chris J., 27, 202, 229
Israel, W., 251
Itzykson, Claude., 94–5, 122, 126, 131, 344

Jackiw, Roman, 32, 68, 134–5, 143, 148–60, 163–4, 207, 282, 295–7, 310–13, 341, 372–4
Jaffe, Arthur, 136–47, 164–5, 374
Jona-Lasinio, G., 117, 129, 281
Jones, Martin, 314, 320
Jordan, Pascual, 7, 241–2, 250, 253, 259
Jost, R., 44, 46

Kadanoff, Leo P., 15, 89, 91–3, 95, 102–4, 106–7, 111–15, 118–19, 123, 125–6, 131, 273, 277–8, 280
Kaiser, David, 47, 248, 343–56
Kastler, Daniel, 44
Kauffman, L., 203
Kaufman, B., 92, 97, 103, 131
Kawasaki, K., 111, 126
Kepler, Johannes, 227–8
Khalatnikov, Isaak M., 92
Klein, Oskar, 167–8, 253
Koch, R. H., 131
Kogut, J., 91, 98, 103, 105, 122, 124–5, 133
Krause, Décio, 5, 324–42
Kuhn, 277

La Rivière, Patrick, 37
Ladyman, J., 324, 338, 340
Lai, S.-N., 115, 133
Landau, Lev D., 13, 36, 40, 52–4, 57, 91–2, 96–7, 131, 264, 267, 269, 272, 334–5, 340, 343, 345, 347–55
Landshoff, P.V., 348–9
Lange, L., 213, 229
Langer, J. S., 111
Langevin, Helene, 172
Lee, B. W., 247, 251
Lee, T. D., 76, 88, 311
Leibniz, 28, 212, 216, 218, 227, 238, 319–20
Lepage, 374
Lewandowski, Jerzy, 187–206, 224, 229, 240
Lewis, D., 303, 309
Lieb, Eliot H., 111, 115, 132, 279, 281
Lifshitz, E. M., 91–2, 131, 334–5, 340
Lorentz, Hendric A., 11, 13, 37, 239

Low, Francis E., 14, 115, 117–18, 131, 251, 268–9, 278–9, 281, 284, 343, 345, 347–8, 350

Ma, S.-K., 94–5, 126, 130–2
Mach, Ernst, 75, 212
Mack, G., 126
Mackey, George, 324, 326, 340, 355–6
MacKinnon, E., 364
MacLane, S., 122, 132, 134
Maidens, Anna, 324–5, 328
Malament, David, 306
Mandelstam, Stanley, 47, 343, 349–50, 352–3, 355
Manin, Yu. I., 324, 340
Maxwell, J. C., 11, 29, 117, 176, 209, 216
McMullin, E., 309
Mehra, J., 159
Mermin, N. David, 128, 265
Merzbacher, Eugen, 315, 323, 340
Mills, R. L., 251
Minkowski, 20, 35, 140
Misner, 174, 189
Mott, N. F., 177

Nambu, Yochiro, 151, 255, 345
Nelson, David R., 13, 16, 26, 114, 118, 126–7, 132, 143, 264–7, 270–1, 274–6, 281, 283, 285
Newton, Isaac, 28–9, 75, 85–6, 96, 137, 176, 214, 218–19
Nickel, B. G., 95, 129–30, 132
Nielson, Holga B., 72, 158, 250–1
Niemeijer, Th., 94, 111, 119, 126, 132
Norton, J., 214–15, 229

Oehme, Rheinhard, 255, 259
Okun, L. B., 348
Oleson, A., 92
Oleson, P., 250–1
Olive, D. I., 348
Omnès, R., 181, 345
Onsager, Lars, 92, 97, 101, 103, 110, 115, 131–2, 279
Oppenheimer, 242, 254–5, 259, 263
Ordóñez, C., 251
Osterwalder, Konrad, 140, 146–7, 258, 362, 364

Parisi, G., 126, 132
Park, T.-S., 251
Patashinskii, A. Z., 91–2, 104, 111, 115, 123, 127, 132
Pauli, Wolfgang, 84–6, 168–70, 173, 242, 250, 259, 263
Peierls, 279
Penrose, 174, 236, 289
Petermann, A., 14, 117, 268–9
Pfeuty, P., 91, 95, 107, 130, 132
Pickering, Andrew, 343
Pinch, Trevor, 355
Poincaré, Henri, 30, 134
Pokrovskii, Valeri L., 91–2, 111, 115, 123, 127, 132
Polchinski, Joseph, 16, 27, 48, 54, 104
Politzer, H. D., 128, 132
Polkinghorne, J. C., 348–9
Pomeranchuck, I., 269
Poole, C. P., 94, 129
Post, H., 325, 332, 337, 340, 342
Preskill, John, 173

Redhead, Michael L. G., 32–40, 211, 231, 250, 301, 309, 315, 323, 325, 332–5, 337–8, 340–2, 363, 365
Reed, M., 42, 46
Reeh, H., 40
Reichenback, H., 225, 229
Riemann, Bernhard, 197, 231, 306
Rimini, A., 289, 364
Rohrlich, Fritz, 81, 208, 252, 254, 256–60, 263, 312–13, 357–67
Rosenfeld, Leon, 168, 170–4, 322–3
Rouet, A., 251
Rovelli, Carlo, 20, 175, 188–9, 202–3, 205, 207–232, 262–3, 285–6, 297
Rowland, Henry A., 75, 82
Rowlinson, J. S., 102, 111, 132
Rubbia, Carlo, 74, 79
Ruelle, 43
Rugh, Svend E., 10
Rújula, A. De, 82
Rushbrooke, G. S., 102, 111, 132
Russell, Bertrand, 30

Sakuri, J. J., 293–4
Salam, Abdus, 76, 79, 82, 152
Schnitzer, Howard J., 161–5, 282
Schrader, R., 258, 364
Schrödinger, Erwin, 2, 8, 42, 84, 169, 241, 325–6, 332, 340, 372
Schroer, Bert, 27
Schultz T. D., 115, 132
Schweber, Silvan S., 27, 38–40, 47, 76, 83, 158, 344, 346
Schwinger, Julian, 11, 38, 47, 57, 70, 76–7, 82, 173, 254, 259, 311, 314, 323
Screaton, G. R., 345, 347
Segal, I., 146
Seiberg, N., 18, 381
Serber, Robert, 255, 259
Shankar, Ramamurti, 26, 47–55, 69, 82, 90, 120, 128, 132, 207, 246, 264, 280–1, 283
Shimony, Abner, 27, 86, 324, 341
Shirkov, D. V., 13, 117, 129
Simon, Barry, 42, 46, 143, 147
Simon, Saunders, 285, 382–3
Smolin, Lee, 189, 202–3, 223–4, 227, 229
Sommerfeld, Arnold, 7
Spencer, Thomas, 140, 146
Stachel, John, 5, 19, 22, 71, 166–75, 202, 211, 214, 226, 229–30, 232–41, 272, 281, 284, 312–13, 323
Stanley, Gene, 271
Stanley, H. E., 98, 104, 121, 123, 126, 132
Stauffer, D., 111, 126, 131
Stein, Howard, 29, 46
Stone, M., 41, 158
Stora, R., 251
Streater, R. F., 44, 46, 362, 365
Strominger, A., 294
Stueckelberg, Ernest C. G., 14, 117, 268–9
Suppes, P., 334–5, 340
Suzuki, M., 95, 126, 132
Symanzik, Kurt, 126, 132, 143, 147, 277

Taylor, J. C., 232, 348
Teller, Paul, 19, 26, 38, 40, 82, 314–23, 324–5, 334–5, 337–8, 340, 342, 366
Thorne, Kip S., 173

Ting, S. C. C., 78
Tisza, Laszlo, 83–6, 111, 384
Tomonaga, Sin-itiro, 38, 57, 252
Toraldo di Francia, G., 325, 327, 339, 340
Toulouse, G., 91, 132
Treiman, Sam B., 18, 68–73, 129, 158, 259
Tyutin, I. V., 247, 251, 311

Uhlenbeck, G.E, 111

Vafa, Cumrun, 145, 147
Valentini, A., 293–4
Van de Waals, 96, 102
Van Fraassem, B., 325, 335–7, 341
Van Leeuwen, J. M. J., 94, 111, 119, 126, 129, 132
Velo, G., 146
Veltman, Martinus, 344
Von Neumann, 229

Wagner, William G., 173
Wald, Robert M., 9
Wallace, David J., 94–5, 122, 127–9, 132–3
Weber, T., 289, 364
Wegner, Franz J., 91, 93, 95, 104, 106–7, 125–6, 131, 133, 271
Weinberg, Steven, 13, 16–17, 31–2, 38, 48, 54, 71, 79, 152, 163–4, 209, 234, 241–56, 258–63, 266, 272, 278–9, 285, 311, 341, 357, 368–86
Weingard, Robert, 186, 223, 287–97, 341, 363, 365
Weisskopf, V. F., 84, 242, 259, 263

Weyl, H., 325, 329, 337, 341
Wheeler, John A., 174, 254, 283, 364–5
Widom, Benjamin, 91–2, 103, 111, 115–16, 126, 128, 133
Wightman, Arthur S., 41–6, 69, 93, 129, 138, 146, 185, 202, 237, 246, 362, 364–5, 368–86
Wigner, Eugene Paul, 7, 136–7, 147, 157, 159, 237, 242, 250, 253, 257, 259
Wilczek, F., 128, 131, 251
Wilson, Kenneth G., 15, 72, 88–9, 91–3, 95, 98, 103, 108–11, 114–18, 120–9, 132–5, 158, 249, 251, 265, 267–70, 272–3, 277–8, 281, 285
Witten, Edward, 48, 145, 147, 158, 164, 290
Wu, T. T., 309

Yamada, Yasuhiko, 146
Yamawaki, K., 247, 251
Yang, C. N., 111, 251, 300, 309, 311
Yukawa, Hideki, 252

Zakharov, V. I., 251
Zee, A., 126
Zemelo, E., 327, 334, 341
Zimmermann, 371
Zinkernagel, Henrik, 10, 384–5
Zinn-Justin, Jean, 115, 126, 129, 131, 230, 247, 251, 377
Zuber, Jean-Bernard, 344
Zuckerman, Daniel M., 128
Zumino, B., 158
Zurek, W. H., 360, 365

Subject index

*-algebra, 209–10, 224
3-metrics g_{ab}, 189–90

acoustics, 78
affine connection, 239
Aharonov–Bohm effect, 159, 298–313
algebra, 8, 14, 35, 37
  field, 14, 45
  Grassmann, 141
  local, 36, 45, 211
  operator, 44, 131
  spinor, 86, 169
American Academy of Arts and Science, 32
analyticity, 248, 350
Anderson–Higgs bosons, *see* Higgs bosons
angular distribution, 346
angular momentum, 53, 188
anharmonic oscillators, 149
anomalies, 155, 158–60
  Adler–Bell–Jackiw, 158
  axial, 160
  chiral, 156, 158
  scale, 159
anomalons, 74
Anti-Matter Spectrometer in Space, 78, 82
anticommutators, 35, 44, 282
antimatter, 76
antiquarks, 79
Aristotelian continuum, 39
asymptotic freedom, 18, 30, 48, 57, 60–1, 128, 135,
  150, 160, 249, 268, 274, 283
asymptotic region, 9, 72, 89, 105–6, 108, 112–13,
  124, 133–4, 363
asymptotic symmetry, 99, 101
Atiyah–Patodi–Singer spectral flow, 158
Atiyah–Singer index theorem, 158
atomism, 39
automorphism, 229
autonomy, 343, 358
axioms, 324, 327–30, 377

bare parameters, 12–13, 57
BCS channel, 52
BCS instability, 53
Bekenstein–Hawking entropy, 24, 204
Bell inequality, 36, 40, 298, 301–2
Bell theorem, *see* Bell inequality
Berry phase, 312
Big Bang, 80, 180

binary pulsar, 215
binding energy, 47, 79
black boxes, 34, 142
black hole entropy, 158, 188, 203, 230
black holes, 9, 24, 75, 80, 150, 179, 204
Bohm's theory, 261, 287–97
Bohm–Vigier formulation, 258
Bohr magneton, 103
Bohr's principle of complementarity, 253, 382
Boltzmann factor, 109, 118
Bondi–Metzer–Sachs Group, 236
bootstrap, 352–5
Borchers' theory, 8
Borchers' Selection Criterion, 45
bosons, 363–4
BRST invariance, 247, 311
Buchholz–Fredenhagen Criterion, 45

Cabibbo angle, 80
Callan–Symanzik equations, 95, 126, 129, 132, 161,
  264, 267
Cantor's set theory, 5
cardinality, 329–30, 333
Cartesian relationism, 212
Casimir effect, 10, 178
causality, 23, 25, 45–6, 58, 63–4, 228, 234, 238, 244,
  248, 377
chaos, 94, 108
charge density, 171
charge operators, 153
charges
  bare, 6
  chiral, 158
  electric, 13, 79, 171
  fractional, 149, 158, 161
  renormalized, 13, 39, 272
chemical potenital, 49
Chern–Pontryagin density, 158
Chern–Simons three-form, 290, 292, 310
chiral invariance, 263
Christoffel symbols, 171
chromodynamic flux, 60
classical limit, 7–8, 21–2
classical mechanics, 62, 66, 75, 78, 170, 209
  of point masses (CMP), 84, 86
cluster decomposition, 17, 69, 234, 243–6, 250, 377
cluster expansion, 93, 141
cohomology, 247, 306
Coleman–Mandula theorem, 282

colours, 275
commutation relation, 44, 137, 144, 170, 245, 296, 349
  canonical (CCR), 7, 241, 362
  space-like, 37
commutators, 35, 282, 377
complex angular momentum plane, 47
compressibility, 101, 103
Compton effect, 167, 360
Compton wavelength, 274, 284, 359
condensation, 18
condensed matter physics, 13–15, 47–55, 77, 79, 86, 89–90, 92, 95, 149, 155–8, 176, 260, 265–71, 273, 280–1, 285–6
configuration space, 177, 190–2, 200, 295
confinement, 18, 23, 30, 57–8, 61–2, 68, 143, 268, 278
conformal invariance, 145
constructive program, 12–13, 17–18, 32, 58, 94, 140–3, 163, 188, 210
constructivism, 34
contiguity, 227
conventions, 3
convergence factors, 13
Cooper pair, 53
Copenhagen complementarity, 21, 35
correlation coefficient, 37
correlation functions, 97, 101, 103, 105, 111, 113, 115, 130, 231, 273
correlation length, 101, 103, 112, 120
correlations
  maximal, 36–7
  vacuum, 36
correspondence principle, 7
cosmic rays, 78
cosmological constant, 24, 62, 77, 373, 384–6
cosmology, 1, 75, 77–80, 149, 170, 180, 230, 236, 263, 295, 383
Coulomb interaction, 48, 55, 70–1, 156, 244–5, 311, 377
countability, 333
coupling, 13, 25, 60–1, 66, 114, 375
  local, 11–12
  strong, 23, 26, 56–7, 61, 64, 71, 224
  weak, 23, 26
coupling constant, 38, 51, 80, 95, 97, 104, 108–9, 112, 119, 122, 132–3, 139–40, 249, 266, 273–5, 280, 345–6, 370, 374, 380, 383
coupling function, 51, 53
covariance, 5, 30
CP violation, 24, 80–1
CPT theorem, 43, 46, 69
critical density, 100
critical exponents, 89, 91–2, 96–7, 100, 102, 106–8, 121, 126, 128–33
critical indices, *see* critical exponents
critical phases, 95
critical phenomena, 69, 89, 90, 95, 97–8, 102, 105–6, 115, 126–33, 135, 155, 158, 161, 264, 268, 272–3
critical point, 92, 94, 96, 99–100, 104–11, 116, 120–5, 129–31, 133, 264, 270, 273
critical temperature, 99, 105
critical trajectories, 120, 124
criticality, 97, 100, 103, 112–13
crossover, 105–7, 124
Cumulative Hierarchy, 328

Curie point, 96, 132
Curie temperature, 264, 377
current algebra, 158
curvature, 178–9, 189–90, 225, 231, 258, 292, 384
Cutkosky graphs, 351–3
cutoff, 38–9, 50–1, 60, 66, 86, 105, 110, 122, 146, 250, 285, 370–1, 375
cyl, *see* cylindrical functions
cylindrical functions, 193, 195–9

D-branes, 24
D-function, 170
δ-function, 156–7, 159, 244, 317–8, 351
de Broglie wave, 85, 299
de Broglie's neutrino theory of light, 169
decay amplitude, 154
decoherence, 181
decoupling, 13, 15–17, 26, 59, 143, 277
decoupling theorem, 16, 31
Dedekind cut, 185
deep inelastic lepton scattering, 79, 135
density functions, 350–1
diffeomorphism invariance, 20–2, 58, 187, 189, 194–5, 198–9, 202, 204–6, 214, 220–6, 228–30
dilatons, 66
dimensionality, 95, 102, 113, 128–30, 155, 160, 264
dimensionless ratio, 101
dimensions
  anomalous, 89, 96–7, 123
  canonical, 97–8
dipole–dipole couplings, 103, 106–7
Dirac phase factor, 300, 304, 307–8
Dirac's equation, 76, 136, 156, 242, 252, 257, 293, 307
Dirac–Weyl equation, 153
dispersion relations, 69, 248, 255, 345, 348–50
  double, 343, 349–50, 356
divergences, 31, 77, 98, 139, 161, 254
  cancellations of, 76
  infrared, 149, 152
  ultraviolet, 11–13, 23, 56–7, 61, 64, 143, 149–50, 152, 154–9, 162–3, 268
Doplicher–Hagg–Robertson Criterion, 45
duality, 18–20, 23–6, 63–6, 145, 147, 242
  wave–particle, 35, 253, 259

effective action, 242
  quantum, 179
electrodynamics, 56, 62, 83–4, 257, 312, 372–3
electromagnetic potential, 288, 298–301, 305–9, 312
electromagnetism, 23, 56, 75, 79, 148, 169, 298–302, 304–9
electron's trajectories, 241, 305–7
electron–positron pair production, 179
electrons, 6, 39, 47–8, 57, 78, 81, 85, 136, 156, 166, 177, 217, 258, 265, 272, 291, 299–308, 326, 345, 363, 366, 374, 382
electroweak theory, 1, 58
elementary particle physics, 1
emergence, 3, 66, 177, 291
  epistemological, 4
emergentism, 32
emission and absorption of radiation, 40
energy barrier, 152
energy density, 10, 152, 178, 384–5

energy gap, 8, 363
energy shift, 142
energy–momentum conservation, 154, 260
entities, 4–5, 8–10, 20, 24, 34, 158, 178, 187, 215, 217–8, 225, 272, 284, 326, 328, 338
entropy, 261
epiphenomena, 4
epistemology, 34, 256
EPR, 36–8
epsilon expansion, 89, 93, 95, 107, 133
equations of motion, 151, 213–15, 224, 231, 288, 292–3, 348–9
equations of state, 103, 105, 133
equilibrium phenomena, 108
equivalence classes, 43, 338
equivalence principle, 245
ether, 11, 22, 178
Euler–Heisenberg action, *see* effective action
Everett and post-Everett formalisms, 181
excitations, 1, 19, 23–5
    of the vacuum, 11, 14
excited states, 9–10
expectation values, 11, 43, 46, 68, 361, 377
exponent relations, 97, 100, 102–5, 110, 113, 126
exponents
    correlation, 102, 108
    decay, 101
    thermodynamic, 102, 105, 108

Feigenbaum's period-doubling cascades, 94
Fermi liquids, 53–4, 94
Fermi surface, 49–50, 52, 54, 94
Fermi theory of nuclear beta decay, 254, 259
Fermi theory of weak interactions, 60
Fermi velocity, 50
Fermilab, 361
fermion model, 335–6
fermions, 6, 7, 49, 51, 53–5, 77, 79–80, 363–4
    Weyl, 158
ferromagnet, 92, 96, 101–3, 106–7, 110–11, 130–2
Feynman diagrams, 49, 64, 68, 94–5, 122, 129, 132, 140, 248, 316, 343–56, 363, 386
Feynman rules, 311, 344
Feynman's functional integral, *see* path integral
Feynman–Dyson series, 243
fiber bundle, 179, 240, 308
field equations, 141, 293–4, 308
    Einstein, 168–70, 174, 178, 189, 203, 214, 258, 263, 288
    gravitational, 22, 167–8, 172–4, 239
    non-linear, 173
field operators, 83, 153, 157, 361
field quanta, 7–9, 338
field theories, 19
    axiomatic, 11, 19, 32, 36, 138, 188, 282
    conformal, 62, 145–7, 161, 245
    effective, 16–18, 24–6, 31–3, 38–40, 47–55, 59, 60, 62, 64, 66, 69, 86, 90, 161–5, 234, 246–50, 259, 262, 267–8, 285
    formal, 87
    local, 60, 72, 83
    Minkowskian, 192, 194–5, 199, 201–2
field–particle relationship, 6
fields, 209, 216–20, 242, 359–62
    background, 178–9, 188, 190–1, 194
    boson, 35

classical, 8, 29, 35, 241, 360
Coulomb, 359, 361, 366
Dirac, 55, 293
electric, 188, 191
electromagnetic, 28, 45, 66, 68, 85, 106, 151, 153–4, 165, 169–72, 176, 235, 241, 296, 300, 306–7, 361, 382
Euclidean, 94
fermion, 35, 153–5, 293, 295
fluctuating, 11, 108
free, 5–7, 9–10, 22, 35, 38, 45–6, 139–40, 193, 209, 362
gauge, 23, 287–97, 308, 378
gravitational, 19–22, 24, 68, 166–8, 171–2, 176–9, 207, 214, 217–19, 224, 226, 235–6, 287, 296
interacting, 8, 13, 38, 46
intermediate vector boson, 361
interpolating, 8
Killing vector, 179, 180, 237
Klein–Gordon, 242
local, 5–6, 8, 11, 20, 29, 32, 35, 43–4, 148, 164, 268
magnetic, 101, 103, 105–6, 110–12, 115, 298–301
Majorana, 55
massive, 35, 362
matter, 201
Maxwell, 177, 216, 242, 287
metric, 5, 218, 226, 239, 288–9
point-like, 11, 24, 32
scalar, 43, 45–6, 58, 176–7, 180, 375–6
spinor, 153, 177, 180
string, 291–3
thermal, 125
vector, 177, 180, 287
velocity, 305–6
Yang–Mills, *see* gauge fields
fine structure constant, 57, 376, 380
fixed points, 25, 71–3, 88–9, 95, 102, 108, 116, 120, 123–6, 265–7, 272, 278
fluctuations, 7–8, 10, 13, 91–2, 95–7, 108, 114, 143, 174, 264, 273, 277, 283–4
    density, 130
    quantum, 3, 11, 13, 15, 25, 225
    vacuum, 25–6, 29, 36
fluid dynamics, 78
fluid percolation, 95
Fock space representation, 6, 8–10, 22, 24, 35, 179, 194, 200–1, 315, 332–7, 361–3
four–Fermi interaction, 50
Fourier analysis, 38
Fourier series, 94
fractionization, 159
Fredenhagen's theorem, 37
free energy, 90, 103, 105, 109, 122–3, 125, 134, 270
Friedman–Robertson–Walker solution, 263
functional equations, 14, 179
functional integral, 2, 13, 48

galaxy, 78, 215
Galilean invariance, 360
gas–liquid coexistence curve, 99–102
gauge
    Coulomb, 244
    global, 13, 300
    local, 13

gauge bosons, 1, 18, 62, 154
gauge conditions, 29
gauge dependence, 299, 305–6, 308
gauge fixing, 143, 225, 247, 311
gauge groups, 14, 24–5, 48, 61, 70, 77, 81, 168
gauge invariance, 6, 17–18, 30, 153–4, 193, 200,
    245–7, 289–90, 300, 304–8, 311
gauge potentials, 143, 308
gauge theory, non-Abelian, 1, 30, 58, 66, 76–9,
    127–8, 136, 145, 158, 274, 287–313
Gaussian integral, 50, 311
Gaussian model, 121
general theory of relativity (GTR), 5, 17–28, 58,
    66, 71, 75, 150, 166–7, 170–4, 180, 187–91, 194,
    203–7, 210–12, 215–21, 224–233, 235–40,
    245–6, 250, 287–92, 295, 312
geometrodynamics, 189–90, 201, 283
geometry, 22, 161, 212
    differential, 220, 283, 312
    non-commutative, 144–6, 223
    quantum, 187–206, 224, 229, 235
    Riemannian, 172, 201, 205
ghost, 6, 30, 180, 292, 311
    Landau, 12, 371
gluons, 14, 26, 30, 60, 262, 275–9, 363, 365
Goldstone–Nambu boson, 149
grand unified theory, 81, 167
gravitation, 58, 71, 166–74, 208, 231–2, 235, 240,
    247–9, 263
gravitational constant, 169
gravitational potentials, 168
gravitational radius, 171
graviton–graviton scattering, 234, 294
graviton–photon scattering, 234
gravitons, 62, 169, 171, 193, 236, 291, 294–5, 363
gravity, 17, 23–8, 56, 70–2, 77, 85, 181, 215, 226,
    283, 287–97, 386
Green's functions, 9, 141, 269, 370
Gross–Neveu model, 55, 189
ground states, 10–11, 25, 45, 49, 58, 140, 142, 148
groups
    continuous, 16, 90
    discrete, 90

Haag's theorem, 140, 316
Haag–Kastler axioms, 140
Haag–Ruelle axioms, 140
hadamard form, 46
hadron–hadron interactions, 279
hadrons, 26, 57
Hall effect, 265
Halperin–Nelson–Kosterlitz–Thouless–Young
    theory of two-dimensional melting, 107
Hamiltonian stability, 94
Hamiltonian–Jacobi equation, 293
Hamiltonians, 14–16, 20, 35, 41–2, 49, 54, 62,
    89, 92, 104–5, 108–10, 116–19, 123–5, 142,
    153–7, 168, 190, 193, 201, 207–8, 221, 243–5,
    254, 261, 267, 290, 293, 311, 344–5, 348–9, 352,
    378, 385
harmonic oscillators, 311, 314
Hartle–Hawking vacuum, 179
Hawking radiation, 178–9
Heisenberg chain, 55
Heisenberg condition, 138–9
Heisenberg model, 102, 132

Heisenberg's uncertainty relations, *see* uncertainty
    relations
helicity, 153–5
helium atoms, 79
Higgs bosons, 1, 3, 18, 30, 69, 79, 81, 144, 242, 361,
    370, 375, 385
Higgs model, 141
Higgs parameters, 80
Hilbert problems, 324
Hilbert space, 35–6, 41–4, 140, 160, 180, 189, 192–8,
    200, 204, 209, 223, 295, 297, 302, 335–7, 360,
    362
Hodge decomposition theorem, 305–6, 308
hole argument, 5, 320
Holonomy, 191, 312
homogeneity, 276, 283–4
horizon, 204, 232
hydrodynamics, 94, 264–6
hydrogen atoms, 47
hyperbolic differential equations, 46
hypothetico-deductive methodology, 6

ideal gas laws, 94
incommensurability, 257, 358
incompatibility, 24, 220–1
indistinguishability, 324–32, 338, 342, 373
individuality, 325, 333, 338, 342
infinite subtractions, 38–9, 374
infinities, 1, 2
instantons, 55
instrumentalism, 2, 3, 34–5, 284
integrable models, 62
interactions
    effective, 13
    electromagnetic, 59, 151, 242–3, 249, 305
    electroweak, 18, 30, 56–7, 61, 177, 242–3, 248–9,
        287
    g-dependent, 272
    gauge, 71
    gravitational, 61, 149, 171, 173, 249
    local, 11, 25, 300
    nonrenormalizable, 31, 60–1, 189, 199–200, 203,
        246, 249, 288, 369, 371
    renormalizable, 16, 48, 76, 290, 371
    strong, 18, 30, 48, 56–8, 62, 242–3, 248–9, 268,
        287, 344, 347–8, 350–4
interfaces, 95, 99
interference pattern, 298–300, 309
inverse operation, 15–16
irrelevance, 89, 91, 102–5, 126
Ising model, 55, 92, 97, 101–3, 110–11, 115, 119,
    129–33, 273
isospin, 79, 151, 247

Kaluza–Klein theory, 19, 23, 180, 287–8, 294, 313,
    375
KAM (Kolmogorov–Arnold–Moser) theory, 93
Kantian metaphysics, 5
kaons, 80
Klein–Gordon equation, 177, 252
knot, 192, 202, 222, 230
Kondo problem, 95, 158
Kosh–Lestalitz transition, 279
Kosterlitz–Thouless theory of two-dimensional
    superfluidity, 107, 135
Kretschmann–Komar coordinates, 225

Lagrangians, 2, 12, 16, 41, 43, 47–8, 51, 66, 118, 145, 169, 207–8, 243–7, 259, 261, 274, 277, 280, 307, 311–12, 346, 376, 386
Landau graphs, 351–3
Landau–Ginzburg model, 121, 123
Lange–Neumann material systems, 225
large N methods, 62
large-scale structure, 80
lattice gauge theories, 72, 88, 143, 146, 192–3, 266, 278–9, 374–5
lattice spacing, 102–3, 111–12, 115
law of conservation, 29, 151, 158–9, 167, 180
laws of motion, 228
Leibniz–Clarke correspondence, 219
LEP collaborations at CERN, 74, 81, 361
lepton number, 86
leptons, 1, 18, 56, 80, 154, 174
Lie groups, 79, 192, 197
light cone, 238, 292–4
local action, 301–9
locality, 2, 11, 13, 19, 23, 37, 58–9, 86, 90, 156, 164, 211, 282, 298, 301–4, 309
localizability, 24, 209–11, 220–1, 224
localization, 12–13, 20, 21, 44–5, 95, 141–2, 207–32, 315
logic of quanta, 324–42
loop, 222
loop diagrams, 246
loop space representation, 189, 224, 230
Lorentz groups, 177, 244–5
Lorentz invariance, 25, 48, 68, 86, 163–4, 243–8, 250, 263, 289–90, 311, 318, 360, 377
low energy, 16–17, 49–50, 55, 59–62, 66, 82, 140–1, 149, 162, 194, 234–5, 246–8, 256, 262, 274, 276–7, 353, 381
    approximation, 31
low energy theorem, 378

M-atoms, 327–8
m-atoms, 327–31, 338
M-theory, 32, 162
magnetic flux, 299, 309
magnetic impurities, 95
magnetic materials, 101, 106
magnetic moment, 86, 136
magnetic monopoles, 307
magnetism, 58
magnetization, 96, 102–3
    spontaneous, 99, 101, 106
Mandelstam diagram, 248, 344, 352
manifolds
    differential, 45, 213, 216, 220–1, 239
    hyperbolic, 45
    metric, 211, 213, 220
    space–time, 220
Manin problem, 326
many-particle theory, 35
marginality, 89, 105–7, 126
mass, 14, 38–9, 57, 169, 245, 310, 370
    gravitational, 78
    Higgs, 370, 375, 385
    inertial, 78
    quark, 61, 80
mass gap, 43–4
mass hierarchy, 18, 70
mass shell, 346–7, 349, 351

mass spectra, 141
mathematics, 72, 89–135, 326–32
matrix mechanics, 7
matter, 216–17, 252–3
Maxwell equations, 136, 177–8, 252
Maxwell theory, 28, 169
mean-field theories, 97, 99, 101
measurability, 19, 171–2, 235
measure, 192–3, 200
    Gaussian, 190, 194
    Haar, 192, 196
measurement, 28, 57, 64, 155, 181–3, 203, 211, 214, 231–2, 236–7, 288, 296, 302, 321–3, 360, 366–7, 382
membranes, 95
mesons, 57, 59, 60
metaphysics, 9
methodology, 40, 256
metric, 215, 217, 288–9, 306, 320
    background, 201
    invariant, 197
    Minkowski, 190, 289
    space–time, 187, 287, 294
metric space, 221
microcausality, 377–8
Minkowski space, 45, 58, 62, 108, 190, 193, 233, 236, 288, 306, 320
moment of inertia, 263
momentum conservation, 51
momentum transfer, 53, 345–7, 350
monojets, 74
Monte Carlo method, 94–5, 132
muons, 81

Navier–Stokes equations, 94, 260, 264–6, 271
negative energy, 9, 155, 315
negative energy sea, 155–8
neutral currents, strangeness-changing, 79
neutrino mass, 18, 69, 81
neutrino oscillation, 18, 69, 81
neutrinos, 169
neutron star, 78
neutrons, 265, 344
Newton's gravitation theory, 208–9, 216, 239, 312, 379
Newton's gravitational constant, 62–3, 71
Newton–Wigner states, 317
Newtonian mechanics, 208–9, 219, 227–30, 237, 384
no-signalling theorems, 37
Noether's theorem, 151, 245
non-individuals, 326–31
non-perturbative theory, 223, 226, 370, 374
nonlocality, 298, 304, 309
nuclear democracy, 352–4
nucleon–nucleon scattering, 346
nucleons, 57, 59, 78, 258
number operator, 8, 35

O(1) groups, 287
observability, 6
observables, 14, 22, 35–7, 57, 68–70, 209–11, 216, 222–5
    Dirac, 201
    geometrical, 201
occupation number, 35
ontological synthesis, 24–7

ontology, 2–6, 9, 24–31, 34, 217, 238, 256, 284, 307, 357–67
  field, 10, 314, 324–42
  particle, 10, 314, 324–56
operationalism, 38
operator
  Batalin–Vilkovisky, 247
  Dirac, 223
  geometric, 198–200
  Hermitian, 42, 289
  Laplacian, 190
  local, 125
  quantum, 22, 198
  self-adjoint, 42, 196–8, 209, 224
operator formalism, 7, 10–11
operator products, 11, 22
operator-product expansion, 368
order parameter, 91–2, 95–8, 101–2
order–disorder transition, 101, 132
Osterwalder–Schrader axioms, 140

ϕ^4 field theory, 121, 142, 146, 150, 375
π–π scattering, *see* pion–pion scattering
Padé approximation, 91, 129–30
Palatini action, 290
parastatistics, 45
parity, 151
particle annihilation, 10, 76, 176, 181, 243–5, 261, 362
particle creation, 10, 76, 176, 181, 243–5, 261, 362
particle production, 43
particle–hole conjecture, 347–8
particles, 6, 8, 22–3, 179, 242, 315, 359–62
  charged, 257
  composite, 243, 352, 363
  elementary, 363
  free, 3, 258, 348
  massless, 9, 55, 96, 149, 153–5, 258, 287, 291
  point, 84, 86
  virtual, 3, 10, 148, 179, 258, 316, 344, 346, 349, 363, 365
partition functions, 48–9, 55, 92, 109, 111, 117–19
path integral, 7, 9, 50, 62, 66, 93, 155, 240–1, 261, 344, 362
Pauli exclusion principle, 253
Pauli matrices, 86, 293
Pauli term, 243
perturbation, 106–7, 235
perturbation theory, 12, 57, 61, 66, 94, 98, 121–2, 136, 140, 158, 188, 234, 242, 245, 249, 261, 268, 344–8, 355
perturbative expansion, 3, 12, 38, 49, 57, 64, 93, 257, 290–1, 348–9, 370, 375, 378
phase cell, 12–13, 141–2
phase shift, 156–7, 299
phase space, 50, 120, 191, 193–4
phase transitions, 31, 92, 96, 116, 127–33, 141, 146
phonons, 265
photoelectric effect, 241
photon–electron scattering, 345
photons, 47, 85, 148, 177, 241–2, 291, 360, 363, 366, 381
  zero energy, 9
pion–nucleon interactions, 277, 345–7, 350
  low-energy, 30

pion–pion scattering, 248, 347, 353
pions, 254, 262, 344–6
Planck length, 193
Planck mass, 56, 60, 62, 65–6, 77, 80–1, 247, 256
Planck scale, 62–3, 70, 77, 188, 203, 222–4, 372
Planck's constant, 138, 167, 257
Planck's quantum of action, 63, 65
Poincaré group, 9, 42–5, 127
Poincaré invariance, 42, 58, 72, 174, 205, 209, 263
Poisson bracket, 191, 200
polarization, 241, 274
polology, 344–9
positivism, 9, 35
positronium, 79
potential function, 139
power-law decay, 96–7, 99, 113–14, 124
probabilities, 36–7, 45, 57, 109, 180, 183–6, 362
probability distribution, 224–5
probability interpretation, 181, 186, 209, 232, 295
probability wave, 8, 20
Proca equation, 177
propagator, 244, 349, 351
protons, 6, 39, 48, 78, 265, 344, 363

quadrupole radiation formula, 171
quanta, 5–7, 29, 315, 324–42
  gravitational, *see* gravitons
  light, 168
quantization, 2, 6, 8, 22–6, 63–5, 144, 152, 155, 167–9, 172–4, 176, 181, 188–9, 200, 204, 236, 238–9, 359
  canonical, 7, 223, 378
  loop, 223
quantization condition, 7, 66
quantum chromodynamics (QCD), 1, 11–14, 18, 26, 30, 48, 55, 59–63, 69, 71, 88, 136, 155, 159–60, 165, 223, 256, 261–2, 266, 278, 368–72, 374–5
quantum electrodynamics (QED), 11–14, 47–8, 56–7, 60–2, 66, 71, 76, 136, 158, 162–5, 168, 171, 205, 234, 241, 254–8, 268–9, 272, 291, 343, 352, 370–1
quantum field, 8–11, 19–21, 29, 41–6, 96, 99, 108, 137–8, 144, 169, 211, 214, 220, 242, 248, 349
quantum field theory (QFT), 1–386
quantum gravity, 5, 17, 20–1, 27, 30, 58, 62–3, 70–2, 77, 81, 162–3, 166–74, 177–81, 187–90, 193, 200–3, 215, 222–4, 226, 229–30, 234–6, 288–91, 295–6, 364
quantum jumps, 176
quantum mechanics, 12, 17, 28, 35–8, 57–8, 62–70, 73–9, 84–8, 96, 107–8, 124, 148, 152–3, 156, 159, 160, 163, 167–70, 178, 181, 185–8, 207–10, 222–9, 232–3, 236–8, 241, 244–6, 253, 258, 288, 294, 296–8, 302–5, 308–11, 323, 359–64, 377, 382
quark confinement, *see* confinement
quark mixing, 80
quark–gluon interactions, 1
quarks, 1, 3, 14, 18, 26, 30, 39, 55–60, 68, 78, 154–5, 174, 217, 254, 262, 265, 275–9, 363–5
  down, 78
  top, 81, 374
  up, 78
quasi-autonomous domains, 3, 31, 39
quasi-cardinality, 329–33, 338
quasi-pluralism, 26
quasi-sets, 326–31, 338, 341

radiation, 217
  gravitational, 21, 166, 215
radiation theory, 35, 45
radiative corrections, 56, 71
realism, 2–6, 10, 38
reality, 2–6, 9, 14–5, 19–20, 35–7, 156, 209, 229–31, 277, 285, 309, 338
recursion relations, 113, 119, 121–2
reducibility, 3
reductionism, 24–7, 29, 31–2, 255, 260, 358, 384
Reeh–Schlieder theorem, 36, 43
reference-systems, 213–15, 229, 318–19
regularization, 22–3, 38, 62, 144, 155, 187, 194–8, 200
relationalism, 225, 316, 319–22
relativism, 34
relativity, 17, 19, 57, 63–4, 166, 170, 189, 222, 225, 238
relevance, 89, 91, 105–7, 126
renormalizability, 17, 30–1, 249, 256, 290, 376
renormalizable theories, 12, 60, 140, 150, 242–3, 247, 250, 259, 369
renormalization, 10, 14–16, 29, 38–40, 49, 53, 56–7, 61, 98, 121–4, 139–40, 143, 157, 160, 180, 242–3, 255–6, 268–9, 274, 314–16, 323, 369, 375, 378
  perturbative, 12–13, 368
renormalization constants, 199
renormalization factor, 113
renormalization group, 11–16, 25, 29–31, 48–9, 57–60, 66, 69, 72–3, 89–136, 142, 146, 155, 158, 161, 164, 249, 264–286, 314, 316
  equation, 14, 31, 249, 269, 271–2, 283
  flow, 15, 17, 62, 89–90, 98, 105, 108, 123–6, 142, 267
  Gell-Mann–Low version, 13–16, 95, 126, 135, 249–50, 266–9, 272, 275
  Kadanoff–Wilson version, 13–16, 249–50
representation
  interaction, 272
  irreducible, 42, 45
  unitary, 42
rescaling factor, 112, 118, 120, 122
response functions, 53
Reynold's numbers, 265
Royal Society of London, 40, 98

S-matrix, 8–9, 17, 44, 47–8, 68, 156, 243–8, 253–5, 259–61, 311, 316, 343–56, 365, 378
saddlepoints, 66
scale
  energy, 14–17, 31, 39, 59–61, 63, 66, 70, 99, 162, 246, 249, 265, 268, 277, 371–2
  Laplacian, 156
  length, 13, 29, 48, 63, 72, 92, 97, 156, 264, 266, 277
    of Angstroms, 265
  mass, 16, 31, 61–2, 161
  unification, 62
scale dependence, 13, 15, 30, 89, 96–8, 123, 269–77, 283, 285
scale invariance, 89, 96–7, 154–9, 268, 274, 278, 371
scaling, 46, 79, 89, 91–2, 102–6, 108, 111–14, 123–6, 130–2, 135
  Kadanoff, 268
scaling function, 103–6, 110, 114
scaling law, 103, 110, 129, 131, 273

scattering, 43, 231, 338
scattering amplitudes, 62–5, 141, 346–50, 363
Schrödinger equation, 35–6, 41, 124, 156–7, 160, 176, 289, 292, 299, 360
Schrödinger logics, 326–7
Schrödinger picture, 288, 295
Schrödinger's wave function, *see* wave function
Schrödinger's wave mechanics, 148, 266
Schwinger's source theory, 10
Schwinger's vacuum persistence amplitude, 178–9
Schwinger–Tomonaga formulation, 258
screening effects, 13, 273–4
  anti-, 13
second quantization, 35, 44, 176–7, 216, 242, 315, 335, 361
self-energy, 168, 373
semiclassical approximation, 64, 73, 200
semi-group, 15, 89, 90, 108, 122, 125, 131, 134, 269–70, 273, 281
separability, 298, 301–6, 309
series expansion, 103, 113, 115, 137, 139
set theory, 324, 327, 333–4
short distance, 12–14, 17–18, 38, 57, 63–5, 157, 160–2, 165, 265, 268, 277, 283–4
sigma model, nonlinear, 55
simultaneity, 220, 237–8
singleton, 331
singularity
  critical, 101, 104
  infrared, 277
  light-cone, 22–3
singularity (pole), 344, 346–9, 351
singularity theorems, 234
smearing functions, 195
SO(3) groups, 190
soft pions, 246
solar quadrupole moment, 208
solenoid, 299, 302, 305–8
solitons, 24, 158, 161
space, 1, 74
  absolute, 11
  compact, 192
space-time, 5–13, 19–31, 35–7, 43–6, 58–66, 70, 72, 86, 139–46, 166–75, 187–90, 202–33, 239, 241, 245–6, 258, 287–94, 316–23
space-time points, 5, 11, 19, 25, 77, 220, 231, 244, 298, 300–3, 307, 316, 319, 360
space-translation, 43, 151, 154
spatio-temporal relations, 5, 19, 21, 24
special theory of relativity (STR), 5, 7, 11, 19–20, 23–4, 37, 71, 75–6, 137, 150, 170, 174, 208–9, 213, 216, 233–4, 237, 239, 253
specific heat, 101, 103–4
spectral condition, 9, 14, 42, 45–6
spin, 14, 43, 46, 55, 69, 72, 85–6, 128, 133, 153, 169, 245
  block, 112, 114
  Ising, 111–12, 120, 123
spin connection, 290
spin-glass, 127
spin network state, 203–4, 223
standard model, 1, 16–18, 23–6, 29–31, 56–8, 61–3, 69, 72–6, 78–83, 144, 148–9, 154, 163, 207, 210, 216, 242–3, 245–50, 256, 370, 379
Stanford Linear Accelerator Center (SLAC), 48, 361

statistical mechanics, 28, 49, 75–7, 80, 89–135, 161, 165, 176, 203, 229, 270–3, 275
statistics, 43, 46, 69, 89–135
  Bose–Einstein, 35, 176, 253
  Fermi–Dirac, 253
  quantum, 324, 326
Stone–Von Neumann theorems, 41–2
stress-energy density, 178
string theory, 1, 17–18, 23–7, 32–3, 58–6, 70–3, 82, 86–7, 127, 136–8, 146, 150, 158–9, 162–3, 176, 180, 223, 226, 234–5, 241–2, 249–50, 261–2, 290–5, 368, 371, 379–81, 386
structure
  causal, 4, 11, 19
  causal–hierarchical, 3
  classical, 6, 7
  conceptual, 2–4
  formal, 12
  geometric, 303–4
  hierarchical, 1, 29
  logic, 4, 6
  mathematical, 2
  metric, 19, 20–2
  physical, 38
  theoretical, 1–5, 19, 24–5, 28, 30, 83, 117
$SU(2)$ groups, 79, 188, 190–4, 197–8, 202, 216, 288, 290, 381
$SU(3)$ groups, 61, 70, 127, 216, 288, 381
$SU(5)$ groups, 379
$SU(6)$ groups, 79
subquasi-sets, 330
substantiality, 10
substantivalism, 23–4, 320, 322–3
substratum, 7–8, 10–11, 24–5, 208
supercharge, 145
superconductivity, 58, 91, 131, 246, 249
superconductivity models, 18
superconductor, 54
  high-T_c, 79
superfluid helium, 91, 107
superfluidity, 91, 107
supergravity, 71–2, 180, 190, 201, 386
supernovae, 80
superposition principle, 171–2
superselection rules, 14, 44–5
superstring theory, 74–5, 77
supersymmetry, 17–19, 23, 48, 58, 62, 66, 69–73, 81–2, 141, 144–7, 247, 250, 282, 371–2
susceptibility, 101, 103, 333
symmetry, 72–3, 91, 130, 151, 159, 261, 263
  chiral, 61, 154–5, 158, 276
  color, 14, 276
  gauge, 18, 23, 56, 60, 69, 151, 163, 243, 247
  global, 58, 69, 247
  internal, 282
  local, 58, 60, 247
  rotational, 58, 69, 151, 263
  scale, 155, 157–9
  space-time, 282
  unitary, 79
symmetry breaking, 31–2, 61–2, 68–9, 73, 141, 150–1, 247, 263, 276
  anomalous (quantum mechanical), 30, 152–5, 157–8, 274
  spontaneous, 16–17, 25, 30–1, 57–8, 81, 149, 151–2, 161, 163, 262, 277, 375, 383–6

symmetry group, 9, 25, 70, 127, 236

tau leptons, 374
Taylor series, 51, 97
tensor
  energy momentum, 215
  metric, 230
tensor-product state-vector, 302
Tevatron Collider, 81
thermodynamic properties, 92, 103, 111
thermodynamics, 9, 78–9, 102, 109, 111, 132, 229, 261, 373
Thirring model, 268
time, 225–6, 237–8
  absolute, 84
time-ordered products, 243
time-translation, 43, 151, 154, 243
TOE, 39, 86, 364
topological defects, 24
topological quantum field theory (TQFT), 222, 227
topology, 22, 161, 204, 290, 306
  Gel'fand, 192
  quantum, 202
transformation, 15–16
  canonical, 191
  coordinate, 169
  Fourier, 54, 241, 244
  gauge, 68, 169, 191–2, 202, 288, 290–2, 299–300, 302, 304–8
  Lorentz, 43, 244, 289
  renormalization group, 14–15, 26, 31, 95, 108, 118–20, 122–5, 134, 269–70, 272–6, 280
  scale, 154
  self-similarity, 14, 111, 276
  unitary, 383
transition matrix, *see* S-matrix
translation invariance
  time, 36, 237–8
tree approximation, 246
turbulence, 265–6
two-component spinor, 293
two-photon decay of the neutral pion, 154

$U(1)$ groups, 79, 216, 287–8, 381
uncertainty relations, 7, 11, 22–5, 35, 171, 235, 237, 273, 283–4, 296, 384–5
unified field theories, 170, 233
unitarity, 20, 25, 30, 57, 89, 180, 221, 248, 350
universality, 89, 94, 101–2, 104–6, 108, 114, 120, 123–6, 131, 267, 270
universe, 80, 263, 288–9
  early, 62, 149
Unruh effect, 9, 179, 366

vacuum energy, 386
vacuum polarization, 13, 178–9
vacuum states, 6, 8–10, 22–5, 29, 35–6, 42–5, 58, 68, 73, 99, 142, 156, 209, 290, 377, 386
Van der Waals approximation, 99
Van der Waarden equation, 293
variables
  classical, 96
  conjugate, 245
  dynamical, 64, 189, 232
  Grassmann, 295
  hidden, 28, 114